AA003885

International Workshop on Quantum Chromodynamics - Theory and Experiment (QCD@Work 2024)

EPJ Web of Conferences Volume 314 (2024)

Trani, Italy
18-21 June 2024

Editors:

Giuseppe E. Bruno
Gabriele Chiodini
Pietro Colangelo
Claudio Corianò
Donato M. Creanza
Fulvia De Fazio
Floriana Giannuzzi
Salvatore My
Eugenio Nappi
Stefano Nicotri

ISBN: 979-8-3313-1058-5

Printed from e-media with permission by:

Curran Associates, Inc.
57 Morehouse Lane
Red Hook, NY 12571

Some format issues inherent in the e-media version may also appear in this print version.

For additional information, please contact EDP Sciences – Web of Conferences at the address below.

EDP Sciences – Web of Conferences
17, Avenue du Hoggar
Parc d'Activité de Courtabœuf
BP 112
F-91944 Les Ulis Cedex A
France

Phone: +33 (0) 1 69 18 75 75
Fax: +33 (0) 1 69 28 84 91

contact-edps@webofconferences.org

Additional copies of this publication are available from:

Curran Associates, Inc.
57 Morehouse Lane
Red Hook, NY 12571 USA
Phone: 845-758-0400
Fax: 845-758-2633
Email: curran@proceedings.com
Web: www.proceedings.com

TABLE OF CONTENTS

Editors

Giuseppe E. Bruno
Dipartimento Interateneo di Fisica "M. Merlin", Università and Politecnico di Bari
Via Orabona 4, 70126 Bari, Italy - **e.mail:** giuseppe.bruno@ba.infn.it

Gabriele Chiodini
Istituto Nazionale di Fisica Nucleare – Sezione di Lecce
Via Arnesano, 73100 Lecce, Italy - **e.mail:** gabriele.chiodini@le.infn.it

Pietro Colangelo
Istituto Nazionale di Fisica Nucleare – Sezione di Bari
Via Orabona 4, 70126 Bari, Italy - **e.mail:** pietro.colangelo@ba.infn.it

Claudio Corianò
Dipartimento di Matematica e Fisica "E. De Giorgi", Università del Salento
Via Arnesano, 73100 Lecce, Italy - **e.mail:** claudio.coriano@le.infn.it

Donato M. Creanza
Dipartimento Interateneo di Fisica "M. Merlin", Università and Politecnico di Bari
Via Orabona 4, 70126 Bari, Italy - **e.mail:** donato.creanza@ba.infn.it

Fulvia De Fazio
Istituto Nazionale di Fisica Nucleare – Sezione di Bari
Via Orabona 4, 70126 Bari, Italy - **e.mail:** fulvia.defazio@ba.infn.it

Floriana Giannuzzi
Istituto Nazionale di Fisica Nucleare – Sezione di Bari
Via Orabona 4, 70126 Bari, Italy - **e.mail:** floriana.giannuzzi@ba.infn.it

Salvatore My
Dipartimento Interateneo di Fisica "M. Merlin", Università and Politecnico di Bari
Via Orabona 4, 70126 Bari, Italy - **e.mail:** salvatore.my@ba.infn.it

Eugenio Nappi
Istituto Nazionale di Fisica Nucleare – Sezione di Bari
Via Orabona 4, 70126 Bari, Italy - **e.mail:** eugenio.nappi@ba.infn.it

Stefano Nicotri
Istituto Nazionale di Fisica Nucleare – Sezione di Bari
Via Orabona 4, 70126 Bari, Italy - **e.mail:** nicotri@infn.it

Scientific Advisory Committee

Elke Caroline Aschenauer (Brookhaven National Laboratory)
Albert de Roeck (CERN Geneva)
Laura Fabbietti (TUM Munchen)
Fernando Ferroni (GSSI – L'Aquila and INFN)
Michelangelo Mangano (CERN Geneva)
Guido Martinelli (Accademia dei Lincei, INFN and Sapienza U, Roma)
Mikhail A. Shifman (Minnesota University)
George Sterman (SUNY Stony Brook)

Local Organizing Committee

Giuseppe E. Bruno (Bari Polytechnic)
Gabriele Chiodini (INFN-Lecce)
Pietro Colangelo (INFN-Bari)
Claudio Corianò (Università del Salento)
Donato M. Creanza (Bari Polytechnic)
Fulvia De Fazio (INFN-Bari)
Floriana Giannuzzi (INFN-Bari)
Salvatore My (Bari University)
Eugenio Nappi (INFN-Bari)
Stefano Nicotri (INFN-Bari)

Participants

Anna Rita	Altamura	INFN - Bari
Fabio	Anulli	INFN - Roma I
Emre	Askan	Middle East Technical University - Ankara
Francesco	Becattini	University of Florence
Federico	Betti	University of Edinburgh
Francesco	Bigazzi	INFN - Florence
Andrea	Bizzeti	University of Modena and Reggio Emilia and INFN
Marco	Bochicchio	INFN - Rome I
Ginevra	Braga	Gran Sasso Science Institute (GSSI)- L'Aquila
Giuseppe Eugenio	Bruno	Politecnico and INFN - Bari
Andrzej	Buras	TUM Institute for Advanced Study - Munchen
Hande	Cetin	Middle East Technical University - Ankara
Xun	Chen	University of South China
Gabriele	Chiodini	INFN - Lecce
Giovanni Antonio	Chirilli	University of Regensburg and Università del Salento
Pietro	Colangelo	INFN - Bari
Angelo	Colelli	University of Bari and INFN
Eimear	Conroy	University of Oxford
Gennaro	Corcella	INFN - Lab. Nazionali di Frascati
Claudio	Coriano'	University of Salento and INFN - Lecce
Aldo	Cotrone	University of Florence and INFN
Donato Maria	Creanza	Politecnico and INFN - Bari
Silvia	Dalla Torre	INFN - Trieste
Fulvia	De Fazio	INFN - Bari
Luigi	Delle Rose	Università della Calabria and INFN
Mitsutoshi	Fujita	University of South China
Isabella	Garzia	University of Ferrara and INFN Ferrara
Martina	Giachello	Gran Sasso Science Institute (GSSI)- L'Aquila
Floriana	Giannuzzi	INFN - Bari
Giovanni	Grilli di Cortona	Laboratori Nazionali del Gran Sasso - INFN
Marco	Guzzi	Kennesaw State University (USA)
László	Gyulai	HUN- REN Wigner RCP - Budapest
Stefano	Lionetti	University of Salento and INFN - Lecce
Francesco	Loparco	INFN - Bari
Nicola	Losacco	University of Bari and INFN
Wolfgang	Lucha	Austrian Academy of Sciences
Kiminad	Mamo	William and Mary University Williamsburg (USA)

Manoj Kumar	Mandal	University of Padova and INFN
Miguel Angel	Martin Contreras	University of South China
Vincenzo	Mastrapasqua	INFN - Bari
Annalisa	Mastroserio	Università di Foggia and INFN - Bari
Barbara	Mele	INFN - Roma I
Dario	Melle	INFN - Lecce
Davide	Milillo	University of Bari and INFN
Shubham	Mishra	IIT Hyderabad (India)
Maitreyee	Mukhejee	Panjab University Chandigarh (India)
Francesco	Murgia	INFN - Cagliari
Salvatore	My	University of Bari and INFN
Eugenio	Nappi	INFN - Bari
Fernando	Navarra	Institute of Physics , University of São Paulo (Brazil)
Madeeha	Nazish	CHEP Punjabi university Pakistan - Lahore
Stefano	Nicotri	INFN - Bari
Nicola	Nicassio	University of Bari and INFN - Bari
Antimo	Palano	INFN - Bari
Alessandro	Papa	University of Calabria and INFN
Marco	Pappagallo	University of Bari and INFN
Luca	Passarella	Politecnico di Torino and INFN
Antonio	Passeri	INFN - Roma Tre
Rajendra Nath	Patra	INFN - Bari
Gil	Paz	Wayne State University - Detroit
Fulvio	Piccinini	INFN - Pavia
Maimoona	Razzaq	University of Calabria and INFN
Edilson Alfonso	Reyes Rojas	Universidad de Pamplona (Colombia)
Giovanni	Ridolfi	University of Genova and INFN
Hagop	Sazdjian	Saclay University and IJCLab Paris
Andreas	Schmitt	University of Southampton
Irfan	Siddique	University of Chinese Academy of Sciences
Akash	Singh	IISER Mohali (India)
Aditya	Srivastav	Indian Institute of Technology Hyderabad (India)
Mikhail	Stephanov	University of Illinois - Chicago
Seyed Farid	Taghavi	Technical University of Munich (TUM)
Riccardo	Tommasi	University of Salento and INFN - Lecce
Aldo	Vera	Universidad Major - Santiago (Chile)
Kun	Xu	University of Chinese Academy of Sciences
Jingxu	Zhang	Lanzhou University (China)

Preface

The 2024 edition of **QCD@Work - International Workshop on Quantum Chromodynamics - Theory and Experiment** was held in Trani, Italy, from 18 to 21 June 2024. The venue was the magnificent Palazzo Delle Arti Beltrani. It was the 11th edition of a series started in 2001 in Martina Franca and continued in Conversano (2003 and 2005), Martina Franca (2007, 2010, 2016), Lecce (2012, 2022), Giovinazzo (2014) and Matera (2018).

The aim of this workshop is to bring together senior and young researchers, theorists and experimentalists putting QCD to work in different frameworks, for discussions, dissemination of the results, promotion of new collaborations, education of the young scientists. The main topics discussed in this edition were:

- low energy and non-perturbative aspects of QCD
- progress in perturbative QCD
- physics of heavy quarks
- QCD in extreme conditions of baryon density and temperature
- new holographic methods for strong interactions.

This edition was the occasion to commemorate two eminent Italian QCD scientists, Stefano Catani and Marcello Ciafaloni, who sadly passed away in 2023.

About 80 researchers (with a large fraction of young scientists) actively participated with presentations and discussions. There was a lively poster session: the prizes for the best posters were awarded to Drs. Maimona Razzaq (Cosenza) and Akash Sing (IISER Mohali).

In this volume we publish the workshop proceedings.

We acknowledge the Institutions and individuals who have contributed to the successful realization of the event. For financial support we thank Istituto Nazionale di Fisica Nucleare (INFN), Commissione Scientifica Nazionale Teorica INFN, Groups I and IV of INFN-Bari, Sezione INFN and Groups I, IV - Lecce, Dipartimento di Matematica e Fisica "Ennio De Giorgi" of Unisalento. We thank Dr. Vito Manzari and Prof. Daniele Martello, Directors of the Bari and Lecce INFN Units. We thank Dr. Niki Battaglia and the Trani "Associazione delle Arti", who manage Palazzo Delle Arti Beltrani. We are particularly grateful to Dr. Mary Lonigro and the staff of Intermeeting Sud for the great help and kindness in the organization of the meeting and in the assistance to participants. We thank all young people of Locanda Del Giullare - Trani, for the warm hospitality and cheerfulness. We acknowledge the help of Dr. Francesca Assisi and of the staff of INFN Bari administration service. Finally, we thank Prof. Francesco Pepe for having allowed us to use two of his pictures of Trani as logos of the workshop.

Giuseppe E. Bruno, Gabriele Chiodini, Pietro Colangelo, Claudio Corianò, Donato M. Creanza, Fulvia De Fazio, Floriana Giannuzzi, Salvatore My, Eugenio Nappi and Stefano Nicotri

Statement of Peer review

In submitting conference proceedings to *Web of Conferences*, the editors of the proceedings certify to the Publisher that

1. They adhere to its **Policy on Publishing Integrity** in order to safeguard good scientific practice in publishing.
2. All articles have been subjected to peer review administered by the proceedings editors.
3. Reviews have been conducted by expert referees, who have been requested to provide unbiased and constructive comments aimed, whenever possible, at improving the work.
4. Proceedings editors have taken all reasonable steps to ensure the quality of the materials they publish and their decision to accept or reject a paper for publication has been based only on the merits of the work and the relevance to the journal.

QCD @ Work 2024 - International Workshop on Quantum Chromodynamics - Theory and Experiment
18-21 June 2024- Trani (Italy)

Proceedings editor(s):

Giuseppe E. Bruno, Gabriele Chiodini, Pietro Colangelo, Claudio Corianò, Donato M. Creanza, Fulvia De Fazio, Floriana Giannuzzi, Salvatore My, Eugenio Nappi, Stefano Nicotri

Bari, 31 October 2024

EPJ Web of Conferences **314**, 00001 (2024) https://doi.org/10.1051/epjconf/202431400001
QCD@Work 2024

Stefano and Marcello @ Work

A personal selection of results obtained by Stefano Catani and Marcello Ciafaloni

Giovanni Ridolfi[1,*]

[1]Dipartimento di Fisica, Università di Genova and INFN Sezione di Genova, Via Dodecaneso 33, 16146 Genova, Italy

Abstract. To honor the memory of Stefano Catani and Marcello Ciafaloni, I describe part of their scientific results in the field of Quantum ChromoDynamics during their long and fruitful careers.

1 Introduction

Stefano Catani and Marcello Ciafaloni (Stefano and Marcello henceforth) have given important contributions to many aspects of quantum field theory and its applications to high energy physics. Reviewing all of them in a short talk is an impossible task. I will limit myself to a selection of topics, largely influenced by the subject of this Workshop (Quantum Chromodynamics) and by my personal interests.

2 Soft gluon resummation

Stefano's and Marcello's results I am most familiar with are those obtained in the context of the all-order resummation of large logarithmic contributions in the perturbative expansion of cross sections close to the threshold energy for the production of heavy systems. I summarize here the basic ideas of this subject following a simple and beautiful review presented by Stefano at a workshop in Montpellier in 1996. [1]

Physical cross sections are always inclusive over arbitrarily soft particles in the final state, because of finite detector resolution. On the other hand, inclusiveness plays a crucial role in QCD calculations, because infrared divergences from virtual corrections are cancelled by radiation of undetected real gluons. The finite left-over of these cancellations give large contributions if the tagged final state is forced to take most of the available energy.

Let us consider a generic process, e.g. the production of some heavy system accompanied by unobserved radiation, which carries a fraction $1 - z$ of the available energy $\sqrt{s}$. Loop corrections and real emission contributions to the cross section at order α_s are separately infrared divergent, and must be regularized by an infrared cut-off ϵ:

$$\frac{dw_{\text{virtual}}}{dz} = -2C\,\alpha_s\,\delta(1-z)\int_0^{1-\epsilon}\frac{dy}{1-y}\log\frac{1}{1-y} + \text{finite terms} \tag{1}$$

$$\frac{dw_{\text{real}}}{dz} = +2C\,\alpha_s\,\frac{1}{1-z}\log\frac{1}{1-z}\,\theta(1-\epsilon-z) + \text{finite terms} \tag{2}$$

*e-mail:giovanni.ridolfi@ge.infn.it

because of the bremsstrahlung energy spectrum $d\omega/\omega$ and the spectrum for collinear emission, dk_T^2/k_T^2. The sum of the two contributions is finite as $\epsilon \to 0$, and can be written

$$\frac{dw(z)}{dz} = \frac{dw_{\text{virtual}}}{dz} + \frac{dw_{\text{real}}}{dz} = 2\,C\,\alpha_s \left[\frac{1}{1-z}\,\log\frac{1}{1-z}\right]_+ \tag{3}$$

with the usual definition of the + distributions

$$\int_0^1 dz\,[D(z)]_+ f(z) = \int_0^1 dz\,D(z)\,[f(z) - f(1)]\,. \tag{4}$$

The contribution of virtual corrections is concentrated at $z = 1$, while the real emission contribution is spread over the range $x < z < 1$, where x is the fraction of total energy carried by the observed final state. Hence, the contribution of soft emission to the cross section is proportional to

$$\int_x^1 dz\,\frac{dw}{dz} = -C\,\alpha_s\,\log^2(1-x), \tag{5}$$

a finite left-over of the cancellation of infrared divergences. When observations are restricted to the region of x close to 1, the phase space for real emission is suppressed, and the finite left-over becomes large. The same mechanism is replicated at higher orders; at order n, at most two powers of $\log(1-x)$ for each power of α_s appear in the perturbative coefficients:

$$C_n(x)\alpha_s^n = \alpha_s^n \sum_{m=1}^{2n} c_{nm}\,\log^m(1-x) + \text{ non singular terms.} \tag{6}$$

The perturbative expansion becomes unreliable in the limit $x \to 1$, and therefore logarithmically enhanced contributions must be resummed to all orders. Typical examples of processes where resummation of soft gluons plays a role are lepton-nucleon scattering in the quasi-elastic limit ($x = x_{\text{Bj}} \to 1$), the production of systems with a large invariant mas Q^2 (Drell-Yan pairs, Higgs) close to threshold ($x = \frac{Q^2}{s}$ for $s \gtrsim Q^2$), transverse momentum q_T spectra in the small-q_T region ($x = 1 - \frac{q_T^2}{Q^2}, q_T^2 \ll Q^2$)).

Soft gluon resummations are typically performed after a suitable integral tranform. In the case of threshold production, one takes Mellin moments of the cross section with respect to x:

$$\hat{F}(N) = \int_0^1 dx\,x^{N-1} F(x); \qquad F(N) = \frac{1}{2\pi i}\int_{\bar{N}-i\infty}^{\bar{N}+i\infty} dN\,x^{-N}\hat{F}(N). \tag{7}$$

(in the case of transverse momentum the appropriate transform is two-dimensional Fourier transform with respect to $\vec{q}_T$.) In the space of the Mellin variable N, the region $x \to 1$ is mapped in the large N region; in this limit, the perturbative expansion of a coefficient function takes the form

$$C(N) = \sum_{n=0}^{\infty} \alpha_s^n \sum_{k=1}^{2n} c_{nk}\,\log^k N + \text{ non singular terms} \tag{8}$$

with at most two powers of $\log N$ for each power of α_s. Terms with $n+1 \le k \le 2n$ are leading logs, while terms with $k = n$ are next-to-leading logs.

Resummation of soft emission contributions at the leading logarithmic approximation can be performed easily in QED on the basis of the eikonal approximation. However, its generalization to QCD, and to higher logarithmic accuracy, is highly non trivial. Both Stefano and Marcello have given central contribution in this direction. I summarize them below.

1. Eikonal emission exponentiates in QED because soft photons are emitted independently of each other. This is not the case in QCD, because of the presence of three-gluon and four-gluon vertices. However, gluon correlations can be shown to cancel in the soft limit, therefore leading to the exponentiation of the single-emission cross section in the leading log approximation. A proof of the exponentiation at leading log in QCD was given by Stefano and Marcello in ref. [2].

2. In QCD, running coupling effects are especially relevant. It was shown by Marcello and collaborators (see e.g. refs. [3], [4]) that the appropriate scle choice for the running coupling is the transverse momentum k_T of emitted gluons with respect to the direction of the emitting parton. On the basis of kinematics, k_T^2 is bounded from above by the hard scale of the process Q^2 times a power of $1-x$. Hence, the upper bound on the transverse momentum can be very different from Q^2 in the threshold limit. This is an important point, because by this choice leading log terms of order $\alpha_s^k \log^{k+1} N$ are resummed.

3. Also next-to-leading logarithmic terms can be shown to take the form of an exponential of a function of the conjugated variable N. A proof was given by Stefano and Luca Trentadue in ref. [5].

In the case of deep-inelastic scattering and Drell-Yan pair production, the resummed coefficient functions in the conjugate space take the form

$$\frac{C_N(Q^2)}{C_N^{\rm LO}(Q^2)} = g_0(Q^2) \exp G_N(Q^2) + O\left(\frac{\log^k N}{N}\right) \tag{9}$$

where $C_N^{\rm LO}(Q^2)$ is the leading-order result, and

$$\begin{aligned} G_N^{\rm DIS} Q^2) &= \log \Delta_q(Q^2,\mu^2) + \log J_q(Q^2) + \log \Delta_{\rm int}^{\rm DIS}(Q^2) \\ G_N^{\rm DY} Q^2) &= 2\log \Delta_q(Q^2,\mu^2) + \log \Delta_{\rm int}^{\rm DY}(Q^2) \end{aligned}$$

with

$$\begin{aligned} \Delta_q(Q^2,\mu^2) &= \int_0^1 dz \frac{z^{N-1}-1}{1-z} \int_{\mu^2}^{Q^2(1-z)^2} \frac{dq^2}{q^2} A(\alpha_s(q^2)) \\ J_q(Q^2,\mu^2) &= \int_0^1 dz \frac{z^{N-1}-1}{1-z} \left[\int_{Q^2(1-z)^2}^{Q^2(1-z)} \frac{dq^2}{q^2} A(\alpha_s(q^2)) + B(\alpha_s(Q^2(1-z)))\right] \\ \Delta_{\rm int}(Q^2,\mu^2) &= \int_0^1 dz \frac{z^{N-1}-1}{1-z} D(\alpha_s(Q^2(1-z)^2)) \end{aligned}$$

and A, B, D analytic functions of their argument. After performing the integrals, the exponent takes the form of an expansion in powers of α_s with fixed $\alpha_s \log N$.

A difficulty immediately arises. The ratio $\frac{C_N(Q^2)}{C_N^{\rm LO}(Q^2)}$ is a function of $\alpha_s(Q^2/N^a), a = 1,2$ which can be expanded in powers of $\alpha_s(Q^2)$. To next-to-leading log we have

$$\alpha_s\left(\frac{Q^2}{N^a}\right) = \frac{\alpha_s(Q^2)}{1+L}\left[1-\alpha_s(Q^2)\frac{\beta_1}{\beta_0}\frac{\log(1+L)}{1+L}\right]; \qquad L = \alpha_s(Q^2)\beta_0 \log\frac{1}{N^a} \tag{10}$$

which has a simple pole at $L=-1$, or

$$N = N_L \equiv e^{\frac{1}{\bar\alpha}}; \qquad \bar\alpha = a\beta_0\alpha_s(Q^2) \tag{11}$$

(the so-called Landau singularity) and a branch cut on the real positive N axis for $\mathrm{Re}\, N > N_L$. Hence, the inverse Mellin transform of $C_N(Q^2)/C_N^{\mathrm{LO}}(Q^2)$ does not exist. This has to do with non-perturbative contributions, which are not under control in perturbation theory.

A solution, now widely adopted, is presented in a paper [6] by Stefano, in collaboration with M. Mangano, P. Nason and L. Trentadue. Their suggestion (named the minimal prescription) is in fact very simple: they suggest to define

$$\sigma(x, Q^2) = \frac{1}{2\pi i} \int_{c-i\infty}^{c+i\infty} dN\, x^{-N}\, \mathcal{L}(N, Q^2)\, C(N, \alpha_s(Q^2)) \tag{12}$$

with the constant c taken to the right of all singularities of $C(N)$, but to the left of the Landau pole, as shown in fig. 1. Clearly, this is not a true inverse Mellin: the integrand is not analytical

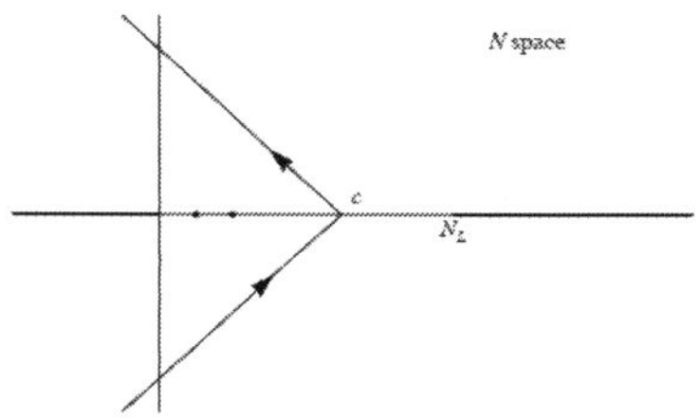

Figure 1. The inversion contour of the minimal prescription.

in any right half-plane, because of the branch cut due to the Landau pole. However, it is shown in ref. [6] that the minimal prescription, which provides a well-defined cross section for all values of x, is an asymptotic sum of the original (divergent) perturbative expansion. Furthermore, the difference between the original series, truncated at the best-approximation term, and the minimal prescription, is suppressed more strongly than any power of Λ^2/Q^2.

The minimal prescription allows one to obtain resummed predictions for physical observables, and since its formulation an impressive amount of phenomenological results have been produced by Stefano and collaborators. An important example is the calculation of the cross section for the production of a Higgs boson at hadron colliders, presented in ref. [7]. Figs. 2 and 3 are taken from that paper.

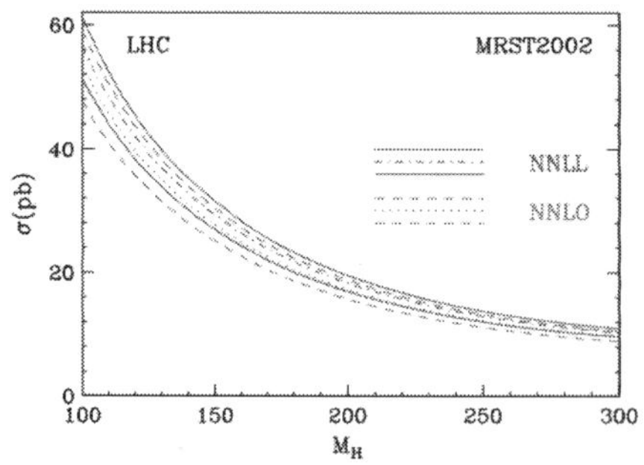

Figure 2. Higgs production at the LHC.

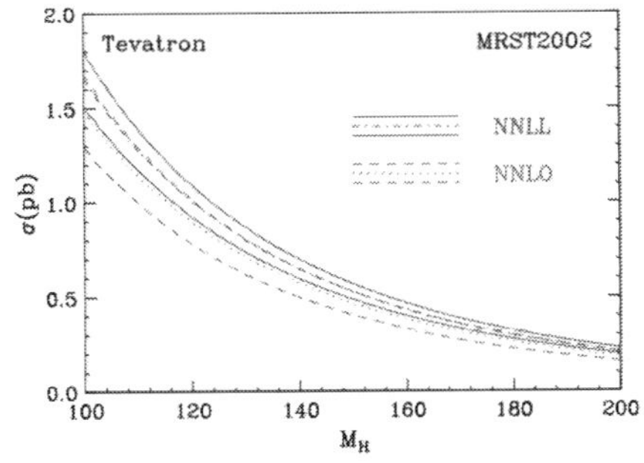

Figure 3. Higgs production at the Tevatron.

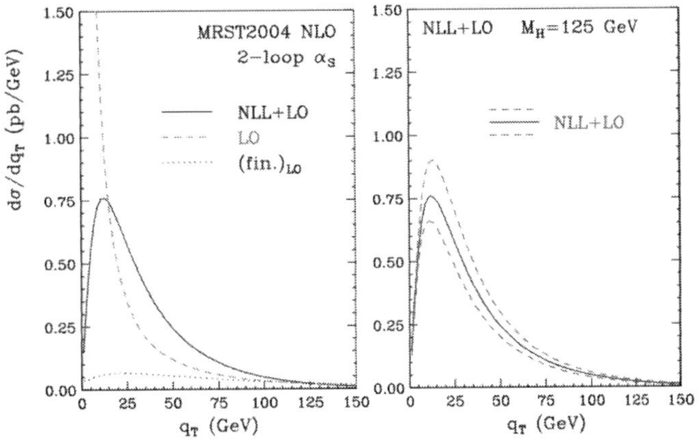

Figure 4. Left: NLL+LO compared with the LO spectrum. Right: uncertainty band from scale variations.

The uncertainty bands are obtained by varying the renormalization and factorization scales around a central value. We note that resummation has a sizable effect on central values, and considerably reduces the theoretical uncertainty.

In some cases, cross section are affected not only quantitatively, but also qualitatively, by the resummation of soft logarithms. This is the case for transverse momentum distributions, where the inclusion of large logs of q_T^2/Q^2 produces a suppression of the cross section in the low-q_T region. The q_T spectrum of Higgs production at the LHC was computed in ref. [8], from which figs. 4 and 5 are taken.

Many other applications of the resummation formalism outlined above were performed by Stefano and a long list of collaborators. [10]-[25].

The subject of soft gluon resummation was addressed later by Stefano with Pino Marchesini and Bryan Webber from a different point of view, which proves to be suitable for implementation in shower Monte Carlo Codes. [9]

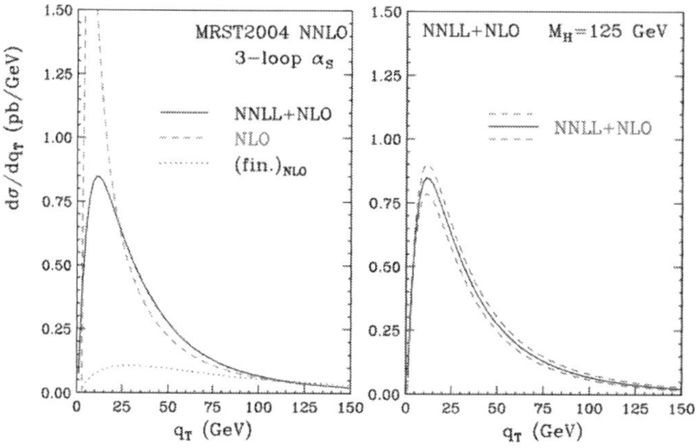

Figure 5. Left: NNLL+NLO compared with the LO spectrum. Right: uncertainty band from scale variations.

3 High-energy resummation

A relevant part of the collaboration between Stefano and Marcello is the study of perturbative QCD in the large energy regime. These subject was addressed in a series of papers in collaboration with Francesco Hautmann ([26]-[32]).

Perturbative QCD predictions for hadronic processes at squared energy s and large transverse momentum $p_T \sim s$ are remarkably accurate in the regime $p_T \gg \Lambda$, where Λ is the QCD energy scale, of order a few hundred MeV. Indeed, non-perturbative contributions suppressed by powers of Λ/p_T, while logarithmic corrections to the naive parton model are systematically computable as a power series in $\alpha_s(p_T^2) \sim (\beta_0 \log \frac{p_T^2}{\Lambda^2})^{-1} \ll 1$. However, in the regime $s \gg p_t^2 \gg \Lambda^2$, powers of $\log x = \log \frac{p_t^2}{s}$ (usually referred to as small-x, or high-energy logarithms) appear in the perturbative coefficients and spoil the accuracy of perturbative predictions. A similar situation arises whenever the process is characterized by two (or more) hard scales, sizably different from each other; an example is the production of heavy quarks.

The leading high-energy logarithmic contributions to total cross sections are powers of $\alpha_s \log x$, which is of order one in the small-x limit. Stefano, Marcello and F. Hautmann have shown in refs. [26]-[32] that these large logarithmic contributions can be summed to all orders for hard processes directly coupled to gluons. To this purpose, they prove a high-energy (or k_T) factorization property of the cross section, valid in the small x limit, where by a universal k_T depending parton density is convoluted with a partonic cross section with an off-shell gluon in the initial state.

4 All-order calculations and Monte Carlo codes

In a beautiful series of papers in collaboration with Mike Seymour [33]-[35], Stefano has introduced a generalization of the subtraction method for the computation of high-order perturbative contributions to QCD observables

The cross section for a process with m partons in the final state at leading order receives next-to-leading order corrections from emission of one extra parton, and from virtual corrections to the m parton process:

$$d\sigma^{\text{NLO}} = d\sigma^{\text{R}}_{\text{m+1partons}} + d\sigma^{\text{v}}_{\text{m partons}}. \tag{13}$$

Both contributions are divergent in the infrared region of the momentum of virtual or real gluons, and the divergence is cancelled in the sum.

The cancellation of infrared singularities is not easy to implement numerically, because it takes place between processes with different final states. In most calculations, the so-called subtraction method is employed, which amounts to add and subtract an arbitrary term with $m + 1$ partons in the final state:

$$d\sigma^{\rm NLO} = \left[d\sigma^{\rm R}_{\rm m+1partons} - d\sigma^{\rm A}_{\rm m+1partons}\right] + d\sigma^{\rm A}_{\rm m+1partons} + d\sigma^{\rm v}_{\rm m\,partons}. \tag{14}$$

The cross section $d\sigma^{\rm A}$ is chosen in such a a way that it cancels the singularities in $d\sigma^{\rm R}_{\rm m+1partons}$ and can be integrated analytically in the singular region. Furthermore, it is convenient to choose it so that it is independent of the observable one wants to compute.

The Catani-Seymour formalism, named the dipole formalism by the authors, is a choice of $d\sigma^{\rm A}$ which is completely general, i.e. not only observable-independent for a given process, but also process independent. The dipole implementation of the subtraction method proves to be especially useful for the implementation of QCD corrections in a Monte Carlo event generator.

5 Non-QCD research

Although not directly related to QCD, some other achievements in Marcello's scientific work are worth mentioning. In the early stage of his scientific career Marcello gave significant contributions in the study of composite models and relativistic bound states, mainly in collaboration with Pietro Menotti at Scuola Normale Superiore in Pisa, where he was a student. [39]-[36]

More recently, Marcello has transferred his knowledge in QCD in an interesting series of papers about the role of large logarithmic corrections arising in the electroweak theory. Most of this work was performed in collaboration with his son Paolo. [40]-[47]

In the latest stage of his scientific activity, Marcello devoted his attention to gravitational scattering at transplanckian energies. [48]

6 Personal recollections

I would like to conclude with some personal reflections on my relationships with Marcello and Stefano. I had the opportunity to meet Marcello a few times between 1986 and 1988, when I was a student in Pisa. Later, in 2001, we were both visiting the CERN Theory Division for an extended period, which gave me many chances to spend time with him. I was always struck by Marcello's profound and vast knowledge of quantum field theory and advanced mathematics, as well as by his calm and pleasant way of discussing both physics and other topics.

Stefano, on the other hand, was a close friend. We first met at an International School organized by the Italian Physical Society in Varenna, on Lake Como, in 1984, and we stayed in close contact ever since. Our scientific interests overlapped significantly, yet for various reasons, we never had the chance to collaborate. However, I was fortunate to have many conversations with Stefano, which were often illuminating. In those moments, I always felt that Stefano's understanding of the subject went far deeper than mine, and that, in some way, I could not reach the level of insight he possessed. I cherish those conversations as a precious gift. For that, and for all the time we shared, I will miss Stefano deeply.

References

[1] S. Catani, Nucl. Phys. B Proc. Suppl. **54** (1997), 107
[2] S. Catani and M. Ciafaloni, Nucl. Phys. B **236** (1984), 61
[3] D. Amati, A. Bassetto, M. Ciafaloni, G. Marchesini and G. Veneziano, Nucl. Phys. B **173** (1980), 429
[4] M. Ciafaloni and G. Curci, Phys. Lett. B **102** (1981), 352
[5] S. Catani and L. Trentadue, Nucl. Phys. B **327** (1989), 323
[6] S. Catani, M. L. Mangano, P. Nason and L. Trentadue, Nucl. Phys. B **478** (1996), 273
[7] S. Catani, D. de Florian, M. Grazzini and P. Nason, JHEP **07** (2003), 028
[8] G. Bozzi, S. Catani, D. de Florian and M. Grazzini, Nucl. Phys. B **737** (2006), 73
[9] S. Catani, B. R. Webber and G. Marchesini, Nucl. Phys. B **349** (1991), 635
[10] S. Catani, M. Grazzini and H. Sargsyan, JHEP **11** (2018), 061
[11] S. Catani, D. de Florian, G. Ferrera and M. Grazzini, JHEP **12** (2015), 047
[12] S. Catani, M. Grazzini and A. Torre, Nucl. Phys. B **890** (2014), 518
[13] S. Catani, L. Cieri, D. de Florian, G. Ferrera and M. Grazzini, Nucl. Phys. B **888** (2014), 75
[14] S. Catani, L. Cieri, D. de Florian, G. Ferrera and M. Grazzini, Nucl. Phys. B **881** (2014), 414
[15] S. Catani, M. Grazzini and A. Torre, Nucl. Phys. B **874** (2013), 720
[16] S. Catani and M. Grazzini, Nucl. Phys. B **845** (2011), 297
[17] G. Bozzi, S. Catani, G. Ferrera, D. de Florian and M. Grazzini, Phys. Lett. B **696** (2011), 207
[18] G. Bozzi, S. Catani, G. Ferrera, D. de Florian and M. Grazzini, Nucl. Phys. B **815** (2009), 174
[19] G. Bozzi, S. Catani, D. de Florian and M. Grazzini, Nucl. Phys. B **791** (2008), 1
[20] R. Bonciani, S. Catani, M. L. Mangano and P. Nason, Phys. Lett. B **575** (2003), 268
[21] M. Cacciari and S. Catani, Nucl. Phys. B **617** (2001), 253
[22] S. Catani, M. L. Mangano, P. Nason, C. Oleari and W. Vogelsang, JHEP **03** (1999), 025
[23] S. Catani, M. L. Mangano and P. Nason, JHEP **07** (1998), 024
[24] R. Bonciani, S. Catani, M. L. Mangano and P. Nason, Nucl. Phys. B **529** (1998), 424 [erratum: Nucl. Phys. B **803** (2008), 234]
[25] S. Catani, L. Trentadue, G. Turnock and B. R. Webber, Nucl. Phys. B **407** (1993), 3
[26] S. Catani, M. Ciafaloni and F. Hautmann, Phys. Lett. B **242** (1990), 97-102
[27] S. Catani, M. Ciafaloni and F. Hautmann, Nucl. Phys. B Proc. Suppl. **18** (1991), 220-225
[28] S. Catani, M. Ciafaloni and F. Hautmann, Nucl. Phys. B **366** (1991), 135-188
[29] S. Catani, M. Ciafaloni and F. Hautmann, Nucl. Phys. B Proc. Suppl. **27** (1992), 33-38
[30] S. Catani, M. Ciafaloni and F. Hautmann, Nucl. Phys. B Proc. Suppl. **29** (1992), 182-191
[31] S. Catani, M. Ciafaloni and F. Hautmann, Phys. Lett. B **307** (1993), 147-153
[32] S. Catani and F. Hautmann, Nucl. Phys. B **427** (1994), 475
[33] S. Catani and M. H. Seymour, Phys. Lett. B **378** (1996), 287
[34] S. Catani and M. H. Seymour, Nucl. Phys. B **485** (1997), 291 [erratum: Nucl. Phys. B **510** (1998), 503-504]
[35] S. Catani, S. Dittmaier, M. H. Seymour and Z. Trocsanyi, Nucl. Phys. B **627** (2002), 189

[36] M. Ciafaloni and P. Menotti, Phys. Rev. **140** (1965), B929
[37] M. Ciafaloni and P. Menotti, Nuovo Cim. **46** (1966), 162
[38] M. Ciafaloni and P. Menotti, Phys. Rev. **173** (1968), 1575
[39] M. Ciafaloni and P. Menotti, Nucl. Phys. B **54** (1973), 483
[40] M. Ciafaloni, P. Ciafaloni and D. Comelli, Phys. Rev. Lett. **84** (2000), 4810
[41] M. Ciafaloni, P. Ciafaloni and D. Comelli, Nucl. Phys. B **589** (2000), 359
[42] M. Ciafaloni, P. Ciafaloni and D. Comelli, Phys. Lett. B **501** (2001), 216
[43] M. Ciafaloni, P. Ciafaloni and D. Comelli, Nucl. Phys. B **613** (2001), 382
[44] M. Ciafaloni, P. Ciafaloni and D. Comelli, Phys. Rev. Lett. **87** (2001), 211802
[45] M. Ciafaloni, P. Ciafaloni and D. Comelli, Phys. Rev. Lett. **88** (2002), 102001
[46] M. Ciafaloni, P. Ciafaloni and D. Comelli, JHEP **05** (2008), 039
[47] M. Ciafaloni, P. Ciafaloni and D. Comelli, JHEP **03** (2010), 072
[48] M. Ciafaloni, D. Colferai and G. Veneziano, Phys. Rev. D **99** (2019) no.6, 066008

EPJ Web of Conferences **314**, 00002 (2024) https://doi.org/10.1051/epjconf/202431400002
QCD@Work 2024

Hunting new animalcula with rare K and B decays

Andrzej J. Buras[1,2,*]

[1]TUM Institute for Advanced Study, Lichtenbergstr. 2a, D-85748 Garching, Germany
[2]Physik Department, TUM School of Natural Sciences, TU München, James-Franck-Straße, D-85748 Garching, Germany

Abstract. We summarize the recent strategy for an efficient hunting of new animalcula with the help of rare K and B decays that avoids the use of the $|V_{cb}|$ and $|V_{ub}|$ parameters that are subject to tensions between their determinations from inclusive and exclusive decays. In particular we update the values of the $|V_{cb}|$-independent ratios of various K and B decay branching ratios predicted by the Standard Model. We also stress the usefulness of the $|V_{cb}|-\gamma$ plots in the search for new physics. We select *the magnificant seven* among rare K and B decays that should play a leading role in the search for new physics due to their theoretical cleanness: $B^+ \to K^+(K^*)\nu\bar{\nu}$ and $K^+ \to \pi^+\nu\bar{\nu}$ measured recently by Belle II and NA62, respectively, $K_L \to \pi^0\nu\bar{\nu}$ investigated by KOTO and also $B_{s,d} \to \mu^+\mu^-$ and $K_S \to \mu^+\mu^-$ measured by the LHCb, CMS and ATLAS experiments at CERN.

1 Introduction

The year 1676 was a very important year for the humanity. In this year Antoni van Leeuvenhoek (1632-1723) discovered the empire of bacteria. He called these small creatures *animalcula* (small animals). This discovery was a mile stone in our civilization for at least two reasons:

- He discovered invisible to us creatures which over thousands of years were systematcally killing the humans, often responsible for millions of death in one year. While Antoni van Leeuvanhoek did not know that bacteria could be dangerous for humans, his followers like L. Pasteur (1822-1895), Robert Koch (1843-1910) and other *microbe hunters* not only realized the danger coming from this tiny creatures but also developed weapons against this empire.
- He was the first human who looked at short distance scales invisible to us, discovering thereby a new *underground world*. At that time researchers looked mainly at large distances, discovering new planets and finding laws, like Kepler laws, that Izaak Newton was able to derive from his mechanics.

While van Leeuvanhoek could reach the resolution down to roughly 10^{-6}m, over the last 350 years this resolution could be improved by many orders of magnitude. On the way down to shortest distance scales scientists discovered *nanouniverse* (10^{-9}m), *femtouniverse* (10^{-15}m) relevant for nuclear particle physics and low energy elementary particle physics and finally *attouniverse* (10^{-18}m) that is the territory of contemporary high energy elementary particle physics.

In this decade and the coming decades we will be able to improve the resolution of the short distance scales by at least an order of magnitude, extending the picture of fundamental physics down

*Talk given at QCD@Work 2024, Trani, June 2024.

to scales $5 \cdot 10^{-20}$m with the help of the high energy processes at the Large Hadron Collider (LHC). Further resolution, down to scales as short as 10^{-21}m (*zeptouniverse*) or even shorter scales, should be possible with the help of high precision experiments in which flavour violating processes played a prominent role for decades [1]. These notes deal with the latter route to the short distance scales which is an indirect route based solely on quantum fluctuations.

The main goal of these efforts is not only the curiosity of whether new animalcula beyond the ones of the Standard Model (SM), already discovered, exist. The main motivation is the hope that finding them will help us to answer a number of very important questions that the SM cannot answer. They are well known and will not be listed here.

The main strategy is at first sight very simple. One calculates various observables like branching ratios of various decays of mesons and leptons within the SM and compares them with the experimental data. Finding the departures from SM predictions, called these days *anomalies*, signals the presence of new animalcula that through quantum fluctuations affect SM predictions. In order to identify the nature of these animalcula it is crucial to measure many of these anomalies as precisely as possible.

In a given *measurement* of an anomaly both experimentalists and theorists are involved simply because in order to find departures from SM predictions one needs both precise experimental data and precise theory. Once these two requirements are satisfied, there are several routes one can follow. The most common one found in the literature are global fits in concrete new physics (NP) models in which these anomalies could be explained.

Another route is to study first the patterns of anomalies observed in the the data and compare them with the patterns of deviation from SM predictions in a given NP scenario. Such patterns, that expose suppressions and enhancements of various observables relative to SM predictions, can be considered as DNAs of the animalcula hunted by us. In particular the correlations between various enhancements and suppressions can rule out some NP scenarios already before any global fit is performed. This is the strategy proposed by Jennifer Girrbach-Noe and myself in 2013 [2]. It has been documented in several subsequent papers, in particular in my book [3] and recently in [4] with many colourful plots. Therefore I will not discuss it here.

The present notes deal with a strategy for obtaining most precise SM predictions for rare K and B meson decays to date, a very importat ingredient in the indirect search for NP. It has been developed this time in collaboration with Elena Venturini [5, 6].

The strategy in question, to be called **BV-strategy** for simplicity, has been motivated by the problems in the determination of the CKM parameters that play a very important role in SM predictions for rare K and B decays. Their values are usually obtained from global fits dominated by UTfitter [7] and CKMfitter [8]. However as stressed in [5, 6] and later in [9] this determination is presently problematic for the following reasons:

- In a global fit which contains processes that could be infected by NP the resulting CKM parameters are also infected by it and consequently the resulting branching ratios cannot be considered as genuine SM predictions. Consequently the resulting deviations from SM predictions obtained in this manner (the pulls) are not the deviations one would find if the CKM parameters were not infected by NP.
- Tensions in the determinations of $|V_{cb}|$ and $|V_{ub}|$ from inclusive and exclusive tree-level decays [10–12]. Using these results lowers the precision with which CKM parameters can be determined and consequently also lowers the precision of SM predictions for many observabies as illustrated soon. Therefore the inclusion of these determinations in the fit should be avoided until theorists agree what the values of $|V_{cb}|$ and $|V_{ub}|$ are.

- Hadronic uncertainties in some observables included in the fit are much larger than in many rare K and B decays. Even if they can be given a lower weight in the fit, they lower the precision and should be presently avoided.

2 BV-strategy

In what follows I want to summarize the BV strategy developed in two papers with Elena Venturini [5, 6] which generalized my 2003 strategy used for $B_{s,d} \to \mu^+\mu^-$ decays [13] to all K and B decays. This strategy deals with the second and the third item above but as I realized in [9] it solves also the first problem. It consists of five steps.

Step 1

Remove CKM dependence from observables as much as possible by calculating suitable ratios of decay branching ratios to the mass differences ΔM_s and ΔM_d in the case of B_s and B_d decays, respectively and to the parameter $|\varepsilon_K|$ in the case of Kaon decays. By suitable we mean for instance that in order to eliminate the $|V_{cb}|$ dependences in the branching ratios for $K^+ \to \pi^+\nu\bar{\nu}$ and $K_L \to \pi^0\nu\bar{\nu}$, the parameter $|\varepsilon_K|$ has to be raised, as given later, to the power 0.82 and 1.18, respectively. For $B_{s,d}$ decays one just divides the branching ratios by $\Delta M_{s,d}$, respectively. In this manner CKM dependence can be fully eliminated for all B decay branching ratios. For K decays only the dependence on the angle β in the Unitarity Triangle (UT) remains. The dependence on the angle γ in the UT is practically absent so that future improvements on the measurements of γ by LHCb and Belle II collaborations will not have any impact on these particular ratios although they will be very important for other ratios discussed by us. On the other hand improved measurements of β and improved values of hadronic parameters will reduce the uncertainties in these ratios. It should be emphasized that already these ratios constitute very good tests of the SM, simply because they are significantly more precise than the individual branching ratios.

Step 2

Set ΔM_s, ΔM_d, $|\varepsilon_K|$ and the mixing induced CP asymmetries $S_{\psi K_S}$ and $S_{\psi\phi}$ to their experimental values. This is done usually in global fits as well but here we confine the fit to these observables. The justification for this step is the fact that within a good approximation all these observables can be simultaneously described within the SM without any need for NP contributions [6] and the theoretical and experimental status of these $\Delta F = 2$ observables is exeptionally good. In turn this step not only avoids the tensions in the determinations of $|V_{cb}|$ and $|V_{ub}|$ in tree level decays, but also provides in Step 5 SM predictions for numerous rare K and B branching ratios that are most accurate to date [5, 6, 9].

Step 3

In order to be sure that the $\Delta F = 2$ archipelago is not infected by NP a rapid test has to be performed with the help of the $|V_{cb}| - \gamma$ plot [5, 6]. This test turns out to be negative dominantly thanks to the 2+1+1 HPQCD lattice calculations of $B_{s,d} - \bar{B}_{s,d}$ hadronic matrix elements [14][1]. The superiority of the $|V_{ub}| - \gamma$ plots over UT plots in this context has been emphasized in [18]. We will stress it again below.

[1] Similar results for ΔM_d and ΔM_s hadronic matrix elements have been obtained within the HQET sum rules in [15] and [16, 17], respectively.

Step 4

As the previous step has lead to a negative rapid test we can now determine the CKM parameters without NP infection on the basis of $\Delta F = 2$ observables alone. It should be noted that this step can be considered as a reduced global fit of CKM parameters in which only $\Delta F = 2$ observables have been taken into account.

Step 5

All the problems listed above are avoided in this manner and having CKM parameters at hand one can make rather precise SM predictions for the observables outside the $\Delta F = 2$ archipelago and compare them with the experimental data. In particular one can predict suitable $|V_{cb}|$-independent ratios between various branching ratios not only within a given meson system but also involving different meson systems. Such ratios depend generally on β and γ with β already precisely known and γ determined in Step 4 more precisely than presently by the LHCb and Belle II experiments.

In order to motivate this strategy in explicit terms, let us recall the values of $|V_{cb}|$ extracted from inclusive and exclusive tree-level semi-leptonic $b \to c$ decays [11, 12]

$$|V_{cb}|_{\rm incl} = (41.97 \pm 0.48) \cdot 10^{-3}, \qquad |V_{cb}|_{\rm excl} = (39.21 \pm 0.62) \cdot 10^{-3}. \tag{1}$$

As rare K and B decays and mixing parameters are sensitive functions of $|V_{cb}|$, varying it from $39 \cdot 10^{-3}$ to $42 \cdot 10^{-3}$ changes $\Delta M_{s,d}$ and B-decay branching ratios by roughly 16%, $K^+ \to \pi^+ \nu\bar\nu$ branching ratio by 23%, ε_K by 29% and $K_L \to \pi^0 \nu\bar\nu$ and $K_S \to \mu^+\mu^-$ branching ratios by 35%.

These uncertainties are clearly a disaster for those like me, my collaborators and other experts in NLO and NNLO calculations who spent decades to reduce theoretical uncertainties in basically all important rare K and B decays and quark mixing observables down to $(1-2)\%$ [19]. It is also a disaster for lattice QCD experts who for quark mixing observables and in particular meson weak decay constants achieved the accuracy at the level of a few percents [12].

3 The $|V_{cb}|$-independent ratios

One constructs then in Step 1 a multitude of $|V_{cb}|$-independent ratios R_i not only of branching ratios to quark mixing observables but also of branching ratios themselves. Those which involve branching ratios from different meson systems depend generally on β and γ. Once β and γ will be precisely measured, this multitude of $R_i(\beta,\gamma)$ will provide very good tests of the SM. However, using Step 5 of our strategy it is possible to make SM predictions for these ratios already today and we will list several of them below.

The details of the execution of this strategy can be found in [5, 6, 9]. In particular analytic expressions for sixteen ratios $R_i(\beta,\gamma)$ and plots for them can be found in [5]. A guide to these relations can be found in Section 4 of the latter paper. Additional ratios, predictions for all ratios considered and for 26 individual branching ratios resulting from our strategy are presented in [9]. Here we just list most interesting results obtained in these papers and update some of them.

We begin with the most interesting $|V_{cb}|$-independent ratios that involve both branching ratios and $\Delta F = 2$ observables. They read

$$\boxed{\frac{\overline{\mathcal{B}}(B_s \to \mu^+\mu^-)}{\Delta M_s} = (2.13 \pm 0.07) \times 10^{-10}\,{\rm ps}}\,, \tag{2}$$

$$\boxed{\frac{\mathcal{B}(B_d \to \mu^+\mu^-)}{\Delta M_d} = (2.02 \pm 0.08) \times 10^{-10}\,{\rm ps}}\,, \tag{3}$$

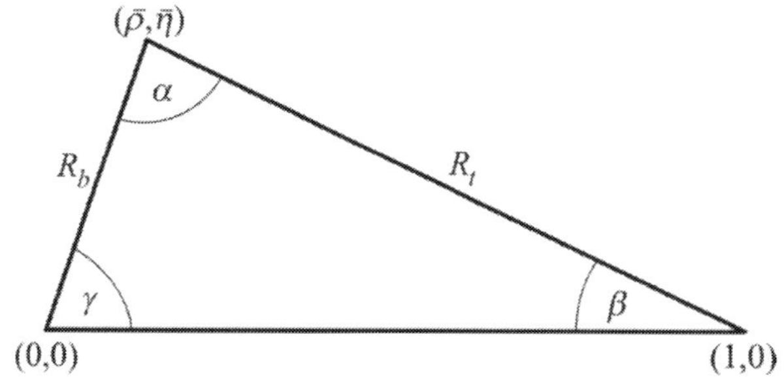

Figure 1. The Unitarity Triangle.

$$\boxed{\frac{\mathcal{B}(K^+\to\pi^+\nu\bar\nu)}{|\varepsilon_K|^{0.82}} = (1.31\pm 0.05)\times 10^{-8}\left(\frac{\sin\gamma}{\sin 64.6^\circ}\right)^{0.015}\left(\frac{\sin 22.2^\circ}{\sin\beta}\right)^{0.71},} \tag{4}$$

$$\boxed{\frac{\mathcal{B}(K_L\to\pi^0\nu\bar\nu)}{|\varepsilon_K|^{1.18}} = (3.87\pm 0.06)\times 10^{-8}\left(\frac{\sin\gamma}{\sin 64.6^\circ}\right)^{0.03}\left(\frac{\sin\beta}{\sin 22.2^\circ}\right)^{0.98}.} \tag{5}$$

The negligible dependence on γ should be noticed so that the angle β plays more important role than γ in this context. Using the experimental values of ΔM_s, ΔM_d, $|\varepsilon_K|$ and β these ratios imply the most accurate predictions for the four branching ratios in question in the SM to date [5, 6]. Moreover, they are independent of the value of $|V_{cb}|$. We will present them in Section 6.

The ratios $R_i(\beta,\gamma)$ within the SM are given below. The explicit expressions for these ratios as functions of β and γ are given in [5]. Here we just list the final results using the CKM parameters in (19) which were not given there. Moreover in the case of the ratios R_5 and R_7 we use the most recent results for the formfactors entering $\mathcal{B}(B^+\to K^+\nu\bar\nu)$ from the HPQCD collaboration [20, 21]. Here we go.

$$\boxed{R_0(\beta) = \frac{\mathcal{B}(K^+\to\pi^+\nu\bar\nu)}{\mathcal{B}(K_L\to\pi^0\nu\bar\nu)^{0.7}} = (2.03\pm 0.11)\times 10^{-3},} \tag{6}$$

$$\boxed{R_{\rm SL} = \frac{\mathcal{B}(K_S\to\mu^+\mu^-)_{\rm SD}}{\mathcal{B}(K_L\to\pi^0\nu\bar\nu)} = (6.29\pm 0.52)\times 10^{-3},} \tag{7}$$

$$\boxed{R_1(\beta,\gamma) = \frac{\mathcal{B}(K^+\to\pi^+\nu\bar\nu)}{\left[\overline{\mathcal{B}}(B_s\to\mu^+\mu^-)\right]^{1.4}} = 53.23\pm 2.75,} \tag{8}$$

$$\boxed{R_2(\beta,\gamma) = \frac{\mathcal{B}(K^+\to\pi^+\nu\bar\nu)}{[\mathcal{B}(B_d\to\mu^+\mu^-)]^{1.4}} = (8.37\pm 0.47)\times 10^{3},} \tag{9}$$

$$\boxed{R_3(\beta,\gamma) = \frac{\mathcal{B}(K_L\to\pi^0\nu\bar\nu)}{\left[\overline{\mathcal{B}}(B_s\to\mu^+\mu^-)\right]^{2}} = (2.06\pm 0.16)\times 10^{6},} \tag{10}$$

$$\boxed{R_4(\beta,\gamma) = \frac{\mathcal{B}(K_L \to \pi^0 \nu\bar{\nu})}{[\mathcal{B}(B_d \to \mu^+\mu^-)]^2} = (2.83 \pm 0.24) \times 10^9\,,} \tag{11}$$

$$\boxed{R_5(\beta,\gamma) = \frac{\mathcal{B}(K^+ \to \pi^+ \nu\bar{\nu})}{[\mathcal{B}(B^+ \to K^+\nu\bar{\nu})]^{1.4}} = (2.32 \pm 0.13) \times 10^{-3}\,,} \tag{12}$$

$$\boxed{R_6(\beta,\gamma) = \frac{\mathcal{B}(K^+ \to \pi^+ \nu\bar{\nu})}{[\mathcal{B}(B^0 \to K^{0*}\nu\bar{\nu})]^{1.4}} = (8.43 \pm 1.21) \times 10^{-4}\,,} \tag{13}$$

$$\boxed{R_7 = \frac{\mathcal{B}(B^+ \to K^+ \nu\bar{\nu})}{\overline{\mathcal{B}}(B_s \to \mu^+\mu^-)} = (1.30 \pm 0.07) \times 10^3\,,} \tag{14}$$

$$\boxed{R_8 = \frac{\mathcal{B}(B^0 \to K^{*0} \nu\bar{\nu})}{\overline{\mathcal{B}}(B_s \to \mu^+\mu^-)} = (2.68 \pm 0.25) \times 10^3\,.} \tag{15}$$

4 New look at the CKM matrix and $|V_{cb}| - \gamma$ plots

Let us emphasize that our strategy uses as basic CKM parameters [5, 6, 9, 22]

$$|V_{us}| = \lambda, \qquad |V_{cb}|, \qquad \beta, \qquad \gamma, \tag{16}$$

with β and γ being the two angels of the UT shown in Fig. 1.

This differs from the standard parametrization of the CKM matrix that involves three real parameters and one complex phase

$$s_{12} = |V_{us}|, \qquad s_{13} = |V_{ub}|, \qquad s_{23} = |V_{cb}|, \qquad \gamma, \tag{17}$$

that can be determined separately in tree-level decays. Consequently, basically all flavour phenomenology in the last three decades used this set of parameters. In particular the determination of the UT was dominated by the measurements of its sides R_b and R_t through tree-level B decays and the $\Delta M_d/\Delta M_s$ ratio, respectively, with some participation of the measurements of the angle β through the mixing induced CP-asymmetries like $S_{\psi K_S}$, the parameter ε_K and much less precise angle γ. This is the case not only of global analyses by UTfitter[23] and CKMfitter [8] but also of less sophisticated determinations of the CKM matrix and of the UT.

However, as pointed out in [5, 6, 9], the most powerful strategy appears eventually to be the one which uses as basic CKM parameters the ones in (16), that is two mixing angles and two phases. This choice is superior to the one in which β is replaced by $|V_{ub}|$ for several reasons:

- The known tensions between exclusive and inclusive determinations of $|V_{cb}|$ and $|V_{ub}|$ [11, 12] are represented only by $|V_{cb}|$ which can be eliminated efficiently by constructing suitable ratios of flavour observables $R_i(\beta,\gamma)$, see previous section, which are free of the tensions in question.
- As pointed out already in 2002 [24], the most efficient strategy for a precise determination of the apex of the UT, that is $(\bar{\varrho},\bar{\eta})$, is to use the measurements of the angles β and γ. Indeed, among any pairs of two variables representing the sides and the angles of the UT that are assumed for this exercise to be known with the same precision, the measurement of (β,γ) results in the most accurate values of $(\bar{\varrho},\bar{\eta})$. The second best strategy would be the measurements of R_b and γ. However, in view of the tensions between different determinations of $|V_{ub}|$ and $|V_{cb}|$, that enter R_b, the (β,γ) strategy will be a clear winner once LHCb and Belle II collaborations will improve the measurements of these two angles.

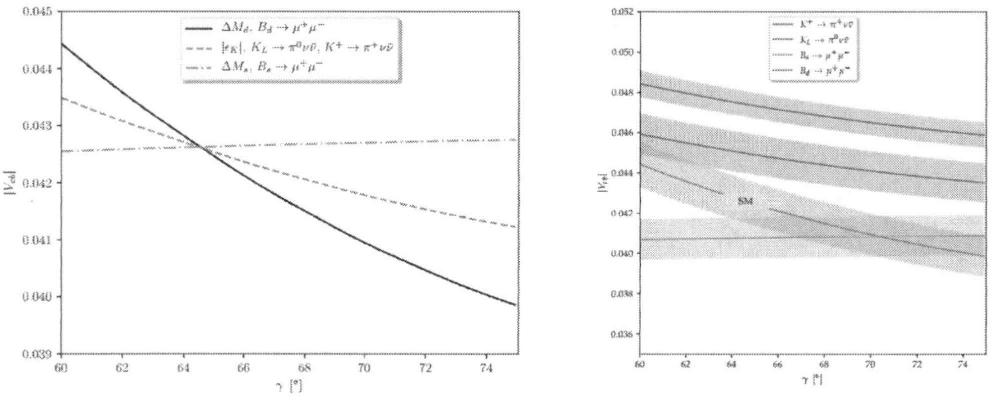

Figure 2. Left: Schematic illustration of the action of the seven observables in the $|V_{cb}| - \gamma$ plane in the context of the SM. We set $\beta = 22.2°$ and all uncertainties to zero. Right: The impact of hypothetical future measurements of the branching ratios for $K^+ \to \pi^+ \nu\bar{\nu}$, $K_L \to \pi^0 \nu\bar{\nu}$, $B_d \to \mu^+\mu^-$ and $B_s\, \mu^+\mu^-$ on the $|V_{cb}| - \gamma$ plane. All uncertainties are included. The yellow disc represents the SM as obtained in (19). From [18].

- The $|V_{cb}| - \gamma$ plots for fixed β, proposed in [5, 6] are, as emphasized in [18], useful companions to common unitarity triangle fits because they exhibit better possible inconsistences between $|V_{cb}|$ and (β, γ) determinations than the latter fits. We will demonstrate this below.

In this context let us present two simple formulae that are central in the (β, γ) strategy as they allow to calculate the appex of the UT in no time, but to my knowledge they have been presented only recently for the first time [1]. They read[2]

$$\boxed{\bar{\varrho} = \frac{\sin\beta\cos\gamma}{\sin(\beta+\gamma)}, \qquad \bar{\eta} = \frac{\sin\beta\sin\gamma}{\sin(\beta+\gamma)}.} \tag{18}$$

Evidently they can be derived by high-school students, but the UT is unknown to them and somehow no flavour physicist got the idea to present them in print until its first appearance in [1].

The superiority of the $|V_{cb}| - \gamma$ plots with respect to $|V_{cb}|$ and γ over UT plots has been emphasized in [18]. Indeed,

- They exhibit $|V_{cb}|$ and its correlation with γ determined through a given observable in the SM, allowing thereby monitoring the progress on both parameters expected in the coming years. Violation of this correlation in experiment will clearly indicate NP at work.
- They utilize the strong sensitivity of rare K decay processes to $|V_{cb}|$ thereby providing precise determination of $|V_{cb}|$ even with modest experimental precision on their branching ratios.
- They exhibit, as shown in Fig. 2, the action of ΔM_s and of B_s decays, like $B_s \to \mu^+\mu^-$ which is not possible in the common UT-plot.
- Once the accuracy of γ measurements will approach 1° it will be easier to monitor this progress on the $|V_{cb}| - \gamma$ plot than on the UT-plot.

[2] This year one can celebrate the thirties birthday of these parameters [25], with their version without bars presented earlier by Wolfenstein [26].

Table 1. Present most accurate SM estimates of the branching ratios for the *the magnificant seven* obtained using the BV-strategy. The experimental result on $K^+ \to \pi^+\nu\bar{\nu}$ from NA62 includes the recent data presented in a seminar at CERN.

Decay	SM Branching Ratio	Data
$B_s \to \mu^+\mu^-$	$(3.78^{+0.15}_{-0.10}) \cdot 10^{-9}$	$(3.45 \pm 0.29) \cdot 10^{-9}$ [27]
$B_d \to \mu^+\mu^-$	$(1.02^{+0.05}_{-0.03}) \cdot 10^{-10}$	$\leq 2.05 \cdot 10^{-10}$ [28]
$B^+ \to K^+\nu\bar{\nu}$	$(4.92 \pm 0.30) \cdot 10^{-6}$	$(13 \pm 4) \cdot 10^{-6}$ [29]
$B^0 \to K^{0*}\nu\bar{\nu}$	$(10.13 \pm 0.92) \cdot 10^{-6}$	$\leq 1.5 \cdot 10^{-5}$ [30]
$K^+ \to \pi^+\nu\bar{\nu}$	$(8.60 \pm 0.42) \cdot 10^{-11}$	$(13.0 \pm 3.1) \cdot 10^{-11}$ [31]
$K_L \to \pi^0\nu\bar{\nu}$	$(2.94 \pm 0.15) \cdot 10^{-11}$	$\leq 2.0 \cdot 10^{-9}$ [32]
$(K_S \to \mu^+\mu^-)_{\rm SD}$	$(1.85 \pm 0.12) \cdot 10^{-13}$	$\leq 2.1 \cdot 10^{-10}$ [33]

In Fig. 2 we show examples of such plots that can be used in the search for new animalcula. See figure caption for explanations.

5 CKM parameters

As the rapid test for the $\Delta F = 2$ observables turned out to be negative we can now determine the CKM parameters using these observables without NP infection. We find [6]

$$\boxed{|V_{cb}| = 42.6(4) \times 10^{-3}, \quad \gamma = 64.6(16)^\circ, \quad \beta = 22.2(7)^\circ, \quad |V_{ub}| = 3.72(11) \times 10^{-3}} \tag{19}$$

and consequently

$$\boxed{|V_{ts}| = 41.9(4) \times 10^{-3}, \qquad |V_{td}| = 8.66(14) \times 10^{-3}, \qquad {\rm Im}\lambda_t = 1.43(5) \times 10^{-4},} \tag{20}$$

$$\boxed{\bar{\varrho} = 0.164(12), \qquad \bar{\eta} = 0.341(11),} \tag{21}$$

where $\lambda_t = V_{ts}^* V_{td}$. Note that our result for γ agrees perfectly with the most recent LHCb measurement: $\gamma = 64.6(28)^\circ$.

6 The magnificant seven

Until now we dealt only with the issue of the CKM parameters but the choice of the observables that are particularly theoretically clean is also very important. In my view the seven ones listed in Table 1 are particulary promising. We give there their SM values obtained using our strategy together with the present experimental results. Clearly other decays, in particular $B \to K\ell^+\ell^-$ and $B \to K^*\ell^+\ell^-$, that remain still central in the search for NP, are very important. Unfortunately, there are different views among theorists on the role of long distance QCD effects in these decays and the interpretation of the observed anomalies in these decays is difficult at present.

The strong suppression of NP in $\Delta F = 2$ prosesses imposes stringent constraints on the parameter space of Z' models and limit their potential effects in rare decays. However, as emphasized in [34–36], the NP contributions to $\Delta F = 2$ processes in these models can be suppressed for a particular pattern of left-handed and right-handed flavour-violating Z' couplings so that NP effects in rare K and B decays can be sizable.

The implications of the suppression of NP contributions to ε_K on rare K decays within Z' models have been studied in [37]. In this case it is sufficient to choose the relevant left-handed couplings to be purely imaginary implying strong correlations between three Kaon decays listed in Table 1 and the ratio ε'/ε as seen in the left plot in Fig. 3. The strict correlation between $K^+ \to \pi^+\nu\bar{\nu}$ and $K_L \to \pi^0\nu\bar{\nu}$ decays in the presence of strong suppression of NP to ε_K seen in this figure is an example of a general correlation between these decays pointed out by Monika Blanke long time ago [38]. Its version in Z' models is presented in the right plot in Fig. 3 taken from [39].

In this context the recent NA62 result for $K^+ \to \pi^+\nu\bar{\nu}$, listed in Table 1, implies

$$\boxed{R^+_{\nu\bar{\nu}} = 1.51 \pm 0.35 \qquad (\mathrm{SM}: \; R^+_{\nu\bar{\nu}} = 1).} \tag{22}$$

making hopes that the branching ratios for $K_L \to \pi^0\nu\bar{\nu}$ and $K_S \to \mu^+\mu^-$ could be strongly enhanced over the SM values. However, to this end the experimental error on $K^+ \to \pi^+\nu\bar{\nu}$ branching ratio should be decreased.

In the case of B decays sufficient suppresion of NP contributions to ΔM_s and ΔM_d requires the presence of both left-handed and right-handed Z' couplings to quarks. The corresponding analysis for rare B decays has been performed in [40] finding again strong correlations between B-decay branching ratios in Table 1 and the ones for $B \to K\ell^+\ell^-$ and $B \to K^*\ell^+\ell^-$ allowing for the explanation of the anomalies in the latter ones despite the suppression of NP contributions to $\Delta F = 2$ processes. In 331 models, in which only left-handed Z' couplings are involved, some amount of NP contributions at the level of 5% has to be allowed to explain the anomalies in question [41].

In the case of decays with neutrinos in the final state the SM predictions are obtained assuming the Dirac nature of neutrinos. However they could be of Majorana type implying different predictions. Moreover, as neutrinos are invisible, what is really measured are the decays $K \to \pi + E_{\mathrm{miss}}$ and $B \to K(K^*) + E_{\mathrm{miss}}$ with E_{miss} standing for the missing energy. This means that instead of neutrinos there could be different neutral particles like scalars, fermions and vectors in the final state including dark particles. Strategies for disentangling these possibilities through kinematic distributions in the missing energy E_{miss} have been developed recently in [42]. See also earlier papers listed there. In particular a recent paper on the anatomy of such distributions in $K^+ \to \pi^+\nu\bar{\nu}$ [43]. In fact such a distinction is possible although in many cases it will be a challenge for experimentalists.

7 Summary and shopping list

Despite the presence of various anomalies in the existing data, it is clearly not evident which animalcula could be responsible for them. Possibly, the main candidates are Z' vector bosons, leptoquarks and vector-like quarks and vector-like leptons but to find out without any doubts what they are we need more data, in particular for theoretically clean decays as the ones listed in Table 1. Precise measurements of branching ratios for these decays and of various kinematical distributions should allow in this decade to find out which animalcula are responsible for them. The strategies presented here

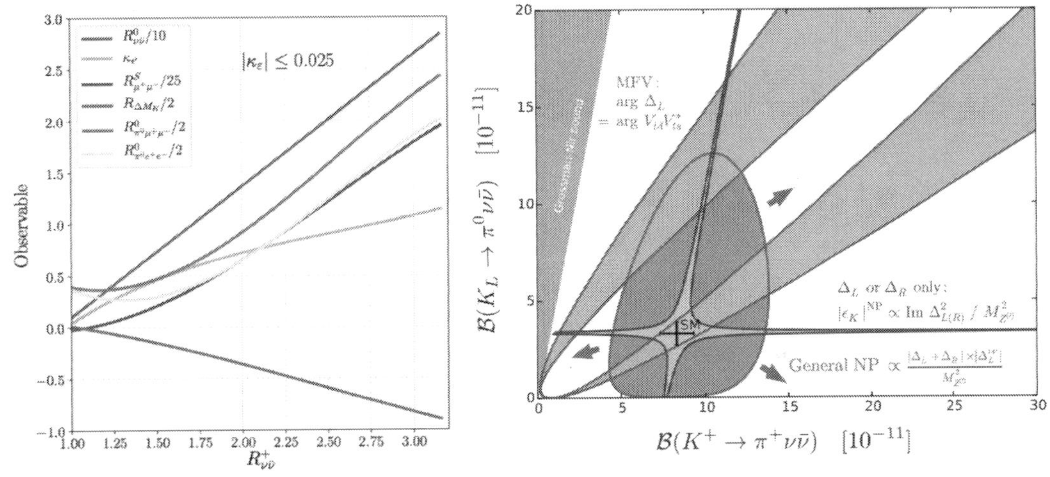

Figure 3. Left: Correlations between the observable $R^+_{\nu\bar\nu}$ and various other Kaon observables in a Z' model. All ratios $R_i = 1$ in the SM. Note that for $K_L \to \pi^0\nu\bar\nu$ and $K_S \to \mu^+\mu^-$ the ratios are divided by 10 and 25 respectively. From [37]. Right: Correlations between $K^+ \to \pi^+\nu\bar\nu$ and $K_L \to \pi^0\nu\bar\nu$ in general Z' models with green region representing MFV models and blue one coming from ε_K constraint with only LH or RH couplings [38]. From [39].

should be helpful in this respect, in particular the ratios $R_i(\beta,\gamma)$ listed in Section 3 independenty of whether NP effects are strongly suppressed in $\Delta F = 2$ processes or not.

Let me finish this writing with my shopping list for the coming years. I list here only the entries related to flavour physics.

1. Precise measurements of the branching ratios for *the magnificant seven* and of missing energy distributions for four decays among them. Presently on a forefront are the $B \to K^{(*)}\nu\bar\nu$ decays studied intensively by the Belle II experiment [29], giving some hints for NP contributions. For selected recent analyses of these data see [44–57]. The same applies to $K^+ \to \pi^+\nu\bar\nu$ for which a very interesting result has been provided very recently by the NA62 experiment listed in Table 1. Taking the present experimental result for $B_s \to \mu^+\mu^-$ allows us to determine the present $|V_{cb}|$-independent experimental values for the the ratios R_1 and R_7 in (8) and (14) that read

$$\boxed{R_1 = 91.4 \pm 22.8, \ (\mathrm{SM}: 53.23 \pm 2.75)}\ \boxed{R_7 = (3.8 \pm 1.2)\cdot 10^{-3}, \ (\mathrm{SM}: (1.30 \pm 0.07)\cdot 10^{3}).} \tag{23}$$

   They differ significantly from very precise SM values. Let us hope that the experimental errors on these ratios will decrease in the coming years.

2. Precise measurements of the branching ratios and of other observables in $B \to K\ell^+\ell^-$ and $B \to K^*\ell^+\ell^-$ decays and the clarification of the anomalies in them.

3. Precise measurements of $|V_{cb}|$ and γ in tree-level decays that would provide additional tests of the BV-strategy.

4. Clarification of anomalies in $B \to \pi K$ decays [58–63].

5. The $\Delta I = 1/2$ rule for $K \to \pi\pi$ is almost 70 years old and we do not yet know whether it is fully explained by the SM. The basic QCD dynamics behind this rule - contained in the hadronic matrix elements of current-current operators - has been identified analytically first in 1986 in the framework of the Dual QCD (DQCD) [64] with some improvements in 2014 [65]. This has been confirmed more than 30 years later by the RBC-UKQCD collaboration [66] although the modest accuracy of both approaches still allows for some NP contributions. See [67] for the most recent summary. It would be important to have two lattice calculations with 2 + 1 + 1 flavours.

6. Similar comments apply to ε'/ε for which the only Lattice QCD calculation is from the RBC-UKQCD with 2 + 1 flavours and without isospin breaking effects [66]. Hopefully it will be improved in the coming years and also performed by another QCD lattice collaboration. A stressed in [1] Jean-Marc Gérard and myself expect on the basis of DQCD developed with Bill Bardeen [64, 68, 69], significant NP contributions to ε'/ε that as seen in the left plot in Fig. 3 are consistent with the recent result on $K^+ \to \pi^+\nu\bar{\nu}$ of the NA62 collaboration.

7. After 37 years of waiting [65, 70, 71] Lattice QCD confirmed the DQCD claim that the $\hat{B}_K \leq 0.75$ with the most accurate result from the RBC-UKQCD collaboration $\hat{B}_K = 0.7436(82)$ [72] to be compared with $\hat{B}_K = 0.73 \pm 0.02$ from DQCD [65]. As the ETM collaboration found $\hat{B}_K = 0.717(18)(16)$ [73] it is likely that the central value of $\hat{B}_K$ is very close to the DQCD one. This would imply that if there is some small NP contribution to ε_K it must be of the same sign as the SM one. Let us hope that further improvements on $\hat{B}_K$ and NP matrix elements [72–75] will be made in the coming years.

8. Calculation of hadronic matrix elements relevant for ΔM_s and ΔM_d with 2 + 1 + 1 flavours by other lattice QCD collaborations in order to check HPQCD results for these matrix elements.

9. Further search for lepton flavour violation.

We should have then great time in the rest of this decade but unfortunately this does not depend entirely on flavour physicists as we have recently seen in the case of CERN decision on the HIKE experiment. In view of the recent NA62 result on $K^+ \to \pi^+\nu\bar{\nu}$ this could turn out to be a big mistake.

It is a pleasure to thank the organizers of this workshop for inviting me to this interesting event and for an impressive hospitality. Particular thanks go to Elena Venturini for fantastic time we spent together developing our strategy. Thanks go also to Jason Aebischer, Jacky Kumar and Peter Stangl for collaborations on the implications of the suppressed NP contributions to $\Delta F = 2$ processes and to Julia Harz and Martin Mojahed for developing strategies for disentangling NP in $K \to \pi\nu\bar{\nu}$ and $B \to K(K^*)\nu\bar{\nu}$ decays. Financial support from the Excellence Cluster ORIGINS, funded by the Deutsche Forschungsgemeinschaft (DFG, German Research Foundation), Excellence Strategy, EXC-2094, 390783311 is acknowledged.

References

[1] A.J. Buras, *Kaon Theory: 50 Years Later* (2023), 2307.15737
[2] A.J. Buras, J. Girrbach, Rept. Prog. Phys. **77**, 086201 (2014), 1306.3775

[3] A.J. Buras, *Gauge Theory of Weak Decays* (Cambridge University Press, 2020), ISBN 978-1-139-52410-0, 978-1-107-03403-7
[4] A.J. Buras (2024), 2403.02387
[5] A.J. Buras, E. Venturini, Acta Phys. Polon. B **53**, A1 (2021), 2109.11032
[6] A.J. Buras, E. Venturini, Eur. Phys. J. C **82**, 615 (2022), 2203.11960
[7] M. Bona et al. (UTfit), Rend. Lincei Sci. Fis. Nat. **34**, 37 (2023), 2212.03894
[8] J. Charles et al. (CKMfitter Group), Eur. Phys. J. **C41**, 1 (2005), hep-ph/0406184
[9] A.J. Buras, Eur. Phys. J. C **83**, 66 (2023), 2209.03968
[10] M. Bordone, B. Capdevila, P. Gambino, Phys. Lett. B **822**, 136679 (2021), 2107.00604
[11] G. Finauri, P. Gambino, JHEP **02**, 206 (2024), 2310.20324
[12] Y. Aoki et al. (Flavour Lattice Averaging Group (FLAG)), Eur. Phys. J. C **82**, 869 (2022), 2111.09849
[13] A.J. Buras, Phys. Lett. **B566**, 115 (2003), hep-ph/0303060
[14] R.J. Dowdall, C.T.H. Davies, R.R. Horgan, G.P. Lepage, C.J. Monahan, J. Shigemitsu, M. Wingate, Phys. Rev. D **100**, 094508 (2019), 1907.01025
[15] M. Kirk, A. Lenz, T. Rauh, JHEP **12**, 068 (2017), [Erratum: JHEP 06, 162 (2020)], 1711.02100
[16] D. King, A. Lenz, T. Rauh, JHEP **05**, 034 (2019), 1904.00940
[17] D. King, M. Kirk, A. Lenz, T. Rauh (2019), [Addendum: JHEP 03, 112 (2020)], 1911.07856
[18] A.J. Buras, Eur. Phys. J. C **82**, 612 (2022), 2204.10337
[19] A.J. Buras, Phys. Rept. **1025**, 1 (2023), 1102.5650
[20] W.G. Parrott, C. Bouchard, C.T.H. Davies (HPQCD), Phys. Rev. D **107**, 014511 (2023), [Erratum: Phys.Rev.D 107, 119903 (2023)], 2207.13371
[21] W.G. Parrott, C. Bouchard, C.T.H. Davies (HPQCD), Phys. Rev. D **107**, 014510 (2023), 2207.12468
[22] M. Blanke, A.J. Buras, Eur. Phys. J. C **79**, 159 (2019), 1812.06963
[23] M. Bona et al. (UTfit), JHEP **0803**, 049 (2008), 0707.0636
[24] A.J. Buras, F. Parodi, A. Stocchi, JHEP **0301**, 029 (2003), hep-ph/0207101
[25] A.J. Buras, M.E. Lautenbacher, G. Ostermaier, Phys. Rev. **D50**, 3433 (1994), hep-ph/9403384
[26] L. Wolfenstein, Phys. Rev. Lett. **51**, 1945 (1983)
[27] Y. Amhis et al. (HFLAV) (2022), 2206.07501
[28] R. Aaij et al. (LHCb), Phys. Rev. D **105**, 012010 (2022), 2108.09283
[29] I. Adachi et al. (Belle-II) (2023), 2311.14647
[30] J. Grygier et al. (Belle), Phys. Rev. **D96**, 091101 (2017), [Addendum: Phys. Rev.D97,no.9,099902(2018)], 1702.03224
[31] M. Zamkovský et al. (NA62), PoS **DISCRETE2020-2021**, 070 (2022)
[32] J. Ahn et al. (KOTO), Phys. Rev. Lett. **122**, 021802 (2019), 1810.09655
[33] R. Aaij et al. (LHCb), Phys. Rev. Lett. **125**, 231801 (2020), 2001.10354
[34] A.J. Buras, F. De Fazio, J. Girrbach, Eur. Phys. J. **C74**, 2950 (2014), 1404.3824
[35] A.J. Buras, D. Buttazzo, J. Girrbach-Noe, R. Knegjens, JHEP **1411**, 121 (2014), 1408.0728
[36] A. Crivellin, L. Hofer, J. Matias, U. Nierste, S. Pokorski, J. Rosiek, Phys. Rev. **D92**, 054013 (2015), 1504.07928
[37] J. Aebischer, A.J. Buras, J. Kumar, Eur. Phys. J. C **83**, 368 (2023), 2302.00013
[38] M. Blanke, Acta Phys.Polon. **B41**, 127 (2010), 0904.2528
[39] A.J. Buras, D. Buttazzo, R. Knegjens, JHEP **11**, 166 (2015), 1507.08672

[40] A.J. Buras, P. Stangl (2024), 2411.XXXXX
[41] A.J. Buras, F. De Fazio, JHEP **03**, 219 (2023), 2301.02649
[42] A.J. Buras, J. Harz, M.A. Mojahed, JHEP **10**, 087 (2024), 2405.06742
[43] M. Gorbahn, U. Moldanazarova, K.H. Sieja, E. Stamou, M. Tabet, Eur. Phys. J. C **84**, 680 (2024), 2312.06494
[44] R. Bause, H. Gisbert, M. Golz, G. Hiller, JHEP **12**, 061 (2021), 2109.01675
[45] X.G. He, G. Valencia, Phys. Lett. B **821**, 136607 (2021), 2108.05033
[46] R. Bause, H. Gisbert, M. Golz, G. Hiller, Eur. Phys. J. C **83**, 419 (2023), 2209.04457
[47] D. Bečirević, G. Piazza, O. Sumensari, Eur. Phys. J. C **83**, 252 (2023), 2301.06990
[48] R. Bause, H. Gisbert, G. Hiller (2023), 2309.00075
[49] L. Allwicher, D. Becirevic, G. Piazza, S. Rosauro-Alcaraz, O. Sumensari (2023), 2309.02246
[50] H.K. Dreiner, J.Y. Günther, Z.S. Wang (2023), 2309.03727
[51] W. Altmannshofer, A. Crivellin, H. Haigh, G. Inguglia, J. Martin Camalich (2023), 2311.14629
[52] E. Gabrielli, L. Marzola, K. Müürsepp, M. Raidal, Eur. Phys. J. C **84**, 460 (2024), 2402.05901
[53] B.F. Hou, X.Q. Li, M. Shen, Y.D. Yang, X.B. Yuan, JHEP **06**, 172 (2024), 2402.19208
[54] X.G. He, X.D. Ma, M.A. Schmidt, G. Valencia, R.R. Volkas, JHEP **07**, 168 (2024), 2403.12485
[55] P.D. Bolton, S. Fajfer, J.F. Kamenik, M. Novoa-Brunet, Phys. Rev. D **110**, 055001 (2024), 2403.13887
[56] D. Marzocca, M. Nardecchia, A. Stanzione, C. Toni (2024), 2404.06533
[57] L. Allwicher, M. Bordone, G. Isidori, G. Piazza, A. Stanzione (2024), 2410.21444
[58] A.J. Buras, R. Fleischer, S. Recksiegel, F. Schwab, Phys. Rev. Lett. **92**, 101804 (2004), hep-ph/0312259
[59] A.J. Buras, R. Fleischer, S. Recksiegel, F. Schwab, Nucl. Phys. B **697**, 133 (2004), hep-ph/0402112
[60] R. Fleischer, R. Jaarsma, E. Malami, K.K. Vos, Eur. Phys. J. C **78**, 943 (2018), 1806.08783
[61] R. Berthiaume, B. Bhattacharya, R. Boumris, A. Jean, S. Kumbhakar, D. London (2023), 2311.18011
[62] A. Datta, J. Kumar, S. Kumbhakar, D. London (2024), 2408.03380
[63] A. Szabelski (2024), 2410.16396
[64] W.A. Bardeen, A.J. Buras, J.M. Gérard, Phys. Lett. **B192**, 138 (1987)
[65] A.J. Buras, J.M. Gérard, W.A. Bardeen, Eur. Phys. J. **C74**, 2871 (2014), 1401.1385
[66] R. Abbott et al. (RBC, UKQCD), Phys. Rev. D **102**, 054509 (2020), 2004.09440
[67] A.J. Buras, *ε'/ε in the Standard Model and Beyond: 2021*, in *11th International Workshop on the CKM Unitarity Triangle* (2022), 2203.12632
[68] W.A. Bardeen, A.J. Buras, J.M. Gérard, Phys. Lett. **B180**, 133 (1986)
[69] W.A. Bardeen, A.J. Buras, J.M. Gérard, Nucl. Phys. **B293**, 787 (1987)
[70] W.A. Bardeen, A.J. Buras, J.M. Gérard, Phys. Lett. **B211**, 343 (1988)
[71] J.M. Gérard, JHEP **1102**, 075 (2011), 1012.2026
[72] P.A. Boyle, F. Erben, J.M. Flynn, N. Garron, J. Kettle, R. Mukherjee, J.T. Tsang (RBC, UKQCD), Phys. Rev. D **110**, 034501 (2024), 2404.02297
[73] N. Carrasco, P. Dimopoulos, R. Frezzotti, V. Lubicz, G.C. Rossi, S. Simula, C. Tarantino (ETM), Phys. Rev. **D92**, 034516 (2015), 1505.06639
[74] A.J. Buras, J.M. Gérard, Acta Phys. Polon. B **50**, 121 (2019), 1804.02401
[75] B.J. Choi et al. (SWME), Phys. Rev. **D93**, 014511 (2016), 1509.00592

Precision measurements with kaon and pion decays at CERN

Andrea Bizzeti[1,2,*] for the NA48/2 and NA62 Collaborations

[1]Dipartimento di Scienze Fisiche, Informatiche e Matematiche, Università di Modena e Reggio Emilia, Modena, Italy
[2]INFN, Sezione di Firenze, Sesto Fiorentino, Italy

Abstract.
The NA62 measurement of the branching ratio and the di-photon mass spectrum of the $K^+ \to \pi^+\gamma\gamma$ decay is presented, together with a preliminary result on the branching fraction of the rare decay $\pi^0 \to e^+e^-$.
The first observation of the decay $K^\pm \to \pi^0\pi^0\mu^\pm\nu(K^{00}_{\mu4})$ by the NA48/2 experiment at the CERN is also reported, together with the measurement of the branching ratio.

1 The NA62 experiment

Tha NA62 experiment at CERN is a fixed-target experiment designed to measure the branching fraction of the $K^+ \to \pi^+\nu\bar{\nu}$ very rare decay[1–3] with a precision similar to the theoretical predictions[4]. Its experimental layout is described in [5] and schematically shown in Figure 1.

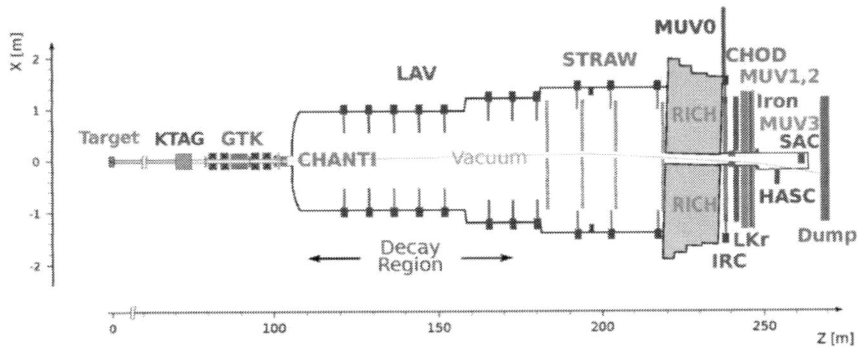

Figure 1. Layout of the NA62 experiment at CERN SPS.

An unseparated 75 GeV/c hadron beam containing $\pi^+(70\%)$, $p(24\%)$ and $K^+(6\%)$ is produced by a 400 GeV/c proton beam from the CERN Super Proton Synchrotron impinging on a beryllium target. Kaons in this secondary beam are identified by a differential Čerenkov

*e-mail: andrea.bizzeti@fi.infn.it

counter (KTAG) with $\sigma_t \sim 70$ ps; trajectories and momenta of all beam particles are measured by a three-station Si-pixel beam spectrometer (GTK), which is followed by an evacuated fiducial volume. Trajectories and momenta of charged particles from K^+ decays are reconstructed with a spectrometer consisting of a dipole magnet and four tracking chambers made of STRAW tubes. Trigger time with $\sigma_t \sim 70$ ps is provided by a Ring Image Cherenkov detector (RICH)[8], together with identification of charged particles[9]. The presence of signals in the quadrants of a scintillation hodoscope (CHOD) is used as a trigger condition. A quasi-homogeneous liquid kripton electromagnetic calorimeter (LKr), a hadron calorimeter (MUV1,2) and muon detectors (MUV3) are also used for particle identification. Large ring-shaped lead-glass detectors (LAV) and small angle (IRC, SAC) photon detectors are used to veto photons outside the LKr acceptance.

Thanks to auxiliary trigger lines [6, 7], the NA62 physics program includes most of the K^+ decay channels and the possibility of precisely measuring decays as $K^+ \to \pi^+\gamma\gamma$ and $K^+ \to \pi^+\pi^0$ followed by the prompt $\pi^0 \to e^+e^-$ decay.

2 NA62 measurement of the $K^+ \to \pi^+\gamma\gamma$ decay

Measurements of radiative Kaon decays allow to test predictions from Chiral Perturbation Theory (ChPT), an effective theory describing low-energy QCD processes. The branching fraction and $\gamma\gamma$ invariant mass spectrum of the $K^+ \to \pi^+\gamma\gamma$ decay have been calculated at both leading and next-to-leading orders in ChPT expansion[10–12], using additional external parameters which are determined from other kaon decays and an unknown real parameter $\hat{c}$ specific for this decay[13].

A sample of $K^+ \to \pi^+\gamma\gamma$ decays was collected by the NA62 experiment in 2017–2018. The abundant $K^+ \to \pi^+\pi^0$ decay followed by the prompt $\pi^0 \to \gamma\gamma$ decay, whose final state contains the same set of particles as the signal, was collected concurrently using the same trigger line and is used as normalization channel.

A common set of cuts is used for the first step in the selection of signal and normalization events, requiring a single positive track in STRAW identified as a π^+ and matched with a K^+ beam track in GTK, two energy clusters in the LKr and kinematic constraints on the $\pi^+\gamma\gamma$ invariant mass (Fig. 2) and total momentum.

The differential decay width of the $K^+ \to \pi^+\gamma\gamma$ decay is measured as a function of the normalized $\gamma\gamma$ invariant mass squared $z = m^2_{\gamma\gamma}/m^2_K$, computed from the reconstructed K^+ and π^+ four-momenta as $z = (P_K - P_\pi)^2/m^2_K$. Different regions in z are used to select signal ($0.20 < z < 0.51$) and normalization ($0.04 < z < 0.12$) candidates. The background is estimated from simulations and validated with data in control regions. The signal region of the z spectrum contains 3984 candidates, with an estimated background contamination of 291 ± 14 events[14]. The main background comes from the $K^+ \to \pi^+\pi^0\gamma$, $\pi^0 \to \gamma\gamma$ decay chain with γ energy clusters overlapping in the LKr.

The next-to leading order $O(p^6)$ contribution in ChPT is found to be necessary to properly describe the measured z spectrum (Fig. 3). The $\hat{c}$ parameter is then extracted by fitting the observed z distribution using the ChPT $O(p^6)$ description, obtaining: $\hat{c} = 1.144 \pm 0.069_{\text{stat}} \pm 0.034_{\text{syst}}$.

The branching fraction is measured by integrating the ChPT $O(p^6)$ differential branching fraction in the full kinematic range, resulting in: $B(K^+ \to \pi^+\gamma\gamma) = (9.61 \pm 0.15_{\text{stat}} \pm 0.07_{\text{syst}}) \times 10^{-7}$, with an improvement of more than a factor 3 over the previous best measurement[15]. A measurement of the model-independent partial branching fraction is also performed in bins of the z observable and then integrated, obtaining: $B_{\text{MI}}(K^+ \to \pi^+\gamma\gamma\ ,\ z > 0.20) = (9.46 \pm 0.19_{\text{stat}} \pm 0.07_{\text{syst}}) \times 10^{-7}$.

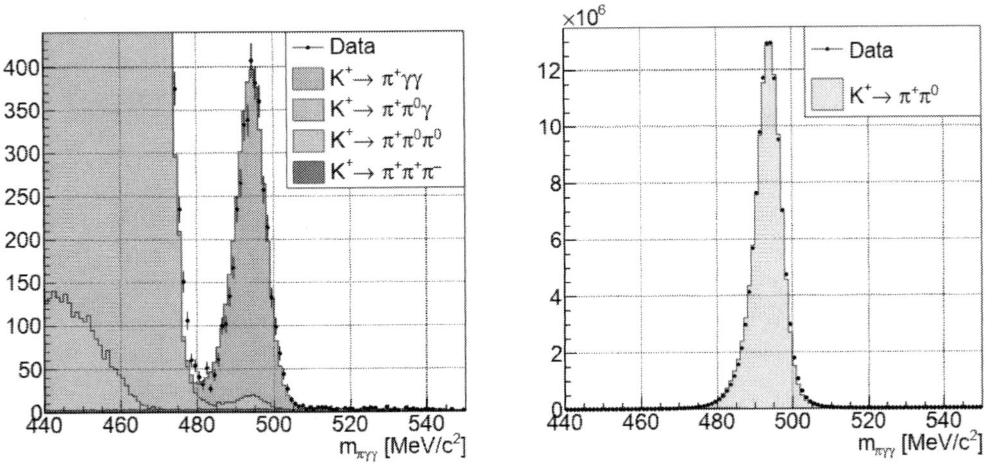

Figure 2. Left: measured $\pi^+\gamma\gamma$ invariant mass distribution for $K^+ \to \pi^+\gamma\gamma$ candidates (markers) and for simulated signal and background distributions (histograms). The signal is observed as a peak in this distribution. Histograms are listed from top to bottom. Right: measured (markers) and simualted (histogram) $\pi^+\gamma\gamma$ invariant mass distribution for $K^+ \to \pi^+\pi^0$ candidates (normalization channel).

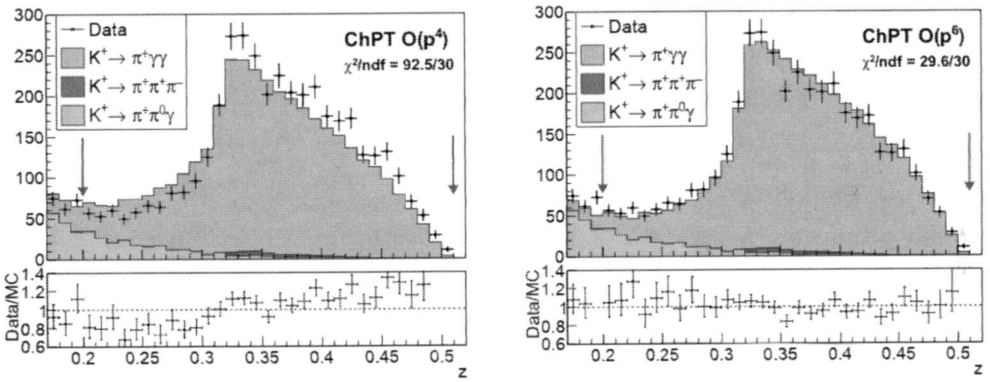

Figure 3. Reconstructed z spectrum of the $K^+ \to \pi^+\gamma\gamma$ candidates (markers), estimated background contributions and simulated signal spectrum (histograms) generated using ChPT $O(p^4)$ (left) and $O(p^6)$ (right) descriptions. The signal region used in the fit is indicated with arrows. Histograms are listed from top to bottom.

More details about this analysis can be found in[14], together with a description of the first search for production and prompt decay of an axion-like particle a with gluon coupling in the process $K^+ \to \pi^+ a,\ a \to \gamma\gamma$.

3 The NA62 measurement of the $\pi^0 \to e^+e^-$ decay

In the measurement of the $\pi^0 \to e^+e^-$ branching fraction the possible presence of radiative photons in the final state must be properly addressed by applying radiative corrections to the observable branching fraction, which is inclusive of final-state radiation.

Using the normalized e^+e^- invariant mass squared $x = m_{ee}^2/m_{\pi^0}^2$ partial branching fractions $B(\pi^0 \to e^+e^-(\gamma)\ ,\ x > x_{\rm cut})$ can be measured. A high value $x_{\rm cut} = 0.95$ is chosen to suppress the contribution of Dalitz decays $\pi^0 \to \gamma e^+e^-$, dominating the low-x region, to about 3.3% of the number of observed $\pi^0 \to e^+e^-(\gamma)$ decays.

The best past measurement of this branching fraction, performed by KTeV[16], is $B_{\rm KTeV}(\pi^0 \to e^+e^-(\gamma), x > 0.95) = (6.44 \pm 0.25 \pm 0.22) \times 10^{-8}$. By applying the latest radiative corrections[17, 18], the measured branching fraction can be extrapolated to the branching fraction without final-state radiation, $B_{\rm KTeV}(\pi^0 \to e^+e^-$, no-rad$) = (6.84 \pm 0.35) \times 10^{-8}$. The comparison with the latest theoretical predictions[19] $B_{\rm theory}(\pi^0 \to e^+e^-$, no-rad$) = (6.25 \pm 0.03) \times 10^{-8}$ shows a disagreement of about two standard deviations.

Data collected by NA62 in 2017 and 2018 were used in this analysis, selecting $K^+ \to \pi^+\pi^0$ decays followed by the prompt decay $\pi^0 \to e^+e^-$ $(K^+ \to \pi^+\pi^0_{ee})$ as signal and $K^+ \to \pi^+e^+e^-$ events as normalization channel. As signal and normalization events contain the same set of particles in the final state, they were both collected using the same multi-track electron trigger line, downscaled by a factor 8, and the following common selection criteria are used in the analysis, leading to cancellation of most systematic effects. Three STRAW tracks forming a vertex are required, with kinematic constraints on their total and transverse momentum; The ratio E/p between the energy deposited in the LKr and the momentum measured by the STRAW spectrometer is used to identify tracks as $\pi^+(E/p < 0.9)$ or $e^\pm(0.9 < E/p < 1.1)$. The total invariant mass $m_{\pi ee}$ (Fig. 4-Left) is required to be in the range $(480,\ 510)\,{\rm MeV}/c^2$. The di-electron invariant mass (Fig. 4-Right) is required to be in the range $(130,\ 140)\,{\rm MeV}/c^2$ for the signal events and in the range $(140,\ 360)\,{\rm MeV}/c^2$ for the normalization events.

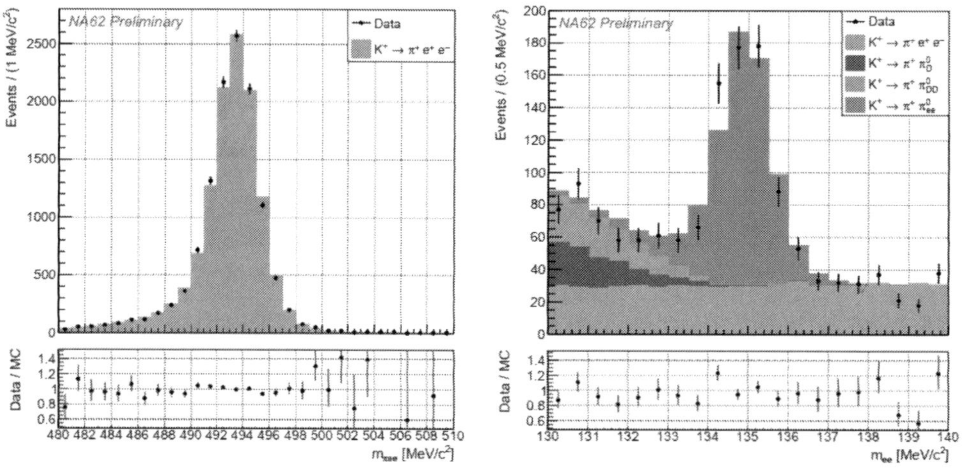

Figure 4. Left: measured (markers) and simulated (histograms) $\pi^+e^+e^-$ invariant mass distribution for the $K^+ \to \pi^+e^+e^-$ normalization channel candidates. Right: e^+e^- invariant mass distribution for $K^+ \to \pi^+\pi^0_{ee}$ candidates (markers) and for simulated signal and background distributions (histograms). Histograms are listed from bottom to top.

The main background sources for the signal are the irreducible $K^+ \to \pi^+e^+e^-$ decay, flat in the π^0 mass region, and the $K^+ \to \pi^+\pi^0$, $\pi^0 \to e^+e^-\gamma$ $(K^+ \to \pi^+\pi^0_{\rm D})$ decay chain when the gamma is converted or lost. The $K^+ \to \pi^+\pi^0$, $\pi^0 \to e^+e^-e^+e^-$ $(K^+ \to \pi^+\pi^0_{\rm DD})$ with two undetected $e^\pm$ also contributes to the background.

The number of signal events $N_s = 597 \pm 9$ is obtained from a maximum likelyhood fit of the m_{ee} spectrum to the combination of simulated distributions for signal and background channels, with the signal branching fraction as a free parameter and backgroun branching fractions taken as external inputs[20]. The resulting preliminary branching fraction is: $B_{\text{NA62}}(\pi^0 \to e^+e^-(\gamma),\ x > 0.95) = (5.86 \pm 0.30_{\text{stat}} \pm 0.11_{\text{syst}} \pm 0.19_{\text{ext}}) \times 10^{-8} = (5.86 \pm 0.37) \times 10^{-8}$.

Using the latest radiative corrections[17, 18], this result is extrapolated to obtain the preliminary branching fraction without final-state radiation: $B_{\text{NA62}}(\pi^0 \to e^+e^-, \text{no-rad}) = (6.22 \pm 0.39) \times 10^{-8}$. More details about this analysis can be found in [21]. This result is compatible with the KTeV measurement[16] and has a similar precision. However, due to its lower central value, the NA62 preliminary result is in agreement with the theoretical expectations[19].

As the precision of the NA62 measurement is presently limited by the signal sample size, improvements are expected by collecting a larger sample with an improved trigger and a reduced downscaling factor; moreover, a new measurement of $B(\pi^0 \to e^+e^-)$ by NA62 will also reduce the dominating external uncertainty.

4 The NA48/2 first observation of the $K^\pm \to \pi^0\pi^0\mu^\pm\nu$ decay

The NA48/2 experiment has been taking data in the years 2003 and 2004, detecting in-flight decays of charged kaons to search for direct CP violation in $K^\pm \to 3\pi$ decays [22]. It used two simultaneous oppositely charged beams of 60 GeV/c momentum produced by 400 GeV/c primary CERN SPS protons impinging on a beryllium target. Charged particles in the beams were measured by a the KAon BEam Spectrometer (KABES) with a spatial, momentum and time resolutions of about 800 μm, 1% and 600 ps, respectively. The two beams were superimposed in the decay volume inside a 114 m long vacuum tank.

Downstream of the decay volume, a magnetic spectrometer (DCH), consisting of a dipole magnet and four drift chamber stations, measured trajectories of charged particles with a spatial resolution of 100 μm, achieving a momentum resolution $\Delta p/p = (1.0 \oplus 0.044p)\%$, with p expressed in GeV/c. The spectrometer was followed by a scintillator hodoscope (CHOD) with $\sim$ 150 ps time resolution, whose signals were used for triggering.

Photons and electrons were precisely measured by a quasi-homogeneous Liquid Krypton electromagnetic calorimeter (LKr). The ratio E/p between the energy deposited in the LKr and the momentum measured by the spectrometer is used for particle identification. An iron-scintillator hadronic calorimeter (HCAL), three planes of scintillators for muon detection (MUV) and several photon veto detectors completed the experimental setup, a detailed description of which can be found in [23].

The kinematics of the K_{l4} decays ($l = e, \mu$), $K^+ \to \pi^+\pi^-l^+\nu$ (K^{+-}_{l4}) and $K^+ \to \pi^0\pi^0l^+\nu$ (K^{00}_{l4}), is conveniently described by the five Cabibbo-Maksymowicz variables [24, 25]: the angles θ_π, θ_e and ϕ shown in Fig. 5-Left, and the squares of the dipion and dilepton invariant masses $S_\pi = m^2_{\pi\pi}$ and $S_l = m^2_{l\nu}$.

Differential rates of K_{l4} decays depends on 4 form factors F, G, H and R, whose values can be calculated in the framework of ChPT[26]. In the case of $\pi^0\pi^0$ dipion, S-wave approximation leads to $G = H = 0$. The F form factor measured in $K^+ \to \pi^0\pi^0e^+\nu_e$ decays[27] can be used in the present analysis thanks to lepton universality, while the R form factor, not contributing to the K_{e4} decays due to the smallness of the electron mass, can be estimated by ChPT-based theoretical calculations[26].

$K^+ \to \pi^0\pi^0\mu^+\nu$ ($K^{00}_{\mu 4}$) signal events are selected from data collected by NA48/2 in 2003 and 2004. The $K^+ \to \pi^0\pi^0\pi^+$ ($K^{00}_{3\pi}$) decay is chosen as normalization channel. The first

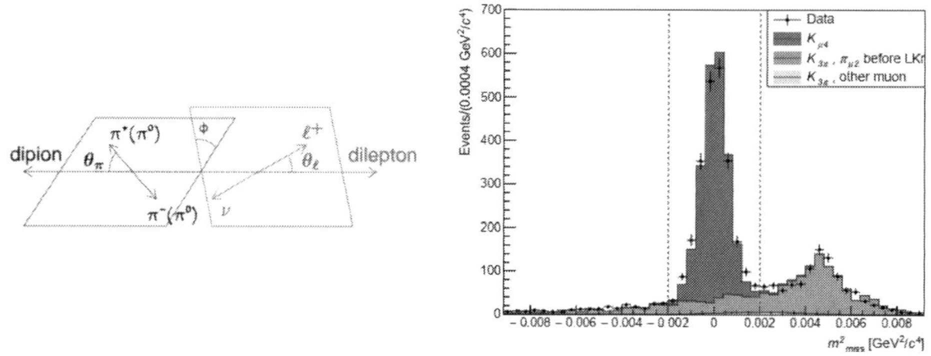

Figure 5. Left: kinematics of K_{l4} decays and definition of Cabibbo-Maksymowicz angles θ_π, θ_l and ϕ. Arrows indicating dipion and dilepton momenta are defined in the kaon rest frame, while those indicating pion (lepton) momenta are defined in the dipion (dilepton) "center-of-mass" frame, i.e. the reference frame in which the total momentum of the two pions (leptons) is zero. Right: distribution of $M^2_{\rm miss}(K^{00}_{\mu4})$ for the selected data events (markers), simualated background and signal contributions (histograms). Vertical lines show the signal region. Histograms are listed from top to bottom.

part of the event selection is common to signal and normalization, leading to a significant cancellation of systematic effects due to detector and trigger inefficiencies.

At least one DCH track and one KABES track are required, whose extrapolations must be consistent with a kaon decay occurring at a specific point (decay vertex). Each combination of four photon candidates, consistent in time and depositing in the LKr more than 3 GeV each, is considered as a possible result of two π^0 decays occurring at the decay vertex if the photon energies and positions are compatible with the π^0 decay kinematics. In events containing more than one combination, the one which better fits the decay kinematics is chosen.

Normalization ($K^{00}_{3\pi}$) decay candidates are then required to have a DCH track momentum larger than 5 GeV/c; moreover, in the plane of the 3π invariant mass $M_{3\pi}$ vs. the transverse momentum P_T of the DCH track the event must lie within an ellipse centered at the kaon PDG mass $M_{3\pi} = M_K^{\rm PDG}$[20] and $P_T = 5$ MeV/c, with semi-axes $\Delta M_{3\pi} = 10$ MeV/c^2 and $\Delta P_T = 20$ MeV/c. About 73 million $K_{3\pi}$ normalization events have been selected, with a negligible background.

From the four-momenta $P_{\rm particle}$ of the reconstructed particles two missing squared masses, $m_{\rm miss} = m^2_{\rm miss}(K^{00}_{\mu4}) = (P_K - P_{\pi^0_1} - P_{\pi^0_2} - P_{\mu^+})^2$ and $m^2_{\rm miss}(K^{00}_{3\pi}) = (P_K - P_{\pi^0_1} - P_{\pi^0_2} - P_{\pi^+})^2$ are calculated using different mass hypothesis (muon or pion mass) for the charged particle detected by the DCH.

Signal candidates are selected from the events rejected by the $K^{00}_{3\pi}$ elliptic cut, with the additional requirement that the DCH track must have a momentum larger than 10 GeV/c and an associated muon detected in the MUV. Kinematic cuts $\cos\theta_l < 0.6$ and $m_{\rm miss}(K^{00}_{3\pi}) < 0.5\, m_{\rm miss}(K^{00}_{\mu4}) - 0.0008\,{\rm GeV}^2/{\rm c}^4$ are applied to further suppress the $K^{00}_{3\pi}$ background. Finally, $S_l = m^2(\mu\nu) > 0.03\,{\rm GeV}^2/{\rm c}^4$ is required to exclude the very large background from the $K^\pm \to \pi^0\pi^0\pi^\pm\,;\pi^\pm \to \mu^\pm\nu$ decay chain.

Fig. 5-Right shows the $m^2_{\rm miss}(K^{00}_{\mu4})$ missing squared mass distribution of the 3718 selected $K^{00}_{\mu4}$ signal candidates, 2437 of which lie within the signal region $[-0.002, +0.002]$ GeV^2/c^4. The $K^{00}_{\mu4}$ signal is observed as a peak in this distribution, providing the first observation of this decay mode. The expected background contribution in the signal region is estimated to be $354 \pm 33_{\rm stat} \pm 62_{\rm syst}$ by performing a fit in the sidebands.

The resulting branching fraction for the restricted phase space ($S_l > 0.03$ GeV^2/c^4) is $B(K^{\pm} \to \pi^0\pi^0\mu^{\pm}\nu,\ S_l > 0.03\,{\rm GeV}^2/{\rm c}^4) = (0.65 \pm 0.02_{\rm stat} \pm 0.02_{\rm syst} \pm 0.01_{\rm ext}) \times 10^{-6}$ =, with an uncertainty dominated by the statistics of the signal sample and the uncertainty in the background evaluation. The extrapolated full phase space branching fraction is: $B(K^{\pm} \to \pi^0\pi^0\mu^{\pm}\nu) = (3.45 \pm 0.10_{\rm stat} \pm 0.11_{\rm syst} \pm 0.05_{\rm ext}) \times 10^{-6}$. More details about this analysis can be found in[28].

References

[1] E. Cortina Gil *et al.* (NA62 Collaboration), First search for $K^+ \to \pi^+\nu\bar{\nu}$ using the decay-in-flight technique. Phys. Lett. B **791**, 156–166 (2019). https://doi.org/10.1016/j.physletb.2019.01.067

[2] E. Cortina Gil *et al.* (NA62 Collaboration), An investigation of the very rare $K^+ \to \pi^+\nu\bar{\nu}$ decay. J. High Energ. Phys. **2020**, 042 (2020). https://doi.org/10.1007/JHEP11(2020)042

[3] E. Cortina Gil *et al.* (NA62 Collaboration), Measurement of the very rare $K^+ \to \pi^+\nu\bar{\nu}$ decay. J. High Energ. Phys. **2021**, 093 (2021). https://doi.org/10.1007/JHEP06(2021)093

[4] A. J. Buras, D. Buttazzo, J. Girrbach-Noe and R. Knegjens, $K^+ \to \pi^+\nu\bar{\nu}$ and $K_L \to \pi^0\nu\bar{\nu}$ in the Standard Model: status and perspectives. J. High Energ. Phys. **2015**, 33 (2015). https://doi.org/10.1007/JHEP11(2015)033

[5] E. Cortina Gil *et al.* (NA62 Collaboration), The beam and detector of the NA62 experiment at CERN. J.INST **12**, P05025 (2017). https://doi.org/10.1088/1748-0221/12/05/P05025

[6] R. Ammendola *et al.*, The integrated low-level trigger and readout system of the CERN NA62 experiment. Nucl. Instr. Meth. A **929**, 1-22 (2019) https://doi.org/10.1016/j.nima.2019.03.012

[7] E. Cortina Gil *et al.* (NA62 Collaboration), Performance of the NA62 trigger system. J. High Energ. Phys. **2023**, 122 (2023). https://doi.org/10.1007/JHEP03(2023)122

[8] G. Anzivino *et al.*, Light detection system and time resolution of the NA62 RICH. J. Inst. **15**, P10025 (2020). https://doi.org/10.1088/1748-0221/15/10/P10025

[9] G. Anzivino *et al.*, Precise mirror alignment and basic performance of the RICH detector of the NA62 experiment at CERN. J. Inst. **13**, P07012 (2018). https://doi.org/10.1088/1748-0221/13/07/P07012

[10] G. Ecker, A. Pich, E. De Rafael, Radiative Kaon decays and CP violation in chiral perturbation theory. Nucl. Phys. B **303**, 665–702 (1988). https://doi.org/10.1016/0550-3213(88)90425-7

[11] G. D'Ambrosio and J. Portolés, Unitarity and vector meson contributions to $K^+ \to \pi^+\gamma\gamma$. Phys. Lett. B **386**, 403–412 (1996). https://doi.org/10.1016/0370-2693(96)00970-7

[12] J. M. Gérard, C. Smith, S. T. Trine, Radiative kaon decays and the penguin contribution to the $\Delta I = 1/2$ rule. Nucl. Phys. B **730**, 1–36 (2005). https://doi.org/10.1016/j.nuclphysb.2005.09.040

[13] G. D'Ambrosio, M. Knecht and S. Neshatpour, Determination of the structure of the $K \to \pi\pi\pi$ amplitudes from recent data. Phys. Lett. B **835**, 137594 (2022). https://doi.org/10.1016/j.physletb.2022.137594

[14] E. Cortina Gil *et al.* (NA62 Collaboration), Measurement of the $K^+ \to \pi^+\gamma\gamma$ decay. Phys. Lett. B **850**, 138513 (2024). https://doi.org/10.1016/j.physletb.2024.138513

[15] C. Lazzeroni *et al.* (NA62 Collaboration), Study of the $K^\pm \to \pi^\pm\gamma\gamma$ decay by the NA62 experiment. Phys. Lett. B **732**, 65–74 (2014). http://dx.doi.org/10.1016/j.physletb.2014.03.016

[16] E. Abouzaid *et al.* (KTeV Collaboration), Measurement of the rare decay $\pi^0 \to e^+e^-$. Phys. Rev. D **75**, 012004 (2007). https://doi.org/10.1103/PhysRevD.75.012004

[17] P. Vaško and J. Novotný, Two-loop QED radiative corrections to the decay $\pi^0 \to e^+e^-$: the virtual corrections and soft-photon bremsstrahlung. J. High Energ. Phys. **2011**, 122 (2011). https://doi.org/10.1007/JHEP10(2011)122

[18] T. Husek, K. Kampf and J. Novotný, Rare decay $\pi^0 \to e^+e^-$: on corrections beyond the leading order. Eur. Phys. J. C **74**, 3010 (2014). https://doi.org/10.1140/epjc/s10052-014-3010-4

[19] M. Hoferichter and T. Teubner, Mixed Leptonic and Hadronic Corrections to the Anomalous Magnetic Moment of the Muon. Phys. Rev. Lett. **128**, 112002 (2022). https://doi.org/10.1103/PhysRevLett.128.112002

[20] R. L. Workman *et al.* (Particle Data Group), *The Review of Particle Physics*. Prog. Theor. Exp. Phys. **2022**, 083C01 (2022). https://doi.org/10.1093/ptep/ptac097

[21] F. Brizioli (NA62 Collaboration), New results from analyses of rare kaon and pion decays at the NA62 experiment. Submitted to *Proc. 58th Rencontres de Moriond – Electroweak Interactions & Unified Theories, La Thuile (IT), 24–31 March 2024*.

[22] J. R. Batley *et al.* (NA48/2 Collaboration), Search for direct CP violating charge asymmetries in $K^\pm \to \pi^\pm\pi^+\pi^-$ and $K^\pm \to \pi^\pm\pi^0\pi^0$ decays. Eur. Phys. J. C **52**, 875–891 (2007). https://doi.org/10.1140/epjc/s10052-007-0456-7

[23] V. Fanti *et al.* (NA48 Collaboration), The beam and detector for the NA48 neutral kaon CP violation experiment at CERN. Nucl. Instr. Meth. A **584**, 433 (2007). https://doi.org/10.1016/j.nima.2007.01.178

[24] N. Cabibbo and A. Maksymowicz, Angular Correlations in K_{e4} Decays and Determination of Low-Energy $\pi - \pi$ Phase Shifts. Phys. Rev. **137**, B438 (1965). https://doi.org/10.1103/PhysRev.137.B438

[25] N. Cabibbo and A. Maksymowicz, Angular Correlations in K_{e4} Decays and Determination of Low-Energy $\pi - \pi$ Phase Shifts (Erratum). Phys. Rev. **168**, 1926 (1968). https://doi.org/10.1103/PhysRev.168.1926

[26] J. Bijnens, G. Colangelo, J. Gasser, K_{l4} decays beyond one loop. Nucl. Phys. B **427**, 427–454 (1994). https://doi.org/10.1016/0550-3213(94)90634-3

[27] J. R. Batley *et al.* (NA48/2 Collaboration), Detailed study of the $K^\pm \to \pi^0\pi^0e^\pm\nu(K_{e4}^{00})$ decay properties. J. High Energ. Phys. **2014**, 159 (2014). http://doi.org/10.1007/JHEP08(2014)159

[28] J. R. Batley *et al.* (NA48/2 Collaboration), First observation and study of the $K^\pm \to \pi^0\pi^0\mu^\pm\nu$ decay. J. High Energ. Phys. **2024**, 137 (2024). https://doi.org/10.1007/JHEP03(2024)137

FCNC decays of spin-1/2 double heavy baryons to spin-3/2 single heavy baryons

Tahmasib Aliev[1,*], *Emre* Askan[1,**], *Altug* Ozpineci[1,***], and *Yasemin* Sarac[2,****]

[1]Middle East Technical University, Physics Department
[2]Atilim University, Electrical and Electronics Engineering Department

Abstract. In recent years, there have been many discoveries in the spectroscopy of hadrons containing heavy quarks. Almost all baryons containing a single heavy quark are discovered in experiments. The quark model predicts the existence of many baryons with double heavy quarks. Among the possible double heavy baryons, only Ξ_{cc}^{+} and Ξ_{cc}^{++} have been experimentally observed in LHCb. Flavor-changing neutral current (FCNC) processes represent a promising platform for precise testing of SM and looking for new physics beyond the SM. In this study, the weak decays of spin-1/2 double heavy baryons to spin-3/2 single heavy baryons induced by FCNC are studied within the light cone QCD sum rules method. First, the transition form factors of $\Xi_{bb}^{0,-} \to \Xi_{b}^{*0,-}\bar{\ell}\ell$, $\Xi_{bb}^{0,-} \to \Sigma_{b}^{*0,-}\bar{\ell}\ell$, $\Omega_{bb}^{-} \to \Xi_{b}^{*-}\bar{\ell}\ell$, $\Omega_{bb}^{-} \to \Sigma_{b}^{*-}\bar{\ell}\ell$ decays are calculated. Then, the corresponding decay widths are estimated using the results for the form factors.

1 Introduction

Development of the experimental technique allowed the discovery of baryons containing a single heavy quark. The quark model predicts the existence of the baryons containing two heavy quarks. Despite the enormous efforts of experimentalists, only Ξ_{cc}^{+} and Ξ_{cc}^{++} baryons are observed [1][2]. Remaining doubly heavy baryons expected their own discoveries. Therefore, a comprehensive theoretical study of their properties, strong, electromagnetic, and weak decay, receives special attention. Among these decays, weak decays governed by flavor-changing neutral current (FCNC) represent promising ones; this is due to the fact that in the Standard Model (SM), at tree level, these decays are forbidden and take place only at the loop level. Therefore, these decays are very sensitive to the gauge structure of the effective Hamiltonian. At the same time, these decays are good places for searching for new physics. At present work, we study the weak FCNC decays of spin-1/2 doubly heavy baryons to spin-3/2 single heavy baryons within light-cone sum rules.

*e-mail: taliev@metu.edu.tr
**e-mail: easkan@metu.edu.tr
***e-mail: ozpineci@metu.edu.tr
****e-mail: yasemin.sarac@atilim.edu.tr

2 Light-cone sum rules of the decay form factors

The flavor-changing neutral transitions ($b \to (d \text{ or } s)\bar{\ell}\ell$) are governed by the weak effective Hamiltonian [3]:

$$\mathcal{H}^{eff} = -4\frac{G_F}{\sqrt{2}}\left[C_7^{eff}O_7 + C_9^{eff}O_9 + C_{10}O_{10}\right] \tag{1}$$

where

$$\begin{aligned} O_7 &= \frac{e}{16\pi^2}\sum_q m_Q\left(\bar{q}_L\sigma_{\mu\nu}Q_R\right)F^{\mu\nu} \\ O_9 &= \frac{e^2}{16\pi^2}\sum_q\left(\bar{q}_L\gamma_\mu Q_L\right)\left(\bar{\ell}\gamma^\mu\ell\right) \\ O_{10} &= \frac{e^2}{16\pi^2}\sum_q\left(\bar{q}_L\gamma_\mu Q_L\right)\left(\bar{\ell}\gamma^\mu\gamma_5\ell\right) \end{aligned} \tag{2}$$

Wilson coefficients in the eq. 1 are given in[3]. To examine the FCNC decays of $\left(\frac{1}{2}\right)_{QQ} \to \left(\frac{3}{2}\right)_Q\bar{\ell}\ell$, the matrix element $\left\langle\left(\frac{3}{2}\right)_Q\middle|\mathcal{H}^{eff}\middle|\left(\frac{1}{2}\right)_{QQ}\right\rangle$ is needed. The matrix elements contain $\left\langle\left(\frac{3}{2}\right)_Q\middle|\bar{q}\gamma_\mu(\gamma_5)Q\middle|\left(\frac{1}{2}\right)_{QQ}\right\rangle$, $\left\langle\left(\frac{3}{2}\right)_Q\middle|\bar{q}\sigma_{\mu\nu}Q\middle|\left(\frac{1}{2}\right)_{QQ}\right\rangle$ which can be written in terms of the form factors as follow[4]:

$$\begin{aligned} \left\langle B^*_{Q^{(\prime)}}(p)\middle|\bar{q}\gamma_\mu(\gamma_5)Q^{(\prime)}\middle|B_{QQ^{(\prime)}}(p')\right\rangle = \bar{u}^\lambda_{Q^{(\prime)}}(p)(\gamma_5)\Big[&(q_\lambda p_\mu - pq g_{\lambda\mu})F_1^V + (q_\lambda\gamma_\mu - \not{q} g_{\lambda\mu})F_2^V \\ &+ (q_\lambda q_\mu - q^2 g_{\lambda\mu})F_3^V + q_\lambda q_\mu F_0^V\Big]u_{QQ^{(\prime)}}(p') \end{aligned} \tag{3}$$

$$\begin{aligned} \left\langle B^*_{Q^{(\prime)}}(p)\middle|\bar{q}i\sigma_{\mu\nu}Q^{(\prime)}\middle|B_{QQ^{(\prime)}}(p')\right\rangle = \bar{u}^\lambda_{Q^{(\prime)}}(p)\Big[&-iq_\lambda\sigma_{\mu\nu}F_1^T + q_\lambda\left(\gamma_\mu p_\nu - \gamma_\nu p_\mu\right)F_2^T \\ &+ q_\lambda\left(\gamma_\mu q_\nu - \gamma_\nu q_\mu\right)F_3^T + q_\lambda\left(p_\mu q_\nu - p_\nu q_\mu\right)F_4^T \\ &+ \left(g_{\mu\lambda}\gamma_\nu - g_{\nu\lambda}\gamma_\mu\right)F_5^T + \left(g_{\mu\lambda}p_\nu - g_{\nu\lambda}p_\mu\right)F_6^T \\ &+ \left(g_{\mu\lambda}q_\nu - g_{\nu\lambda}q_\mu\right)F_7^T\Big]\gamma_5 u_{QQ^{(\prime)}}(p') \end{aligned} \tag{4}$$

Relation between the form-factors with the contracted q^ν $\left(T_i^{V(A)}\right)$ and generic form-factors $\left(F_i^T\right)$ can be found in [4]. To calculate the form factors given in eqs 3 and 4 using light-cone sum rules, we use the following correlation function:

$$\Pi_\Gamma = i\int d^4x e^{iqx}\langle 0|\mathcal{T}\left\{J_\Gamma^\dagger(x)j_{QQ}(0)\right\}|B^*(p)\rangle \tag{5}$$

where j_{QQ} is the interpolating current of the spin-1/2 double heavy baryons, J_Γ is the transition current, namely $\bar{q}\gamma_\mu Q$, $\bar{q}\gamma_\mu\gamma_5 Q$ and $\bar{q}\sigma_{\mu\nu}Q$, and $B^*(p)$ denotes the spin-3/2 single heavy baryon state with momentum p.

2.1 Phenomenological side of the correlation function

Saturating the correlation function with states carrying the same quantum numbers of B_{QQ} states and separating the ground state contribution, we get

$$\Pi_\Gamma = \frac{\langle 0|j_{QQ}(0)|B_{QQ}(p')\rangle\left\langle B_{QQ}(p')\middle|J_\Gamma^\dagger\middle|B_Q^*(p)\right\rangle}{p'^2 - M_h^2} + \ldots \tag{6}$$

where ... means higher state and continuum contributions.

The matrix element $\langle 0 | j_{QQ}(0) | B_{QQ}(p') \rangle$ is defined as

$$\langle 0 | j_{QQ}(0) | B_{QQ}(p') \rangle = \lambda_{QQ} u_{QQ}(p') \tag{7}$$

where λ_{QQ} is the residue of the spin-1/2 baryon. The residues of the doubly heavy spin-1/2 baryons are calculated in [5]. Using the general definition $\left\langle B_{QQ}(p') \middle| J_\Gamma^\dagger \middle| B_Q^*(p) \right\rangle = \bar{u}_{QQ}(p') O_\Gamma^\lambda u_Q^\lambda(p)$ and performing summation over the spins of B_{QQ} baryons we get the expressions of the hadronic part of the correlation functions:

$$\Pi_\mu^V = \frac{\lambda_{QQ}}{p'^2 - m_{1/2}^2} \Big[(F_0^V + F_3^V) q_\lambda q_\mu \not{q} + F_1^V p_\mu q_\lambda \not{q} + \not{q} g_{\mu\lambda} (-q^2 F_3^V)$$
$$- F_2^V \gamma_\mu q_\lambda \not{q} + (\gamma_\mu q_\lambda - \not{q} g_{\mu\lambda})(m_{1/2} + m_{3/2}) F_2^V + \cdots \Big] \gamma_5 u_Q^\lambda(p) \tag{8}$$

$$\Pi_\mu^A = \frac{\lambda_{QQ}}{p'^2 - m_{1/2}^2} \Big[(F_0^A + F_3^A) q_\lambda q_\mu \not{q} + F_1^A p_\mu q_\lambda \not{q} + \not{q} g_{\mu\lambda} (-q^2 F_3^A)$$
$$- F_2^A \gamma_\mu q_\lambda \not{q} + (\gamma_\mu q_\lambda - \not{q} g_{\mu\lambda})(m_{1/2} - m_{3/2}) F_2^A + \cdots \Big] u_Q^\lambda(p) \tag{9}$$

$$\Pi_{\mu\nu}^T = \frac{\lambda_{QQ}}{p'^2 - M_1^2} \Big[F_1^T q_\lambda (\gamma_\mu \gamma_\nu - g_{\mu\nu}) \not{q} - F_2^T q_\lambda (\gamma_\mu p_\nu - \gamma_\nu p_\mu) \not{q} - F_3^T q_\lambda (\gamma_\mu q_\nu - \gamma_\nu q_\mu) \not{q}$$
$$+ F_4^T q_\lambda (p_\nu q_\mu - p_\mu q_\nu) \not{q} - F_5^T (\gamma_\nu g_{\mu\lambda} - \gamma_\mu g_{\nu\lambda}) \not{q}$$
$$+ F_6^T (p_\nu g_{\mu\lambda} - p_\mu g_{\nu\lambda}) \not{q} + F_7^T (q_\nu g_{\mu\lambda} - q_\mu g_{\nu\lambda}) \not{q} + \cdots \Big] \gamma_5 u_Q^\lambda(p) \tag{10}$$

where $m_{1/2}$ denotes the mass of the spin-1/2 baryon and $m_{3/2}$ denotes the mass of the spin-3/2 baryon. In the last line of the above equations, only the structures that have been used are written explicitly, and the three dots denote other structures.

2.2 Theoretical part of the correlation function

From eq. 5 it follows that for the calculation of the theoretical part of the correlation function, the expressions of the interpolating currents are needed. The interpolating current of the doubly heavy baryon is defined as:

$$j_{QQq}(x) = \sqrt{2} \varepsilon_{abc} \left[\left(Q^{aT}(x) C q^b(x) \right) \gamma_5 Q^c(x) + \beta \left(Q^{aT}(x) C \gamma_5 q^b(x) \right) Q^c(x) \right] \tag{11}$$

where Q denotes the heavy quark and q denotes the light quark. The matrix elements $\left\langle 0 \middle| q_{1\beta}(x) q_{2\eta}(0) Q_\lambda(0) \middle| B_Q^* \right\rangle$ in terms of the single heavy baryon distribution amplitudes are given as: (see [6] and [7])

$$\langle 0 | q_{1\alpha}(t_1 n) q_{2\beta}(t_2 n) Q_\gamma(0) | B_Q^* \rangle = \sum_i (\Gamma_i)_{\beta\alpha} A_\gamma^i(t_1, t_2) \tag{12}$$

where

$$\Gamma_1 = -\frac{1}{4} C \quad \Gamma_2 = -\frac{1}{4} \gamma_\mu C \quad \Gamma_3 = -\frac{1}{8} \sigma_{\mu\nu} C \quad \Gamma_4 = -\frac{1}{4} \gamma_\mu \gamma_5 C \tag{13}$$

$$
\begin{aligned}
A^1 &= -\frac{1}{\sqrt{3}} f^{(2)} \psi_{\parallel}^{\mathbb{1}}(t_1, t_2) \bar{v}_\mu u_Q^\mu(v) \\
A^2 &= \frac{1}{\sqrt{3}} f^{(1)} \left[-\bar{v}_\nu \frac{v_+}{2} \psi_{\parallel}^{n}(t_1, t_2) \bar{n}_\mu + \psi_{\perp}^{\mathbb{1}}(t_1, t_2) \left(g_{\mu\nu} - \bar{v}_\mu \bar{v}_\nu \right) + \frac{\bar{v}_\nu}{2v_+} \psi_{\parallel}^{\bar{n}}(t_1, t_2) n_\mu \right] u_Q^\nu(v) \\
A^3 &= -\frac{i}{\sqrt{3}} f^{(2)} \left[-\left(v_+ \bar{n}_\nu \psi_{\perp}^{n}(t_1, t_2) + \frac{1}{v_+} n_\nu \psi_{\perp}^{\bar{n}}(t_1, t_2) \right) \left(g_{\mu\alpha} - \bar{v}_\mu \bar{v}_\alpha \right) - n_\mu \bar{n}_\nu \psi_{\parallel}^{n\bar{n}}(t_1, t_2) \bar{v}_\alpha \right] u_Q^\alpha(v) \\
A^4 &= -\frac{i}{2\sqrt{3}} f^{(1)} \epsilon^{\mu\nu\alpha\beta} n_\alpha \bar{n}_\beta \psi_{\perp}^{n\bar{n}}(t_1, t_2) u_Q^\nu(v)
\end{aligned}
\tag{14}
$$

Here, $u_Q^\nu(v)$ denotes the Rarita-Schwinger spinor of the spin-3/2 baryon with velocity v and n_α, $\bar{n}_\alpha$, and $\bar{v}_\alpha$ are defined as:

$$
n_\alpha = \frac{1}{vx} x_\alpha \qquad \bar{n}_\alpha = 2v_\alpha - \frac{1}{vx} x_\alpha \qquad \bar{v}_\alpha = n_\alpha - v_\alpha, \tag{15}
$$

3 Numerical analysis

In this section, we perform the numerical analysis of the form factors. Then, using the form factors, we will estimate the decay widths. The parameters entering the sum rules are the masses of the heavy quarks, the mass of the doubly heavy baryon and its residue, decay constants $f^{(1)}$ and $f^{(2)}$ of the single heavy baryon. For the masses of the heavy quarks, we used their values in the $\overline{MS}$ scheme, $\overline{m}_b(\overline{m}_b) = (4.18 \pm 0.03)$ GeV and $\overline{m}_c(\overline{m}_c) = (1.275 \pm 0.025)$ GeV [8], residues of the doubly heavy baryons are calculated in [5], and the masses can be found in [2] and[8] and the decay constants $f^{(1)}$ and $f^{(2)}$ can be found in[9]

The non-perturbative contribution to the light-cone sum rules comes from the distribution amplitudes (DA's) of the corresponding baryons. Forms of the DA's are given in [6]. [1]

In addition, we also have three parameters, namely the continuum threshold s_0, the Borel mass M^2, and the β parameter. To determine the working region of M^2, there are two requirements. M^2 should be large enough to ensure that the leading twist contribution is larger than the higher twist contributions, and it also should be small enough to suppress the higher states and continuum contributions. To examine the stability of the results with respect to Borel mass and the continuum threshold, we present the dependency of the results to M^2 with fixed values of β and s_0 and dependency of the results to $\cos(\theta)$ parameter where $\beta = \tan(\theta)$ with fixed values of M^2 and s_0 values in figure-1.

To extend the obtained results in the physical region of the q^2, which is $4m_\ell^2 \leq q^2 \leq (m_{1/2}^2 - m_{3/2}^2)$, we used the double pole parametrization.

$$
F_i(q^2) = \frac{F_i(0)}{1 - \dfrac{q^2}{m_i^2} + \alpha_i \left(\dfrac{q^2}{m_i^2} \right)^2} \tag{16}
$$

Using the results of the form factors, we present the decay widths in the table below.

4 Conclusion

In this work, we presented the FCNC decay widths for the transitions of the spin-1/2 double heavy baryons to spin-3/2 single heavy baryons. We applied the light-cone sum rules method

[1]In [6], the authors expand the model functions in terms of Gegenbauer polynomials and chose $A = 1/2$ for the calculation of the model functions in the DA's. Since the value $A = 1/2$ causes some of the integral diverge, we simply excluded the last term in the Gegenbauer polynomial expansion.

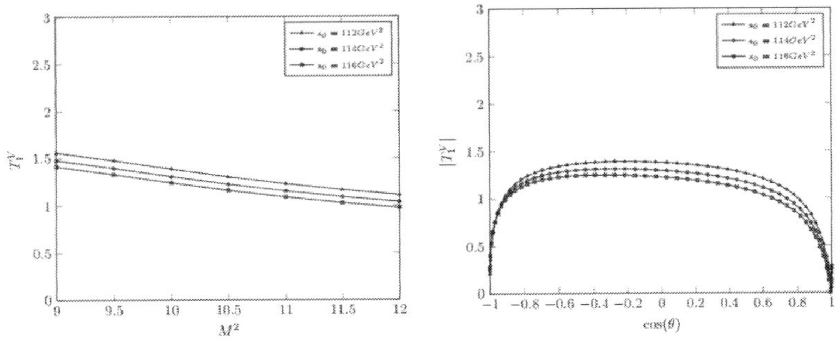

(a) (b)

Figure 1. **a)** M^2 dependence of the T_1^V for $\Xi_{bb}^- \to \Sigma_b^{*-}$ transitions with different s_0 values where $\beta = 5$ **b)** $\cos(\theta)$ dependence of the $|T_1^V|$ for $\Xi_{bb}^- \to \Sigma_b^{*-}$ transitions for fixed M^2(10 GeV^2) with different s_0 values where $\beta = \tan(\theta)$.

Table 1. Decay widths for the $B_{QQ} \to B_Q^* e^+ e^-$ transitions.

Transition	Γ / GeV	Γ / GeV [10]
$\Omega_{bb}^- \to \Omega_b^{*-} e^+ e^-$	1.94×10^{-19}	2.71×10^{-19}
$\Omega_{bb}^- \to \Xi_b^{*-} e^+ e^-$	9.35×10^{-21}	3.06×10^{-21}
$\Xi_{bb}^- \to \Sigma_b^{*-} e^+ e^-$	9.49×10^{-21}	6.52×10^{-21}
$\Xi_{bb}^- \to \Xi_b^{*-} e^+ e^-$	1.98×10^{-19}	1.43×10^{-19}
$\Xi_{bb}^0 \to \Sigma_b^{*0} e^+ e^-$	9.49×10^{-21}	3.27×10^{-21}
$\Xi_{bb}^0 \to \Xi_b^{*0} e^+ e^-$	1.98×10^{-19}	1.45×10^{-19}

to the FCNC decays of doubly heavy baryons and calculated the form factors. Using the form factors, we estimated the decay widths. Our predictions on the decay widths of the spin-1/2 double heavy baryon decays to the spin-3/2 single heavy baryon can be checked in further experiments in LHC.

This research has been supported by TUBITAK (The Scientific and Technological Research Council of Türkiye) under grant no 122F385.

References

[1] M. Mattson *et al.* [SELEX], Phys. Rev. Lett. **89**, 112001 (2002) doi:10.1103/PhysRevLett.89.112001

[2] R. Aaij *et al.* [LHCb], Phys. Rev. Lett. **119**, no.11, 112001 (2017) doi:10.1103/PhysRevLett.119.112001

[3] G. Buchalla, A. J. Buras and M. E. Lautenbacher, Rev. Mod. Phys. **68**, 1125-1144 (1996) doi:10.1103/RevModPhys.68.1125

[4] G. Hiller and R. Zwicky, JHEP **11**, 073 (2021) doi:10.1007/JHEP11(2021)073

[5] T. M. Aliev, K. Azizi and M. Savci, Nucl. Phys. A **895**, 59-70 (2012) doi:10.1016/j.nuclphysa.2012.09.009
[6] A. Ali, C. Hambrock, A. Y. Parkhomenko and W. Wang, Eur. Phys. J. C **73**, no.2, 2302 (2013) doi:10.1140/epjc/s10052-013-2302-4
[7] A. Ali, C. Hambrock and A. Y. Parkhomenko, Theor. Math. Phys. **170**, 2-16 (2012) doi:10.1007/s11232-012-0002-0
[8] R. L. Workman *et al.* [Particle Data Group], PTEP **2022**, 083C01 (2022) doi:10.1093/ptep/ptac097
[9] S. Groote, J. G. Korner and O. I. Yakovlev, Phys. Rev. D **55**, 3016-3026 (1997) doi:10.1103/PhysRevD.55.3016
[10] X. H. Hu, R. H. Li and Z. P. Xing, Eur. Phys. J. C **80**, no.4, 320 (2020) doi:10.1140/epjc/s10052-020-7851-8

Exclusive $c \to u\gamma$ transitions of B_c meson

Nicola Losacco[1,2,*]

[1]Istituto Nazionale di Fisica Nucleare, Sezione di Bari, Via Orabona 4, I-70126 Bari, Italy
[2]Dipartimento Interateneo di Fisica "M. Merlin", Università e Politecnico di Bari, via Orabona 4, 70126 Bari, Italy

Abstract. We study the rare decays of the B_c meson induced by the flavour changing neutral current $c \to u\gamma$ transition. In the Standard Model they are strongly suppressed by the Glashow-Iliopoulos-Maiani mechanism, therefore they are sensitive to new physics. The difficulty is to get rid of long-distance contributions. We study such effects in radiative B_c transitions both to B^* and to the axial-vector B_1' meson.

1 Introduction

The decays induced by the flavour changing neutral current $c \to u$ transitions play an important role in the search for new physics (NP) phenomena [1]. In the Standard Model (SM) the relevant weak Hamiltonian involves small Wilson coefficients resulting from the efficient GIM cancellation [2]: the corresponding hadronic amplitudes are highly suppressed for $c \to u\gamma$, $c \to u\ell^+\ell^-$ and $c \to u\nu\bar{\nu}$ [3]. Studies devoted to $B_c^+ \to B^{*+}\gamma$ concluded that in this channel the long-distance (LD) contributions do not have the overwhelming size as in the case of D, D_s and Λ_c decays [4]. We reconsider the issue, extending the analysis to the $B_c^+ \to B_1'^+\gamma$ mode, with $B_1'^+$ the lightest axial-vector beauty meson. We find that for B_c decays to spin-1 positive-parity B_u excitations, the LD contributions are comparable to the B^{*+} case. However, a hadronic suppression in the short-distance (SD) amplitude reduces the role of this channel for searching NP.

2 Effective weak Hamiltonian

To describe the $c \to u\gamma$ transition we consider the effective weak Hamiltonian [3]

$$\mathcal{H}_{eff} = 4\frac{G_F}{\sqrt{2}}\left[\sum_{q=d,s} V_{cq}^* V_{uq}\left(C_1 O_1^{(q)} + C_2 O_2^{(q)}\right) + \sum_{i=3}^{6} C_1 O_i + \sum_{i=7}^{8}(C_i O_i + C_i' O_i')\right] \quad (1)$$

expressed in terms of current-current operators $O_{1,2}^{(q)}$, QCD penguins operators $O_{3,...,6}$ and electromagnetic and gluon dipole operators $O_{7,8}^{(\prime)}$. The effective coefficient C_7^{eff} including two-loop QCD matrix elements of the operators $O_{3,...,6}$, is in the range [3]

$$C_7^{\text{eff}} \in [-0.00151 - i0.00556|_s + i0.00005|_{CKM}, -0.00088 - i0.00327|_s + i0.00002|_{CKM}], \quad (2)$$

*e-mail: nicola.losacco@ba.infn.it

where the subscripts indicate the contributions to the imaginary part from the strong phases and the phases of the CKM matrix. C_7' turns out to be small, $C_7' \sim m_u/m_c$, therefore the operator O_7' can be neglected in SM. Eq. (2) shows the GIM suppression in the SM for the dipole O_7, since $C_7 \sim 10^{-3}$. This coefficient can be significantly enhanced in a SM extension (BSM) scenario. General bounds can be established [5]:

$$|C_7|, |C_7'| \lesssim 0.5. \tag{3}$$

The amplitude of the transition $B_c(p) \to A(p', \epsilon)\gamma(q, \lambda)$, where A is a 1^+ state is given by

$$\mathcal{A}(B_c(p) \to A(p', \epsilon)\gamma(q, \lambda)) = \left\{A_{PC}\left[p \cdot q g^{\alpha\beta} - q^\alpha p^\beta\right] + iA_{PV}\varepsilon^{\alpha\beta\mu\nu} p_\mu q_\nu\right\} \epsilon^*_\alpha \lambda^*_\beta. \tag{4}$$

3 Short-distance amplitude

The SD amplitude comes from the O_7' and $O_{1,2}$ operators Fig. 1. The weak annihilation (WA) topology is doubly Cabibbo-suppressed, therefore it can be ignored in the evaluation of the SD.

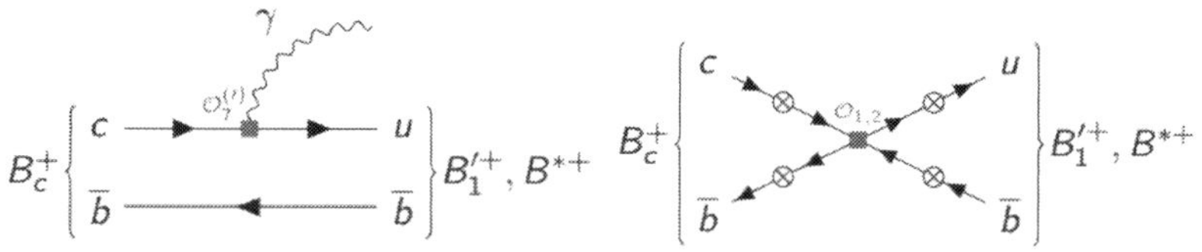

Figure 1. Electromagnetic dipole (left) and weak annihilation amplitudes (right) for $B_c \to B_1'(B^*)\gamma$. The crosses in the WA diagram correspond to the photon emission.

Focusing on the positive parity channel, the hadronic matrix element is parametrized as

$$\begin{aligned}\langle B_1'(p', \epsilon)| \bar{u}\sigma_{\mu\nu}c |B_c(p)\rangle = &\frac{\epsilon^* \cdot q}{\left(m_{B_c} + m_{B_1'}\right)^2}\left(p_\mu p'_\nu - p_\nu p'_\mu\right) T_0'(q^2) \\ &+ \left(p_\mu \epsilon^*_\nu - p_\nu \epsilon^*_\mu\right) T_1'(q^2) + \left(p'_\mu \epsilon^*_\nu - p'_\nu \epsilon^*_\mu\right) T_2'(q^2),\end{aligned} \tag{5}$$

with $q = p - p'$, λ and ϵ the photon and B_1' polarization vectors. Since the $\bar{u}\sigma_{\mu\nu}\gamma_5 c$ matrix element is obtained using $\sigma_{\mu\nu}\gamma_5 = -\frac{i}{2}\varepsilon_{\mu\nu\alpha\beta}\sigma^{\alpha\beta}$, with $\varepsilon^{0123} = -1$, the parity-conserving and parity-violating SD amplitudes in Eq. (4) are given by

$$A_{PC}^{SD} = i\frac{G_F}{(2\pi)^{3/2}} m_c \alpha^{1/2}(C_7^{eff} + C_7')(T_1'(0) + T_2'(0)), \tag{6}$$

$$A_{PV}^{SD} = -i\frac{G_F}{(2\pi)^{3/2}} m_c \alpha^{1/2}(C_7^{eff} - C_7')(T_1'(0) + T_2'(0)). \tag{7}$$

To determine the form factors T_i we use an approach based on the heavy quark spin symmetry [6–9]. For a generic Dirac matrix Γ, the matrix elements of quark currents for negative and positive parity doublets, invariant under rotations of the heavy quark spins, can be obtained using the trace formalism [10]:

$$\langle B_1'(v, k)| \bar{u}\Gamma c |B_c(v)\rangle = -\mathrm{Tr}\left[\bar{S}\,\Omega'(v, a_0 k)\,\Gamma H\right], \tag{8}$$

where S and H are the positive and negative parity spin doublet fields representation of the states (B_c, B_c^*) and (B_0, B_1'). The form factor Ω' encoding the nonperturbative dynamics is written as [6, 7]

$$\Omega'(v, a_0 k) = \Omega_1'(v, a_0 k) + \not{k} a_0 \Omega_2'(v, a_0 k)\,. \tag{9}$$

The dimensionful parameter a_0 represents the length scale involved in the process: we use the same value for both matrix elements. Since $m_c \ll m_b$, the b quark remain almost unaffected in the B_c transition, and the velocity of the final meson is nearly unchanged. Using Eq. (8) we obtain for B_1'

$$\langle B_1'(v, k; \epsilon)|\, \bar{u}\sigma_{\mu\nu} c\, |B_c(v)\rangle = -i\sqrt{\frac{m_{B_c}}{m_{B_1'}}}\left[\Omega_1'\left(k_\mu \epsilon_\nu^* - k_\nu \epsilon_\mu^*\right) + 2a_0\Omega_2' m_{B_1'}\epsilon^*\cdot v\left(k_\mu v_\nu - k_\nu v_\mu\right)\right]. \tag{10}$$

The form factors $\Omega_{1,2}'$ can be obtained using a determination of matrix elements in a non perturbative approach. In [11] using a covariant light-front approach the form factors for the $B_c \to (B_{sJ}, B_{dJ})$ decays in SM have been determined. We can use these form factors assuming isospin symmetry.

The heavy quark spin symmetry produces relations allowing to determine the tensor form factors. We obtain the finite mass form factors in term of Ω_1' and $a_0\, \Omega_2'$:

$$T_0'(q^2) = 2i\frac{(m_{B_c} + m_{B_1'})^2\,\sqrt{m_{B_1'}}}{m_{B_c}^{3/2}} a_0\, \Omega_2', \tag{11}$$

$$T_1'(q^2) = -\frac{m_{B_1'}}{m_{B_c}} T_2'(q^2), \tag{12}$$

$$T_2'(q^2) = -i\sqrt{\frac{m_{B_c}}{m_{B_1'}}}\Omega_1'. \tag{13}$$

Eq. (12) together with Eqs. (6)-(7) shows that the SD contribution for the positive-parity final state is suppressed due to a cancellation between two terms.

4 Long-distance amplitude

LD contributions correspond to processes involving intermediate hadrons. There are two kind of terms. The first one comes from the weak annihilation (WA) amplitude in Fig. 2 (top panel) with the photon radiated by any quark. The second type of contributions are pole terms induced by Eq. (1) with intermediate neutral vector mesons, as in Fig. 2 (bottom panel). For $B_c \to B_1'\gamma$ the intermediate ρ^0, ω and ϕ mesons are far from the kinematical range, therefore their contribution is suppressed respect to the case for $B_c \to B_u^*\gamma$. Let us focusing on B_1'.

4.1 WA with intermediate hadrons

The first weak annihilation diagram in Fig. 2 involves two terms. In the first one the B_c has weak transition to B, B_0 and B^*, coupled to B_1' and a photon:

$$\mathcal{A}^{B_c\to B_1'\gamma} = \mathcal{A}^{B_c\to B_{\rm res}}\frac{i}{\tilde{p}^2 - m_{B_{\rm res}}^2}\mathcal{A}^{B_{\rm res}\to B_1'\gamma}. \tag{14}$$

$B_{\rm res}$ is one of the off-shell states B, B_0 and B^*, the momentum $\tilde{p}$ is such that $\tilde{p}^2 = m_{B_c}^2$. $\mathcal{A}^{B_c\to B_{\rm res}}$ can be computed from the Hamiltonian (1) using factorization:

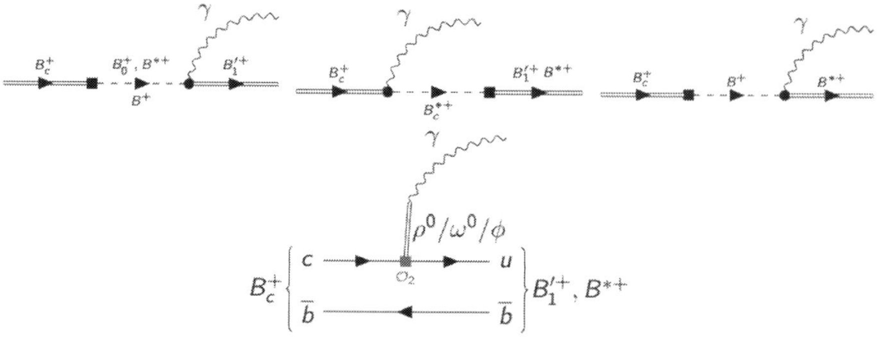

Figure 2. LD contribution to $B_c \to B'_1\gamma$ and B^*. Top panel: Weak annihilation contribution, where the box represent the insertion of a weak operator. Bottom panel: long-distance pole contribution for the B'_1 and B^* cases.

$$\mathcal{A}^{B_c \to B_{\rm res}} = \frac{G_F}{\sqrt{2}} V^*_{cb} V_{ub} a_1 \langle B_{\rm res}| \bar{u}\gamma^\mu (1-\gamma_5) b |0\rangle \langle 0| \bar{b}\gamma_\mu (1-\gamma_5) c |B_c\rangle \tag{15}$$

where $a_1 = C_1 + \frac{C_2}{3}$. The matrix elements in the amplitude involves the decay constants of mesons in the process, f_{B_c},$f_{B_{\rm res}}$. The matrix elements of two mesons and the photon can be written in terms of effective couplings $g_{1,2,3}$, which can be determined by light-cone QCD sum rules (LCSR) in HQET [12]. Using Eq. (14), the LD contributions are given by

$$\mathcal{A}(B_c \to B \to B'_1\gamma) = i\frac{G_F}{\sqrt{2}} V^*_{cb} V_{ub} a_1 f_{B_c} f_B \frac{m^2_{B_c}}{m^2_{B_c} - m^2_B} e g_1 \left[(\lambda^* \cdot \epsilon^*)(p' \cdot q) - (\lambda^* \cdot p')(\epsilon^* \cdot q)\right], \tag{16}$$

$$\mathcal{A}(B_c \to B^* \to B'_1\gamma) = 0, \tag{17}$$

$$\mathcal{A}(B_c \to B_0 \to B'_1\gamma) = \frac{G_F}{\sqrt{2}} V^*_{cb} V_{ub} a_1 f_{B_c} f_{B_0} \frac{m^2_{B_c}}{m^2_{B_c} - m^2_{B_0}} e g_3 \varepsilon_{\alpha\beta\sigma\tau} \lambda^{*\alpha} \epsilon^{*\beta} \tilde{p}^\sigma q^\tau. \tag{18}$$

For the contribution of the second diagram in Fig. 2 the radiative emission from the B_c to the B^*_c is followed by the annihilation to B'_1. The photon emission amplitude can be parametrized as

$$\mathcal{A}^{B_c \to B^*_c\gamma} = \langle B^*_c(\tilde{p}, \eta)\gamma(q, \lambda)|B_c(p)\rangle = ie g_4 \varepsilon_{\alpha\beta\sigma\tau} \lambda^{*\alpha} \eta^{*\beta} \tilde{p}^\sigma q^\tau, \tag{19}$$

with g_4 obtained from the prediction $\Gamma(B^*_c \to B_c\gamma) = 33$ eV [13]:

$$g_4^2 = \frac{24 m^3_{B^*_c} \Gamma(B^*_c \to B_c\gamma)}{\alpha \left(m^2_{B^*_c} - m^2_{B_c}\right)^3}. \tag{20}$$

For the transition to B'_1 the amplitude reads:

$$\mathcal{A}^{B^*_c \to B'_1} = \frac{G_F}{\sqrt{2}} V^*_{cb} V_{ub} a_1 \langle 0| \bar{b}\gamma_\mu (1-\gamma_5) c |B^*_c\rangle \langle B'_1| \bar{u}\gamma^\mu (1-\gamma_5) b |0\rangle, \tag{21}$$

with the decay constant $f_{B_c^*}$ and $f_{B_1'}$ parametrizing the matrix elements. The amplitude for the whole process is

$$\mathcal{A}(B_c \to \gamma B_c^* \to B_1') = \frac{G_F}{\sqrt{2}} V_{cb}^* V_{ub} a_1 f_{B_1} f_{B_c^*} \frac{m_{B_1} m_{B_c^*}}{m_{B_1}^2 - m_{B_c^*}^2} e g_4 \varepsilon_{\alpha\beta\sigma\tau} \lambda^{*\alpha} \eta^{*\beta} \tilde{p}^\sigma q^\tau. \quad (22)$$

4.2 Pole contribution

Even though the pole contributions should be suppressed, we examine their impact on the B_1' channel.

The amplitude for $B_c \to B_1' V \to B_1' \gamma$, where V is one of the resonances, reads

$$\mathcal{A} = \mathcal{A}^{B_c \to B_1' V} \frac{i}{q^2 - m_V^2 + i m_V \Gamma_V} \mathcal{A}^{V \to \gamma}, \quad (23)$$

where

$$\mathcal{A}^{B_c \to B_1' V} = i \frac{G_F}{\sqrt{2}} a_2 V_{cD}^* V_{uD} m_V f_V \epsilon_V^{*\mu} \langle B_1'(p', \epsilon) | \bar{u} \gamma_\mu (1 - \gamma_5) c | B_c(p) \rangle , \quad (24)$$

$$\mathcal{A}^{V \to \gamma} = -ie \tilde{Q}_D m_V f_V \epsilon_V^\mu \lambda_\mu^*. \quad (25)$$

D refers to the kind of down quark in the process. For $D = d$ we have $\tilde{Q}_d = \frac{Q_d}{\sqrt{2}}$, for $D = s$ we have $\tilde{Q}_s = Q_s$. To evaluate the full amplitude in Eq. (23), we consider that from gauge invariance the longitudinal helicity amplitude must be discarded [14]. This condition is translated into a constraint on the form factor $V_0'(0) = 0$.

5 $B_c \to B_u^* \gamma$

To test NP in the radiative decay of B_c through the rare transition $c \to u\gamma$, the process $B_c \to B^*\gamma$ is the first one to study. The SD amplitude is given in Eqs. (6)-(7), after switching the parity conserving with parity violating expression and with suitable substitutions, $m_{B_1'} \to m_{B^*}$ and appropriate tensor form factors parametrizing the matrix element. The LD contributions for this channel are depicted in Fig. 2. Their evaluation follows the same steps as for the LD case of the B_1' channel. The LD WA contribution with the B^+ resonance is evaluated following [4]. All matrix elements are parametrized as in [15], with the form factors for the $B_c \to B_d$ transition constructed using lattice QCD results and heavy quark spin symmetry. Isospin symmetry is also used.

6 Results

In Fig. 3 we compare the effect of the LD contributions and of the SD one in the B^* and B_1' modes.
The LD contributions dominate in the small region of $|C_7^{eff}|$, which corresponds to the SM prediction for the Wilson coefficient. If some NP effect increases C_7^{eff}, it would be better observed in the 1^- channel.

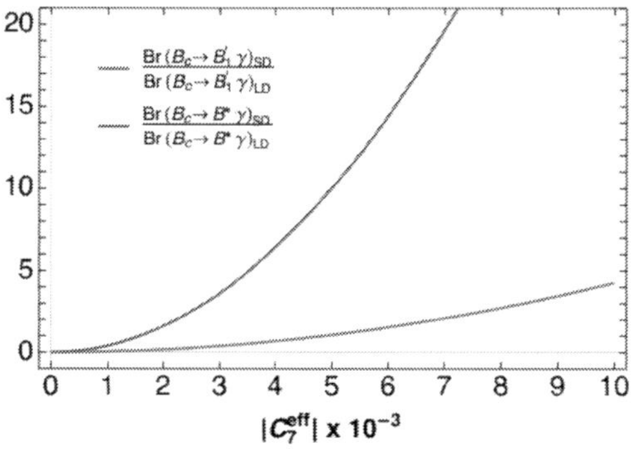

Figure 3. Comparison between the ratio due to SD and LD contributions to the rates of channels with B_1' and B^* as final states.

7 Conclusions

Motivated by previous analysis on the rare transition $c \to u\gamma$ in the B_c decays to B^*, we focus on the B_1' channel to study if it is less polluted by LD contributions.

From a model independent analysis based on heavy quark spin symmetry, we found the relation in Eq. (12) between the tensor form factor entering in the SD amplitude for the B_1' channel. Their combination in the amplitude produces a suppression of the SD contribution to the process. Therefore, while LD contributions affect more the B^* channel, the hadronic suppression in the B_1' case makes it less suitable for accessing NP.

I thank P. Colangelo and F. De Fazio for collaboration. This study has been carried out within the INFN project (Iniziativa Specifica) SPIF. The research has been partly funded by the European Union – Next Generation EU through the research Grant No. P2022Z4P4B "SOPHYA - Sustainable Optimised PHYsics Algorithms: fundamental physics to build an advanced society" under the program PRIN 2022 PNRR of the Italian Ministero dell'Università e Ricerca (MUR).

References

[1] H. Gisbert, M. Golz, D.S. Mitzel, Mod. Phys. Lett. A **36**, 2130002 (2021), 2011.09478
[2] A. Buras, *Gauge Theory of Weak Decays* (Cambridge University Press, 2020), ISBN 978-1-139-52410-0, 978-1-107-03403-7
[3] S. de Boer, G. Hiller, JHEP **08**, 091 (2017), 1701.06392
[4] S. Fajfer, S. Prelovsek, P. Singer, Phys. Rev. D **59**, 114003 (1999), [Erratum: Phys.Rev.D 64, 099903 (2001)], hep-ph/9901252
[5] N. Adolph, G. Hiller, A. Tayduganov, Phys. Rev. D **99**, 075023 (2019), 1812.04679
[6] E.E. Jenkins, M.E. Luke, A.V. Manohar, M.J. Savage, Nucl. Phys. B **390**, 463 (1993), hep-ph/9204238
[7] P. Colangelo, F. De Fazio, Phys. Rev. D **61**, 034012 (2000), hep-ph/9909423
[8] P. Colangelo, F. De Fazio, F. Loparco, N. Losacco, M. Novoa-Brunet, JHEP **09**, 028 (2022), 2205.08933
[9] P. Colangelo, F. De Fazio, N. Losacco, F. Loparco, M. Novoa-Brunet, Phys. Rev. D **106**, 094005 (2022), 2208.13398

[10] A.F. Falk, Nucl. Phys. B **378**, 79 (1992)
[11] Y.J. Shi, W. Wang, Z.X. Zhao, Eur. Phys. J. C **76**, 555 (2016), 1607.00622
[12] P. Colangelo, F. De Fazio, A. Ozpineci, Phys. Rev. D **72**, 074004 (2005), hep-ph/0505195
[13] D. Ebert, R.N. Faustov, V.O. Galkin, Phys. Rev. D **67**, 014027 (2003), hep-ph/0210381
[14] E. Golowich, S. Pakvasa, Phys. Rev. D **51**, 1215 (1995), hep-ph/9408370
[15] P. Colangelo, F. De Fazio, F. Loparco, Phys. Rev. D **103**, 075019 (2021), 2102.05365

EPJ Web of Conferences **314**, 00006 (2024) https://doi.org/10.1051/epjconf/202431400006
QCD@Work 2024

Recent developments in HQET

Gil Paz[1,*]

[1]Department of Physics and Astronomy, Wayne State University, Detroit, Michigan 48201, USA

Abstract. In this talk we review recent perturbative and non-perturbative developments in Heavy Quark Effective Theory (HQET).

1 Introduction

Heavy Quark Effective Theory (HQET) is an effective field theory for heavy quarks. It is most useful for cases where the quark mass m_Q is much larger than the QCD scale, namely, $m_Q \gg \Lambda_{\rm QCD}$. Using well-known transformations, see, e.g., [1, 2], we can obtain from the QCD Lagrangian the following Lagrangian:

$$\mathcal{L} = \bar{h}_v iv\cdot D h_v + \bar{h}_v i\not{D}_\perp \frac{1}{2m_Q + iv\cdot D} i\not{D}_\perp h_v\,, \tag{1}$$

where h_v is the heavy quark field and $D^\mu_\perp = D^\mu - (v\cdot D)v^\mu$. For $v = (1, \vec{0})$, $D^\mu_\perp = \vec{D}$. Expanding in powers of $iv\cdot D/2m_Q$ gives the dimension-five HQET Lagrangian:

$$\mathcal{L}_{\rm HQET} = \bar{h}_v iv\cdot D h_v - c_2 \bar{h}_v \frac{D_\perp^2}{2m_Q} h_v - c_F \bar{h}_v \frac{\sigma_{\alpha\beta}G^{\alpha\beta}}{4m_Q} h_v + O\left(\frac{1}{m_Q^2}\right). \tag{2}$$

Expanding to higher orders in $1/m_Q$ would give the HQET Lagrangian with tree-level Wilson coefficients. Starting at dimension seven there are other operators with $O(\alpha_s)$ Wilson coefficients, see [3], where the Lagrangian up to dimension seven was presented. The dimension-eight Lagrangian was presented in [4] and [5]. Ref. [4] also presented a method to easily construct local HQET operators of, in principle, any dimension.

Using HQET, observables can be written schematically as a series

$$\text{Observable} = \sum_{n=0}^{\infty}\sum_j c_n^j(\mu)\frac{\langle\, O_n^j(\mu)\,\rangle}{m_Q^n}\,, \tag{3}$$

where $\langle\, O_n^j(\mu)\,\rangle \sim \Lambda^n_{\rm QCD}$ and $\mu \sim m_Q$. The Wilson coefficients $c_n^j(\mu)$ are perturbative and the matrix elements $\langle\, O_n^j(\mu)\,\rangle$ are non-perturbative. Since $\alpha_s(\mu)$ becomes smaller for large μ and $\Lambda_{\rm QCD}/m_Q$ is small, we expect to achieve good precision with just a few terms. To improve the precision we can calculate $c_n^j(\mu)$ to higher powers in α_s and/or include $\langle\, O_n^j(\mu)\,\rangle$ with larger n, assuming we can extract them from data or calculate them using Lattice QCD. What $\langle\, O_n^j(\mu)\,\rangle$ do we usually encounter?

*e-mail: gilpaz@wayne.edu

Strong interaction operators are made of quarks and gluons. These can be local, e.g., $\bar{q}(0)\cdots q(0)$, or non-local, e.g., $\bar{q}(0)\cdots q(tn)$, where n is a light-cone vector ($n^2 = 0$). The general matrix element is of the form, $\langle f(p_f)|O_n^j(\mu)|i(p_i)\rangle$, where $O_n^j(\mu)$ can be local or non-local and p_i, p_f can be independent or not. Let us look at the most common options, ordered by increased complexity.

A local operator between the vacuum and a state gives rise to a number: decay constant, e.g., $\langle 0|\bar{q}\gamma^\mu\gamma_5 h_v|P(v)\rangle = -i\sqrt{m_P} f_P v^\mu$. Decay constants will appear in section 2. A diagonal matrix element of a local operator gives rise to a number: HQET parameter, e.g., $\langle \bar{B}|\bar{b}\,\vec{\boldsymbol{D}}^2\,b|\bar{B}\rangle = 2M_B\mu_\pi^2$. HQET parameters will appear in section 2. A non-diagonal matrix element of a local operator gives rise to a function: form factor, e.g., $\langle D(p_f)|\bar{c}\gamma^\mu b|\bar{B}(p_i)\rangle = f_+(q^2)(p_i + p_f)^\mu + f_-(q^2)(p_i - p_f)^\mu$, where $p_f - p_i = q$. Form factors will appear in section 4. A non-local operator between the vacuum and a state gives rise to a function, e.g., LCDA (Light-Cone Distribution Amplitude): $\langle H_v|\bar{h}_v(0)\not{n}\gamma_5\,[0,tn]\,q_s(tn)|0\rangle = -iF(\mu)\int_0^\infty d\omega\, e^{i\omega t}\phi_+(\omega,\mu)$. LCDAs will appear in section 3. A diagonal matrix element of a non-local operator gives rise to a function, e.g., shape function: $S(\omega) = \int_{-\infty}^{\infty} dt\, e^{i\omega t}\langle \bar{B}(v)|\bar{b}(0)\,[0,tn]b(tn)|\bar{B}(v)\rangle/4\pi M_B$. Finally, non-diagonal matrix element of a non-local operator, e.g. $\langle K^{(*)}(p_f)|\bar{s}_L(0)\gamma^\rho \cdots \tilde{G}_{\alpha\beta} b_L(tn)|B(p_i)\rangle$, gives rise to a non-local form factor, see [6]. We will not discuss shape functions or non-local form factors here.

In the following we review recent perturbative and non-perturbative developments in HQET. Recent is defined to be the period of Spring 2022-Spring 2024. To allow for a broad overview, only some aspects of the papers are highlighted. More details can be found in the original papers. The rest of the talk is structured as follows. Section 2 discusses recent perturbative developments. Section 3 discusses recent developments related to LCDAs. Section 4 discuses recent developments related to form factors. Section 5 discusses other recent developments. We conclude in section 6.

2 Recent developments in HQET: Perturbative

As mentioned in the introduction, we can improve theoretical predictions by calculating c_n^j to higher orders in α_s. Moreno calculated $c_3^{\rm Darwin}$ to $\mathcal{O}(\alpha_s)$ for $\bar{B} \to X_u\,\ell\,\bar{\nu}_\ell$ total rate and leptonic invariant mass in [7]. In some cases the "technology" improved to $\mathcal{O}(\alpha_s^4)$. For example, Takaura used the four-loop relation between the pole and $\overline{\rm MS}$ masses to extracted HQET parameters from B and D meson masses [8]. Lee and Pikelner calculated the four-loop HQET propagator in [9]. This four-loop calculation was used by Grozin to find the four-loop HQET heavy to light anomalous dimension [10]. Looking at [10] in more detail, the anomalous dimension can be used to calculate the ratio of the B meson and D meson decay constants f_B/f_D. Including the four-loop calculation Ref. [10] finds

$$\begin{aligned} \frac{f_B}{f_D} &= \sqrt{\frac{m_D}{m_B}}\left(\frac{\alpha_s^{(4)}(m_c)}{\alpha_s^{(4)}(m_b)}\right)^{-\frac{\tilde{\gamma}_{j0}}{2\beta_0^{(4)}}}\left\{1 + \cdots\alpha_s + \cdots\alpha_s^2 + \cdots\alpha_s^3 + [\sim 1\text{GeV}]\left(\frac{1}{m_c} - \frac{1}{m_b}\right) + \cdots\right\} \\ &= 0.669\cdot(1 + 0.039 + 0.029 + 0.032 + [\sim 0.46]), \end{aligned} \tag{4}$$

where the last term includes an estimate of power corrections. Ref. [10] comments on this equation that "Convergence of the perturbative series is questionable [...]". Numerically, without power corrections $f_B/f_D = 0.736$, and with power corrections $f_B/f_D = 1.04$. Comparing these to the lattice QCD value, $f_B/f_D = 0.896 \pm 0.009$, Ref. [10] concludes that "The effect of the (poorly known) $1/m_{c,b}$ correction is large."

3 Recent developments in HQET: Non-local matrix elements

Non local matrix elements arise in many processes, e.g., the proton parton distribution function in hard QCD processes. In B decays such as $B \to K^*\gamma$ we encounter the B-meson LCDA. It is the Fourier transform of $\langle B|\bar{b}(0)\dots[0,tn]q_s(tn)|0\rangle$. B-meson LCDA also arises when the B meson is in the *final* state, see, e.g., in the decay $W, Z \to B + \gamma$ [11]. Such processes were recently considered in [12] ($W^\pm \to B \pm \gamma$) and [13] ($W^+ \to B^+\ell^+\ell^-$).

Beneke, Finauri, Vos, and Wei considered the process $W^\pm \to B \pm \gamma$ in [12] . It has three scales: hard scale $Q \gg$ heavy quark scale $m_Q \gg$ QCD scale $\Lambda_{\rm QCD}$. In [12] the QCD LCDA:

$$\langle H(p_H)|\bar{Q}(0)\not{n}\gamma^5[0,tn]q(tn)|0\rangle = -if_H n\cdot p_H \int_0^1 du\, e^{iutn\cdot p_H}\phi(u;\mu), \quad (5)$$

was matched to a perturbative function convoluted with HQET LCDA:

$$\langle H_v|\bar{h}_v(0)\not{n}\gamma^5[0,tn]q_s(tn)|0\rangle = -iF_{\rm stat}(\mu)\, n\cdot v \int_0^\infty d\omega\, e^{i\omega tn\cdot v}\varphi_+(\omega;\mu). \quad (6)$$

Such a factorization allows to resum large logs between $\Lambda_{\rm QCD}$ and m_Q and m_Q and the hard scale Q. This paper also consider the evolution of the LCDA. Starting with HQET LCDA at soft scale $\mu_s = 1$ GeV, it is evolved in HQET to the matching scale μ and matched to $\phi(u)$. It is then evolved in QCD to the hard scale m_W, see figure 6 of [12]. The branching ratio they find is [12]

$$\text{Br}(W \to B\gamma) = (2.58 \pm 0.21_{\rm in}\,{}^{+0.05}_{-0.08}{}_{\mu_h}\,{}^{+0.05}_{-0.08}{}_{\mu_b}\,{}^{+0.18}_{-0.13}{}_{\delta}\,{}^{+0.61}_{-0.34}{}_{\beta}\,{}^{+2.95}_{-0.98}{}_{\lambda_B})\cdot 10^{-12}. \quad (7)$$

The uncertainty from the low-scale HQET LCDA parameters λ_B, β is large.

Ishaq, Zafar, Rehman, and Ahmed considered the process $W^+ \to B^+\ell^+\ell^-$ in [13]. Using the scale hierarchy $m_W \sim m_b \gg \Lambda_{\rm QCD}$ they factorize the amplitude as

$$\mathcal{M} = e\,\bar{\ell}\gamma^\mu\ell \int_0^\infty d\omega\, T_\mu(\omega, m_b, q^2, \mu_F)\Phi_B^+(\omega, \mu_F) + O\left(m_b^{-1}\right), \quad (8)$$

where T_μ is the perturbative hard-scattering kernel. Calculating T_μ at $O(\alpha_s)$ they find "[...] the scale dependence in the NLO decay rates [...] gets largely reduced, particularly for relatively large μ_F." See figure 5 of [13]. The theoretical prediction for the branching ratio is $\sim 10^{-11}$ for electrons and muons. It is sensitive to the hadronic parameter λ_B defined as

$$\frac{1}{\lambda_B} \equiv \int_0^\infty \frac{d\omega}{\omega}\phi_B^+(\omega). \quad (9)$$

See figure 6 of [13]. Ref. [13] uses $\lambda_B = 0.35 \pm 0.15$ GeV. Searching for $B^+ \to \ell^+\nu_\ell\gamma$ Belle got $\lambda_B = 0.36^{+0.25}_{-0.09}$ GeV [14]. Ref. [13] concludes that observing $W^+ \to B^+\ell^+\ell^-$ at the LHC could help constrain λ_B.

4 Recent developments in HQET: Local non-diagonal matrix elements

In the SM, $B \to D$ transitions are described by two form factors, and $B \to D^*$ transitions are described by four form factors. At leading power in the heavy quark symmetry *all* of these form factors are described by one universal Isgur-Wise function ξ. Including $1/m_c$ & $1/m_b$ power corrections there are three additional functions and the number grows rapidly at higher powers. Including $1/m_c^2$ corrections, there are additional 20 functions and including $1/m_c^2$ & $1/m_b^2$ there are additional 32 functions, see Table 1 of [15].

Bernlochner, Ligeti, Papucci, Prim, Robinson, and Xiong, suggested in [16] supplemental power-counting that reduces these numbers. The QCD Lagrangian before the $1/m_Q$ expansion is given in equation (1). The postulated power counting is in powers of $i\not{D}_\perp$: currents involve one $i\not{D}_\perp$, Lagrangian insertions involves two $i\not{D}_\perp$'s. Many subleading contributions arise from Lagrangian insertions. Ref. [16] conjectures that terms entering at third order or higher should be suppressed and calls this residual chiral (RC) expansion. Under RC expansion including $1/m_c^2$ corrections, there is one additional function and including $1/m_c^2$ & $1/m_b^2$ there are additional 3 additional functions, see Table 1 of [15]. Even more recently, Bernlochner, Papucci, and Robinson, applied the same method to $\Lambda_b \to \Lambda_c l \nu$ decay [17].

Turning to phenomenology recall that

$$R(D^{(*)}) \equiv \mathrm{Br}(\bar{B} \to D^{(*)}\, \tau\, \bar{\nu}_\tau)/\mathrm{Br}(\bar{B} \to D^{(*)}\, \ell\, \bar{\nu}_\ell), \quad \ell = e, \mu \tag{10}$$

Ref. [16] obtained $R(D) = 0.288(4)$. This should be compared to HFLAV 2024 Standard Model (SM) prediction of $R(D) = 0.298(4)$, and value from experiment of $R(D) = 0.342(26)$ [18]. Ref. [16] obtained $R(D^*) = 0.249(3)$. This should be compared to HFLAV 2024 SM prediction of $R(D^*) = 0.254(5)$, and value from experiment of $R(D^*) = 0.287(12)$ [18]. Clearly the tension between experiment and theory remains for both theoretical predictions.

5 Recent developments in HQET: Other topics

Manzari and Robinson suggested a new theoretical framework for heavy quark resonances in [19]. The new framework uses on-shell recursion techniques to express resonant amplitude as a product of on-shell sub-amplitudes. In figure 5 of [19] the authors present a toy example calculation in this framework and a fixed-width Breit-Wigner side by side with Belle data for D_2^* resonance. From the figure the preliminary results of Ref. [19] seem promising.

Garg and Upadhyay studied F-wave B mesons in HQET in [20]. F-wave B mesons have angular momentum $L = 3$. Adding $L = 3$ to the heavy-quark spin $s_Q = 1/2$ gives 7/2 and 5/2 angular momenta. Adding the light quark spin 1/2 to 5/2 gives a $J = 2$ and $J = 3$ doublet. Adding the light quark spin 1/2 to 7/2 gives a $J = 3$ and $J = 4$ doublet. Thus the spectrum contains two doublets. Ref. [20] used information from, e.g., D mesons, to calculate B meson properties. For example, in table 2 of [20] the authors compare the calculated masses to two quark models.

Vishwakarma and Upadhyay presented an analysis of 2S singly-heavy baryons in HQET in [21]. They used information from measured 2S baryons: $\Xi_c(2970)$ and $\Lambda_b(6070)$, and HQET to calculate 2S baryon properties. For example, in table 7 of [21] the authors compare the calculated masses to two quark models.

6 Conclusions

Arguably, this year marks the 35th anniversary of HQET, based on the date of the publication of "Weak Decays of Heavy Mesons in the Static Quark Approximation," by Isgur and Wise [22]. This paper includes the line "This logarithm can be displayed explicitly by going over to an effective theory where the heavy quark is treated as a static color source." Note the words "effective theory".

HQET is a mature field where some perturbative corrections are known to fourth order, and some non-perturbative power corrections are known to fourth order. In this talk we reviewed recent developments in HQET both on the perturbative side and the non-perturbative side. Theoretical progress in the field mirrors the experimental progress. It is clear from several of the papers discussed in this talk that non-perturbative effects often dominate the uncertainties. Controlling them might require further developments in HQET.

I thank the organizers for the invitation to give this talk. This work was supported by the U.S. Department of Energy grant DE-SC0007983.

References

[1] M. Neubert, Phys. Rept. **245**, 259-396 (1994) [arXiv:hep-ph/9306320 [hep-ph]]
[2] A. V. Manohar and M. B. Wise, Camb. Monogr. Part. Phys. Nucl. Phys. Cosmol. **10**, 1-191 (2000)
[3] A. V. Manohar, Phys. Rev. D **56**, 230-237 (1997) [arXiv:hep-ph/9701294 [hep-ph]]
[4] A. Gunawardana and G. Paz, JHEP **07**, 137 (2017) [arXiv:1702.08904 [hep-ph]]
[5] A. Kobach and S. Pal, Phys. Lett. B **772**, 225-231 (2017) [arXiv:1704.00008 [hep-ph]]
[6] A. Khodjamirian, T. Mannel, A. A. Pivovarov and Y. M. Wang, JHEP **09**, 089 (2010) [arXiv:1006.4945 [hep-ph]]
[7] D. Moreno, Phys. Rev. D **109**, no.7, 074030 (2024) [arXiv:2402.13805 [hep-ph]]
[8] H. Takaura, EPJ Web Conf. **274**, 03003 (2022) [arXiv:2212.02874 [hep-ph]]
[9] R. N. Lee and A. F. Pikelner, JHEP **02**, 097 (2023) [arXiv:2211.03668 [hep-ph]]
[10] A. Grozin, JHEP **02**, 198 (2024) [arXiv:2311.09894 [hep-ph]]
[11] Y. Grossman, M. König and M. Neubert, JHEP **04**, 101 (2015) [arXiv:1501.06569 [hep-ph]]
[12] M. Beneke, G. Finauri, K. K. Vos and Y. Wei, JHEP **09**, 066 (2023) [arXiv:2305.06401 [hep-ph]]
[13] S. Ishaq, S. Zafar, A. Rehman and I. Ahmed, PTEP **2024**, no.6, 063B05 (2024) [arXiv:2404.01696 [hep-ph]]
[14] M. Gelb *et al.* [Belle], Phys. Rev. D **98**, no.11, 112016 (2018) [arXiv:1810.12976 [hep-ex]]
[15] F. U. Bernlochner, Z. Ligeti, M. Papucci, M. T. Prim, D. J. Robinson and C. Xiong, PoS **ICHEP2022**, 758 (2022)
[16] F. U. Bernlochner, Z. Ligeti, M. Papucci, M. T. Prim, D. J. Robinson and C. Xiong, Phys. Rev. D **106**, no.9, 096015 (2022) [arXiv:2206.11281 [hep-ph]]
[17] F. U. Bernlochner, M. Papucci and D. J. Robinson, [arXiv:2312.07758 [hep-ph]]
[18] https://hflav-eos.web.cern.ch/hflav-eos/semi/moriond24/r_dtaunu/RDRDst_hflav_moriond24_68.pdf
[19] C. A. Manzari and D. J. Robinson, [arXiv:2402.12460 [hep-ph]]
[20] R. Garg and A. Upadhyay, PTEP **2022**, no.9, 093B08 (2022) [arXiv:2207.02498 [hep-ph]]
[21] K. K. Vishwakarma and A. Upadhyay, [arXiv:2208.02536 [hep-ph]]
[22] N. Isgur and M. B. Wise, Phys. Lett. B **232**, 113-117 (1989)

Inclusive rare Λ_b decays to photon

Francesco Loparco[1,*]

[1]Istituto Nazionale di Fisica Nucleare, Sezione di Bari, Via Orabona 4, I-70126 Bari, Italy

Abstract. Inclusive $H_b \to X_s \gamma$ decays, with H_b a beauty baryon, are treated exploiting an expansion up to the third order in the inverse of the b quark mass m_b and to LO in α_s, keeping the dependence on the hadron spin. Double distribution $\frac{d^2\Gamma}{dy\, d\cos\theta_P}$ is computed for polarized baryons, with $y = 2\,E_\gamma/m_b$, E_γ the photon energy and θ_P the angle between the baryon spin vector and the photon momentum in the H_b rest-frame. Modifications to the photon polarization asymmetry can probe effects of physics beyond the Standard Model. A systematic method to treat the singular terms in the photon energy spectrum obtained by the OPE is proposed.

1 Introduction

Processes induced at the quark level by the $b \to s\gamma$ transition occur at loop level in the Standard Model (SM) and are therefore sensitive to physics beyond it [1–3]. Upon integration of the heavy quanta, an effective Hamiltonian is obtained in terms of local operators and Wilson coefficients [4, 5]. Physics beyond the Standard Model can induce new operators and/or modify the Wilson coefficients [6–13].

The radiative $b \to s$ transition has been analyzed in theory [14]. As for experiment, after the first observation of the $B \to K^*(892)\gamma$ mode [15] other exclusive processes have been observed [16, 17]. For baryons, the rate and the photon polarization of $\Lambda_b \to \Lambda\gamma$ have been measured [18, 19], and an upper bound has been put to $\mathcal{B}(\Xi_b^- \to \Xi^-\gamma)$ [20].

Considering the inclusive mode $H_b \to X_s\gamma$, several analyses have been carried out for B mesons. In this paper we focus on the case of beauty baryons, in particular Λ_b [21]. Invoking quark-hadron duality, it is possible to exploit the Operator Product Expansion (OPE) and Heavy Quark Effective Theory (HQET) [22] to express inclusive decay widths as an expansion in $1/m_b$. Input quantities are the hadronic matrix elements of local operators.

Several issues in inclusive modes induced by the $b \to s\gamma$ transition need to be considered. The actual expansion parameter is the inverse of the energy released in the process. In some regions of the phase space it is not small. Signals about the reliability of the method show up as singularities in differential distributions. Singular terms appear at higher orders in the Heavy Quark Expansion (HQE) in form of the delta distribution and its derivatives. The origin of such terms can be related to the Fermi motion of the heavy quark in the decaying hadron, and can be accounted for introducing a shape function which encodes information on the distribution of the b quark residual momentum in the hadron [23–25]. In case of B, measurements have been exploited to constrain its moments [26].

Another source of uncertainty in inclusive $b \to s\gamma$ processes are the resolved photon contributions [27], that appear at $O(m_b^{-1})$ [28, 29].

*e-mail: francesco.loparco1@ba.infn.it

The photon energy spectrum can be measured above an energy threshold. For $\bar{B}$ the HFLAV Collaboration provides $\mathcal{B}(\bar{B} \to X_s \gamma) = (3.49 \pm 0.19) \times 10^{-4}$ for $E_\gamma > 1.6\,\text{GeV}$ to be compared to the SM result: $\mathcal{B}(\bar{B} \to X_s \gamma) = (3.36 \pm 0.23) \times 10^{-4}$ for the same threshold [5].

Baryon modes, such as $\Lambda_b \to X_s \gamma$ provide the further possibility to investigate observables sensitive to the spin of the decaying hadron. Heavy baryons with a b-quark produced in Z^0 and top-quark decays are expected to have a sizable polarization, as observed at LEP [30–32]. The application of HQE to baryons requires new information, namely the operator matrix elements for specified hadron spin. These have been determined in [33]. Here we also investigate a method to treat the singular terms in the inclusive photon spectrum to reconstruct the Λ_b shape function, a method which can be systematically applied when higher order terms in the $1/m_b$ expansion are computed.

The plan of the paper is the following. Section 2 describes the $b \to s\gamma$ low-energy Hamiltonian both in SM and in extensions. In section 3 we review the application of the HQE to the inclusive $H_b \to X_s \gamma$ process with H_b a baryon, keeping the dependence on the baryon spin. The correlation between the photon and Λ_b polarizations is studied. A treatment of the singular terms is discussed in section 4. The last section summarizes.

2 Generalized $b \to s\gamma$ effective Hamiltonian

The low-energy Hamiltonian governing $b \to s\gamma$ transition can be written as

$$H_{\text{eff}}^{b\to s\gamma} = -4\,\frac{G_F}{\sqrt{2}}\,V_{tb}\,V_{ts}^*\,\sum_i\,[C_i(\mu)\,O_i + C_i'(\mu)\,O_i'] \ , \tag{1}$$

G_F is the Fermi constant, $V_{tb(s)}$ the CKM matrix elements.[1] Eq. (1) comprises the magnetic penguin operators $O_7 = \frac{e}{16\pi^2}\,[\bar{s}\,\sigma^{\mu\nu}\,(m_s\,P_L + m_b\,P_R)\,b]\,F_{\mu\nu}$ and $O_8 = \frac{g_s}{16\pi^2}\left[\bar{s}_\alpha\,\sigma^{\mu\nu}\left(\frac{\lambda^a}{2}\right)_{\alpha\beta}(m_s\,P_L + m_b\,P_R)\,b_\beta\right]G^a_{\mu\nu}$, with $P_{R,L} = \frac{1\pm\gamma_5}{2}$ helicity projectors, α,β colour indices, λ^a the Gell-Mann matrices. $F_{\mu\nu}$ and $G^a_{\mu\nu}$ are the electromagnetic and gluonic field strengths, e and g_s the electromagnetic and strong coupling constants, m_b and m_s the b and s quark mass. Current-current operators $O_1 = (\bar{s}_\alpha\,\gamma^\mu\,P_L\,c_\beta)\,(\bar{c}_\beta\,\gamma_\mu\,P_L\,b_\alpha)$ and $O_2 = (\bar{s}\,\gamma^\mu\,P_L\,c)\,(\bar{c}\,\gamma_\mu\,P_L\,b)$ and the QCD penguin operators $O_3 = (\bar{s}\,\gamma^\mu\,P_L\,b)\,\sum_q(\bar{q}\,\gamma^\mu\,P_L\,q)$, $O_4 = (\bar{s}_\alpha\,\gamma^\mu\,P_L\,b_\beta)\,\sum_q(\bar{q}_\beta\,\gamma^\mu\,P_L\,q_\alpha)$, $O_5 = (\bar{s}\,\gamma^\mu\,P_L\,b)\,\sum_q(\bar{q}\,\gamma^\mu\,P_R\,q)$ and $O_6 = (\bar{s}_\alpha\,\gamma^\mu\,P_L\,b_\beta)\,\sum_q(\bar{q}_\beta\,\gamma^\mu\,P_R\,q_\alpha)$, where the sum runs over $q = \{u, d, s, c, b\}$, are also comprised. The remaining operators, absent in SM, are analogous to the QCD penguins but have a scalar or tensor structure [34]. The primed operators have opposite chirality with respect to the unprimed ones.

In SM the process $b \to s\gamma$ is described by photon penguin diagrams, with the photon coupled either to the intermediate fermion or to the $W^\pm$, giving rise to the magnetic operator O_7, the only operator contributing at lowest order in QCD. The renormalization group evolution to the scale $\mu_b \simeq O(m_b)$ also involves O_8 and $O_{1,\dots,6}$. Large logarithms producing a strong enhancement of the rate are generated by the mixing of these operators into O_7. The anomalous dimension matrix governing the mixing is regularization scheme dependent. Such dependence is taken into account by defining an effective coefficient $C_7^{\text{eff}}(\mu_b)$ which includes contributions of $O_{1,\dots,6}$ [35]. Therefore, O_7 represents the dominant contribution to $b \to s\gamma$, with the SM Wilson coefficients known at NNLO in QCD [14, 36].

In this study we describe a calculation at LO in α_s so that only the operators O_7 and O_7' have been considered. Hence, the effective Hamiltonian at the scale μ_b can be written as

$$H_{\text{eff}}^{b\to s\gamma} = -4\,\frac{G_F}{\sqrt{2}}\,V_{tb}\,V_{ts}^*\,\{C_7^{\text{eff}}\,O_7 + C_7'^{\text{eff}}\,O_7'\} = -4\,\frac{G_F}{\sqrt{2}}\,\lambda_t\,\frac{e}{16\pi^2}\,\sum_{i=7,7'}\,C_i^{\text{eff}}\,J^i_{\mu\nu}\,F^{\mu\nu} \ , \tag{2}$$

[1] Doubly Cabibbo suppressed terms proportional to $V_{ub}\,V_{us}^*$ have been neglected in (1).

where $\lambda_t = V_{tb} V_{ts}^*$, $J_{\mu\nu}^i = [\bar{s}\,\sigma_{\mu\nu}\,(m_s\,(1-P_i) + m_b\,P_i)\,b]$ and $P_7 = P_R$ and $P_{7'} = P_L$.

3 Inclusive decay width

The differential inclusive decay width can be written as

$$d\Gamma = [dq]\,\frac{G_F^2\,|\lambda_t|^2}{8\,m_{H_b}}\,\frac{\alpha}{\pi^2}\sum_{i,j=7,7'} C_i^{\text{eff}*}\,C_j^{\text{eff}}\,W_{MN}^{ij}\,\mathcal{F}^{MN}\,, \tag{3}$$

with $[dq] = \frac{d^3 q}{(2\pi)^3\,2\,q^0}$ and $\mathcal{F}^{MN} = -4\,q^\nu\,q^{\nu'}\,g^{\mu\mu'}$. By the optical theorem, the hadronic tensor W_{MN}^{ij} is related to the discontinuity of the forward scattering amplitude depicted in Fig. 1

$$T_{MN}^{ij} = i\int d^4x\,e^{-i\,q\cdot x}\,\langle H_b(p,s)|T[J_M^{i\dagger}(x)\,J_N^j(0)]|H_b(p,s)\rangle\,, \tag{4}$$

across the cut corresponding to the process $H_b(p,s) \to X_s(p_X)\,\gamma(q,\epsilon)$: $W_{MN}^{ij} = \frac{1}{\pi}\,\text{Im}\,[T_{MN}^{ij}]$.

The range of the invariant mass p_X^2 of the states produced in B and Λ_b decays (with $p_X = p - q$) is $p_X^2 \in [m_{K^*}^2, m_B^2]$ and $p_X^2 \in [m_\Lambda^2, m_{\Lambda_b}^2]$, respectively. For $m_b \to \infty$, p_X^2 is almost always large enough to exploit the short-distance limit $x \to 0$ in Eq. (4), thus allowing a computation of T^{ij} and W^{ij} by an OPE with expansion parameter $1/m_b$ [37, 38], which can be constructed by expressing the hadron momentum $p = m_H\,v$, where the v the four-velocity, in terms of m_b and of a residual momentum k: $p = m_b\,v + k$. The QCD b quark field is rescaled $b(x) = e^{-i\,m_b v\cdot x}\,b_v(x)$, with $b_v(x)$ and satisfies the equation of motion $b_v(x) = (P_+ + \frac{i\,\not{D}}{2\,m_b})\,b_v(x)$ with $P_+ = \frac{1+\not{v}}{2}$. In terms of $b_v(x)$ Eq. (4) becomes:

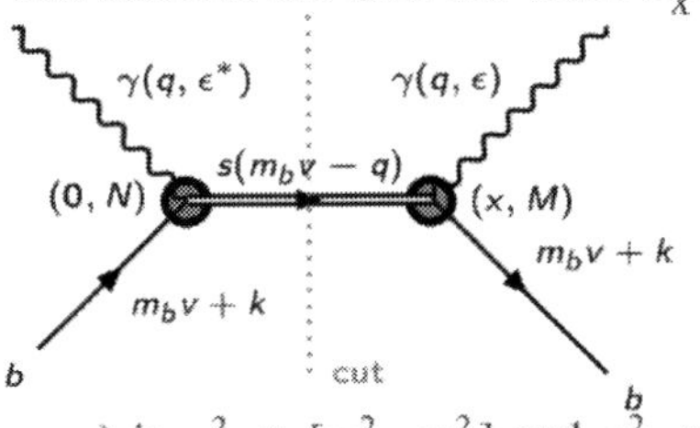

Figure 1. Discontinuity of the forward amplitude across the cut of the radiative process.

$$T_{MN}^{ij} = i\int d^4x\,e^{i\,(m_b v - q)\cdot x}\,\langle H_b(v,s)|T[\hat{J}_M^{i\dagger}(x)\,\hat{J}_N^j(0)]|H_b(v,s)\rangle\,. \tag{5}$$

$\hat{J}^i$ contains the field b_v. The HQE is obtained from

$$T_{MN}^{ij} = \langle H_b(v,s)|\bar{b}_v(0)\,\bar{\Gamma}_M^i\,\frac{1}{m_b\,\not{v} + \not{k} - \not{q} - m_s}\,\Gamma_N^j\,b_v(0)|H_b(v,s)\rangle\,, \tag{6}$$

with $\bar{\Gamma}_M^i = \gamma^0\,\Gamma_M^{i\dagger}\,\gamma^0$ and $\Gamma_M^7 = \sigma_{\mu\nu}\,(m_b\,P_R + m_s\,P_L)$, $\Gamma_M^{7'} = \sigma_{\mu\nu}\,(m_b\,P_L + m_s\,P_R)$. Replacing $k \to i\,D$, with D the QCD covariant derivative, and considering $|k| \sim O(\Lambda_{\text{QCD}})$ we have

$$T_{MN}^{ij} = \sum_{n=0}^{+\infty}\langle H_b(v,s)|\bar{b}_v(0)\,\bar{\Gamma}_M^i\,(\not{p}_s + m_s)\,[i\,\not{D}\,(\not{p}_s + m_s)]^n\,\Gamma_N^j\,b_v(0)|H_b(v,s)\rangle\,\frac{(-1)^n}{\Delta_0^{n+1}}\,, \tag{7}$$

where p_s is the s quark momentum and $\Delta_0 = p_s^2 - m_s^2$. Using the trace formalism [39], we can write the n-th term in the series as

$$\langle H_b(v,s)|\bar{b}_v(0)\,\bar{\Gamma}_M^i\,(\not{p}_s + m_s)\,\underbrace{i\,\not{D}\,(\not{p}_s + m_s)\ldots i\,\not{D}\,(\not{p}_s + m_s)}_{\text{n times}}\,\Gamma_N^j\,b_v(0)|H_b(v,s)\rangle = \tag{8}$$

$$= \left[\bar{\Gamma}_M^i\,(\not{p}_s + m_s)\prod_{k=1}^{n}\left[\gamma_{\mu_k}\,(\not{p}_s + m_s)\right]\Gamma_N^j\right]_{ab}\underbrace{\langle H_b(v,s)|\bar{b}_v(0)\,iD^{\mu_1}\ldots iD^{\mu_n}\,b_v(0)|H_b(v,s)\rangle_{ba}}_{(\mathcal{M}^{\mu_1\ldots\mu_n})_{ba}}\,,$$

with a and b Dirac indices. For any order in the expansion, all the matrix elements $(\mathcal{M}^{\mu_1\ldots\mu_n})_{ba}$ can be written in terms of nonperturbative parameters

$$O(m_b^{-n})\ldots\begin{cases} O(m_b^{-3})\begin{cases} O(m_b^{-2})\begin{cases} -2\,M_H\,\hat{\mu}_\pi^2 = \langle H_b|\bar{b}_v\,iD^\mu\,iD_\mu\,b_v|H_b\rangle \\ 2\,M_H\,\hat{\mu}_G^2 = \langle H_b|\bar{b}_v\,(-i\sigma_{\mu\nu})\,iD^\mu\,iD^\nu\,b_v|H_b\rangle \end{cases} \\ 2\,M_H\,\hat{\rho}_D^3 = \langle H_b|\bar{b}_v\,iD^\mu\,(iv\cdot D)\,iD_\mu\,b_v|H_b\rangle \\ 2\,M_H\,\hat{\rho}_{LS}^3 = \langle H_b|\bar{b}_v\,(-i\sigma_{\mu\nu})\,iD^\mu\,(iv\cdot D)\,iD^\nu\,b_v|H_b\rangle \end{cases} \\ \ldots \end{cases} \quad . \tag{9}$$

$\mathcal{M}^{\mu_1\ldots\mu_n}$ have been derived at $O(m_b^{-3})$ for a polarized baryon in [33], extending the previous results [40–44].

The double distribution, with respect to y and $\cos\theta_P$, using the methods proposed in [39, 45], comprises two terms

$$\frac{d^2\Gamma}{dy\,d\cos\theta_P} = \tilde{\Gamma}_1 + \cos\theta_P\,\tilde{\Gamma}_2\,. \tag{10}$$

Integrating (10) in $\cos\theta_P$, (θ_P being the angle between the hadron spin s and the photon momentum q), one has $\tilde{\Gamma}_1 = \frac{1}{2}\frac{d\Gamma}{dy}$ and the photon energy spectrum

$$\begin{aligned}\frac{1}{\Gamma_0}\frac{d\Gamma}{dy} = &\left[1 - \frac{\hat{\mu}_\pi^2}{2m_b^2} - \frac{\hat{\mu}_G^2}{2m_b^2}\frac{3+5z}{1-z} - \frac{10\hat{\rho}_D^3}{3m_b^3}\frac{1+z}{1-z}\right]\delta(1-z-y) \\ &+\left[\frac{\hat{\mu}_\pi^2}{2m_b^2}(1-z) - \frac{\hat{\mu}_G^2}{6m_b^2}(3+5z) - \frac{4\hat{\rho}_D^3}{3m_b^3}(1+2z) + \frac{2\hat{\rho}_{LS}^3}{3m_b^3}(1+z)\right]\delta'(1-z-y) \\ &+\left[\frac{\hat{\mu}_\pi^2}{6m_b^2}(1-z)^2 - \frac{\hat{\rho}_D^3}{3m_b^3}(1-z)(1+2z) + \frac{\hat{\rho}_{LS}^3}{6m_b^3}(1-z^2)\right]\delta''(1-z-y) \\ &- \frac{\hat{\rho}_D^3}{18m_b^3}(1-z)^2(1+z)\,\delta'''(1-z-y)\,,\end{aligned} \tag{11}$$

where $z = \frac{m_s^2}{m_b^2}$, $\Gamma_0 = \frac{\alpha G_F^2 |\lambda_t|^2}{32\pi^4}\,m_b^5\,(1-z)^3\left[|C_+^{\rm eff}|^2 + |C_+^{\prime\,\rm eff}|^2\right]$, $C_+^{\rm eff} = C_7^{\rm eff} + \sqrt{z}\,C_7^{\prime\,\rm eff}$ and $C_+^{\prime\,\rm eff} = \sqrt{z}\,C_7^{\rm eff} + C_7^{\prime\,\rm eff}$. $\tilde{\Gamma}_2$ is given by:

$$\begin{aligned}-\frac{2}{\Gamma_0}\frac{|C_+^{\rm eff}|^2 + |C_+^{\prime\,\rm eff}|^2}{|C_+^{\rm eff}|^2 - |C_+^{\prime\,\rm eff}|^2}\tilde{\Gamma}_2 = &\left[1 - \frac{13\hat{\mu}_\pi^2}{12m_b^2} - \frac{3\hat{\mu}_G^2}{4m_b^2}\frac{5+3z}{1-z} - \frac{\hat{\rho}_D^3}{6m_b^3}\frac{31+9z}{1-z}\right]\delta(1-z-y) \\ &+\left[\frac{\hat{\mu}_\pi^2}{2m_b^2}(1-z) - \frac{\hat{\mu}_G^2}{2m_b^2}(3+z) - \frac{2\hat{\rho}_D^3}{m_b^3}(1+z)\right]\delta'(1-z-y) \\ &+\left[\frac{\hat{\mu}_\pi^2}{6m_b^2}(1-z)^2 - \frac{\hat{\rho}_D^3}{3m_b^3}(1-z)(1+2z)\right]\delta''(1-z-y) \\ &- \frac{\hat{\rho}_D^3}{18m_b^3}(1-z)^2(1+z)\,\delta'''(1-z-y)\,.\end{aligned} \tag{12}$$

The angular differential distribution also comprises two terms:

$$\frac{d\Gamma(H_b \to X_s\,\gamma)}{d\cos\theta_P} = A + B\,\cos\theta_P\,, \tag{13}$$

with $A = \frac{1}{2}\Gamma(H_b \to X_s\,\gamma)$ and $B = -\frac{\Gamma_0}{2}\frac{|C_+^{\rm eff}|^2 - |C_+^{\prime\,\rm eff}|^2}{|C_+^{\rm eff}|^2 + |C_+^{\prime\,\rm eff}|^2}\left[1 - \frac{13\hat{\mu}_\pi^2}{12m_b^2} - \frac{3\hat{\mu}_G^2}{4m_b^2}\frac{5+3z}{1-z} - \frac{\hat{\rho}_D^3}{6m_b^3}\frac{31+9z}{1-z}\right]$. Integrating in $\cos\theta_P$, the inclusive $H_b \to X_s\,\gamma$ width is given by

$$\Gamma(H_b \to X_s\,\gamma) = \Gamma_0\left[1 - \frac{\hat{\mu}_\pi^2}{2m_b^2} - \frac{\hat{\mu}_G^2}{2m_b^2}\frac{3+5z}{1-z} - \frac{10\hat{\rho}_D^3}{3m_b^3}\frac{1+z}{1-z}\right]. \tag{14}$$

For $C_7' \to 0$ the SM result is recovered.

Another interesting observable is the photon polarization asymmetry $A_P(\cos\theta_P)$ in $b \to s\,\gamma$ transition that quantifies the difference between left and right handed photon produced (the definition can be found in [21]). In SM the photon polarization asymmetry is $A_P(\cos\theta_P) \simeq -1$ for almost all $\cos\theta_P$, it increases only for $\cos\theta_P \to 1$, see Fig. 2 (left panel). Physics beyond SM can produce a deviation of the asymmetry from the SM value with the largest effect for $\cos\theta_P \simeq 1$. Considering both C_7 and C_7' real and varying $C_7'^{\rm eff}/C_7^{\rm eff} \in [-0.3, 0.3]$ we obtain the asymmetry in Fig. 2 (right panel) versus $C_7'^{\rm eff}/C_7^{\rm eff}$ for selected values of $\cos\theta_P$.

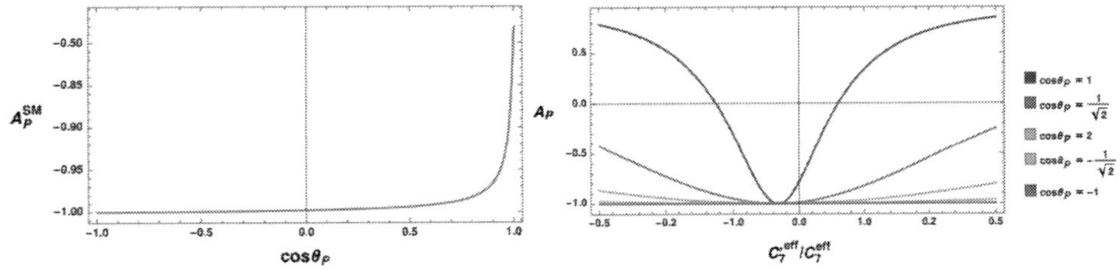

Figure 2. Photon polarization asymmetry $A_P(\cos\theta_P)$ versus $\cos\theta_P$ in SM (left) and photon polarization asymmetry $A_P(\cos\theta_P)$ for several values of $\cos\theta_P$, varying $C_7'^{\rm eff}/C_7^{\rm eff}$ (right).

4 Treatment of the singular terms

The spectrum obtained by the short-distance OPE does not account for possible smearing close to the end point region. In this region, the Fermi motion of the b quark and its soft interactions with the light degrees of freedom in the hadron are relevant. It is possible to resum the singular terms defining the spectral function [23–25, 38, 46]

$$S_s(y) = \sum_{n=0}^{\infty} \frac{M_n}{n!}\,\delta^{(n)}(1-z-y)\,, \tag{15}$$

and introducing the quantity $f(k_+)$ as[2]

$$S_s(y) = \int dk_+\,\delta\left(1 - y - z + \frac{k_+}{m_b}\right)[f(k_+) + O(m_b^{-1})]\,. \tag{16}$$

The photon energy spectrum is obtained by the convolution [24]

$$\frac{d\Gamma}{dy} = \int dk_+\,f(k_+)\,\frac{d\Gamma^*}{dy}\,. \tag{17}$$

In $\frac{d\Gamma^*}{dy}$ the b quark mass m_b is replaced by $m_b^* = m_b + k_+$, an exact substitution at tree level. For $k_+ \in [-m_b, m_{H_b} - m_b]$, replacing $m_b \to m_b^*$ in the variable y, one finds ($m_s = 0$ to simplify the discussion): $y \to \frac{2E_\gamma}{(m_b+k_+)}$. Hence, for $k_+^{\rm max} = m_{H_b} - m_b$ the maximum photon energy is $E_\gamma = \frac{m_{H_b}}{2}$ (physical end-point). The shape function provides an interpretation of the singular terms in the photon energy spectrum obtained in the previous section. Writing Eq. (11) as

$$\frac{1}{\Gamma}\frac{d\Gamma}{dy} = \sum_{n=0}^{3} \frac{M_n}{n!}\,\delta^{(n)}(1-z-y)\,, \tag{18}$$

[2]Perturbative corrections to the shape function and its moments are discussed in [47, 48].

with Γ in Eq. (14) and the moments

$$M_0 = 1\,, \qquad M_1 = \frac{\hat\mu_\pi^2}{2\,m_b^2}\,(1-z) - \frac{\hat\mu_G^2}{6\,m_b^2}\,(3+5\,z) - \frac{4\,\hat\rho_D^3}{3\,m_b^3}\,(1+2\,z) + \frac{2\,\hat\rho_{LS}^3}{3\,m_b^3}\,(1+z)\,, \tag{19}$$

$$M_2 = \frac{\hat\mu_\pi^2}{3\,m_b^2}\,(1-z)^2 - \frac{2\,\hat\rho_D^3}{3\,m_b^3}\,(1-z)(1+2\,z) + \frac{\hat\rho_{LS}^3}{3\,m_b^3}\,(1-z^2)\,,\quad M_3 = -\frac{\hat\rho_D^3}{3\,m_b^3}\,(1-z)^2\,(1+z)\,.$$

Each moment M_n in Eqs. (19), has an expansion $M_n = \sum_{k=n}^{\infty} \frac{M_{n,k}}{m_b^k}$ [24]. M_n in (15) are related to the moments of the photon energy spectrum: $\langle y^k\rangle = \frac{1}{\Gamma}\int_0^{1-z} dy\, y^k\, \frac{d\Gamma}{dy}$. Using (18) we have:

$$\langle y^k\rangle = \sum_{n=0}^{\infty} \frac{M_n}{n!}\int_0^{1-z} dy\, y^k\, \delta^{(n)}(1-z-y) = \sum_{j=0}^{k}\binom{k}{j}(1-z)^{k-j}\,M_j\,, \tag{20}$$

$$\langle y\rangle = (1-z) + M_1\,, \qquad \langle y^2\rangle = (1-z)^2 + 2\,(1-z)\,M_1 + M_2\,, \qquad \sigma_y^2 = \langle y^2\rangle - \langle y\rangle^2\,. \tag{21}$$

The moments of the measured photon energy spectrum can be used to determine the HQET parameters as well as the $\cos\theta_P$ distribution (13).

Let us describe the effect of the Fermi motion. Each order in the $1/m_b$ expansion, the photon energy spectrum obtained by the OPE corresponds to a monochromatic line. At LO, the line is at $y = 1-z$, the next terms correspond to a displacement of this position. The convolution (17) with the shape function provides the smearing of the spectrum.

The shape function is a nonperturbative quantity to be determined by methods such as lattice QCD, or which must be suitably parametrized. We adopt a different point of view. Including infinite terms, the sum in Eq. (18) coincides with $S_s(y)$ in (15). In the sum, the first term corresponds to a monocromatic line at the zero of the δ function, with $\langle y\rangle = 1-z$ and $\sigma_y^2 = 0$, the LO results in Eqs. (21). We write $\delta(b-y) = \lim_{\sigma_y\to 0} \frac{1}{\sqrt{2\pi}\,\sigma_y}\, e^{-\frac{(b-y)^2}{2\sigma_y^2}}$, with $b = (1-z) = \langle y\rangle_{\rm LO}$ and σ_y^2 at each order in $1/m_b$, starting from $1/m_b^2$. For $m_b \to \infty$ the condition $\sigma_y \to 0$ reproduces the partonic result. Our ansatz consists in substituting

$$S_s(y) = \sum_{n=0}^{\infty}\frac{M_n}{n!}\,\delta^{(n)}(1-z-y) \quad\to\quad S_s(y) = \sum_{n=0}^{\infty}\frac{M_n}{n!}\,(-1)^n\,\frac{d^n}{dy^n}\,\frac{1}{\sqrt{2\pi}\,\sigma_y}\,e^{-\frac{(b-y)^2}{2\sigma_y^2}}\,. \tag{22}$$

Using the representation of the Hermite polynomials $H_n(x) = (-1)^n\, e^{x^2}\,\frac{d^n}{dx^n}\,e^{-x^2}$, we obtain:

$$S_s(y) = \frac{1}{\sqrt{2\pi}\,\sigma_y}\sum_{n=0}^{\infty}\frac{M_n}{n!}\left(-\frac{1}{\sqrt{2}\,\sigma_y}\right)^n e^{-\frac{(b-y)^2}{2\sigma_y^2}}\,H_n\!\left(\frac{b-y}{\sqrt{2}\,\sigma_y}\right). \tag{23}$$

Denoting by $\langle y^k\rangle_N$ the moments computed using this expression for the spectral function,

$$\langle y^k\rangle_N = \int_0^{y_{\rm max}} dy\, y^k\, S_s(y)\,, \qquad \text{where} \qquad \lim_{\sigma_y\to 0}\langle y^k\rangle_N = \langle y^k\rangle\,, \tag{24}$$

with $\langle y^k\rangle$ in (20). While b is fixed to the LO result for $\langle y\rangle$, σ_y can be determined at an arbitrary order in the $1/m_b$ expansion. At odds with most models proposed for the shape function, by construction the ansatz (23) can include all moments M_n once they are computed.

In Fig. 3 we plot the spectral function obtained at LO, $O(m_b^{-2})$ and $O(m_b^{-3})$. The maximum photon energy is larger than the partonic result $y = 1-z$. In the same figure we plot the shape function obtained from (16) at $O(m_b^{-2})$ and $O(m_b^{-3})$. As a consequence of broadening the spectrum through the substitution in (22), there is a tail exceeding the physical end-point $\bar y_{\rm max} \simeq \frac{m_{\Lambda_b}}{m_b}\,(1-z)$, and a tail in the shape function exceeding $k_+^{\rm max} = m_{\Lambda_b} - m_b$. This is a spurious effect of the truncation. When higher orders in the HQE are included, the area below such tails approaches to zero.

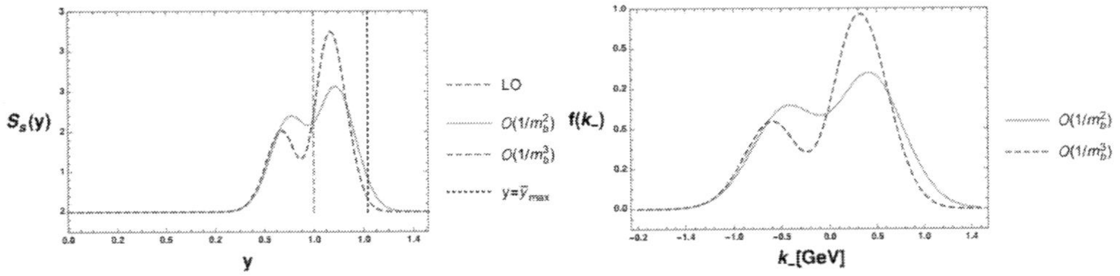

Figure 3. Function $S_s(y)$ (left) and shape function $f(k_+)$ (right) obtained using Eq. (23) up to $n = 3$.

5 Conclusions

The HQE has been exploited to compute the inclusive decay width induced by the $b \to s\gamma$ transition for a beauty baryon. The differential width in the photon energy and in $\cos\theta_P$ allows to construct new observables with respect to mesons. The calculation has been carried out at $O(m_b^{-3})$ for non-vanishing m_s, using the baryon matrix elements determined in [33]. For the singular terms appearing in the spectrum we have proposed a treatment that can be systematically improved including higher order terms in the expansion.

I thank P. Colangelo and F. De Fazio for collaboration. This study has been carried out within the INFN project (Iniziativa Specifica) SPIF.

References

[1] M.A. Shifman, A.I. Vainshtein, V.I. Zakharov, Phys. Rev. D **18**, 2583 (1978), [Erratum: Phys.Rev.D 19, 2815 (1979)]
[2] S. Bertolini, F. Borzumati, A. Masiero, Phys. Rev. Lett. **59**, 180 (1987)
[3] B. Grinstein, R.P. Springer, M.B. Wise, Phys. Lett. B **202**, 138 (1988)
[4] G. Cella, G. Curci, G. Ricciardi, A. Vicere, Phys. Lett. B **325**, 227 (1994), hep-ph/9401254
[5] M. Misiak et al., Phys. Rev. Lett. **114**, 221801 (2015), 1503.01789
[6] B. Grinstein, M.B. Wise, Phys. Lett. B **201**, 274 (1988)
[7] W.S. Hou, R.S. Willey, Phys. Lett. B **202**, 591 (1988)
[8] F. Gabbiani, E. Gabrielli, A. Masiero, L. Silvestrini, Nucl. Phys. B **477**, 321 (1996), hep-ph/9604387
[9] L.L. Everett, G.L. Kane, S. Rigolin, L.T. Wang, T.T. Wang, JHEP **01**, 022 (2002), hep-ph/0112126
[10] F. Borzumati, C. Greub, Y. Yamada, Phys. Rev. D **69**, 055005 (2004), hep-ph/0311151
[11] A.J. Buras, A. Poschenrieder, M. Spranger, A. Weiler, Nucl. Phys. B **678**, 455 (2004), hep-ph/0306158
[12] M. Blanke, A.J. Buras, A. Poschenrieder, C. Tarantino, S. Uhlig, A. Weiler, JHEP **12**, 003 (2006), hep-ph/0605214
[13] M. Misiak, M. Steinhauser, Eur. Phys. J. C **77**, 201 (2017), 1702.04571
[14] A.J. Buras, *Gauge Theory of Weak Decays* (Cambridge University Press, 2020), ISBN 978-1-139-52410-0, 978-1-107-03403-7

[15] R. Ammar et al. (CLEO), Phys. Rev. Lett. **71**, 674 (1993)
[16] Y.S. Amhis et al. (HFLAV) (2022), 2206.07501
[17] R.L. Workman et al. (Particle Data Group), PTEP **2022**, 083C01 (2022)
[18] R. Aaij et al. (LHCb), Phys. Rev. Lett. **123**, 031801 (2019), 1904.06697
[19] R. Aaij et al. (LHCb), Phys. Rev. D **105**, L051104 (2022), 2111.10194
[20] R. Aaij et al. (LHCb), JHEP **01**, 069 (2022), 2108.07678
[21] P. Colangelo, F. De Fazio, F. Loparco, JHEP **10**, 147 (2023), 2306.02748
[22] M. Neubert, Phys. Rept. **245**, 259 (1994), hep-ph/9306320
[23] M. Neubert, Phys. Rev. D **49**, 3392 (1994), hep-ph/9311325
[24] M. Neubert, Phys. Rev. D **49**, 4623 (1994), hep-ph/9312311
[25] I.I.Y. Bigi, M.A. Shifman, N.G. Uraltsev, A.I. Vainshtein, Int. J. Mod. Phys. A **9**, 2467 (1994), hep-ph/9312359
[26] F.U. Bernlochner, H. Lacker, Z. Ligeti, I.W. Stewart, F.J. Tackmann, K. Tackmann (SIMBA), Phys. Rev. Lett. **127**, 102001 (2021), 2007.04320
[27] M. Benzke, S.J. Lee, M. Neubert, G. Paz, JHEP **08**, 099 (2010), 1003.5012
[28] J.F. Donoghue, A.A. Petrov, Phys. Rev. D **53**, 3664 (1996), hep-ph/9510227
[29] S.J. Lee, M. Neubert, G. Paz, Phys. Rev. D **75**, 114005 (2007), hep-ph/0609224
[30] D. Buskulic et al. (ALEPH), Phys. Lett. B **365**, 437 (1996)
[31] G. Abbiendi et al. (OPAL), Phys. Lett. B **444**, 539 (1998), hep-ex/9808006
[32] P. Abreu et al. (DELPHI), Phys. Lett. B **474**, 205 (2000)
[33] P. Colangelo, F. De Fazio, F. Loparco, JHEP **11**, 032 (2020), [Erratum: JHEP 12, 098 (2022)], 2006.13759
[34] F. Borzumati, C. Greub, T. Hurth, D. Wyler, Phys. Rev. D **62**, 075005 (2000), hep-ph/9911245
[35] A.J. Buras, M. Misiak, M. Munz, S. Pokorski, Nucl. Phys. B **424**, 374 (1994), hep-ph/9311345
[36] M. Misiak, Acta Phys. Polon. B **49**, 1291 (2018)
[37] J. Chay, H. Georgi, B. Grinstein, Phys. Lett. B **247**, 399 (1990)
[38] I.I. Bigi, M.A. Shifman, N. Uraltsev, A.I. Vainshtein, Phys. Rev. Lett. **71**, 496 (1993), hep-ph/9304225
[39] B.M. Dassinger, T. Mannel, S. Turczyk, JHEP **03**, 087 (2007), hep-ph/0611168
[40] T. Mannel, A.V. Rusov, F. Shahriaran, Nucl. Phys. B **921**, 211 (2017), 1702.01089
[41] S. Kamali, Int. J. Mod. Phys. **A34**, 1950036 (2019), 1811.07393
[42] Y. Grossman, Z. Ligeti, Phys. Lett. B **332**, 373 (1994), hep-ph/9403376
[43] A.V. Manohar, M.B. Wise, Phys. Rev. **D49**, 1310 (1994), hep-ph/9308246
[44] S. Balk, J. Korner, D. Pirjol, Eur. Phys. J. C **1**, 221 (1998), hep-ph/9703344
[45] T. Mannel, S. Turczyk, N. Uraltsev, JHEP **11**, 109 (2010), 1009.4622
[46] T. Mannel, M. Neubert, Phys. Rev. D **50**, 2037 (1994), hep-ph/9402288
[47] C.W. Bauer, A.V. Manohar, Phys. Rev. D **70**, 034024 (2004), hep-ph/0312109
[48] S.W. Bosch, B.O. Lange, M. Neubert, G. Paz, Nucl. Phys. B **699**, 335 (2004), hep-ph/0402094

EPJ Web of Conferences **314**, 00008 (2024) https://doi.org/10.1051/epjconf/202431400008
QCD@Work 2024

Highlights from the BESIII experiment

Isabella Garzia[1,2,*], on behalf of the BESIII Collaboration

[1]University of Ferrara - Dipartimento di Fisica e Scienze della Terra, Ferrara, Italy
[2]Istituto Nazionale di Fisica Nucleare (INFN), Sezione di Ferrara, Ferrara, Italy

Abstract. Since 2009, the BESIII experiment has been collecting e^+e^- physics data in the energy range between 2.0 up to 4.95 GeV. This allowed the BESIII Collaboration to collect three of the largest data sets of charmonium J/ψ, $\psi(2S)$, and $\psi(3770)$ resonances in the world. In particular, the radiative decay of J/ψ provides a gluon-rich environment and it is therefore regarded as one of the most promising hunting grounds for glueballs, as well as a powerful tool for searching exotic states. In this proceeding, we present the first measurement of the quantum numbers of the $X(2370)$ particle, which shows very promising glueball features, and the first observation of exotic isoscalar state $\eta_1(1855)$ in the $J/\psi \to \gamma\eta\eta'$ decay. Latest studies on the $\eta(1405)/\eta(1474)$ are also reported, together with the observation od a new state, $X(1880)$, observed in the $J/\psi \to \gamma(3\pi^+\pi^-)$ decay. Finally, the normalized differential cross section of inclusive π^0, K^0_S, and η production as a function of hadron momentum are are measured at several center of mass energies from 2.000 to 3.6710 GeV. The results deviates from several theoretical calculation based on existing fragmentation functions and provide new ingredient for global data fits.

1 Introduction

The Beijing Electron Spectrometer III (BESIII) detector [1] is a general purpose magnetic spectrometer installed at the Beijing Electron Positron Collider II (BEPCII) [2]. BEPCII is a symmetric double-ring e^+e^- collider hosted at the Institute of High Energy Physics with a beam energy tuneable in a range from 1.0 GeV to 2.45 GeV. The BESIII detector has a geometrical acceptance of 93% of 4π solid angle and it consists of four main subsystems: a helium-based multilayer drift chamber (MDC), a plastic scintillator time-of-flight (TOF) system, a CsI(Tl) electromagnetic calorimeter (EMC) and a resistive plate muon chamber. The first three subdetectors are enclosed in a superconductive solenoidal magnet which generates a magnetic field of about 1 T. More details can be found in [1].

The BESIII experiment has accumulated about 10 billion J/ψ events, about 2.7×10^9 $\psi(3686)$ events, and a total integrated luminosity of about 20 fb^{-1} at the center of mass energy of 3.773 GeV, that provide a clean laboratory to study light hadron physics and search for light exotic hadron states, such as multiquark states, glueballs and hybrids. Other data has been collected at several center of mass energy from 2 GeV up to 4.95 GeV, which allows to cover a very rich physics program in the so called τ−charm region.

*e-mail: isabella.garzia@unife.it,garzia@fe.infn.it

2 Glueballs and exotic hybrid states

The existence of glueballs has been predicted by Quantum Chromodynamics (QCD). According to Lattice QCD calculation [3–5], the mass of the ground state glueball with quantum number $J^{PC} = 0^{++}$, should be in the range [1.6-1.7] GeV/c^2, while the mass of the first excited state (with $J^{PC} = 2^{++}$) should be around 2.3 GeV/c^2. These mass ranges are accessible at the BESIII experiment. In addition to the high statistic collected, radiative J/ψ decay is a gluon rich process. Therefore, it is regarded as one of the most promising hunting grounds for glueballs.

QCD also allows the existence of hybrid mesons, $q\bar{q}$ states with gluonic degree of freedom. Models and Lattice QCD (LQCD) predict that the exotic $J^{PC} = 1^{-+}$ nonet hybrid mesons is the lightest, with a mass around $1.7 - 2.1$ GeV/c^2 [6]. To date, there are evidence of three isovector 1^{-+} states ($\pi_1(1400)$, $\pi_1(1600)$, and $\pi_1(2015)$) [7–10]. Finding an isoscalar 1^{-+} hybrid state is critical for establishing the hybrid multiplet.

2.1 Observation of exotic isoscalar state $\eta_1(1855)$ in $J/\psi \to \gamma\eta\eta'$

Based on a sample of $(10.09\pm0.04)\times10^9$ J/ψ events collected by BESIII, a Partial Wave Analysis (PWA) of $J/\psi \to \gamma\eta\eta'$ to search for 1^{-+} and investigate the decay property of $f_0(1710)$ was performed [11, 12]. An isoscalar state with exotic quantum number 1^{-+}, denoted as $\eta_1(1855)$, has been observed for the first time with a statistical significance larger than 19σ. Its mass and width are measured to be $(1855 \pm 9^{+6}_{-1})$ MeV/c^2 and $(188 \pm 18^{+3}_{-8})$ MeV [12], in agreement with LQCD calculation [13]. Figure 1 shows the PWA fit projection on the $\eta\eta'$ invariant mass distribution. In addition, the decay $J/\psi \to \gamma f_0(1500) \to \gamma\eta\eta'$ has also been observed with a statistical significance larger that 30σ, while the $J/\psi \to \gamma f_0(1710) \to \gamma\eta\eta'$ is found to be negligible. The suppressed decay rate of the $f_0(1710)$ into $\eta\eta'$ supports the hypothesis that the $f_0(1710)$ has a large overlap with the scalar glueball, and the $f_0(1710)/f_0(2020)$ might be interpreted as flavor singlet [14].

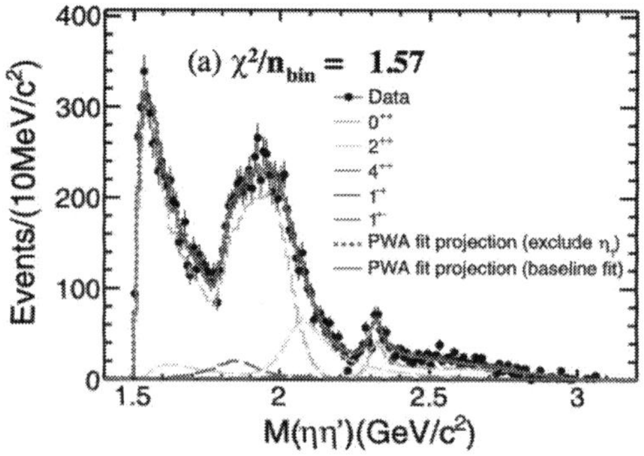

Figure 1. PWA fit projection for the invariant mass distribution of $\eta\eta'$. The red line is the total fit projections from the baseline PWA, while the blue dotted line represents the fit projection from a fit excluding the η_1 component.

2.2 The $\eta(1405)/\eta(1475)$ in $J/\psi \to \gamma K^0_S K^0_S \pi^0$ decay and $J/\psi \to \gamma\gamma\phi$

Radiative J/ψ decay can also been used to investigate the nature of the pseudoscalar structure around 1.4 GeV/c^2, which is often interpreted as the combination of two isoscalar resonant state, i.e., the $\eta(1405)$ and $\eta(1475)$. From a theoretical point of view, the pseudoscalar nonet of ground state is well established, and the $\eta(1295)$ and $\eta(1475)$ are generally assigned to be the first radial excitation of the ground state η and η' [32], and hence it became very difficult to accommodate additional pseudoscalar states. Therefore, the possibility of an additional unconventional state has been widely discussed. To investigate the $\eta(1405/1475)$ nature, BESIII Collaboration performed a PWA analysis of the decay $J/\psi \to \gamma K_S^0 K_S^0 \pi^0$ using the full J/ψ data set, where both mass dependent (MD) and mass independent (MI) approach are performed [16]. In particular, to extract the contributions from different components, a MI PWA is performed: the intermediate states in the $K_S^0 K_S^0 \pi^0$ invariant mass spectrum are parameterized by a separate complex constant in 24 bins of 15 MeV/c^2 width in the region $M(K_S^0 K_S^0 \pi^0) < 1.6$ GeV/c^2 (bin-by-bin analysis). Taking into account the spin-parity, charge conjugation and isospin conservation, all possible decay mode candidates are evaluated, and the $K_S^0 K_S^0 \pi^0$ invariant mass spectra obtained from the MI PWA results are shown in Figure 2. The fit results confirm that the dominant contribution comes from the pseudoscalar component; axial vector component also contribute (peaking at 1.28 GeV/c^2 and 1.42 GeV/c^2); and there is also a small tensor component decaying into $K^*(892)^0 K_S^0$ around 1.52 GeV/c^2 observed for the first time in this decay mode. A mass dependent PWA is also performed, and the results are summarized in Ref. [16]. The nominal solution is in good agreement with the MI PWA results, and it includes two pseudoscalar state, the $\eta(1405)$ and the $\eta(1475)$, two axial vector states, the $f_1(1285)$ and the $f_1(1425)$, and a tensor state, the $f_2(1525)$.

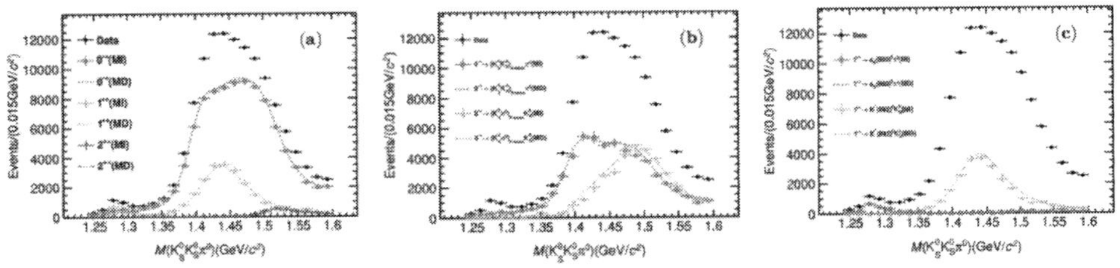

Figure 2. Intensity spectra of $K_S^0 K_S^0 \pi^0$ invariant mass distribution: (a) dominant spin-parity components, (b) dominant decay modes for the pseudoscalar component, and (c) dominant decay modes for the axial vector component. Dots with error bars are the intensity obtained from the MI PWA, while the dotted lines are obtained from MD PWA results.

For the first time, in a recent BESIII analysis was performed a PWA of the decay $J/\psi \to \gamma\gamma\phi$, which serve as flavor filter reaction and play a important role in the investigation of the quark contents of the intermediate resonances [17]. The PWA result shows that the $J/\psi \to \gamma X; X \to \gamma\phi$ process has predominantly 2^{++} components. The $\eta(1405)$, $X(1835)$, and several f-states are observed with a statistical significance greater than 5σ. In particular, the structure with mass around 1.4 GeV/c^2 in the $\gamma\phi$ system is described taking into account the $\eta(1405)$ and $f_1(1420)$ contributions. These results indicate that only one pseudoscalar state is needed in the vicinity of 1.4 GeV/c^2. In addition, the quantum number of the $X(1835)$ are confirmed to be 0^{-+}, and its observation decaying into $\gamma\phi$ final state indicates that this resonance also contains a sizable $s\bar{s}$ component.

2.3 $X(2370)$: Glueball-like particle in $J/\psi \to \gamma K_S^0 K_S^0 \eta'$

The $X(2370)$ resonance was first observed in the $J/\psi \to \gamma\pi^+\pi^-\eta'$ process in the $\pi^+\pi^-\eta'$ invariant mass distribution [18], and it was further observed in a combined measurement of

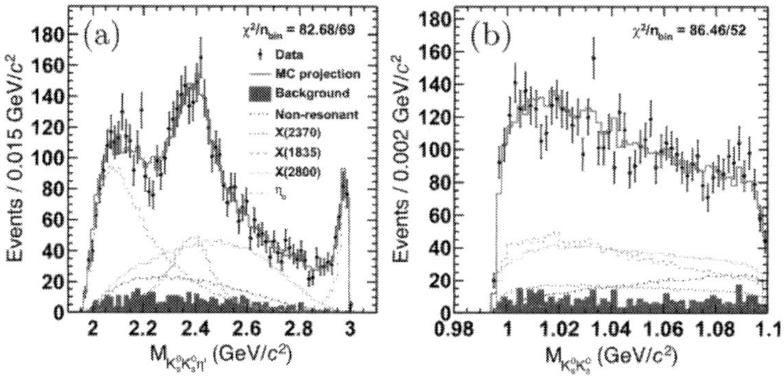

Figure 3. PWA fit projections on the (a) $K_S^0 K_S^0 \eta'$ and (b) $K_S^0 K_S^0$ invariant mass distributions. Dots with error bars refer to data, solid red histograms are the PWA total projections.

$J/\psi \to \gamma K^+ K^- \eta'$ and $J/\psi \to \gamma K_S^0 K_S^0 \eta'$ [19]. These experimental observation stimulated several theoretical interpretations on its nature. Among them, one of the most promising explanation is a pseudoscalar glueball [20–22]. To understand the nature of the $X(2370)$, it is crucial to measure the quantum numbers and the decay modes, and the $J/\psi \to \gamma K_S^0 K_S^0 \eta'$ decay provides a clean environment for the J^{PC} measurement with minimal background contamination due to the fact that the main background processes, $J/\psi \to \pi^0 K_S^0 K_S^0 \eta'$ and $J/\psi \to K_S^0 K_S^0 \eta'$, are forbidden by exchange symmetry and CP conservation. Using the full J/ψ data set collected at BESIII, the spin-parity determination of the $X(2379)$ was obtained for the first time [23]. A strong enhancement near the $K_S^0 K_S^0$ mass threshold from the $f_0(980)$ was observed, together with a clear connection between the $f_0(980)$ and the structure around 2.4 GeV/c^2 in the invariant mass spectra of $K_S^0 K_S^0 \eta'$. After requiring the $K_S^0 K_S^0$ invariant mass to be less that 1.1 GeV/c^2, the structure around 2.4 GeV/c^2 becomes much more prominent in the $K_S^0 K_S^0 \eta'$ mass spectrum. A PWA fit is performed in order to study the properties of the $X(2370)$ state, and the optimal PWA fit results, shown in Figure 3, shows that the data can be well described taking into account he following intermediate states: $X(1835)$, $X(2370)$, η_c, and a broad 0^{-+} structure denoted as $X(2800)$. The PWA fit result indicates a contribution from $X(2370) \to K_S^0 K_S^0 \eta'$ with a statistical significance larger than 14σ, with the mass and width of the $X(2370)$ measured to be $2395 \pm 11(\text{stat})^{+26}_{-94}\text{syst}$ MeV/c^2 and $188^{+18}_{-17}(\text{stat})^{+124}_{-33}\text{syst}$ MeV, respectively. The spin parity of the $X(2370)$ is determined to be 0^{-+} for the first time. In addition, the measured mass is in good agreement with the mass prediction of the lightest pseudoscalar glueball [20], which makes the $X(2370)$ state a promising glueball candidate.

2.4 $X(1880)$: A new state observed in $J/\psi \to \gamma(3\pi^+\pi^-)$

The $X(1835)$ resonance was first observed in the $\pi^+\pi^-\eta'$ invariant mass spectrum [24] and as a $p\bar{p}$ threshold enhancement in the $J/\psi \to \gamma p\bar{p}$ decay [25] at BESIII. Its nature is still controversial and several theoretical models are proposed. The $X(1835)$ was later confirmed by BESIII [26] and CLEO [27] experiments, and the quantum number measured to be 0^{-+} [28]. Meanwhile, a prominent structure, named as $X(1840)$, was observed in the $3(\pi^+\pi^-)$ invariant mass spectrum [29]. Even if it was interpreted as a new decay mode of the $X(1835)$, its width is substantially narrower than that of $X(1835)$.

In an updated analysis of $J/\psi \to \gamma\pi^+\pi^-\eta'$ is observed a significant abrupt change in the slope of the $X(1835) \to \pi^+\pi^-\eta'$ line shape at the $p\bar{p}$ mass threshold [30], which could

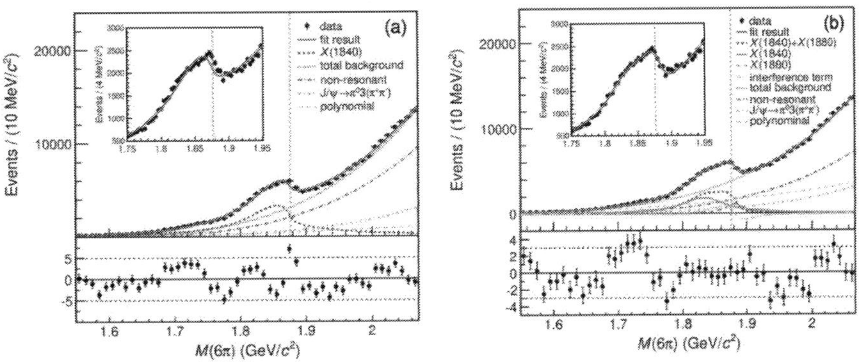

Figure 4. Fit result of the 6π invariant mass distribution ($M(6\pi)$) for model-I (a) and model-II (b).

be originated from the opening of an additional $p\bar{p}$ decay channel or the interference between two different resonance contributions. It is interesting to understand wether a similar phenomenon exist around the $p\bar{p}$ mass threshold in the 6π invariant mass distribution by analysing the $X(1840)$ line shape in the $J/\psi \to \gamma 3(\pi^+\pi^-)$ decay [31]. After the event and track selection, discussed in detail in ref. [31], an anomalous line shape near the $p\bar{p}$ mass threshold is clearly observed, and an unbinned maximum likelihood fit to the 6π invariant mass spectrum ($M(6\pi)$) was performed. When the $X(1840)$ is described as a simple resonant structure, the fit quality is very poor: a simple resonance structure cannot be used to describe the $M(6\pi)$ spectrum. To resolve the discrepancy from data, two different models for the line shape of the structure around the $p\bar{p}$ structure are investigated. In model-I was made the assumption that the change in the line shape of 6π above the $p\bar{p}$ mass threshold is due to the opening of the $X(1840) \to p\bar{p}$ decay, and the anomalous shape is described with a Flattè formula. Figure 4 (a) shows the comparison between the data and the fit result of model-I, where it is clearly visible a tension between data and fit results around the $p\bar{p}$ mass threshold. In order to obtain a better description on data, a second model (model-II) was taken into account: it allows the interference between two resonant structures, denoted as $X(1840)$ and $X(1880)$. As reported in Figure 4 (b), model-II improves the fit quality, and the $X(1880)$ is observed with a statistical significance larger than 10σ. According tho the model-II fit results, the narrow structure $X(1880)$ has a mass of $M_{X(1880)} = 1882.1 \pm 1.7 \pm 0.7$ MeV$/c^2$ and a width of $\Gamma_{X(1880)} = 30.7 \pm 5.5 \pm 2.4$ MeV, while the mass and the width of $X(1840)$ are measured to be $M_{X(1840)} = 1832.5 \pm 3.1 \pm 2.5$ MeV$/c^2$ and $\Gamma_{X(1840)} = 80.7 \pm 5.2 \pm 7.7$ MeV. Compared with the two structure observed in the $\pi^+\pi^-\eta'$ mass spectrum [30], the $X(1840)$ has a mass consistent with that of the $X(1835)$ state but much narrower width, while the mass and width of the $X(1880)$ are in agreement with those reported in [30]. However, to establish the nature of these states the determination of the spin-parity quantum numbers and couple channel amplitude analysis are needed.

3 Unpolarized fragmentation function at BESIII

Fragmentation Functions (FFs) are non-perturbative functions used to describe the hadronization of partons (quarks and gluons) into color-neutral particles. A precise knowledge of FFs will help to understand the color confinement property of the QCD at long distance, as well as they play a fundamental role in constraining the proton spin configuration and the nuclear

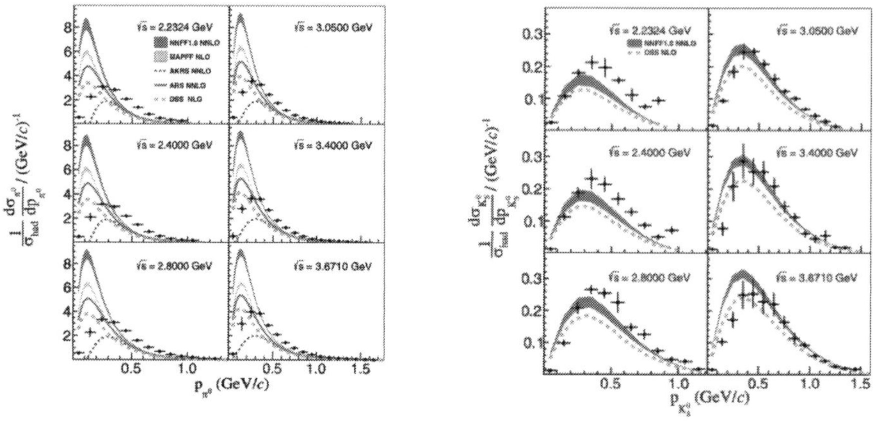

Figure 5. Normalised differential cross section of the $e^+e^- \to \pi^0 + X$ (left) and $e^+e^- \to K_S^0 + X$ (right) processes. Points with error bars refer to data, for which the uncertainties are the quadrature sum of the corresponding statistical and systematic uncertainties. The coloured bands denote several theoretical calculation [33].

parton distribution functions (PDFs). Single inclusive e^+e^- annihilation, $e^+e^- \to h+X$, where h is an identified hadron and X represents everything else, provides a clean laboratory to study the unpolarized FFs. Here, the dimensionless variable $z = 2E_h/\sqrt{s}$ is the relative hadron energy, E_h is the hadron energy, and $\sqrt{s}$ is the center of mass energy. A typical experimental observable is

$$\frac{1}{\sigma(e^+e^- \to \text{hadrons})}\frac{d\sigma(e^+e^- \to h + X}{dp_h},$$

where $\sigma(e^+e^- \to \text{hadrons})$ is the total cross section for e^+e^- annihilation to all possible hadronic final states, and p_h denotes the momentum of the identified hadron h. The observable can be interpreted, in terms of the leading order α_s, as $\sum_q e_q^2[D_q^h(z, \sqrt{s}) + D_{\bar{q}}^h(z, \sqrt{s})]$, where e_q is the fractional charge of the quark q, and $D_{q(\bar{q})}^h(z, \sqrt{s})$ is the unpolarized FF of quark q (anti-quark $\bar{q}$) at $\sqrt{s}$.

As summarised in Ref. [32], several high-precision measurements of unpolarized FFs have been made in e^+e^- annihilation experiments at the center of mass energy above 10 GeV, while few data with huge statistical uncertainties are present in the energy region from 3.6 to 5.2 GeV. Using the BESIII data collected at six center of mass energies from 2.2324 to 3.6710 GeV, the processes $e^+e^- \to \pi^0/K_S^0 + X$ are studied [33]. The normalised differential cross sections for the inclusive π^0 and K_S^0 production are shown in Figure 5 together with various theoretical predictions extrapolated from different FFs determined from existing world data [32]. The results provide broad z coverage from 0.1 to 0.9. For both π^0 and K_S^0, it is clear the discrepancy between data and theoretical prediction, which is observed to depend on both the hadron momentum and $\sqrt{s}$. One possible explanation of this discrepancies is that the theoretical prediction, restricted to the leading twist contribution, it is not sufficient at the BESIII energy scale. It may also be important to consider quark mass and hadron mass correction effects and small-z resummation effects, as well as the fact that the theoretical models are obtained from existing e^+e^- annihilation at high energy scale and then extrapolated to the low-energy scale of BESIII. However, these results provide new ingredient for global data analyses.

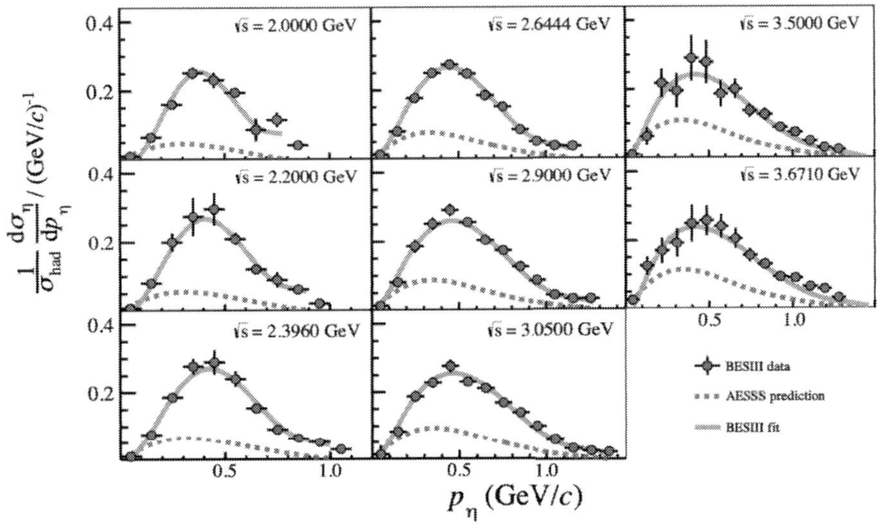

Figure 6. Normalised differential cross sections of the $e^+e^- \to \eta + X$ process. The blue dotted curves denote the prediction by using the AESSS FFs [35], while the grey curves denote the calculations by using the newly extracted η FFs from BESIII.

In addition to π^0/K^0_S, for the first time, BESIII reports the measurement of the process $e^+e^- \to \eta + X$ at eight $\sqrt{s}$ from 2.0000 to 3.6710 GeV, with a z coverage from 0.3 to 0.9 [34]. The normalised differential cross sections for the inclusive η production are shown in Figure 6. The blue dotted curves in Figure 6 represent the theoretical prediction obtained by using the η FF from AESSS parameterization [35] extracted using data of η production in e^+e^- annihilation with $\sqrt{s} > 10$ GeV. The disagreement with BESIII data is evident.

The grey line in Figure 6 shows a new calculation [34] in which also the new η BESIII data are considered. In this case, the model describes the data better. These new results in the relatively low energy region provide special ingredients for FF studies, and they will help to enhance our understanding of the QCD factorization theorem at the leading twist and beyond.

4 Conclusion

The BESIII Collaboration started taking physics data in 2009, and a lot of new physics results are obtained since then. The main focus of this proceeding is the report of the most recent results of the BESIII Collaboration in the QCD field. In particular, thanks to the world largest J/ψ data sample collected by BESIII, a dedicated program on the light hadron spectroscopy and decays has allowed to investigate the nature of the light hybrids, glueball states, and mesonic molecules. In addition to the J/ψ data, data collected at $\sqrt{s}$ from 2.0000 to 3.6710 GeV are used to extract the normalised differential cross section for several inclusive hadrons production. Such studies will be fundamental to a better understanding of fragmentation processes and shed light on the QCD factorization theorem, as well as non-perturbative processes. Finally, new upgrade on the BEPCII collider and on the BESIII detector planned by the end of the 2024 will open new interesting scenario.

References

[1] Ablikim M., et al. (BESIII Collaboration), Nucl. Instrum. Methods Phys. Res., Sect. A **614**, 345 (2010), Design and Construction of the BESIII Detector

[2] Yu C. H., et al. in Proceedings of IPAC2016, Busan, Korea, p.p 1012-1018

[3] Bali G. S., Schilling K., Hulsebos A., Irving A., Michael C., and Stephenson P. W. Phys. Lett. B **309** 378 (1993)

[4] Morningstar C. J. and Peardon M. J., Phys. Rev. D **60**, 034509 (1999)

[5] Gregory E., Irving A., Luciani B., McNeile C., Rago A., Richards C., and Rinaldi E., J. High Energy Phys. **10**, 170 (2012)

[6] Meyer C. A., and Swanson E. S., Prog. Part. Nucl. Phys. **82**, 21 (2015)

[7] Mayer C. A., and Van Haarlem Y., Phys. Rev. C **82**, 025208 (2010)

[8] Klempt E., and Zaitsev A., Phys. Rep. **454**, 1 (2007)

[9] Rodas A., et al., (JPAC Collaboration), Phys. Rev. Lett. **122**, 042002 (2019)

[10] Woss A. J., et al. (Hadron Spectrum Collaboration) Phys. Rev. D **103**, 054503 (2021)

[11] Ablikim M., et al. (BESIII Collaboration), Phys. Rev. L **129**, 192002 (2022)

[12] Ablikim M., et al. (BESIII Collaboration), Phys. Rev. D **106**, 072012 (2022); Phys. Rev. D **107**, 079901 (2023)

[13] Dudek J. J., Edwards R. G., Guo P., Thomas C. E. (Hadron Spectrum Collaboration), Phys. Rev. D **88**, 094505 (2013)

[14] Klempt E., and Sarantev A. V., Phys. Lett. B **826**, 136906 (2022)

[15] Zyla P. A., et al. (Particle Data Group), Prog. Theor. Exp. Phys. **2020**, 083C01 (2020)

[16] Ablikim M., et al. (BESIII Collaboration), JHEP **03**, 121 (2023)

[17] Ablikim M., et al. (BESIII Collaboration), arXiv:2401.00918 https://arxiv.org/pdf/2401.00918

[18] Ablikim M., et al. (BESIII Collaboration), Phys. Rev. L **106**, 072002 (2011)

[19] Ablikim M., et al. (BESIII Collaboration), Eur. Phys. J. C **80**, 746 (2020)

[20] L.-C. Gui, J.-M. Dong, Y. Chen, and Y.-B. Yang, Phys. Rev. D **100**, 054511 (2019)

[21] W. I. Eshraim, S. Janowski, F. Giacosa, and D. H. Rischke, Phys. Rev. D **87**, 054036 (2013)

[22] R. Zhang, W. Sun, Y. Chen, M. Gong, L.-C. Gui, and Z. Liu, Phys. Rev. B **827**, 136960 (2022)

[23] Ablikim M., et al. (BESIII Collaboration), Phys. Rev. Lett. **132**, 181901 (2024)

[24] Ablikim M., et al. (BESIII Collaboration), Phys. Rev. Lett. **95**, 262001 (2005)

[25] J. Z. Bai, et al. (BES Collaboration), Phys. Rev. Lett. **91**, 022001 (2003)

[26] Ablikim M., et al. (BESIII Collaboration), Phys. Rev. Lett. **106**, 072002 (2011)

[27] J. P. Alexander, et al. (CLEO Collaboration), Phys. Rev. D **82**, 092002 (2010)

[28] Ablikim M., et al. (BESIII Collaboration), Phys. Rev. Lett. **115**, 091803 (2015)

[29] Ablikim M., et al. (BESIII Collaboration), Phys. Rev. D **88**, 091502 (2013)

[30] Ablikim M., et al. (BESIII Collaboration), Phys. Rev. Lett. **117**, 042002 (2016)

[31] Ablikim M., et al. (BESIII Collaboration), Phys. Rev. Lett. **132**, 151901 (2024)

[32] Navas S., et al. (Particle Data Group), Phys. Rev. D **110**, 030001 (2024)

[33] Ablikim M., et al. (BESIII Collaboration), Phys. Rev. Lett. **130**, 231901 (2023)

[34] Ablikim M., et al. (BESIII Collaboration), arXiv:2401.17873 https://arxiv.org/pdf/2401.17873

[35] Aidala C. A., et al., Phys. Rev. D **83**, 034002 (2011)

EPJ Web of Conferences **314**, 00009 (2024) https://doi.org/10.1051/epjconf/202431400009
QCD@Work 2024

Various illustrative brief glances at the flavour-cryptoexotic tetraquark states

Wolfgang Lucha[1,*]

[1]Institute for High Energy Physics, Austrian Academy of Sciences, Nikolsdorfergasse 18, A-1050 Vienna, Austria

Abstract. Constituting the largest subset of multiquark states so far observed by experiment, tetraquark mesons carrying overall quark flavour quantum numbers identical to those of the (conventional) quark–antiquark mesons necessitate their particularly deliberate phenomenological analyses. A collection of (more or less recent) insights specific to the latter subclass of exotic hadrons bears the promise to enable even significant improvement in the understanding of these hadrons by theoretical approaches such as, among others, the framework of QCD sum rules.

1 Cryptoexotic tetraquarks as nature's favourite multiquark hadrons

Colour confinement restricts all (observable) bound states of dynamical degrees of freedom of quantum chromodynamics to colour singlets under the action of its gauge group SU(3). These states comprise not only all conventional hadrons — quark–antiquark mesons and three-quark or three-antiquark baryons — but also several variants of exotic hadrons: multiquarks (such as *mesonic* tetraquarks, hexaquarks, *etc.* and *baryonic* pentaquarks, heptaquarks, *etc.*) as well as hybrids composed of quarks and gluons, and purely gluonic bound states denoted as glueballs.

At least at present, the most densely populated *sub*-category of experimentally established multiquark exotic hadrons is the set of tetraquark mesons — constituted by the bound states of two quarks q and two antiquarks $\overline{q}$, with flavour quantum numbers $a, b, c, d \in \{u, d, s, c, b\}$ and masses m_a, m_b, m_c, m_d — of the *flavour-cryptoexotic* sort; for the latter defining characteristic, this subset of exotic hadrons certainly deserves particularly enhanced theoretical attention [1]:

Definition: A *multiquark* hadron state is termed *"flavour-cryptoexotic"* if it does not feature more *open* quark flavours than a respective *conventional* hadron state.

By comprising *at least* one quark–antiquark pair of one and the same flavour, any tetraquark T of flavour-cryptoexotic type exhibits at most two quark flavours, with immediate implications:

$$T = [\overline{q}_a\, q_b\, \overline{q}_b\, q_c] , \qquad a, b, c \in \{u, d, s, c, b\} . \tag{1}$$

- It may therefore undergo *mixing* with all the conventional-meson open-flavour counterparts

$$M_{\overline{a}c} = [\overline{q}_a\, q_c] , \qquad a, c \in \{u, d, s, c, b, t(, \ldots ?)\} . \tag{2}$$

- Its *multiquark* identity cannot be unambiguously told just from its number of *open* flavours.

Extracted from Table 1 of Ref. [2], Table 1 recalls [1] all possible quark-flavour configurations as well as all conceivable antiquark–quark pairings of the flavour-cryptoexotic tetraquarks (1).

*e-mail: Wolfgang.Lucha@oeaw.ac.at

Table 1. Breakdown [2] of the quark-flavour content, $a \neq b \neq c$, of the *flavour-cryptoexotic* tetraquarks (1), in terms of involved and *open* (*viz.*, not counterbalanced by respective *antiflavours*) quark flavours.

Number of Different Flavours	Quark Composition	Number of Open Flavours
3	$\overline{q}_a\, q_b\; \overline{q}_b\, q_c$	2
	$\overline{q}_a\, q_b\; \overline{q}_c\, q_c$	2
2	$\overline{q}_a\, q_a\; \overline{q}_a\, q_b$	2
	$\overline{q}_a\, q_a\; \overline{q}_b\, q_a$	2
	$\overline{q}_a\, q_b\; \overline{q}_b\, q_a$	0
	$\overline{q}_a\, q_a\; \overline{q}_b\, q_b$	0
1	$\overline{q}_a\, q_a\; \overline{q}_a\, q_a$	0

2 Relevance of the tetraquark contributions to four-point correlators

A promising and rather convenient starting point for theoretical explorations of any tetraquark meson is (hardly surprisingly) offered by the time-ordered (T) four-point correlation functions

$$\left\langle \mathrm{T}\Big(j(y)\, j(y')\, j^\dagger(x)\, j^\dagger(x')\Big)\right\rangle \tag{3}$$

of appropriately defined local quark–antiquark bilinear operators (sometimes named currents)

$$j_{\bar a b}(x) \equiv \overline{q}_a(x)\, q_b(x)\ ,$$

wherein all reference to the (for the following irrelevant) spin or parity degrees of freedom has been neglected. The correlation functions (3) govern, among others, all amplitudes describing scatterings of two *conventional* mesons (2), of momenta p_1 and p_2, into two *ordinary* mesons. In these correlation functions, tetraquark mesons manifest in form of poles in the Mandelstam variable $s \equiv (p_1+p_2)^2$. Accordingly, for identifying uniquely *all* genuinely "tetraquark-phile" [3, 4] contributions to any correlation function (3), a selective *criterion* [5] has been proposed:

> **Proposition:** A truly tetraquark-phile contribution to any correlation function (3) depends on s in a non-polynomial fashion and exhibits a four-intermediate-quark *branch cut* that starts at the associated *branch point* at $\hat{s} \equiv (m_a+m_b+m_c+m_d)^2$ [5].

The cut's existence or nonexistence can be established by exploiting the Landau equations [6].

To illustrate the basic ideas underlying this concept, combinatorial simplicity suggests, for definiteness, to focus to that species of tetraquarks among all the configurations categorized in Table 1 that involve precisely three different quark flavours [7, 8], *viz.*, to the tetraquark subset

$$T = [\overline{q}_a\, q_b\, \overline{q}_b\, q_c]\ , \qquad a, b, c \in \{u, d, s, c, b\}\ , \qquad a \neq b \neq c\ . \tag{4}$$

In form of intermediate-state poles, these tetraquarks may show up in the correlation functions

$$\left\langle \mathrm{T}\Big(j_{\bar a b}(y)\, j_{\bar b c}(y')\, j^\dagger_{\bar a b}(x)\, j^\dagger_{\bar b c}(x')\Big)\right\rangle, \qquad \left\langle \mathrm{T}\Big(j_{\bar a c}(y)\, j_{\bar b b}(y')\, j^\dagger_{\bar a c}(x)\, j^\dagger_{\bar b b}(x')\Big)\right\rangle, \tag{5}$$

$$\left\langle \mathrm{T}\Big(j_{\bar a b}(y)\, j_{\bar b c}(y')\, j^\dagger_{\bar a c}(x)\, j^\dagger_{\bar b b}(x')\Big)\right\rangle \tag{6}$$

of either identical (5) or different (6) initial and final distributions of their three quark flavours.

The application of the suggested simple general criterion to the perturbative expansions of these correlation functions demonstrates [5]: for all flavour-cryptoexotic tetraquarks (4), each

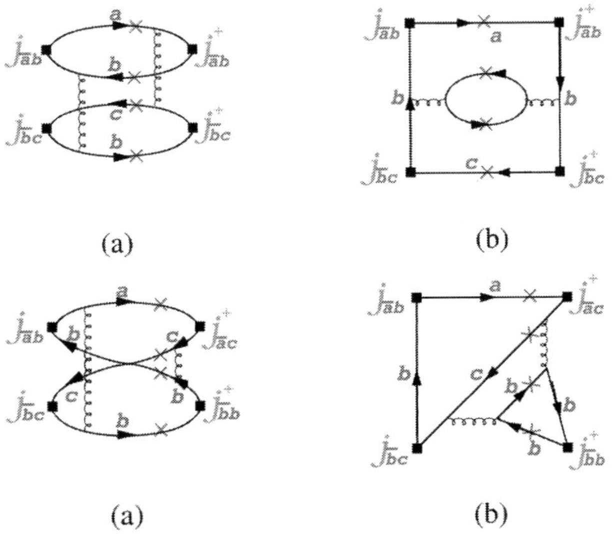

Figure 1. Representative sample of lowest-order tetraquark-phile perturbative contributions to (top row) flavour-preserving correlation functions (5) or (bottom row) flavour-reordering correlation functions (6). In each of the contributions depicted, curly lines represent the (at lowest relevant order) pair of internally exchanged gluons and the foursome of purple crosses indicates a possible location of any four-quark cut.

tetraquark-phile contribution needs at least two internal gluon exchanges of suitable topology. Consequently, tetraquark-phile contributions cannot emerge before the next-to-next-to-lowest perturbative order. Generic examples of the second-order contributions can be found in Fig. 1.

If implemented in analyses of exotic hadron states, multiquark-phile improvements of any adopted framework sharpen one's analytic blades by zooming in on the true multiquark nature of one's bound states of interest. For the bound-state approach provided by the QCD sum-rule formalism [9, 10], the tetraquark-adequate modifications have been sketched in Refs. [11, 12].

3 Multitude of qualitative results and predictions from large-N_c QCD

Essentially *qualitative* insights may be gained by generalizing the gauge group SU(3) of QCD to some special unitary Lie group of degree N_c, SU(N_c), and enabling the number N_c of *colour* degrees of freedom to increase without bound, $N_c \to \infty$, accompanied by a matching decrease to zero of the strong coupling g_s, or the strong fine-structure coupling α_s, at a proper pace [13]:

$$\begin{array}{ccc} g_s \propto \dfrac{1}{\sqrt{N_c}} = O(N_c^{-1/2}) \xrightarrow[N_c\to\infty]{} 0 & & \\ \Updownarrow & \Longrightarrow & N_c\, g_s^2 \propto N_c\, \alpha_s = O(N_c^0) = \text{const}\,. \\ \alpha_s \equiv \dfrac{g_s^2}{4\pi} \propto \dfrac{1}{N_c} = O(N_c^{-1}) \xrightarrow[N_c\to\infty]{} 0 & & \end{array}$$

However, extrapolating the SU(3) quark representation to gauge groups SU(N_c) is, fortunately or unfortunately, far from unique: The by far most common choice is to retain all quarks in the fundamental SU(N_c) representation, of dimension N_c. However, the $\frac{1}{2}\,N_c\,(N_c-1)$-dimensional antisymmetric SU(N_c) representation constitutes another option for the assignment of quarks.

Within the wide realm of exotic hadrons, the *decays* of the tetraquark states of the form (1) and their allowed *mixings* with flavour-compatible ordinary mesons (2) are of utmost interest.

3.1 Large-N_c behaviour of decay width of cryptoexotic tetraquarks: Upper bounds

For every multiquark hadron, on an equal footing with its mass its total *decay width* Γ belongs to the most crucial characteristics of this hadron state. The coupling strength of a tetraquark T to *pairs* of conventional mesons M, encoded in a related transition amplitude $A(T \longleftrightarrow M\,M)$, can be inferred from the respective tetraquark-pole contribution to correlation functions of the sort (3). For all cryptoexotic tetraquark mesons in the current focus, the large-N_c dependences of both categories, (5) and (6), of *tetraquark-phile* correlation functions prove to be the same:[1]

$$\left\langle \mathrm{T}\!\left(j(y)\, j(y')\, j^\dagger(x)\, j^\dagger(x')\right)\right\rangle_{\text{tetraquark-phile}} = O(N_c^2\,\alpha_s^2) = O(N_c^0)\,. \qquad (7)$$

Taking into account the standard [14] N_c dependence of *conventional*-meson decay constants,

$$f_{M_{\bar a b}} \equiv \langle 0|j_{\bar a b}|M_{\bar a b}\rangle \propto \sqrt{N_c} = O(N_c^{1/2})\,,$$

implies for the (potential) cryptoexotic-tetraquark–two-ordinary-meson transition amplitudes

$$\underbrace{A(T \longleftrightarrow M_{\bar a b}\, M_{\bar b c}) = O(N_c^{-1}) \overset{N_c\text{ order}}{\equiv} A(T \longleftrightarrow M_{\bar a c}\, M_{\bar b b}) = O(N_c^{-1})} \qquad (8)$$
$$\Longrightarrow \qquad \Gamma(T) = O(N_c^{-2})\,.$$

If the decays of flavour-cryptoexotic tetraquarks (1) are dominated by any decays into *two* ordinary mesons (2), the squares of the transition amplitudes (8) govern the *total* decay widths $\Gamma(T)$ of any flavour-cryptoexotic tetraquarks (1), whence the latter are of the order $O(N_c^{-2})$ [5].

This prediction, deduced by adopting exclusively tetraquark-phile contributions, however, deviates from the results of similar earlier studies, *cf.* Table 2. The discrepancies can be traced back [1] to either a retainment [15] of contributions to correlation functions that, evidently, are not tetraquark-phile from the point of view of Ref. [5], or a neglect [16] of flavour-rearranging contributions to correlation functions exemplified by the graph (b) in the bottom row of Fig. 1.

Table 2. Large-N_c QCD predictions [2] of *upper bounds* on the decrease of the total decay widths $\Gamma(T)$ of the *flavour-cryptoexotic* tetraquarks (1) with increasing number N_c of the colour degrees of freedom.

Author Collective	**Large-N_c Total Decay Width $\Gamma(T)$**	**References**
Knecht and Peris	$O(1/N_c)$	[15]
Maiani, Polosa and Riquer	$O(1/N_c^3)$	[16]
Lucha, Melikhov and Sazdjian	$O(1/N_c^2)$	[5]

Needless to stress, the tetraquark mesons need not betray their actual existence necessarily already at the *largest* tetraquark-phile order. Consequently, in a strict sense all large-N_c claims of the above kind can provide just an upper bound on the respective factual large-N_c behaviour of tetraquark decay widths. Anyway, in the large-N_c limit the flavour-cryptoexotic tetraquarks are narrow states: for increasing N_c, their total decay widths $\Gamma(T)$ decrease and this even faster than the "well-known" decrease [14] of the decay widths $\Gamma(M)$ of the conventional mesons M,

$$\Gamma(T) = O(N_c^{-2}) \overset{N_c\text{ order}}{<} \Gamma(M) \propto \frac{1}{N_c} = O(N_c^{-1}) \xrightarrow[N_c\to\infty]{} 0\,.$$

Such kind of narrowness definitely promotes the empirical detectability of (unstable) hadrons: in comparison with the *mass* of any such hadron, its total *decay width* should not be too broad.

[1] In contrast to these observations, for tetraquark mesons that involve four mutually unequal quark flavours the two possible sorts of leading-order tetraquark-phile contributions to the related correlation functions differ by one order of N_c [5]; consistency then necessitates the *pairwise* appearance [5] of such genuinely flavour-exotic tetraquark mesons.

3.2 Mixing of cryptoexotic tetraquark and same-open-flavour conventional meson

In contrast to all manifestly flavour-exotic tetraquarks featuring four open quark flavours [17], for every flavour-cryptoexotic tetraquark state the, at least, potential *mixing* with conventional mesons of compatible quantum numbers should be anticipated. Inspecting the contribution of a single intermediate mixing of a flavour-cryptoexotic tetraquark (4) with "flavour-congruent" conventional mesons $M_{\bar{a}c}$ to a correlation function (7) straightforwardly leads to the constraint

$$\underbrace{A(T \longleftrightarrow M_{\bar{a}b}\,M_{\bar{b}c})}_{O(N_c^{-1})}\, g_{TM_{\bar{a}c}}\, \underbrace{A(M_{\bar{a}c} \longleftrightarrow M_{\bar{a}b}\,M_{\bar{b}c})}_{O(N_c^{-1/2})} = O(N_c^{-2})$$

on the products of a tetraquark–two-ordinary-meson transition amplitude $A(T \longleftrightarrow M_{\bar{a}b}\,M_{\bar{b}c})$, a tetraquark–conventional-meson mixing strength $g_{TM_{\bar{a}c}}$ and a three-ordinary-meson coupling $A(M_{\bar{a}c} \longleftrightarrow M_{\bar{a}b}\,M_{\bar{b}c})$, with any latter amplitude established [14] to decrease, for large N_c, like

$$A(M_{\bar{a}c} \longleftrightarrow M_{\bar{a}b}\,M_{\bar{b}c}) \propto \frac{1}{\sqrt{N_c}} = O(N_c^{-1/2}) \xrightarrow[N_c\to\infty]{} 0\,.$$

So, the tetraquark-phile findings (7) yield for the large-N_c behaviour of the mixing strength [5]

$$g_{TM_{\bar{a}c}} \propto \frac{1}{\sqrt{N_c}} = O(N_c^{-1/2}) \xrightarrow[N_c\to\infty]{} 0\,.$$

This observation is by no means surprising: comparing the mixing case with the case of a pure tetraquark-pole contribution betrays just a replacement of the tetraquark–two-ordinary-meson amplitude by the product of a mixing strength and some three-conventional-meson amplitude. The large-N_c decrease of the interchanged intermediate expressions is, qualitatively, identical:

$$A(T \longleftrightarrow M_{\bar{a}b}\,M_{\bar{b}c}) = O(N_c^{-1})$$
$$\Updownarrow$$
$$g_{TM_{\bar{a}c}}\, A(M_{\bar{a}c} \longleftrightarrow M_{\bar{a}b}\,M_{\bar{b}c}) = O(N_c^{-1})\,.$$

Consequently, the mixing of a flavour-cryptoexotic tetraquark with conventional mesons does not possess the capability to alter the large-N_c behaviour of this tetraquark's total decay width.

4 Brief summary of minimally surprising outcomes and conclusions

Almost by definition, the total quark content of each *multiquark* state can be decomposed into (more than one) colour-singlet bound states each of which involves a smaller number of quark constituents, and ultimately into a sequence of conventional hadrons. Consequently, as is well known [18], theoretical analyses of multiquarks have to face the risk of being contaminated or corrupted by contributions providing no information about the multiquark hadrons of interest.

In view of these obstacles, a criterion has been formulated [5] that promises to allow for an (unambiguous) identification of precisely all those contributions to one's preferred framework that bear some relationship to the tetraquark analyzed and for a rigorous rejection of all others.

Application of this *selection tool* to the drastic simplification offered by the generalization of quantum chromodynamics to the set of quantum field theories characterized by the increase beyond bound of their number of colour degrees of freedom, accompanied by the (concurrent) decrease of the strong coupling strength, to the (from an experimental point of view) currently most densely populated multiquark subdivision of all flavour-cryptoexotic tetraquarks proves:

- In this (delicate) limit, the total decay widths of all flavour-cryptoexotic tetraquark states (4) vanish in proportion to the inverse of the square of the number of colour degrees of freedom.
- This decrease to zero proceeds parametrically even faster than that of a conventional meson.

- This decrease is not affected by the potentially occurring mixing of the flavour-cryptoexotic tetraquark state with a conventional meson exhibiting *identical* open quark-flavour content.

Beyond doubt, the considerations of Sects. 2 or 3 and the derived implications, exemplified by the flavour-cryptoexotic tetraquarks (4), ought to be generalized to the entire set of multiquark hadrons and eventually be reflected in future quantitative predictions of multiquark properties.

W. L. thanks both D. Melikhov and H. Sazdjian for pleasant collaboration on the above topics.

References

[1] W. Lucha, Mission target: tetraquark mesons of flavour-cryptoexotic type, Universe **9** (2023) 358, arXiv:2306.15474 [hep-ph]

[2] W. Lucha, D. Melikhov and H. Sazdjian, Exotic tetraquark mesons in large-N_c limit: an unexpected great surprise, EPJ Web Conf. **192** (2018) 00044; arXiv:1808.05519 [hep-ph]

[3] W. Lucha, D. Melikhov and H. Sazdjian, Exotic states and their properties from large-N_c QCD, PoS (EPS-HEP 2017) 390, arXiv:1709.02132 [hep-ph]

[4] W. Lucha, D. Melikhov and H. Sazdjian, Are there narrow flavour-exotic tetraquarks in large-N_c QCD?, Phys. Rev. D **98** (2018) 094011, arXiv:1810.09986 [hep-ph]

[5] W. Lucha, D. Melikhov and H. Sazdjian, Narrow exotic tetraquark mesons in large-N_c QCD, Phys. Rev. D **96** (2017) 014022, arXiv:1706.06003 [hep-ph]

[6] L. D. Landau, On analytic properties of vertex parts in quantum field theory, Nucl. Phys. **13** (1959) 181

[7] W. Lucha, D. Melikhov and H. Sazdjian, Tetraquark and two-meson states at large N_c, Eur. Phys. J. C **77** (2017) 866, arXiv:1710.08316 [hep-ph]

[8] W. Lucha, D. Melikhov and H. Sazdjian, Tetraquarks in large-N_c QCD, Prog. Part. Nucl. Phys. **120** (2021) 103867, arXiv:2102.02542 [hep-ph]

[9] M. A. Shifman, A. I. Vainshtein and V. I. Zakharov, QCD and resonance physics. Theoretical foundations, Nucl. Phys. B **147** (1979) 385

[10] P. Colangelo and A. Khodjamirian, QCD sum rules, a modern perspective, in *At the Frontier of Particle Physics — Handbook of QCD. Boris Ioffe Festschrift*; edited by M. Shifman (World Scientific, Singapore, 2001), Vol. 3, p. 1495, arXiv:hep-ph/0010175

[11] W. Lucha, D. Melikhov and H. Sazdjian, Tetraquark-adequate formulation of QCD sum rules, Phys. Rev. D **100** (2019) 014010, arXiv:1901.03881 [hep-ph]

[12] W. Lucha, D. Melikhov and H. Sazdjian, Tetraquark-adequate QCD sum rules for quark-exchange processes, Phys. Rev. D **100** (2019) 074029, arXiv:1909.06324 [hep-ph]

[13] G. 't Hooft, A planar diagram theory for strong interactions, Nucl. Phys. B **72** (1974) 461

[14] E. Witten, Baryons in the $1/N$ expansion, Nucl. Phys. B **160** (1979) 57

[15] M. Knecht and S. Peris, Narrow tetraquarks at large N, Phys. Rev. D **88** (2013) 036016, arXiv:1307.1273 [hep-ph]

[16] L. Maiani, A. D. Polosa and V. Riquer, Tetraquarks in the $1/N$ expansion and meson–meson resonances, J. High Energy Phys. 06 (2016) 160, arXiv:1605.04839 [hep-ph]

[17] W. Lucha, Mission target: exotic multiquark hadrons — sharpened blades, Universe **9** (2023) 171, arXiv:2302.14439 [hep-ph]

[18] S. Weinberg, Tetraquark mesons in large-N quantum chromodynamics, Phys. Rev. Lett. **110** (2013) 261601, arXiv:1303.0342 [hep-ph]

EPJ Web of Conferences **314**, 00010 (2024) https://doi.org/10.1051/epjconf/202431400010
QCD@Work 2024

Recent results in hadron spectroscopy from CMS

Vincenzo Mastrapasqua[1,*], on behalf of the CMS Collaboration

[1]Istituto Nazionale di Fisica Nucleare - Sezione di Bari, Italy

Abstract. The present report summarizes recent results on the heavy flavor spectroscopy at the CMS experiment obtained using data collected at the Large Hadron Collider during the 2015-2018 data taking with proton-proton collisions at centre-of-mass energy of 13 TeV: the results range from the observation of new structures in the di-charmonium mass spectrum to the observation of new decay modes of beauty baryons such as Ξ_b^- and Λ_b^0.

1 Introduction

The heavy flavour production cross section at the Large Hadron Collider (LHC) is several orders of magnitude larger than at e^+e^- colliders. The CMS experiment [1] at LHC exploits its 4π coverage and high resolution to perform challenging measurements in the Heavy Flavor sector, despite the complex initial state and high background in the proton-proton (*pp*) collision environment. In fact CMS, together with the other LHC experiments, is able to investigate various aspects of the heavy flavour physics, in particular concerning both conventional and exotic spectroscopy.

CMS recently published the study of the invariant mass spectrum of the $J/\psi J/\psi$ final state [2] (2024), following the previous publication from LHCb [3] (2020): CMS reported two structures near the di-J/ψ kinematical threshold, one of which compatible with the resonance $X(6900)$ (a.k.a. $T_{\psi\psi}$) reported by LHCb; in addition, CMS reported evidence of a third structure. The observation of $X(6900)$ has been also confirmed at the LHC by the ATLAS experiment, both in the $J/\psi J/\psi$ and $J/\psi\psi(2S)$ channels [4] shortly before the CMS publication (2023).

The increasing data sample collected by CMS allowed to observe a new decay[1] of the beauty baryon $\Xi_b^- \to \psi(2S)\Xi^-$ [5]. Moreover, the baryon $\Xi_b(5945)^0$ (a.k.a. Ξ_b^{*0}) is reconstructed via $\Xi_b^{*0} \to \Xi_b^-\pi^+$: its mass and width measurements are largely improved with respect to the previous measurement based on the first 5 fb^{-1} of pp collisions data published by CMS [6]. The first observation of Ξ_b^{*0} in fact was published in April 2012 - even before the Higgs discovery - and it represented the first observation of a new particle ever at CMS.

CMS reported recently also on the observation of the new decay $\Lambda_b^0 \to J/\psi\Xi^-K^+$ [7]: this final state is the first multi-body decay with $J/\psi\Xi^-$ ever discovered. This final state represents a promising field for future searches of doubly-strange hidden-charm pentaquarks as intermediate resonance, in analogy to hidden-charm pentaquarks reported by LHCb in the $J/\psi p$ and $J/\psi\Lambda$ systems (e.g $\Lambda_b \to J/\psi pK^-$, $B^+ \to J/\psi p\bar{\Lambda}$) [8–10].

*e-mail: vincenzo.mastrapasqua@ba.infn.it

[1]The inclusion of charge-conjugate decay modes is implied throughout the paper.

2 New structures in the di-J/ψ mass spectrum

The CMS experiment studied the $J/\psi J/\psi(\rightarrow \mu^+\mu^-\mu^+\mu^-)$ invariant mass spectrum using 135 fb^{-1} of pp collisions data at the centre-of-mass energy of $\sqrt{s}$ = 13 TeV collected during the three years 2016 - 2018 [2].

The major sources of background of the di-J/ψ invariant mass spectrum are the Non-Resonant Single Parton Scattering (NRSPS) and the Double Parton Scattering (DPS): the two contributions are modeled with specific parametrizations studied by means of Monte Carlo (MC) simulations. Additional background sources are known and are expected to contribute especially near the di-J/ψ kinematic threshold (e.g. inadequacy of the NRSPS model near threshold, feed-down of partially reconstructed higher mass states, possible coupled-channel interactions, pomeron exchange processes, etc.). In particular a near-threshold enhancement is commonly observed in mass spectra of vector states with same isospin (e.g. observed in $J/\psi\phi$ [11], $J/\psi\omega$ [12]). These contributions are overall described by introducing a Breit-Wigner contribution (BW_0) near the kinematic threshold ($2M^{PDG}_{J/\psi} \simeq 6.2$ GeV).

In order to verify the adequacy of the background model, a fit on the full mass spectrum (up to 15 GeV) is performed using a preliminary signal model, that is the sum of three Breit-Wigner functions, each convoluted with a Gaussian that the describes the detector's resolution effect. The signal resolution, estimated on MC simulations, ranges from 10 MeV (at 6.5 GeV) to 18 MeV (at 7.3 GeV). The background is considered adequately described as the fit probability in the full range is $P(\chi^2) = 98\%$. The fractions for each background component are also extracted from the fit: the NRSPS is dominant (58%), while the DPS (25%) contributes only to the higher part of the spectrum.

Once the background model is validated, the fit features are sequentially added by checking their local statistical significance at each step: a feature is kept if its local statistical significance exceeds 3σ.

An Unbinned Maximum Likelihood (UML) fit is performed on a smaller invariant mass range, namely from 6.2 to 9.0 GeV, using three non-interfering scalar Relativistic Breit-Wigner functions (BW) with Gaussian resolution as signal model: two structures respectively around 6.6 GeV (6.5σ) and 6.9 GeV (9.4σ) are observed (Figure 1). While the latter is compatible within 2σ with the previous observation at the LHCb experiment [3], the former is a new structure, named X(6600). CMS also reports an evidence for a new, third, structure around 7.3 GeV (4.1σ). The overall fit quality is rather poor ($P(\chi^2) = 9\%$) due to the inadequate description of the dips at 6.75 GeV and 7.15 GeV.

The fit quality is improved by allowing the three scalar BW distributions to interfere by introducing complex amplitudes for each state. Therefore the signal model S is written as follows:

$$S = |r_1 e^{i\phi_1} BW_1 + r_2 BW_2 + r_3 e^{i\phi_3} BW_3|^2 \tag{1}$$

where r_i is the modulus of the i-th complex amplitude, and ϕ_i its phase. The phase of the X(6900) is arbitrarily set to zero.

Figure 2 shows the results of the fit of the di-J/ψ invariant mass distribution with a signal model with interference: the fit quality is strongly improved ($P(\chi^2) = 65\%$). Furthermore the proposed signal model is largely favoured with respect to any combination of having only two interfering states out of the three ($P(\chi^2) \simeq 30\%$). Also the local statistical significance for each signal is improved; in particular the structure with the highest mass is the least significant, with 4.7σ. The global statistical significance of the same structure, assessed by means of Monte Carlo pseudo-experiments, slightly exceeds 3σ. The new fit results in a better description of the dips between the structures: the middle structure appears to be

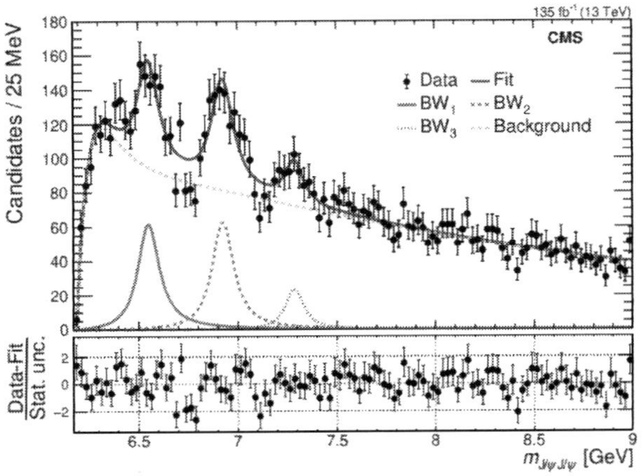

Figure 1. J/ψ-pair invariant mass distribution recorded at CMS using 135 fb^{-1} of pp collisions [2]. The solid blue curve is the result of an UML fit: the dashed yellow curve and the red curves respectively describe the background and the signals. Three structures are established, respectively with a local statistical significance of 6.5σ, 9.4σ, and 4.1σ. The fit probability is rather low (9%) since the model describes poorly the dips between the structures.

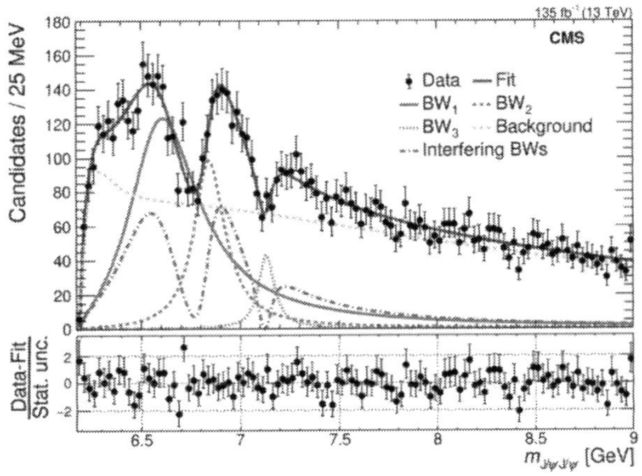

Figure 2. An UML fit is performed on the J/ψ-pair invariant mass distribution using a different model that allows the three signals to interfere with each other [2]. The solid blue curve is the total fit result, the dashed yellow curve is the background contribution, the red curves correspond to the self-interfereing signal components, while the dashed purple curve is the total interference contribution (drawn as positive). This model improves the description of the dips between the structures ($P(\chi^2) = 65\%$).

compatible with the LHCb measurement within 2σ, while the two other structures around 6.6 GeV (observation) and 7.1 GeV (evidence) are respectively named X(6600) and X(7100).

Similar results are obtained if the adopted signal model is borrowed from the LHCb publication [3], yet with a poorer data description: the LHCb model includes interference only between each signal and the NRSPS background, but the states do not interfere with each other. CMS instead considered the interference between the NRSPS contribution and the signal to be less probable, since the former is a mixture of different spin-parity states.

3 Observation of a new Ξ_b^- decay

CMS recently reported the first observation of a new Ξ_b^- decay, $\Xi_b^- \to \psi(2S)\Xi^-$, using 140 fb^{-1} of pp collisions at $\sqrt{s} = 13$ TeV [5]. The branching fraction R of the new decay is measured with respect to the normalization channel $\Xi_b^- \to J/\psi\Xi^-$, a well-known decay with a similar decay chain, as both J/ψ and $\psi(2S)$ are reconstructed in the di-muon channel. The Ξ^- baryon is reconstructed via $\Xi^- \to \Lambda\pi^-$, $\Lambda \to p\pi^-$. The yields for the signal and normalization channels are extracted by means of an UML fit: in each fit the signal is described as the sum of two Gaussians with common mean and different width (to account for the varying resolution across the detector acceptance), while the background is described with a first order polynomial. The yields enter the branching fraction ratio R after being corrected for the acceptance, trigger, and reconstruction efficiencies, which are evaluated on MC. The obtained branching fraction measurement is:

$$R = \frac{\mathcal{B}(\Xi_b^- \to \psi(2S)\Xi^-)}{\mathcal{B}(\Xi_b^- \to J/\psi\Xi^-)} = 0.84^{+0.21}_{-0.19}(\text{stat}) \pm 0.10(\text{syst}) \pm 0.02(\mathcal{B}) \quad (2)$$

where the third uncertainty is related to the uncertainty on the branching fractions of the charmonium states.

The new Ξ_b^- decay mode allows to extend the available sample of Ξ_b^- baryons and to search for excited Ξ_b states decaying into Ξ_b^-. In fact the Ξ_b baryon family is composed of isospin doublets of beauty-strange baryons (bsq): three isospin doublets exist according to total angular momentum ($J^P = \frac{1}{2}^+, \frac{3}{2}^+$) and the total spin of the s and q quarks ($j_{sq} = 0, 1$). The three resulting families are commonly referred to as Ξ_b (ground state), Ξ_b', and Ξ_b^*.

Excited neutral Ξ_b baryons are searched at CMS in the $\Xi_b^-\pi^+$ final state, where the Ξ_b^- baryon is reconstructed in the following decay channels: $\Xi_b^- \to J/\psi\Xi^-$, $\Xi_b^- \to \psi(2S)\Xi^-$, $\Xi_b^- \to J/\psi\Lambda^0K^-$, and $\Xi_b^- \to J/\psi\Sigma^0K^-$ (partial reconstruction), where $\Sigma^0 \to \Lambda^0\gamma_{lost}$, but the photon is too soft to be detected, thus not reconstructed. Charmonia are reconstructed in $J/\psi \to \mu^+\mu^-$, $\psi(2S) \to \mu^+\mu^-$, and $\psi(2S) \to J/\psi\pi^+\pi^-$.

The presented analysis explores the phase-space region of transverse momentum $p_T(\Xi_b^-) > 15$ GeV and rapidity $|y(\Xi_b^-)| < 2.4$, and extends the previous analysis based on 5 fb^{-1} of pp collisions data from Run-1 ($\sqrt{s} = 7$ TeV) [6].

A structure compatible with $\Xi_b(5945)^0 \to \Xi_b^-\pi^+$ is observed, while not present in the same-sign control channel $\Xi_b^-\pi^-$: the signal mass and width are extracted by means of a simultaneous UML fit using the several reconstruction decay chains. The obtained results largely improve the Run-1 results in terms of precision and are in agreement with LHCb [13, 14].

$$m(\Xi_b(5945)^0) = 5952.4 \pm 0.1(\text{stat} + \text{syst}) \pm 0.6(m_{\Xi_b^-})\ \text{MeV} \quad (3)$$

$$\Gamma(\Xi_b(5945)^0) = 0.87^{+0.22}_{-0.20}(\text{stat}) \pm 0.16(\text{syst})\ \text{MeV} \quad (4)$$

Moreover the Ξ_b^{*0}/Ξ_b^- production rate $R_{\Xi_b^{*0}}$ is measured to be approximately 1/4, in agreement with the LHCb result [13].

From considerations on isospin symmetry, 1/9 of the Ξ_b^- are instead produced by the negatively-charged excited Ξ_b^{*-}, thus overall 1/3 of the collected Ξ_b^- is produced from Ξ_b^*: if other higher-mass excited Ξ_b states exist and are sources of Ξ_b^-, then the promptly-produced Ξ_b^- fraction in pp collisions is expected to be less than 2/3.

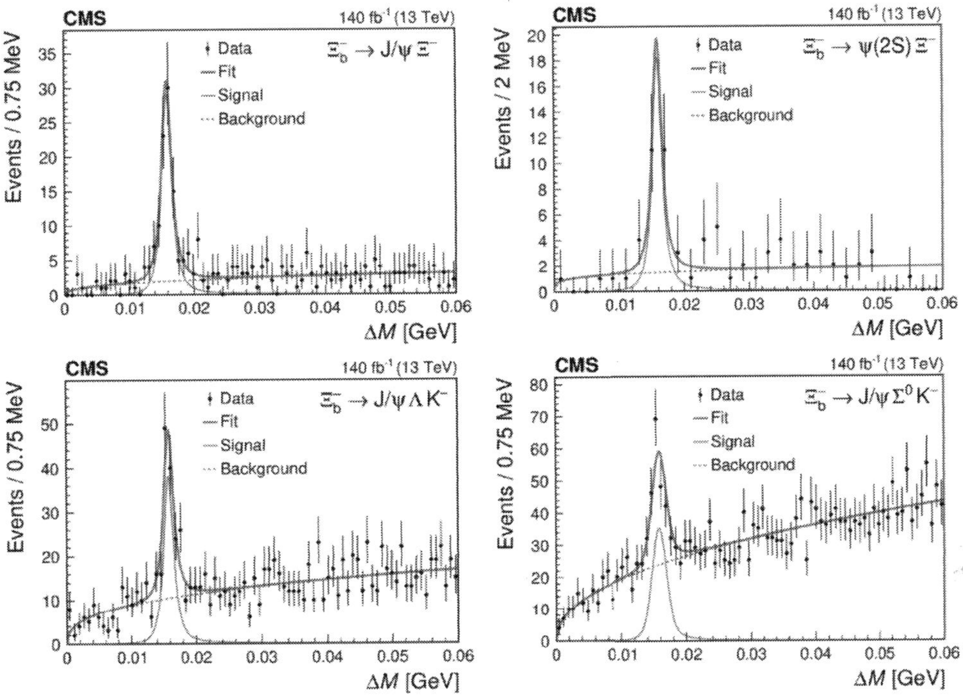

Figure 3. Black dots are the $\Delta M = m(\Xi_b^- \pi^+) - m(\Xi_b^-) + m^{PDG}(\Xi_b^-)$ invariant mass distribution using 140 fb^{-1} of pp collisions data at CMS [5]. The result of a simultaneous UML fit is superimposed (solid red curve): the solid green and dashed blue curves respectively show the signal and background contributions.

4 Observation of a new Λ_b^0 decay

CMS has observed for the first time a new decay of the Λ_b^0 baryon, $\Lambda_b^0 \rightarrow J/\psi \Xi^- K^+$ using 140 fb^{-1} of pp collisions at $\sqrt{s} = 13$ TeV collected during 2016 - 2018 [7].

The decay is reconstructed in a final state with two muons and four tracks, with $J/\psi \rightarrow \mu^+\mu^-$, $\Xi^- \rightarrow \Lambda\pi^-$, and $\Lambda \rightarrow p\pi^-$. The decay $\Lambda_b^0 \rightarrow \psi(2S)(\rightarrow \mu^+\mu^-\pi^+\pi^-)\Lambda(\rightarrow p\pi^-)$ is designated as normalization channel due to the similar kinematics.

The signal yield is extracted by means of an UML fit: the background is modeled with a falling exponential distribution, while the signal contribution is described with a t-Student distribution (the d.o.f. parameter is fixed from MC studies). The extracted signal yield is $N(\Lambda_b^0 \rightarrow J/\psi \Xi^- K^+) = 46 \pm 11$, corresponding to an observation with a significance of 5.8σ.

The signal yield for the normalization channel is analogously extracted: the signal yields enter the branching fraction ratio R after being corrected by an overall efficiency, estimated on MC

$$R = \frac{\mathcal{B}(\Lambda_b^0 \rightarrow J/\psi \Xi^- K^+)}{\mathcal{B}(\Lambda_b^0 \rightarrow \psi(2S)\Lambda)} = [3.38 \pm 1.02(\text{stat}) \pm 0.61(\text{syst}) \pm 0.03(\mathcal{B})]\% \tag{5}$$

where the third error is due to the uncertainty on the $\psi(2S) \rightarrow J/\psi\pi^+\pi^-$ and $\Xi^- \rightarrow \Lambda\pi^-$ branching fractions. The total uncertainty is dominated by the statistics, due to the small signal yield.

A preliminary search for intermediate resonances in the $\Lambda_b^0 \to J/\psi\Xi^- K^+$ is performed, even though it is limited the low signal yield. Background-subtracted distributions are obtained by means of the *sPlot* technique [15]. Figure 4 compares the obtained distributions for any possible two-body decay with the distributions obtained from MC study where the decay is simulated according to a pure phase-space model: no significant difference from the phase-space model is observed with the currently available level of statistics.

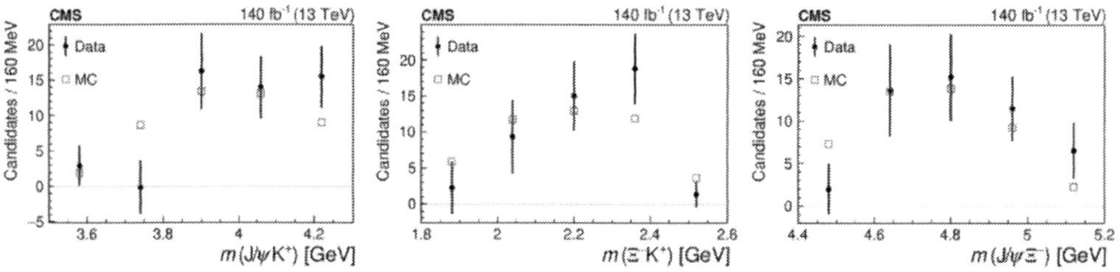

Figure 4. Black dots are the background-subtracted data distribution for the invariant mass of two-body intermediate states in the $\Lambda_b^0 \to J/\psi\Xi^- K^+$ decay [7], obtained with the *sPlot* technique [15]. Data are compared with the distribution of simulated events generated according to a pure phase-space model (red squares). Data are in agreement with the simulations within uncertainties, considering the available level of statistics.

References

[1] CMS Collaboration, The CMS experiment at the CERN LHC. JINST **3** S08004 (2008) https://doi.org/10.1088/1748-0221/3/08/S08004

[2] CMS Collaboration, New structures in the $J/\psi J/\psi$ mass spectrum in Proton-Proton collisions at $\sqrt{s}$ = 13 TeV. Phys. Rev. Lett. **132**, 111901 (2024). https://doi.org/10.1103/PhysRevLett.132.111901

[3] LHCb Collaboration, Observation of structures in the J/ψ-pair mass spectrum. Science Bullettin **65**, 1983 (2020). https://doi.org/10.1016/j.scib.2020.08.032

[4] ATLAS Collaboration, Observation of an excess of di-charmonium events in the four muon final state with the ATLAS detector. Phys. Rev. Lett. **131**, 151902 (2023). https://doi.org/10.1103/PhysRevLett.131.151902

[5] CMS Collaboration, Observation of the $\Xi_b^- \to \psi(2S)\Xi^-$ decay and studies of the Ξ_b^{*0} baryon in proton-proton collisions at $\sqrt{s}$ = 13 TeV. Phys. Rev. D **110**, 012002 (2024). https://doi.org/10.1103/PhysRevD.110.012002

[6] CMS Collaboration, Observation of a new Ξ_b baryon, Phys. Rev. Lett. **108**, 252002 (2012). https://doi.org/10.1103/PhysRevLett.108.252002

[7] CMS Collaboration, Observation of the $\Lambda_b^0 \to J/\psi\Xi^- K^+$ decay. arXiv preprint **2401** 16303 (2024) https://doi.org/10.48550/arXiv.2401.16303, accepted by EPJC

[8] LHCb Collaboration, Observation of $J/\psi p$ resonances consistent with pentaquark states in $\Lambda_b^0 \to J/\psi p K^-$ decays. Phys. Rev. Lett. **115**, 072001 (2015) https://doi.org/10.1103/PhysRevLett.115.072001

[9] LHCb Collaboration, Model-independent evidence for $J/\psi p$ contributions to $\Lambda_b^0 \to J/\psi p K^-$. Phys. Rev. Lett. **117**, 082002 (2016) https://doi.org/10.1103/PhysRevLett.117.082002

[10] LHCb Collaboration, Observation of a narrow pentaquark state, $P_c(4312)^+$, and of two-peak structure of $P_c(4450)^+$. Phys. Rev. Lett. **122**, 222001 (2019) https://doi.org/10.1103/PhysRevLett.122.222001

[11] LHCb Collaboration, Observation of new resonances decaying to $J/\psi K^+$ and $J/\psi\phi$. Phys. Rev. Lett. **127**, 082001 (2021) https://doi.org/10.1103/PhysRevLett.127.082001

[12] BaBar Collaboration, Evidence for the decay $X(3872) \to J/\psi\omega$. Phys. Rev. D **82**, 011101 (2010) https://doi.org/10.1103/PhysRevD.82.011101

[13] LHCb Collaboration, Measurement of the properties of the Ξ_b^{*0} baryon. J. High Energy Phys. **05**, 161 (2016) https://doi.org/10.1007/JHEP05(2016)161

[14] LHCb Collaboration, Observation of new baryons in the $\Xi_b^-\pi^+\pi^-$ and $\Xi_b^0\pi^+\pi^-$ systems. Phys. Rev. Lett. **131**, 171901 (2023) https://doi.org/10.1103/PhysRevLett.131.171901

[15] M. Pivk and F. R. Le Diberder, sPlot: a statistical tool to unfold data distributions. Nucl. Instrum. Meth. A **555**, 356. doi:10.1016/j.nima.2005.08.106

Scattering of mesons and emergence of tetraquarks in two-dimensional QCD

Hagop Sazdjian[1,*]

[1]Université Paris-Saclay, CNRS/IN2P3, IJCLab, 91405 Orsay, France

Abstract. Scattering of two mesons is considered in the framework of two-dimensional QCD in the large-N_c limit with four different quark flavors. The scattering takes place through two coupled channels, corresponding to direct and quark-exchange processes, which are of order $O(N_c^{-2})$ and $O(N_c^{-1})$, respectively. Finiteness of the scattering amplitudes to order N_c^{-2} is pointed out. The theory reduces, at low energies, to an effective theory of mesons, interacting by a quark-exchange process, by means of a contact term. The unitarization of the scattering amplitudes leads to the emergence of a tetraquark bound state, located very close to the lowest two-meson elastic threshold.

1 Introduction

Two-dimensional QCD, in the large-N_c limit, first introduced and analyzed by 't Hooft [1–4], is an efficient tool for probing many properties of hadrons related to the confinement of quarks. In two spacetime dimensions, confinement is a natural property of QCD, due to the strong infrared behavior of the gluon propagator. In the large-N_c limit, inelasticity and pair creation effects are damped, in which case the confinement properties of the theory become more salient. In noncovariant gauges (light-cone, axial), gluon self-interactions and ghosts disappear. Finally, in the many-body case, the interactions between hadronic clusters are free of long-range van der Waals forces.

We investigate, in that framework, the properties of the theory in the four-body sector, made of two quarks and two antiquarks, each with different flavor, and search for the possible existence of tetraquark bound states. The investigation is concentrated on the properties of two-meson scattering amplitudes. Reviews on the large-N_c limit can be found in [5, 6] and more specifically in relation with the tetraquark problem in [7].

2 Quark-antiquark sector

The quark fields are considered in the fundamental representation of the color-gauge group $SU(N_c)$. The light-cone coordinates and momenta are defined as $x^{\pm} = \frac{1}{\sqrt{2}}(x^0 \pm x^1)$, $p_{\pm} = \frac{1}{\sqrt{2}}(p_0 \pm p_1)$. In the light-cone gauge, the gluon field satisfies $A_- = 0$. Because of the particular structure of the quark-gluon vertex in the light-cone gauge, where only the Dirac matrix γ_- appears, the Dirac matrices can be factorized from the N_c-leading expressions of all dynamical quantities.

*e-mail: hagop.sazdjian@ijclab.in2p3.fr

In two dimensions, the coupling constant g of the theory has the dimension of mass. One may introduce the string tension σ according to the definition

$$\sigma \equiv \frac{g^2 N_c}{4}. \tag{1}$$

(g has been defined with the one-index notation of the gluon fields.) In the large-N_c limit, the string tension is assumed to remain finite, which means that the coupling constant squared should decrease like $1/N_c$.

The gluon propagator contributes with its ++ component:

$$D(q) \equiv D_{++}(q) = \frac{i}{q_-^2}. \tag{2}$$

In two dimensions, this propagator is infrared singular. To avoid infrared divergences in intermediate calculations, we introduce a small mass-like parameter λ (behaving like the − component of a momentum) in the propagator, playing the role of an infrared cutoff; to simplify notations, λ will be most often replaced by an inverse quantity, Λ:

$$\frac{i}{q_-^2} \longrightarrow \frac{i}{q_-^2 + \lambda^2}, \qquad \Lambda \equiv \frac{\sigma}{\lambda}. \tag{3}$$

One expects that physical quantities will remain finite when the limits $\lambda \to 0$ or $\Lambda \to \infty$ are taken.

The quark propagator, after taking into account self-energy radiative corrections coming from the large-N_c leading part, represented by the one-gluon contribution in its integral equation, is

$$S(p) = \frac{ip_-}{2p_+p_- - |p_-|\Lambda - m'^2 + i\epsilon}, \qquad m'^2 = m^2 - \frac{2\sigma}{\pi}. \tag{4}$$

One notices that in the limit $\Lambda \to \infty$, the renormalized mass of the quark goes to infinity, indicating that the quark ceases to be an observable free particle with a finite mass.

For quark-antiquark systems in the color-singlet sector, one can write down the corresponding Bethe–Salpeter equation, where the kernel is given, at leading order of N_c, by the one-gluon exchange diagram. Designating by r the total momentum and by p the quark momentum, one introduces the bound state wave function $\phi(r, p)$. Noticing that the interaction in the light-cone gauge is lightlike instantaneous, one further introduces the instantaneous wave function φ defined as

$$\varphi(r, p_-) = \int \frac{dp_+}{2\pi} \phi(r, p). \tag{5}$$

φ satisfies spectral conditions that imply that the momentum component p_- should be positive and bounded by r_-, which itself is positive. Defining then the reduced variable x as

$$x = \frac{p_-}{r_-}, \qquad 0 \leq x \leq 1, \tag{6}$$

the equation satisfied by φ becomes

$$\left[r^2 - \frac{m_1^2}{x} - \frac{m_2^2}{(1-x)}\right]\varphi(x) = -\left(\frac{2\sigma}{\pi}\right)\int_0^1 dy \frac{(\varphi(y) - \varphi(x))}{(y-x)^2}, \tag{7}$$

where m_1 and m_2 are the quark and antiquark masses, respectively. This is the 't Hooft equation [2]. One notices that the infrared cutoff has disappeared, related to the fact that the eigenvalues r^2, which are to be determined, correspond to the physical meson masses.

The spectrum of the eigenfunctions and eigenvalues corresponds to an infinite tower of meson states, characterized by a discrete quantum number n, with a linear increase with n of the higher masses squared, a feature that is typical of a confinement regime.

The quark-antiquark scattering amplitude $\mathcal{T}$ has been calculated in [3]. Its expression is

$$\mathcal{T}(r,x,x') = -\frac{2\sigma}{N_c}\frac{1}{r_-^2(x-x')^2} + \frac{1}{N_c}\sum_n \frac{\widetilde{\phi}_n(r,x)\widetilde{\phi}_n^*(r,x')}{(r^2-r_n^2)}, \tag{8}$$

where $\widetilde{\phi}_n$ is obtained from ϕ_n after factorizing in the latter the free quark-antiquark Green's function. One notices that apart from the one-gluon exchange term (the first term), which cannot be absorbed into products of wave functions, the scattering amplitude is entirely given by the meson wave function contributions. The scattering amplitude $\mathcal{T}$ is cutoff dependent through the complete wave functions ϕ or $\widetilde{\phi}$, which, contrary to the instantaneous wave function φ [Eq. (5], are cutoff dependent. Therefore, $\mathcal{T}$ is not a directly observable quantity. Furthermore, it explicitly contains the long-range one-gluon exchange interaction. In this connection, one of the basic requirements from the theory is that in observable processes, like meson-meson scattering, the cutoff dependences, as well as the long-range contributions, disappear from the final expressions.

3 Two-quark–two-antiquark systems

We now consider systems made of two quarks and two antiquarks. The quark flavors are assumed different from each other; this case avoids mixing problems with ordinary meson states. The quarks are denoted with labels 1 and 3, with masses m_1 and m_3, and the antiquarks with labels $\bar{2}$ and $\bar{4}$, with masses m_2 and m_4. The Green's function, corresponding to the four ingoing and outgoing particles, is schematically represented in Fig. 1.

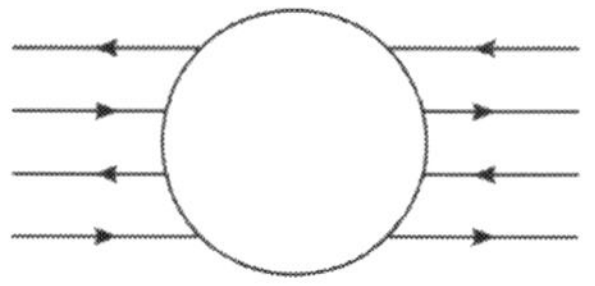

Figure 1. Four-body Green's function, with two quarks, 1 and 3, and two antiquarks, $\bar{2}$ and $\bar{4}$.

We are interested in the color-singlet sectors of the incoming and outgoing particles in the Green's function, describing scattering processes of mesons. We have two sets of mesonic clusters: $(1\bar{2})(3\bar{4})$ and $(1\bar{4})(3\bar{2})$. One can then distinguish four different channels in the scattering process, two of which are called "direct", the ingoing and outgoing clusters being the same, and the two others "recombination", the outgoing clusters having undergone a quark exchange. The four different processes are:

$$(1\bar{2})+(3\bar{4}) \longrightarrow (1\bar{2})+(3\bar{4}), \quad \text{direct channel 1} \quad (D1), \tag{9}$$
$$(1\bar{2})+(3\bar{4}) \longrightarrow (1\bar{4})+(3\bar{2}), \quad \text{recombination channel 1} \quad (R1), \tag{10}$$
$$(1\bar{4})+(3\bar{2}) \longrightarrow (1\bar{4})+(3\bar{2}), \quad \text{direct channel 2} \quad (D2), \tag{11}$$
$$(1\bar{4})+(3\bar{2}) \longrightarrow (1\bar{2})+(3\bar{4}), \quad \text{recombination channel 2} \quad (R2). \tag{12}$$

Integral equations for the Green's functions in the color-singlet sectors at leading orders of N_c can be obtained following the method used in the quark-antiquark case. Their detailed derivation is left to a separate work. We will be content by indicating here some of the qualitative aspects that characterize them. Similar derivations and related technical problems can be found in [8–11].

The integral equations relative to the four channels decouple into two subsets of equations, the first concerning channel 1 and the second, channel 2. Inside each subset, one has two coupled integral equations relative to the direct and recombination processes. Actually, one passes from one subset to the other by appropriate permutations of indices. It is therefore sufficient to concentrate on one subset, channel 1, say.

Specifically, one finds the following large-N_c behaviors for the scattering amplitudes:

$$\mathcal{T}_{R1} = O(N_c^{-1}) + O(N_c^{-3}), \quad (13)$$

$$\mathcal{T}_{D1} = O(N_c^{-2}) + O(N_c^{-4}). \quad (14)$$

The on-mass shell scattering amplitudes are obtained by projecting the off-mass shell scattering amplitudes on the ingoing and outgoing wave functions.

4 Finiteness of the scattering amplitudes to order $1/N_c^2$

An important test for the theory is the verification that the meson-meson scattering amplitudes are finite quantities in the limit $\Lambda \to \infty$. The infrared cutoff is present in the quark propagators, the gluon propagator and the external wave functions. Another test is the verification that the meson-meson scattering amplitudes are free of long-range van der Waals type forces.

The calculations take into account the following features. Each quark propagator carries a damping factor Λ^{-1}. The gluon propagator, upon integration, produces a diverging factor Λ. Each external meson wave function carries a diverging power Λ. Each quark loop integration with respect to the + component of the loop momentum removes one power from the damping factors. Previous calculations of this type have been presented in [3].

Generally, individual diagrams are linearly divergent in Λ. One therefore has to consider combinations of several diagrams having similar structures.

The recombination channel process receives, at order N_c^{-1}, contributions from three diagrams, one of them being free of gluon propagators and the two others containing one gluon propagator. It turns out that the sum of the three diagrams is finite in the limit $\Lambda \to \infty$. The finite part has the structure of an effective four-meson contact term with smooth momentum dependence, excluding any presence of long-range forces. The effective coupling constant of the contact term is provided by an overlapping integral of the four external meson wave functions.

Since the next-to-leading term in the recombination channel, generated by the leading terms in the corresponding Green's function, is of order N_c^{-3}, one should check whether nonleading terms of the $1/N_c$ expansion might produce divergences at order $1/N_c^2$. These arise from quark-loop corrections to the gluon propagator, from vertex corrections and from crossed diagrams.

The quark-loop correction can be calculated using the propagator (4). The result is a modification of the gluon propagator (2)-(3) to the form [3]

$$D(q) = \frac{i}{q_-^2 + \lambda^2 + \lambda |q_-| N_f/(\pi N_c)}, \quad (15)$$

where N_f is the number of quark flavors. The net effect of this change is a renormalization of the cutoff parameter Λ in the following way:

$$\Lambda \longrightarrow \Lambda\Big(1 - \frac{N_f}{\pi^2 N_c}\Big). \quad (16)$$

The same change of Λ also occurs in the quark propagator (4), m'^2 remaining unchanged. Therefore, the cancellation mechanism between gluon and quark propagator divergences remains unaffected.

Vertex corrections and gluon crossed diagrams produce generally leading corrections of order $1/N_c$. However, in the present case, the latter vanish due to the spectral conditions that the meson wave functions should satisfy [Eqs. (5) and (6)].

Therefore, the finiteness of the recombination channel scattering amplitude remains true also at order N_c^{-2}.

The direct channel process receives, at order N_c^{-2}, contributions from 65 diagrams, containing up to four gluon propagators. These diagrams can be be grouped, according to their structure, into five categories, which can be studied independently from each other. The net result is that the divergent parts cancel out in each category and one remains with a finite part which is nothing but the generalized two-meson unitarity diagram generated by the finite part of the recombination process. (The two-meson loops are actually infinite in number, involving all mesons of the quark-antiquark spectra, [Eq. (8)].)

The above results establish the finiteness of the two-meson scattering amplitudes to order N_c^{-2} and provide the equivalent low-energy effective meson theory, which reduces to the recombination process contact term and the generation from it of unitarity diagrams.

5 Unitarized scattering amplitudes

The extension of the above study to higher orders than N_c^{-2} becomes out of reach, because of the increasing number of diagrams and also because of the appearance of new types of diagrams. For the last category, we might observe that even if they produce new types of effective interactions, they could not compete, at large N_c, with those found up to order N_c^{-2}. It is therefore not unreasonable to ignore, for a first study, such possible new effects.

Concerning the higher-order contributions generated by the leading-order diagrams through the Green's functions, we observe that the results obtained above indicated the tendancy of the theory to unitarize the effective contact interaction generated by the recombination process. Assuming that the resulting effective theory satisfies the unitarity property, one is entitled to complete the results obtained up to order N_c^{-2} by unitarizing the corresponding scattering amplitudes.

We undertake that operation using two approximations. First, we keep in the series of intermediate meson states that contribute to the unitarity loop diagrams only the ground state mesons, neglecting their radial excitations. The latter contribute with dampted mass factors and with smaller four-meson couplings, due to the smallness of the overlapping integrals between ground state and radial wave functions. Second, we neglect the momentum dependence of the effective four-meson coupling of the ground-state mesons, by treating it as a constant. With these approximations, the relevant equations are reduced to coupled four-channel equations [Eqs. (9)-(12)], that could be dealt with in matrix form.

Designating by K_R the real coupling constant of the recombination channel (unitarity requiring the equality $K_{R1} = K_{R2}$), the scattering amplitude and the kernel of its integral equation take the following matrix forms:

$$\mathcal{T} = \begin{pmatrix} \mathcal{T}_{D1} & \mathcal{T}_{R2} \\ \mathcal{T}_{R1} & \mathcal{T}_{D2} \end{pmatrix}, \quad K = \begin{pmatrix} 0 & K_R \\ K_R & 0 \end{pmatrix}, \quad J = \begin{pmatrix} J_{1\bar{2},3\bar{4}} & 0 \\ 0 & J_{1\bar{4},3\bar{2}} \end{pmatrix}, \tag{17}$$

where we have also introduced the two-meson loop functions $J_{1\bar{2},3\bar{4}}$ and $J_{1\bar{4},3\bar{2}}$:

$$J_{a\bar{b},c\bar{d}}(s, M_{a\bar{b}}M_{c\bar{d}}) = -\frac{i}{2\sqrt{-\lambda_{a\bar{b},c\bar{d}}}}\Big[1 - \frac{1}{\pi}\arctan\Big(\frac{\sqrt{-\lambda_{a\bar{b},c\bar{d}}}}{s - (M^2_{a\bar{b}} + M^2_{c\bar{d}})}\Big)\Big],$$
$$\lambda_{a\bar{b},c\bar{d}} = \Big(s - (M_{a\bar{b}} + M_{c\bar{d}})^2\Big)\Big(s - (M_{a\bar{b}} - M_{c\bar{d}})^2\Big). \tag{18}$$

When the interaction is represented by a contact term, as is the case here, the unitarization operation reduces to the iteration of the interaction by means of the meson loops. One has

$$\begin{aligned} \mathcal{T} &= (1 - iKJ)^{-1}K \\ &= \frac{1}{\Big(1 + K_R J_{1\bar{4},3\bar{2}} K_R J_{1\bar{2},3\bar{4}}\Big)} \begin{pmatrix} iK_R J_{1\bar{4},3\bar{2}} K_R & K_R \\ K_R & iK_R J_{1\bar{2},3\bar{4}} K_R \end{pmatrix}, \end{aligned} \tag{19}$$

from which one may identify, with (17), the different components of $\mathcal{T}$.

6 Tetraquarks

Tetraquark bound states may occur at the zeros of the denominator of $\mathcal{T}$. This leads to the equation

$$1 + K_R J_{1\bar{4},3\bar{2}} K_R J_{1\bar{2},3\bar{4}} = 0. \tag{20}$$

This equation has generally one bound state solution, which lies very close to the lowest elastic two-meson threshold, with binding energies of the order of 1 MeV or much less, according to the mass configurations of the mesons. The smallness of the binding energy is traced back to the smallness of the coupling constant K_R, which, within a rough estimate, is of the order of $10\,(2\sigma/\pi)$, with respect to some combinations of the meson masses. In the case of two heavy quarks, the latter ratio effectively contributes with a power of 4.

For the particular example of the configuration $(c\bar{s}b\bar{u})$, which leads, in the pseudoscalar sector, to the two two-meson clusters (D^+B^-) and $(D^0\bar{B}^0_s)$, the binding energy comes out of the order of 0.02 MeV with respect to the $(D^0\bar{B}^0_s)$ elastic threshold.

Existence of tetraquark bound states with four different quark flavors, containing b and c quarks, has been observed in Lattice QCD calculations [12–14].

Of particular interest is the comparison of the coupling constants of the tetraquark state to the two meson clusters. Assuming for definiteness that $(M_{1\bar{2}} + M_{3\bar{4}}) > (M_{1\bar{4}} + M_{3\bar{2}})$, and denoting the tetraquark binding energy by E_{TB} and $\Delta M = (M_{1\bar{2}} + M_{3\bar{4}}) - (M_{1\bar{4}} + M_{3\bar{2}})$, one observes from Eqs. (18) and (19), for nonrelativistic systems, that the coupling constant squared of the tetraquark to the clusters $(1\bar{2})(3\bar{4})$ is proportional to $E_{TB}^{-1/2}$, while that to the clusters $(1\bar{4})(3\bar{2})$ is proportional to $(\Delta M)^{-1/2}$. The ratio of these quantities for the example considered above is of the order of 30, which means that the tetraquark couples more strongly to the heavier cluster than to its neigboring one.

The latter property would suggest at first sight that the tetraquark is essentially a molecular system made of the heavier cluster ingredients, here $(1\bar{2})(3\bar{4})$. However, such an interpretation, neglecting the second set of clusters, would not explain in a natural way the particular position of the tetraquark pole in the vicinity of the lighter clusters. The two two-meson clusters seem to play an equal role. This situation is due to the quark-exchange mechanism, which is the basic interaction ingredient of the two-meson clusters, and which needs the presence of the two distinct two-meson clusters to operate.

7 Conclusion

Two-dimensional QCD at large N_c allows a consistent treatment of the confining interactions, with possible control of the infrared divergences. In the sector of four different flavors, the theory reduces, at low energies, to an effective theory of mesons, where the basic interaction is provided by a quark-exchange four-meson contact term. The unitarization of the scattering amplitude in all channels of interaction leads to the prediction of the existence of a tetraquark bound state lying very close to the lightest two-meson elastic threshold.

The author thanks W. Lucha and D. Melikhov for stimulating discussions on the subject. This research has received financial support from the EU research and innovation programme Horizon 2020, under Grant agreement No. 824093.

References

[1] G. 't Hooft, A planar diagram theory for strong interactions, Nucl. Phys. B **72**, 461 (1974). 10.1016/0550-3213(74)90154-0

[2] G. 't Hooft, A two-dimensional model for mesons, Nucl. Phys. B **75**, 461 (1974). 10.1016/0550-3213(74)90088-1

[3] C.G. Callan, Jr., N. Coote, D.J. Gross, Two-dimensional Yang-Mills theory: a model of quark confinement, Phys. Rev. D **13**, 1649 (1976). 10.1103/PhysRevD.13.1649

[4] M.B. Einhorn, Form-factors and deep inelastic scattering in two-dimensional Quantum Chromodynamics, Phys. Rev. D **14**, 3451 (1976). 10.1103/PhysRevD.14.3451

[5] E. Witten, Baryons in the $1/N$ expansion, Nucl. Phys. B **160**, 57 (1979). 10.1016/0550-3213(79)90232-3

[6] S. Coleman, Aspects of Symmetry (Cambridge University Press, Cambridge, U.K., 1985), ISBN 9780521318273, 0521318270

[7] W. Lucha, D. Melikhov, H. Sazdjian, Tetraquarks in large-N_c QCD, Prog. Part. Nucl. Phys. **120**, 103867 (2021), 2102.02542. 10.1016/j.ppnp.2021.103867

[8] W. Heupel, G. Eichmann, C.S. Fischer, Tetraquark bound states in a Bethe-Salpeter approach, Phys. Lett. B **718**, 545 (2012), 1206.5129. 10.1016/j.physletb.2012.11.009

[9] G. Eichmann, C.S. Fischer, W. Heupel, The light scalar mesons as tetraquarks, Phys. Lett. B **753**, 282 (2016), 1508.07178. 10.1016/j.physletb.2015.12.036

[10] G. Eichmann, C.S. Fischer, W. Heupel, N. Santowsky, P.C. Wallbott, Four-quark states from functional methods, Few Body Syst. **61**, 38 (2020), 2008.10240. 10.1007/s00601-020-01571-3

[11] A.N. Kvinikhidze, B. Blankleider, Covariant tetraquark equations in quantum field theory, Phys. Rev. D **106**, 054024 (2022), 2102.09558. 10.1103/PhysRevD.106.054024

[12] C. Alexandrou, J. Finkenrath, T. Leontiou, S. Meinel, M. Pflaumer, M. Wagner, Shallow bound states and hints for broad resonances with quark content $\bar{b}\bar{c}ud$ in $B-\bar{D}$ and $B^*-\bar{D}$ scattering from Lattice QCD, Phys. Rev. Lett. **132**, 151902 (2024), 2312.02925. 10.1103/PhysRevLett.132.151902

[13] M. Padmanath, A. Radhakrishnan, N. Mathur, Bound isoscalar axial-vector $bc\bar{u}\bar{d}$ tetraquark T_{bc} from Lattice QCD using two-meson and diquark-antidiquark variational basis, Phys. Rev. Lett. **132**, 201902 (2024), 2307.14128. 10.1103/PhysRevLett.132.201902

[14] N. Mathur, M. Padmanath, $\bar{b}\bar{c}q_1q_2$ four-quark states from Lattice QCD, PoS **LATTICE2021**, 443 (2022), 2111.01147. 10.22323/1.396.0443

EPJ Web of Conferences **314**, 00012 (2024) https://doi.org/10.1051/epjconf/202431400012
QCD@Work 2024

Study of charmonium decays to $K^0_S K\pi$ in the $B \to (K^0_S K\pi)K$ channels

Antimo Palano[1,*]

[1]Istituto Nazionale di Fisica Nucleare, Sezione di Bari, Italy

Abstract. A study of the $B^+ \to K^0_S K^+ K^- \pi^+$ and $B^+ \to K^0_S K^+ K^+ \pi^-$ decays is performed using proton-proton collisions at center-of-mass energies of 7, 8 and 13 TeV at the LHCb experiment. The $K^0_S K^\pm \pi^\mp$ invariant mass spectra from both decay modes reveal a rich content of charmonium resonances, such as η_c, J/ψ, $\eta_c(2S)$ and χ_{c1}. Precise measurements of their parameters and branching fractions are obtained. Dalitz plot analyses of $\eta_c \to K^0_S K^\pm \pi^\mp$ and $\eta_c(2S) \to K^0_S K^\pm \pi^\mp$ decays are performed.

1 Introduction

Light meson spectroscopy in the strange meson sector is still far to be understood, with the only established strange scalar meson being the $K^*_0(1430)$ resonance, whose parameters are yet to be precisely measured [1]. A widely used method of modeling the $K\pi$ S-wave relies on the results from Ref. [2], which consists of a large threshold enhancement described by a scattering length term and a relativistic Breit–Wigner (BW) function describing the $K^*_0(1430)$ resonance. A similar behavior is observed in the $K\pi$ S-wave measured in D^+ decays [3]. Still unresolved is the possible existence of a broad scalar resonance, $\kappa/K^*_0(700)$, claimed by several experiments [1]; its existence would suggest the possible presence of tetraquark states in the light meson system.

2 B^+ signals and measurement of the charmonium resonances parameters

Candidate events are reconstructed and then background suppressed using a multivariate analysis [4]. The resulting $K^0_S K^+ K^- \pi^+$ and $K^0_S K^+ K^+ \pi^-$ mass spectra are shown in Fig. 1, where B^+ signals over small backgrounds can be seen, yielding 74150 and 58460 events, respectively[1]. The corresponding $K^0_S K^\pm \pi^\mp$ mass spectra are shown in Fig. 2. Above the threshold region a D^0 signal is present and prominent η_c, J/ψ, χ_{c1} and $\eta_c(2S)$ resonances are observed.

The measurements of charmonium-resonance parameters are performed with binned fits to the $K^0_S K^\pm \pi^\mp$ invariant-mass spectra separately in the η_c–J/ψ and the χ_{c1}–$\eta_c(2S)$ mass regions. In the fit to the η_c–J/ψ mass region, the J/ψ width is fixed to the known value [1]. All resonances are described by simple non-relativistic BW functions convolved with the appropriate mass resolution functions. The fit results are summarized in Table 1 and show significant improvements with respect to previous measurements. Figure 3 shows the K^+K^-

*e-mail: antimo.palano@ba.infn.it

[1]The inclusion of charge-conjugate processes is implied throughout the paper.

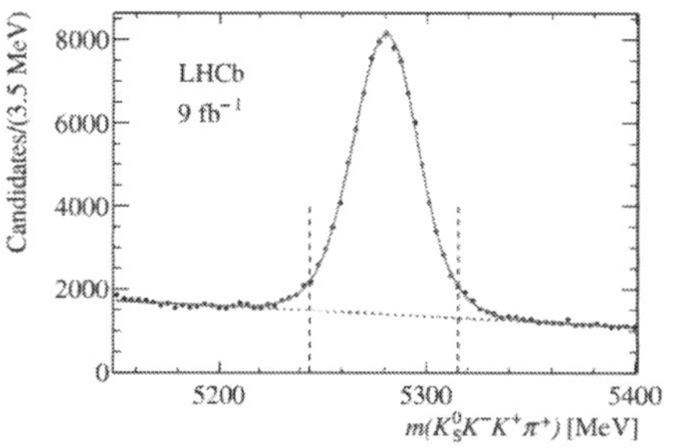

Figure 1. Fitted $K_S^0K^+K^-\pi^+$ and $K_S^0K^+K^+\pi^-$ mass spectra, with indication of signal and background. The vertical lines indicate the B^+ selections.

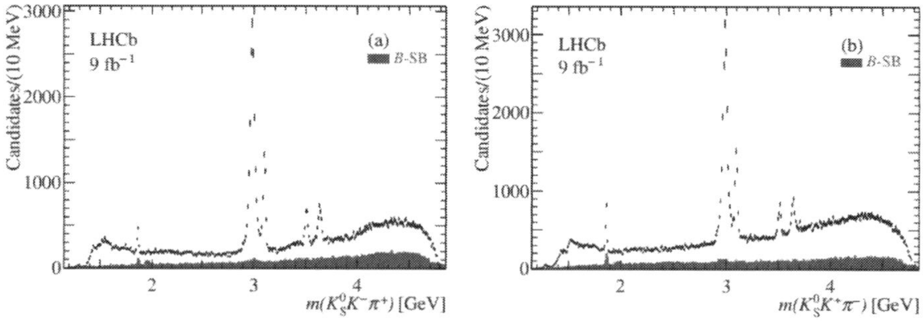

Figure 2. $K_S^0K^\pm\pi^\mp$ mass spectra, from (a) $B^+ \to K_S^0K^+K^-\pi^+$ and (b) $K_S^0K^+K^+\pi^-$. Shaded is the background estimated from B^+ sidebands.

invariant-mass spectrum for $B^+ \to K_S^0K^+K^-\pi^+$ candidates in the χ_{c0}–χ_{c2} mass region, where a prominent χ_{c0} signal can be seen, together with a weaker signal at the χ_{c2} mass. The $B^+ \to \chi_{c0}K$ decay is expected to be zero in simple factorization models. The χ_{c0} resonance is described by a relativistic spin-0 BW convolved with the experimental resolution and interference is allowed with a non-resonant K^+K^- component. The χ_{c2} parameters are fixed to the known values. The fit returns χ_{c0} parameters consistent with PDG averages with χ_{c0} yield of $N_{\chi_{c0}} = 1920 \pm 90$ and relative phase $\phi = -1.290 \pm 0.073$ rad. The χ_{c2} significance is found to be 4.6σ.

3 Measurements of branching fractions

We use the known PDG η_c and J/ψ branching fractions $B^+ \to (\eta_c/J/\psi)K^+$ as reference and obtain the first branching fractions measurements listed in Table 2. The first uncertainty is statistical, the second systematic, the third is due to the PDG uncertainty on the $B^+ \to \eta_c K^+$ or $B^+ \to J/\psi K^+$ branching fractions.

4 η_c Dalitz plot analysis

Figure 4 shows the $K_S^0K^\pm\pi^\mp$ Dalitz plot in the η_c mass region, separately for $B^+ \to K_S^0K^+K^-\pi^+$ and $B^+ \to K_S^0K^+K^+\pi^-$ data which are dominated by uniform constructive hori-

Table 1. Fitted η_c, J/ψ, $\eta_c(2S)$ and χ_{c1} parameters. The known J/ψ mass is subtracted from the fitted value.

Resonance.	mass (MeV)	width (MeV)	Yield
η_c	$2985.01 \pm 0.17 \pm 0.89$	$29.7 \pm 0.5 \pm 0.2$	34910 ± 283
J/ψ	$-0.54 \pm 0.08 \pm 0.45$	0.0929 (fixed)	6696 ± 106
$\eta_c(2S)$	$3637.90 \pm 0.54 \pm 1.40$	$10.77 \pm 1.62 \pm 1.08$	3680 ± 128
χ_{c1}	$3509.84 \pm 0.69 \pm 0.64$	0.88 (fixed)	2760 ± 86

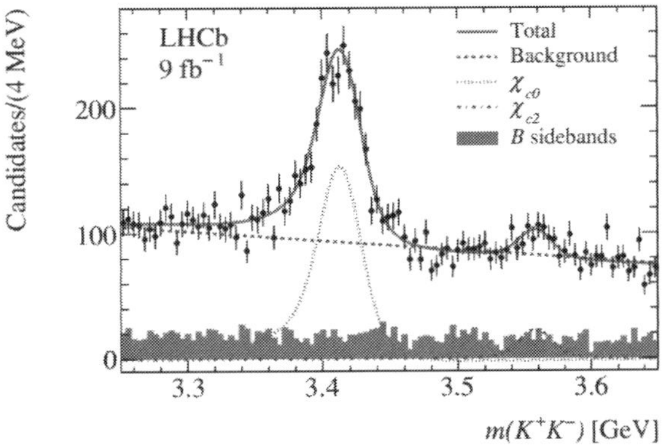

Figure 3. Invariant K^+K^- mass distribution in the χ_{c0}-χ_{c2} mass region for $B^+ \to K^0_S K^+ K^- \pi^+$ decays. The results of the fit are overlayed. Superimposed is the K^+K^- invariant-mass spectrum from the B^+ sidebands.

zontal and vertical bands due to the $K^*_0(1430)$ resonance. G-parity relates the final states $\bar{K}^*_0 K$ and $K^*_0 \bar{K}$ and, for the I=0 $J^{PC} = 0^{-+}$ η_c resonance, G-parity is even and therefore the $\bar{K}^*_0 K$ and $K^*_0 \bar{K}$ bands interfere constructively.

We perform two η_c Dalitz plot analyses. (a) Quasi Model Independent (QMI) measurement of the $K\pi$ S-wave . In this fit, the $K\pi$ invariant-mass spectrum is divided into 37 equally-spaced mass intervals 50MeV wide, and two new fit parameters are added for each interval: the amplitude and phase of the $K\pi$ S-wave. The amplitude is fixed to 1.0 and its phase to $\pi/2$ at an arbitrary point in the mass spectrum, chosen to be interval 16, corresponding to a mass of 1.45 GeV. Other resonant contributions are also added in the fit [4]. Figure 5 shows the fitted $K\pi$ S-wave amplitude and phase obtained from the QMI analysis averaged

Table 2. Measured branching fractions using the η_c (top) or J/ψ (bottom) known branching fractions as reference.

Final state	$\mathcal{B}$ ($\times 10^{-5}$)
	η_c as reference
$B^+ \to K^0 K^+ K^- \pi^+$	$32.28 \pm 0.33 \pm 1.97 \pm 7.17$
$B^+ \to K^0 K^+ K^+ \pi^-$	$26.56 \pm 0.31 \pm 0.68 \pm 5.90$
$B^+ \to \chi_{c0} K^0 \pi^+$	$1.38 \pm 0.07 \pm 0.11 \pm 0.32$
$B^+ \to \chi_{c2} K^0 \pi^+$	$0.87 \pm 0.20 \pm 0.08 \pm 0.20$
	J/ψ as reference
$B^+ \to K^0 K^+ K^- \pi^+$	$34.01 \pm 0.74 \pm 0.91 \pm 3.10$
$B^+ \to K^0 K^+ K^+ \pi^-$	$28.01 \pm 0.68 \pm 1.35 \pm 2.55$
$B^+ \to \chi_{c0} K^0 \pi^+$	$1.45 \pm 0.08 \pm 0.11 \pm 0.16$
$B^+ \to \chi_{c2} K^0 \pi^+$	$0.92 \pm 0.21 \pm 0.08 \pm 0.10$

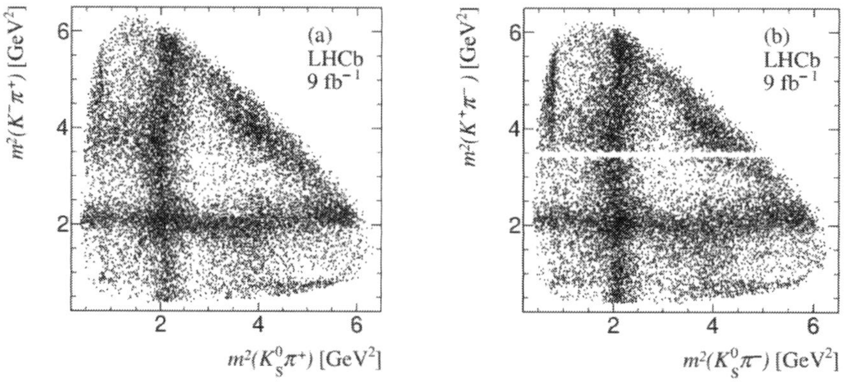

Figure 4. $K^0_S K^{\pm}\pi^{\mp}$ Dalitz plot in the η_c mass region, separately for (a) $B^+ \to K^0_S K^+ K^- \pi^+$ and (b) $B^+ \to K^0_S K^+ K^+ \pi^-$ data. A significant $\bar{D}^0 \to K^+\pi^-$ is removed from the $B^+ \to K^0_S K^+ K^+ \pi^-$ data.

over $B^+ \to K^0_S K^+ K^- \pi^+$ and $B^+ \to K^0_S K^+ K^+ \pi^-$ data, which agree well. The data show a dominance of the $K_0^*(1430)$ resonant contribution followed by a sharp drop and a subsequent signal of the $K_0^*(1950)$ resonance. A contribution in the phase motion not continuous with the rest of the distribution is present in the threshold region of Fig. 5(b). This behavior may be attributed to the inability of the algorithm to obtain a correct phase motion in this region of the phase space due to the absence of additional significant resonant contributions.

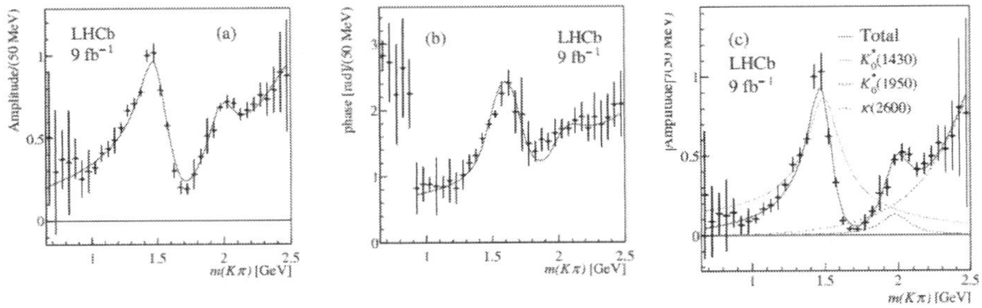

Figure 5. Fit to the (a) amplitude, (b) phase and (c) squared amplitude from the η_c QMI analysis using the results from the isobar-model analysis.

(b) Isobar model. In this fit all the resonances are described by Breit-Wigner functions. The search for the best solution is performed by adding resonances one-by-one, using the $K_0^*(1430)$ resonance as the reference amplitude. It is found that, using only the list of the known resonances, a poor description of the data is obtained. A large improvement (increased likelihood $\Delta(2\log\mathcal{L}) = 1338$) is obtained by adding an additional scalar $K\pi$ resonance with free parameters, labeled in the following as $\kappa(2600)$ and included in the fit using the simple non-relativistic BW function. Resonances having poorly known parameters, namely the $K_0^*(1430)$, $K_0^*(1950)$ and $a_0(1700)$ resonances, are allowed to have free parameters. Table 3 lists the measured resonances parameters where significances are estimated using Wilks' theorem. Table 4 lists the fitted fractional contributions and relative phases obtained by this fitting model and fit projections are shown in Fig. 6. The fit requires the

Table 3. Fitted parameters and significances of the resonances observed in the η_c Dalitz-plot analysis. Significances larger than 10σ are evaluated as $\sqrt{\Delta(2\log\mathcal{L})}$.

Resonance	Mass (MeV)	Γ (MeV)	$\Delta(2log\mathcal{L})$	Significance ($n\sigma$)
$K_0^*(1430)$	$1493 \pm 4 \pm 7$	$215 \pm 7 \pm 4$		
$K_0^*(1950)$	$1980 \pm 14 \pm 19$	$229 \pm 26 \pm 16$	316	17.8
$a_0(1700)$	$1736 \pm 10 \pm 12$	$134 \pm 17 \pm 61$	161	12.7
$\kappa(2600)$	$2662 \pm 59 \pm 201$	$480 \pm 47 \pm 72$	1338	36.6

presence of $a_0(1700)^- \to K_S^0 K^-$, first observed by BaBar ($a_0(1700)^- \to \eta\pi^-$) in the Dalitz plot analysis of $\eta_c \to \eta\pi^+\pi^-$ [5].

We test the consistency between the $K\pi$ S−wave model obtained from the QMI analysis with that obtained from the isobar model analysis. In the isobar model the $K\pi$ S−wave is described in terms of $K_0^*(1430) + K_0^*(1950) + \kappa(2600)$ interfering resonances. We perform a χ^2 fit to the QMI $K\pi$ S−wave amplitude and phase using the model:

$$f(m) = c_1 BW_{K^*(1430)}(m)e^{i\phi_1} + c_2 BW_{K^*(1950)}(m)e^{i\phi_2} + c_3 BW_{\kappa(2600)}(m)e^{i\phi_3}$$

where m indicates the $K\pi$ mass. In this fit, the mass and width of the three resonances are fixed to the results obtained in the isobar model analysis while their coefficients and relative phases are left as free parameters. The fit has a p-value of 63% and is shown in Fig. 5, together with the composition of the fitted resonances. This test shows that the $K\pi$ S-wave obtained using the QMI method can be interpreted as due to the sum of the three interfering K_0^* resonances observed with the isobar model.

Table 4. Results from the Dalitz-plot analysis of the η_c using the isobar model. Inverse-variance-weighted averages from $B^+ \to K^0_S K^+ K^- \pi^+$ and $B^+ \to K^0_S K^+ K^+ \pi^-$ data.

Final state	Fraction (%)	Phase [rad]
$K_0^*(1430)\bar{K}$	$33.4 \pm 0.9 \pm 2.0$	0.
$a_0(980)^-\pi^+$	$5.1 \pm 0.5 \pm 0.8$	$-3.38 \pm 0.06 \pm 0.08$
$K_2^*(1420)\bar{K}$	$14.6 \pm 0.7 \pm 0.8$	$3.54 \pm 0.02 \pm 0.07$
$a_2(1310)^-\pi^+$	$1.1 \pm 0.1 \pm 0.2$	$-2.89 \pm 0.08 \pm 0.18$
$K_0^*(1950)\bar{K}$	$3.7 \pm 0.3 \pm 0.2$	$-0.44 \pm 0.04 \pm 0.45$
$a_0(1700)^-\pi^+$	$1.1 \pm 0.2 \pm 0.2$	$1.05 \pm 0.06 \pm 0.15$
$a_0(1450)^-\pi^+$	$2.6 \pm 0.3 \pm 0.5$	$-4.82 \pm 0.06 \pm 0.13$
$a_2(1750)^-\pi^+$	$0.3 \pm 0.1 \pm 0.1$	$2.33 \pm 0.12 \pm 0.11$
$\kappa(2600)\bar{K}$	$61.8 \pm 2.4 \pm 5.4$	$-0.37 \pm 0.03 \pm 0.09$
Sum	$123.7 \pm 2.7 \pm 4.7$	

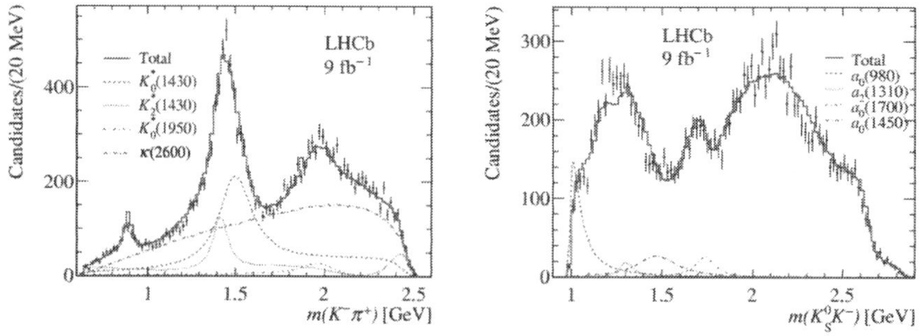

Figure 6. Fit projections on the $K^-\pi^+$, $K^0_S\pi^+$ and $K^0_S K^-$ invariant-mass distributions from the Dalitz-plot analysis of the η_c decay using the isobar model for the $B^+ \to K^0_S K^+ K^- \pi^+$ data.

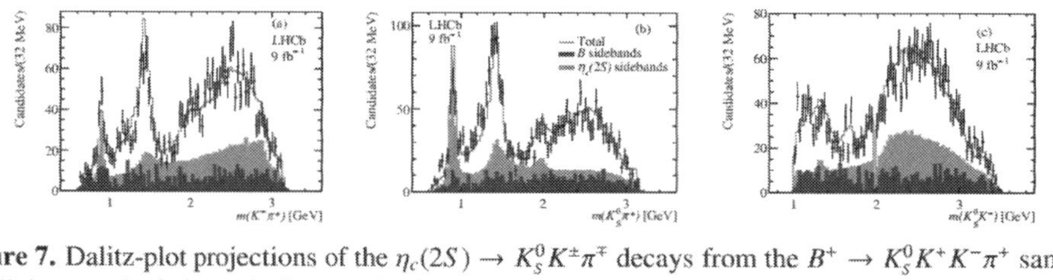

Figure 7. Dalitz-plot projections of the $\eta_c(2S) \to K^0_S K^\pm \pi^\mp$ decays from the $B^+ \to K^0_S K^+ K^- \pi^+$ sample. The light gray shaded area is the contribution from the fitted sidebands and the dark shaded gray area is the contribution from the B^+ sidebands.

5 First Dalitz plot analysis of $\eta_c(2S) \to K^0_S K^\pm \pi^\mp$ in $B^+ \to K^0_S K^+ K^- \pi^+$

The $\eta_c(2S)$ Dalitz plot is similar to that from the η_c but with larger phase space. To describe the data, we only use the isobar model Dalitz plot analysis with all the resonances parameters fixed to that obtained from the η_c analysis. Fit projections are shown in Fig. 7 and results are listed in Table 5. We note large interference effects and similar features as for the η_c analysis, including a small enhancement at the position of the $a_0(1700)$ resonance.

Table 5. Results from the Dalitz-plot analysis of $\eta_c(2S) \to K_S^0 K^\pm \pi^\mp$ in the $B^+ \to K_S^0 K^+ K^- \pi^+$ final state using the isobar model.

Final state	Fraction [%]	Phase [rad]
$K_0^*(1430)K$	$25.5 \pm 3.3 \pm 4.1$	0.
$K_2^*(1430)K$	$24.5 \pm 3.3 \pm 4.4$	$3.10 \pm 0.11 \pm 0.08$
$K_0^*(1950)K$	$3.7 \pm 1.3 \pm 1.1$	$-0.82 \pm 0.17 \pm 0.24$
$a_0(1700)^-\pi^+$	$1.7 \pm 1.1 \pm 0.5$	$1.22 \pm 0.32 \pm 0.90$
$a_0(1450)^-\pi^+$	$7.8 \pm 1.9 \pm 1.0$	$1.86 \pm 0.14 \pm 0.56$
$a_2(1750)^-\pi^+$	$4.9 \pm 1.4 \pm 1.1$	$-1.75 \pm 0.15 \pm 0.39$
$\kappa(2600)K$	$124.2 \pm 9.0 \pm 7.7$	$-0.91 \pm 0.10 \pm 0.08$
Sum	$192.3 \pm 10.9 \pm 10.0$	

Table 6. Inverse-variance-weighted averages of the fractional contributions for χ_{c1} decays to $K^*\bar{K}$ resonances from the $B^+ \to K_S^0 K^+ K^- \pi^+$ and $B^+ \to K_S^0 K^+ K^+ \pi^-$ final states. The reported uncertainties are statistical, systematic and from the uncertainty on the χ_{c1} branching fraction.

Decay mode	Fraction	Branching fraction ($\times 10^{-3}$)
$\mathcal{B}(\chi_{c1} \to K^*(892)^0\bar{K}^0)$	$0.099 \pm 0.012 \pm 0.004$	$1.04 \pm 0.13 \pm 0.04 \pm 0.09$
$\mathcal{B}(\chi_{c1} \to K_2^*(1430)^0\bar{K}^0)$	$0.111 \pm 0.015 \pm 0.005$	$1.17 \pm 0.16 \pm 0.05 \pm 0.10$
$\mathcal{B}(\chi_{c1} \to K^*(892)^+K^-)$	$0.112 \pm 0.016 \pm 0.013$	$1.18 \pm 0.17 \pm 0.14 \pm 0.10$
$\mathcal{B}(\chi_{c1} \to K_2^*(1430)^+K^-)$	$0.143 \pm 0.018 \pm 0.006$	$1.61 \pm 0.19 \pm 0.19 \pm 0.14$

6 Study of the χ_{c1} decay to $K_S^0 K^\pm \pi^\mp$

The χ_{c1} Dalitz plot for $B^+ \to K_S^0 K^+ K^- \pi^+$ data, is dominated by the presence of $K^*(892)$ and $K_2^*(1430)$ resonances. A fit to the $K\pi$ and $K_S^0\pi$ invariant-mass distributions is performed using an empirical background shape in addition to two relativistic BW functions describing the $K^*(892)$ and $K_2^*(1430)$ resonances with parameters fixed to their known values. The fractional $K^*(892)$ and $K_2^*(1430)$ contributions are converted into $\chi_{c1} \to K^*\bar{K}$ branching fractions, listed in Table 6, by multiplying the fractional contributions by the known χ_{c1} branching fraction, $\mathcal{B}(\chi_{c1} \to \bar{K}^0 K^+ \pi^-) = (7.0 \pm 0.6) \times 10^{-3}$ [1] and correcting for unseen K^* decay modes. We obtain branching fractions determinations with improved precision with respect to previous measurements.

References

[1] S. Navas *et al.* [Particle Data Group], "Review of particle physics,", Phys. Rev. D **110** (2024) no.3, 030001. https://doi:10.1103/PhysRevD.110.030001

[2] D. Aston *et al.* "A Study of $K^-\pi^+$ Scattering in the Reaction $K^-p \to K^-\pi^+n$ at 11-GeV/c," Nucl. Phys. B **296** (1988), 493-526 https://doi:10.1016/0550-3213(88)90028-4

[3] E. M. Aitala *et al.* [E791], "Model independent measurement of S-wave $K^-\pi^+$ systems using $D^+ \to K^-\pi^+\pi^+$ decays from Fermilab E791," Phys. Rev. D **73** (2006), 032004 [erratum: Phys. Rev. D **74** (2006), 059901] https://doi:10.1103/PhysRevD.73.032004

[4] R. Aaij *et al.* [LHCb], "Study of charmonium decays to $K_S^0 K\pi$ in the $B \to (K_S^0 K\pi)K$ channels," Phys. Rev. D **108** (2023) no.3, 032010 https://doi:10.1103/PhysRevD.108.032010

[5] J. P. Lees *et al.* [BaBar], "Light meson spectroscopy from Dalitz plot analyses of η_c decays to $\eta' K^+K^-$, $\eta'\pi^+\pi^-$, and $\eta\pi^+\pi^-$ produced in two-photon interactions," Phys. Rev. D **104** (2021) no.7, 072002 https://doi:10.1103/PhysRevD.104.072002

EPJ Web of Conferences **314**, 00013 (2024) https://doi.org/10.1051/epjconf/202431400013
QCD@Work 2024

Hyperon physics at BESIII

Jingxu Zhang[1,*], on behalf of BESIII Collaboration

[1]Lanzhou University

Abstract. The BESIII experiment has been successfully operating at the BEPCII collider since 2008, collecting extensive data samples at the resonances J/ψ and $\psi(2S)$. Recently, BESIII has presented a series of analyses focusing on hyperon physics, particularly the production of hyperon-antihyperon pairs from charmonium states. The polarizations of Ξ^- and Ξ^0 hyperons have been accurately measured at BESIII, providing a new avenue to test the CP symmetry of multi-strange hyperons. Additionally, BESIII has determined the absolute branching fractions and angular distribution parameters α_γ in Λ and Σ^+ radiative hyperon decays with significantly higher accuracy than previous measurements. BESIII has also reported on the interactions of Ξ^0 and Λ hyperons with nuclei. In this proceeding, we present a comprehensive review of the recent advancements in hyperon physics achieved by the BESIII experiment.

1 Introduction

The study of hyperons provides critical insights into the strong interaction, particularly in the transitional region between non-perturbative and perturbative Quantum Chromodynamics (QCD) regimes. The decays of hyperon-antihyperon pairs produced in e^+e^- annihilation offer exceptional laboratories for probing fundamental discrete symmetries, including CP violation. Although the Standard Model (SM) of particle physics has been immensely successful, it does not explain the large matter-antimatter asymmetry in our universe. CP violation is key to understanding this asymmetry [1]. However, CP violation has only been observed experimentally in the meson sector [2–5], not in any baryon sector. Previous experiments constructed CP observables that depend on both strong and weak phases. A novel method from BESIII provides a separation between these phases. By exploiting spin entanglement between the double strange Ξ hyperon and its antiparticle $\bar{\Xi}$, the weak phase difference can be determined directly [6]. This allows us to directly identify the source of CP violation from the SM and beyond. Radiative hyperon decays provide valuable insight into the nature of nonleptonic weak interactions. Generally, the radiative decays of a spin-1/2 hyperon are described by a parity-conserving (P wave) and a parity-violating (S wave) process. The parity-violating amplitude of radiative hyperon decays is predicted to be small in the limit of SU(3) flavor symmetry [7], which implies that α_γ should be approximately zero. The parameter α_γ, which characterizes the mixing of S and P waves, is, however, observed to be significantly large in radiative hyperon decays [8]. To understand and resolve this discrepancy between theory and experiments, new and more precise measurements of all weak radiative hyperon decays are required. To understand baryon-baryon interactions, it is crucial to have

*e-mail: zhangjingxu20@lzu.edu.cn

a detailed comprehension of both nucleon-nucleon and hyperon-nucleon interactions [9]. Although the theory of nucleon-nucleon and antinucleon-nucleon interactions is well developed, our understanding of hyperon-nucleon interaction remains highly uncertain due to the limited number of relevant measurements. The study of hyperon-nucleon interaction is also crucial for determining the equation of state of nuclear matter at supersaturation densities and for understanding the so-called "hyperon puzzle" of neutron stars [10, 11].

With extensive data samples collected at the charmonium resonances of 10^{10} J/ψ [12] and 3×10^9 $\psi(2S)$ [13], BESIII has conducted a series of detailed analyses focused on the production and decay of hyperon-antihyperon pairs. These studies have yielded precise measurements of hyperon properties and interactions, contributing to the broader field of particle physics.

2 Hyperon transverse polarization and CP tests

The production of multi-strange hyperon-antihyperon pairs through the resonances J/ψ and $\psi(2S)$ decay provides the noval method to investigate the spin polarization [6]. The subsequent non-leptonic weak decay of hyperon provide good environment to test the CP symmetry. For a spin-1/2 hyperon decaying into another spin-1/2 baryon and a pion meson, the final states are consist of the parity odd (S-Wava) and parity even (P-Wave) amplitudes. The total decay amplitude is $A = S + P\boldsymbol{\sigma} \cdot \hat{\boldsymbol{n}}$, where $\hat{n}$ is the unit notation along the daughter baryon momentum in the mother rest frame. The hyperon weak decay process can be parameterized by the decay parameters [14]

$$\alpha = \frac{2\mathrm{Re}(S * P)}{|S|^2 + |P|^2}, \ \beta = \frac{2\mathrm{Im}(S * P)}{|S|^2 + |P|^2} = \sqrt{1 - \alpha^2} \sin\phi. \tag{1}$$

Here, the α and ϕ are the CP-odd parameters. The parameter α is the angular distribution asymmetry of the final state baryon

$$dN/d\Omega = \frac{1}{4\pi}(1 + \alpha \mathbf{P} \cdot \hat{\boldsymbol{n}}), \tag{2}$$

where the $\mathbf{P}$ denotes the polarization vector of the mother hyperon. For cascade decay $\Xi^{-/0} \to \Lambda\pi^{-/0}$, the parameter ϕ can be determined by the subsequent $\Lambda \to p\pi^{-/0}$ decay. If CP is conserved, the decay parameters have the same values but opposite sign, like $\alpha = -\bar{\alpha}, \beta = -\bar{\beta}$ and $\phi = -\bar{\phi}$. Then, one can define the CP observables to test the symmetry:

$$A_{CP} = \frac{\alpha + \bar{\alpha}}{\alpha - \bar{\alpha}}, \ \phi_{CP} = \frac{\phi + \bar{\phi}}{2}, \ B_{CP} = \frac{\beta + \bar{\beta}}{\alpha - \bar{\alpha}} = (\xi_P - \xi_S). \tag{3}$$

The $(\xi_P - \xi_S)$ is the weak phase difference. All the non-zero values of three observables indicate the CP violation. The Standard Model predicts the CP violation for Ξ hyperon is around $A_{CP} \sim 10^{-5}$ [6].

To describe the hyperon pairs production from electron and positron annihilation, including the subsequently two-body hyperon weak decay, the joint angular distribution for single decay is [15]

$$\mathcal{W}(\xi; \omega) = \sum_{\mu,\bar{\nu}=0}^{3} C_{\mu\bar{\nu}} \sum_{\mu',\bar{\nu}'=0}^{3} a^{B}_{\mu,\mu'} a^{\bar{B}}_{\bar{\nu},\bar{\nu}'}. \tag{4}$$

The vector $\boldsymbol{\xi}$ denotes the polar and azimuthal helicity angles of hyperons, $\boldsymbol{\xi} = (\theta,\ \theta_\Lambda,\ \varphi_\Lambda,\ \theta_{\bar{\Lambda}},\ \varphi_{\bar{\Lambda}},\ \theta_p,\ \varphi_p,\ \theta_{\bar{p}},\ \varphi_{\bar{p}})$. The eight free parameters are $\boldsymbol{\omega}$ =

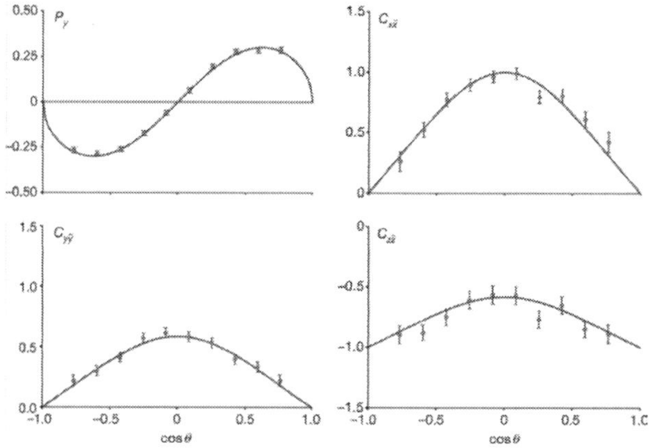

Figure 1. The acceptance correction distributions of transverse polarization P_y and spin correlations C_{xx}, C_{yy} and C_{zz} as functions of $\cos\theta_{\Xi}$ in the $e^+e^- \to J/\psi \to \Xi^-\bar{\Xi}^+$.

(α_ψ, $\Delta\Phi$, α_Λ, $\bar{\alpha}_\Lambda$, α_Ξ, $\bar{\alpha}_\Xi$, ϕ_Ξ, $\bar{\phi}_\Xi$). The such elements of 4×4 spin density matrices, which describe the two-body hyperon weak decay the process, are parameterized in terms of the hyperon decay parameters α and ϕ, and depend on the helicity angles. The hyperon production density matrix $C_{\mu\bar{\nu}}$ is defined by the spin correlation C_{ij} and transverse polarization P_y:

$$C_{\mu\bar{\nu}} = (1+\alpha_\psi\cos^2\theta)\begin{pmatrix} 1 & 0 & P_y & 0 \\ 0 & C_{xx} & 0 & C_{xz} \\ -P_y & 0 & C_{yy} & 0 \\ 0 & -C_{xz} & 0 & C_{zz} \end{pmatrix}. \tag{5}$$

In 2022 and 2023 years, BESIII experiment successively reported the observations of Ξ^- and Ξ^0 polarization in the decay of J/ψ and $\psi(2S)$ [16–19]. Figure 1 shows the transverse polarization and spin correlations in $J/\psi \to \Xi^-\bar{\Xi}^+$. The fitting results for $\psi(2S)$, $J/\psi \to \Xi\bar{\Xi}$ decays are shown in Table 1. Comparing the results of Ξ^- and Ξ^0, the angular distribution α_ψ and spin polarization $\Delta\Phi$ are similar except the polarization in $\psi(2S) \to \Xi^0\bar{\Xi}^0$. For results between J/ψ and $\psi(2S)$, the α_ψ and $\Delta\Phi$ are different. The CP-odd decay parameters α_Ξ and ϕ_Ξ are also observed in these decay processes, demonstrating consistency of the parameters across different charmonium resonances. Based on the measurements of decay parameters, the $A_{CP,\Xi}$ is determined to be $A_{CP,\Xi^-} = (0.6\pm1.3\pm0.6)\times10^{-2}$ and $A_{CP,\Xi^0} = (-5.4\pm6.5\pm3.1)\times10^{-3}$. These results are the very precise tests of the Ξ hyperon CP symmetry. The weak phase difference, which denotes the alternative way to test CP which avoid the influence from strong final states interaction of $\Lambda-\pi$, is determined by the $J/\psi \to \Xi^-\bar{\Xi}^+$ and $J/\psi \to \Xi^0\bar{\Xi}^0$ decays. The results for weak phase difference are the first determination in Ξ even in any baryons. It shows that $(\xi_p-\xi_s)_{\Xi^-} = (1.2\pm3.4\pm0.8)\times10^{-2}$ rad and $(\xi_p-\xi_s)_{\Xi^0} = (0.0\pm1.7\pm0.2)\times10^{-2}$ rad.

3 Hyperon weak radiative decay

Since 1990s, the branching fractions of radiative hyperon decays have not been updated. The radiative hyperon decays $\Lambda \to n\gamma$ [20] and $\Sigma^+ \to p\gamma$ [21] were observed recently at BESIII

Table 1. The number of events of charmonium states (N_ψ) produced by e^+e^- annihilation and the signal events of $\Xi\bar{\Xi}$ pairs ($N_{\Xi\bar{\Xi}}$).

	$J/\psi \to \Xi^-\bar{\Xi}^+$ [16]	$J/\psi \to \Xi^0\bar{\Xi}^0$ [17]	$\psi(2S) \to \Xi^-\bar{\Xi}^+$ [18]	$\psi(2S) \to \Xi^0\bar{\Xi}^0$ [19]
α_ψ	0.586±0.012±0.010	0.514±0.006±0.015	0.693±0.048±0.049	0.665±0.086±0.081
$\Delta\Phi$(rad)	1.213±0.046±0.016	1.168±0.019±0.018	0.667±0.111±0.058	−0.050±0.150±0.020
α_Ξ	−0.376±0.007±0.003	−0.3750±0.0034±0.0016	−0.344±0.025±0.007	−0.358±0.042±0.013
$\bar{\alpha}_\Xi$	0.371±0.007±0.002	0.3790±0.0034±0.0021	0.355±0.025±0.002	0.363±0.042±0.013
ϕ_Ξ(rad)	0.011±0.019±0.009	0.0051±0.0096±0.0018	0.023±0.074±0.003	0.027±0.117±0.011
$\bar{\phi}_\Xi$(rad)	−0.021±0.019±0.007	−0.0053±0.0097±0.0019	−0.123±0.073±0.004	−0.185±0.116±0.017

using 10 billion J/ψ data sets. The Dtag method is used to measure the branching fraction:

$$\mathrm{BF}(B \to b\gamma) = \frac{N_{DT}\epsilon_{ST}}{N_{ST}\epsilon_{DT}}. \tag{6}$$

Here, BF is the branching fraction, B means mother particle and b is daughter baryon, $N_{DT(ST)}$ and $\epsilon_{DT(ST)}$ are the double tag (single tag) yields and the corresponding detection efficiencies. The previous BESIII reported that Λ and Σ^+ from $J/\psi \to \Lambda\bar{\Lambda}$ [22] and $J/\psi \to \Sigma^+\bar{\Sigma}^-$ [23] decays are polarized. The polarization can be used to determine the parameter α_γ in the $\Lambda \to n\gamma$ and $\Sigma^+ \to p\gamma$ from the angular distributions of mother baryons B. The helicity angles are defined from Eq. 4.

Figure. 2 shows the results of BFs and decay asymmetry parameters α_γ in $\Lambda \to n\gamma$ and $\Sigma^+ \to p\gamma$, including the PDG results and theoretical predictions. The measured BFs are both lower than the PDG values by 5.6σ in $\Lambda \to n\gamma$ and 4.2σ in $\Sigma^+ \to p\gamma$. By analyzing the joint angular distribution, the $\alpha_\gamma = -0.016 \pm 0.010 \pm 0.005$ in $\Lambda \to n\gamma$ is determined for the first time, and $\alpha_\gamma = -0.651 \pm 0.056 \pm 0.020$ in $\Sigma^+ \to p\gamma$ is consistent with PDG value. The BF and α_γ results of the charged conjugate modes are consistent within uncertainties, and there is no indication of any CP violation.

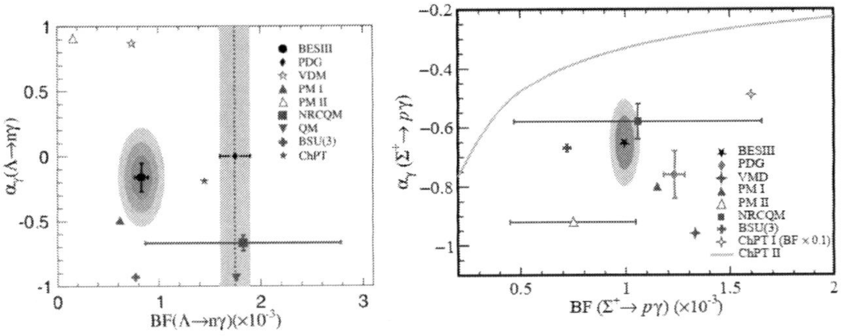

Figure 2. Two dimensional distributions of the BF and α_γ in $\Lambda \to n\gamma$(Left) and $\Sigma^+ \to p\gamma$(Right) decays.

4 Hyperon nucleus interaction

The experimental study of the interaction between hyperons and different target materials began in the 1960s and has lasted for more than half a century. In 2023, BESIII reported on the research of interactions of Ξ^0 [24] hyperon and nucleons. It's the first study of hyperon-nucleon interactions in electron-positron collisions and opens up a new direction for such researches. Based on the large J/ψ data samples, BESIII can provide amount of hyperon beams. Afterwards the hyperon beams can interact with the material in the beam pipe. The material of the beam pipe is composed of gold (^{197}Au), beryllium (^{9}Be), and oil(^{12}C: ^{1}H = 1: 2.13). The cross sections of the hyperons and nucleus reactions are determine with [25]

$$\sigma = \frac{N_{sig}}{\epsilon_{sig}\mathcal{B}\mathcal{L}_{eff}}, \quad (7)$$

where N_{sig} is the signal yields of the reaction, ϵ is the selection efficiency, $\mathcal{B}$ is the product of the branching ratios of all intermediate resonances, and $\mathcal{L}_{eff}$ is the effective luminosity of the hyperon flux. Figure 3 shows the distribution of signal yields of the Ξ^- produced by $\Xi^0 + n$ interaction. The clear signal is observed with a statistical significance of 7.1σ. By analyzing the various decay points in target materials, the cross sections of Ξ^0 and nuclei interactions are determined at the momentum of the Ξ^0 of $P_{\Xi^0} = 0.818$ GeV/c^2, such as $\sigma(\Xi^0 +^9 \text{Be} \rightarrow \Xi^- +^8 \text{Be}) = (22.1 \pm 5.3 \pm 4.5)$ mb.

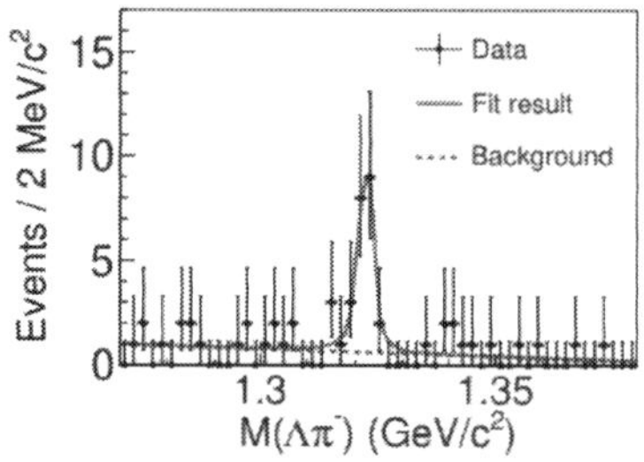

Figure 3. Distribution of the $M_{\Lambda\pi^-}$. The black dots are the data, the red solid curve is the total fit result and blue dashed curve is the background component.

BESIII also reports the scattering of Λ and nucleons with significant observation [26, 27]. The total cross sections in $\Lambda +^9 \text{Be} \rightarrow \Sigma^+ + X$, $\Lambda p \rightarrow \Lambda p$ and $\bar{\Lambda}p \rightarrow \bar{\Lambda}p$. The X refers to any possible particles produced accompanying the Σ^+. This is the first study of Λ-nucleon interactions at an e^+e^- collider and the study of $\bar{\Lambda}p$ is the first observation of antihyperon-nucleon scattering. These measurements will serve as important inputs for the theoretical understanding of the (anti)hyperon-nucleon interaction.

5 Summary

Since its inception in 2009, BESIII has been operating successfully and has collected substantial data samples in the τ-charm physics region. Significant advancements have been made in hyperon physics, including first observations of spin polarization and CP violation in the two-body weak decays of Ξ^- and Ξ^0 hyperons, precise measurements of absolute branching fractions, and studies on angular asymmetries in weak radiative decays involving Λ and Σ^+ hyperons, as well as the determination of hyperon-nucleon scattering processes with Ξ^0, Λ, and $\bar{\Lambda}$ for the first time at an e^+e^- collider.

This work is supported in part by National Key Research and Development Program of China under Contracts Nos. 2020YFA0406400, 2020YFA0406403; National Natural Science Foundation of China under Contracts Nos. 12075107, 12247101; the 111 Project under Grant No. B20063.

References

[1] A.D. Sakharov, Violation of CP Invariance, C asymmetry, and baryon asymmetry of the universe, Pisma Zh. Eksp. Teor. Fiz. **5**, 32 (1967). 10.1070/PU1991v034n05ABEH002497

[2] J.H. Christenson, J.W. Cronin, V.L. Fitch, R. Turlay, Evidence for the 2π Decay of the K_2^0 Meson, Phys. Rev. Lett. **13**, 138 (1964). 10.1103/PhysRevLett.13.138

[3] B. Aubert et al. (BaBar), Observation of CP violation in the B^0 meson system, Phys. Rev. Lett. **87**, 091801 (2001). 10.1103/PhysRevLett.87.091801

[4] K. Abe et al. (Belle), Observation of large CP violation in the neutral B meson system, Phys. Rev. Lett. **87**, 091802 (2001). 10.1103/PhysRevLett.87.091802

[5] R. Aaij et al. (LHCb), Observation of CP Violation in Charm Decays, Phys. Rev. Lett. **122**, 211803 (2019). 10.1103/PhysRevLett.122.211803

[6] N. Salone, P. Adlarson, V. Batozskaya, A. Kupsc, S. Leupold, J. Tandean, Study of CP violation in hyperon decays at super-charm-tau factories with a polarized electron beam, Phys. Rev. D **105**, 116022 (2022), 2203.03035. 10.1103/PhysRevD.105.116022

[7] Y. Hara, Nonleptonic Decays of Baryons and the Eightfold Way, Phys. Rev. Lett. **12**, 378 (1964). 10.1103/PhysRevLett.12.378

[8] M. Kobayashi, J. Haba, T. Homma, H. Kawai, K. Miyake, T.S. Nakamura, N. Sasao, Y. Sugimoto, New Measurement of the Asymmetry Parameter for the $\Sigma^+ \to p\gamma$ Decay, Phys. Rev. Lett. **59**, 868 (1987). 10.1103/PhysRevLett.59.868

[9] R.L. Workman et al. (Particle Data Group), Review of Particle Physics, PTEP **2022**, 083C01 (2022). 10.1093/ptep/ptac097

[10] I. Vidaña, Hyperons and neutron stars, Nucl. Phys. A **914**, 367 (2013). 10.1016/j.nuclphysa.2013.01.015

[11] L. Tolos, L. Fabbietti, Strangeness in Nuclei and Neutron Stars, Prog. Part. Nucl. Phys. **112**, 103770 (2020), 2002.09223. 10.1016/j.ppnp.2020.103770

[12] M. Ablikim et al. (BESIII), Number of J/ψ events at BESIII, Chin. Phys. C **46**, 074001 (2022), 2111.07571. 10.1088/1674-1137/ac5c2e

[13] M. Ablikim et al. (BESIII), Determination of the number of $\psi(3686)$ events taken at BESIII*, Chin. Phys. C **48**, 093001 (2024), 2403.06766. 10.1088/1674-1137/ad595b

[14] T.D. Lee, C.N. Yang, General Partial Wave Analysis of the Decay of a Hyperon of Spin 1/2, Phys. Rev. **108**, 1645 (1957). 10.1103/PhysRev.108.1645

[15] E. Perotti, G. Fäldt, A. Kupsc, S. Leupold, J.J. Song, Polarization observables in e^+e^- annihilation to a baryon-antibaryon pair, Phys. Rev. **D99**, 056008 (2019). 10.1103/PhysRevD.99.056008

[16] M. Ablikim et al. (BESIII), Probing CP symmetry and weak phases with entangled double-strange baryons, Nature **606**, 64 (2022), 2105.11155. 10.1038/s41586-022-04624-1

[17] M. Ablikim et al. (BESIII), Tests of CP symmetry in entangled Ξ0-Ξ̄0 pairs, Phys. Rev. D **108**, L031106 (2023), 2305.09218. 10.1103/PhysRevD.108.L031106

[18] M. Ablikim et al. (BESIII), Observation of Ξ- hyperon transverse polarization in ψ(3686)→Ξ-Ξ̄+, Phys. Rev. D **106**, L091101 (2022), 2206.10900. 10.1103/PhysRevD.106.L091101

[19] M. Ablikim et al. (BESIII), First simultaneous measurement of Ξ0 and Ξ̄0 asymmetry parameters in ψ(3686) decay, Phys. Rev. D **108**, L011101 (2023), 2302.09767. 10.1103/PhysRevD.108.L011101

[20] M. Ablikim et al. (BESIII), Measurement of the Absolute Branching Fraction and Decay Asymmetry of Λ→nγ, Phys. Rev. Lett. **129**, 212002 (2022), 2206.10791.

10.1103/PhysRevLett.129.212002

[21] M. Ablikim et al. (BESIII), Precision Measurement of the Decay Σ+→pγ in the Process J/ψ→Σ+Σ¯-, Phys. Rev. Lett. **130**, 211901 (2023), 2302.13568. 10.1103/PhysRevLett.130.211901

[22] M. Ablikim et al. (BESIII), Precise Measurements of Decay Parameters and CP Asymmetry with Entangled $\Lambda - \bar{\Lambda}$ Pairs, Phys. Rev. Lett. **129**, 131801 (2022), 2204.11058. 10.1103/PhysRevLett.129.131801

[23] M. Ablikim et al. (BESIII), Σ^+ and $\bar{\Sigma}^-$ polarization in the J/ψ and $\psi(3686)$ decays, Phys. Rev. Lett. **125**, 052004 (2020), 2004.07701. 10.1103/PhysRevLett.125.052004

[24] M. Ablikim et al. (BESIII), First Study of Reaction Ξ0n→Ξ-p Using Ξ0-Nucleus Scattering at an Electron-Positron Collider, Phys. Rev. Lett. **130**, 251902 (2023), 2304.13921. 10.1103/PhysRevLett.130.251902

[25] J. Dai, H.B. Li, H. Miao, J. Zhang, Prospects to study hyperon-nucleon interactions at BESIII*, Chin. Phys. C **48**, 073003 (2024), 2209.12601. 10.1088/1674-1137/ad3dde

[26] M. Ablikim et al. (BESIII), First measurement of ΛN inelastic scattering with Λ from e+e−→J/ψ→ΛΛ¯, Phys. Rev. C **109**, L052201 (2024), 2310.00720. 10.1103/PhysRevC.109.L052201

[27] M. Ablikim et al. (BESIII), First Study of Antihyperon-Nucleon Scattering Λ¯p→Λ¯p and Measurement of Λp→Λp Cross Section, Phys. Rev. Lett. **132**, 231902 (2024), 2401.09012. 10.1103/PhysRevLett.132.231902

Low energy description of single flavor baryons

Francesco Bigazzi[1,*], *Aldo L.* Cotrone[1,2,**], and *Andrea* Olzi[1,2,***]

[1]INFN, Sezione di Firenze, Via G. Sansone 1, I-50019 Sesto Fiorentino (Firenze), Italy
[2]Dipartimento di Fisica e Astronomia, Universitá di Firenze, Via G. Sansone 1, I-50019 Sesto Fiorentino (Firenze), Italy

Abstract. In single-flavored large N QCD, the standard low-energy description of baryons in terms of Skyrmions is no longer available. Recently it has been proposed that the correct low-energy description in this case is in terms of a pancake-shaped sheet, a quantum Hall droplet. We will describe how this proposal can be made concrete in Holographic QCD. We present the brane configuration describing the sheet and an approximate solution with the expected physical properties.

1 No Skyrmion with one flavor

The usual low-energy description of pions and baryons for N_f quark flavors is in terms of the pion matrix

$$U = \exp\left(i \sum_{a=1}^{N_f^2-1} \frac{\pi^a(x) T^a}{f_\pi} \right), \tag{1}$$

where π^a are the pion fields, T_a the generators of $SU(N_f)$ and f_π is the pion decay constant. The (effective) Chiral Lagrangian is

$$\mathcal{L}_{\text{eff}} = -\frac{f_\pi^2}{4} Tr[\partial_\mu U \partial^\mu U^\dagger] + \frac{1}{32e^2} Tr\left[U^\dagger \partial_\mu U, U^\dagger \partial_\nu U\right]^2 , \tag{2}$$

where e is a coupling constant. The second piece, called the Skyrme term, allows for the description of baryons as solitons of this Lagrangian [1], e.g. for $N_f = 2$ one has the standard "Skyrmion"

$$U = \exp\left(i \frac{f(r)}{r} x^i \sigma^i \right), \tag{3}$$

where x^i are the target space coordinates, $r^2 = \sum_i (x^i)^2$, σ^i are the Pauli matrices and $f(r)$ is a known function. It can be shown that if $f(0) = \pi k$ with $k \in \mathbb{Z}$, then the solution has winding number k for $\Pi_3(SU(N_f = 2)) = \mathbb{Z}$, where the three-sphere of Π_3 is in coordinate space. This is interpreted as the baryon number.

The problem with this description is that $\Pi_3(U(N_f = 1)) = 0$. In other words, the Skyrme term trivializes if $N_f = 1$. So, there are no Skyrmions with one flavor. Of course, baryons made by one quark flavor do exist, so the problem is with their correct description at low energies.

*e-mail: bigazzi@fi.infn.it
**e-mail: cotrone@fi.infn.it
***e-mail: andrea.olzi@unifi.it

2 A proposal for the solution

Recently Komargodski, following previous studies [2–6], proposed that in single-flavor large N QCD, baryons can be seen as charged sheets [7]. With a single flavor, the low energy Lagrangian is just the Lagrangian for the equivalent of the η' field (which is light when the number of colors N is large) [8]

$$\mathcal{L}_{\text{eff}} = \frac{1}{2}(\partial\eta')^2 - \frac{1}{2}m\Lambda_{\text{QCD}}\cos(\eta') - \frac{1}{2}m_{WV}^2\text{Min}_{k\in\mathbb{Z}}(\eta' + 2\pi k)^2 . \tag{4}$$

In this expression m is the quark mass, Λ_{QCD} the QCD dynamical scale, and m_{WV} the Witten-Veneziano mass. The potential has a cusp singularity behavior at $\eta' = \pi$, where extra light gluonic degrees of freedom are localized. This locus is called the "sheet". The η' has non-trivial monodromy through the sheet. The infinitely extended sheet is similar to the standard QCD domain-wall at $\theta_{YM} = \pi$, but with no associated charge, so it is unstable or, more precisely, metastable at large N. But, crucially, it can be stabilized by the baryonic charge.

In fact, the sheet (as the domain wall) can be argued to host a $U(1)_N$ Chern-Simons theory on its world-volume, so it is a "quantum Hall droplet". Then, if it has a boundary, it can have a chiral edge mode carrying baryonic charge, see figure 1. If this is the case, the charge forbids the complete decay of the "pancake-shaped" sheet, since this is the lightest particle which carries one unit of charge.

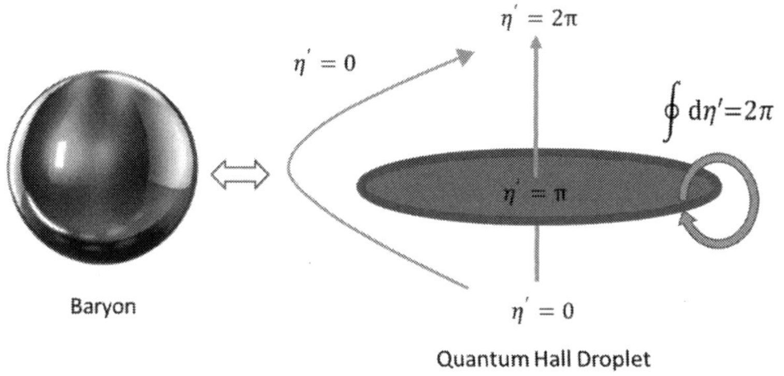

Figure 1. In the case of one flavor, the baryon is really a quantum Hall droplet of pancake shape. The η' field has a non-trivial monodromy across the droplet world-volume.

The crucial question is: can we test the proposal and compute properties of this baryon? In the next section we present some progress in this direction in a theory similar to QCD, known as "Holographic QCD" or the "Witten-Sakai-Sugimoto model".

3 Realization in Holographic QCD

Holographic QCD [9, 10] corresponds to a Type IIA background in string theory, generated by N D4-branes wrapped on a circle, with N_f D8/anti-D8-brane pairs in the probe approximation (that is, not backreacting on the background). At low energies, this configuration is dual to a $SU(N)$ non-supersymmetric four-dimensional Yang-Mills theory coupled to a tower of Kaluza-Klein modes in the adjoint representation and N_f chiral quark flavors. It is similar to (planar) QCD, showing confinement, mass gap, and chiral symmetry breaking. Its vacuum structure is the same as in QCD. Baryons in this theory can be described as non-Abelian instantonic configurations of the gauge field on the D8-branes [11], but only for $N_f \geq 2$. So, again: how do we describe the sheet, that is the baryon in a single-flavored holographic QCD?

In [12] it was realized that the behavior of η' through the sheet is reproduced if a D6-brane describes the latter (see also [13]). In fact, the η' corresponds to a fluctuating mode of the D8-brane gauge field. The action (4) can be derived as the low energy limit of the D8-brane action (following the steps in [14]), and its coupling with a D6-brane implies a 2π monodromy when the field configuration crosses the brane. Moreover, the D6-brane, upon reduction on a four-sphere of the background, hosts a $U(1)_N$ Chern-Simons theory on its world-volume [15].

So, the sheet is realized as a D6-brane in this theory. The quantum Hall droplet with a circular boundary corresponds to a D6-brane attached to the D8-brane, as depicted in figure 2. The D6-brane constitutes the gluonic core of the baryon. This is the equivalent of the string

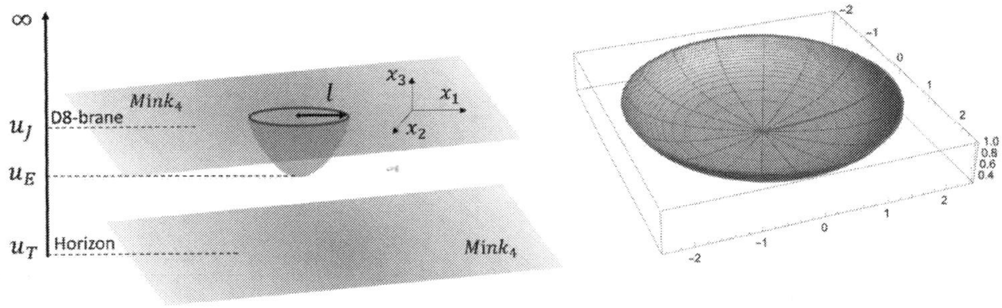

Figure 2. Left: a cartoon of the D6-brane describing the gluonic core of the baryon hanging from the D8-brane in the deconfined, chiral symmetry broken phase. Right: the actual solution for the D6-brane shape from solving the equations of motion coming from the Dirac-Born-Infeld action.

junction at the core of the baryon, where the chromoelectric flux tubes (strings) attached to the quarks meet.

In the following, we are going to describe the baryon in the simplest setting, corresponding to the deconfined phase of the quantum field theory but still in the chiral symmetry breaking phase - a possibility present in Holographic QCD [16]. The presence of a horizon in the background, corresponding to the deconfined phase, implies some simplifications in the equations [17].

We would like to describe the quantum Hall droplet. The profile of the D6-brane ending on the D8 brane in the background can be calculated numerically by solving the equations of motion coming from the Dirac-Born-Infeld action, which describes the low-energy physics of the D-branes. A representative profile is given by the plot on the right of figure 2. The profile is immersed in a three-dimensional space because of the holographic radial direction, keeping track of the energy scale - the dual quantum Hall droplet is two-dimensional.

Calculating the energy of this solution one finds out that it grows linearly for "large" radius l, and quadratically for small l. Nevertheless, the configuration cannot have an arbitrarily large radius in the deconfined phase: if l is too large, the bottom of the D6-brane falls into the horizon, transitioning to a cylinder-shaped configuration. The dual sheet is no longer a pancake, but an annulus. Physically, this is dual to a circular global string, or a so-called "vorton" in the charged case. Its energy grows linearly with l for every value of l. It can be shown that these two embeddings are separated by a first-order phase transition [17].

The D6-brane profile is not the end of the story. A D6-brane ending on a D8-brane is a magnetic source for the D8-brane gauge field F describing the η' and, crucially, the charge allowing for a stable configuration. The gauge field on the D8-brane describes the mesonic content of the baryon. So, the D8-brane gauge field is necessarily turned on. The profile of its components is again dictated by the DBI action, this time for the D8-brane. As customary, at leading order at low energies, the equations of motion reduce to electromagnetism (remember that $N_f = 1$). Since also the D8-brane is wrapped on a four-sphere of the background, the theory at low energies is five-dimensional. Electromagnetism in five dimensions admits a Chern-Simons coupling. Considering that the background is non-trivial, altogether one faces a problem in the five-dimensional Maxwell-Chern-Simons theory on curved space.

The equations for the charged sheet are a coupled system of non-linear partial differential equations. In fact, the symmetry of the problem dictates that the solution will depend on the radial direction ρ in the plane of the sheet, on the transverse direction in Minkowski space, which is denoted as x_3, and on the holographic radial direction, which we will call z. Needless to say, these equations are extremely hard to solve even numerically.

3.1 An approximate solution

An approximate solution, along the lines of [11], can be found if one assumes that the droplet is extremely small, i.e. its radius l scales as a negative power of the large 't Hooft coupling λ.[1]

The D6-brane intersects the D8-brane at its tip at $z = 0$, where it is a good approximation to consider the background as flat. In the flat space limit the symmetry in the x_3, z plane is restored, such that the dependence is just on the radius r in this plane. The set of equations "simplify" to

$$\begin{aligned}
&\rho\partial_r(rF_{tr}) + r\partial_\rho(\rho F_{t\rho}) - \tau(F_{\rho\psi}F_{r\theta} - F_{r\psi}F_{\rho\theta}) = 0\,,\\
&\frac{1}{\rho}\partial_r(rF_{r\psi}) + r\partial_\rho\left(\frac{1}{\rho}F_{\rho\psi}\right) - \tau/2(F_{tr}F_{\rho\theta} - F_{t\rho}F_{r\theta}) = 0\,,\\
&\frac{1}{r}\partial_\rho(\rho F_{\rho\theta}) + \rho\partial_r\left(\frac{1}{r}F_{r\theta}\right) - \tau/2(F_{t\rho}F_{r\psi} - F_{tr}F_{\rho\psi}) = 0\,,\\
&dF = -\sqrt{2}\delta(r)\delta(\rho - l)d\rho \wedge dr \wedge d\theta\,,
\end{aligned} \tag{5}$$

where $\tau \sim 1/\lambda$ is a constant and ϕ, θ are the angular coordinates in the plane of the sheet and the (x_3, z)–plane. The last equation is the violation of the Bianchi identities due to the circular source (the D6-brane boundary on the D8-brane).

Even the flat-space system (5) is not easily solved because the fields diverge close to the source as usual in electromagnetism. However, assuming that the droplet is extremely small, it admits an approximate solution with the desired properties:

- It is an equilibrium configuration, of size

$$l_{\text{eq}} \sim \sqrt{\frac{\Lambda_{QCD}}{\lambda T^3_{\chi SB}}}\,, \tag{6}$$

  where Λ_{QCD} is the dynamical scale of the theory and $T_{\chi SB}$ is the temperature of the chiral symmetry breaking transition. As expected, l_{eq} does not depend on N and scales as a negative power of λ.

[1] More precisely, λ is a 't Hooft-like parameter proportional to the ratio between the confining string tension and the square of the Kaluza-Klein mass scale. The holographic description we are adopting is reliable when $\lambda \gg 1$.

- Its mass is

$$M \sim \frac{\lambda^2 N T_{\chi S B}^3}{\Lambda_{\mathrm{QCD}}^2}\,, \tag{7}$$

with the correct scaling with N for a baryon in the planar limit.

- It has a unit baryon number

$$n_B = \frac{1}{12\pi^2}\int d^4x\,(F_{\rho\phi}F_{r\theta} - F_{r\phi}F_{\rho\theta}) = 1\,. \tag{8}$$

- It has spin $N/2$ (for a single flavor all the quark spins are aligned).

Nevertheless, this solution is based on a scaling assumption about the fields and, most importantly, the cutoff close to the source.

4 Conclusions

In conclusion, we presented a (partial) concrete holographic realization of Komargodski's proposal for the low-energy description of single-flavor baryons and computed its basic properties. Calculating the corresponding numeric solution could corroborate and improve the approximated solution. We would also like to compute other observables associated with this baryon and extend the solution to the confined phase. Finally, it is worth mentioning that this construction has applications to axionic dark matter [18, 19], and that other studies related to the sheet appeared in [20–30].

References

[1] T. H. R. Skyrme, 'A Nonlinear field theory," Proc. Roy. Soc. Lond. A **260**, 127-138 (1961). https://doi.org/10.1098/rspa.1961.0018

[2] I. I. Kogan, "Axions, monopoles and cosmic strings," [arXiv:hep-ph/9305307 [hep-ph]].

[3] G. Gabadadze and M. A. Shifman, "Vacuum structure and the axion walls in gluodynamics and QCD with light quarks," Phys. Rev. D **62**, 114003 (2000) [arXiv:hep-ph/0007345 [hep-ph]]. https://doi.org/10.1103/PhysRevD.62.114003

[4] M. M. Forbes and A. R. Zhitnitsky, "Domain walls in QCD," JHEP **10**, 013 (2001) [arXiv:hep-ph/0008315 [hep-ph]]. https://doi.org/10.1088/1126-6708/2001/10/013

[5] D. T. Son, M. A. Stephanov and A. R. Zhitnitsky, "Domain walls of high density QCD," Phys. Rev. Lett. **86**, 3955-3958 (2001) [arXiv:hep-ph/0012041 [hep-ph]]. https://doi.org/

[6] G. Gabadadze and M. Shifman, "QCD vacuum and axions: What's happening?," Int. J. Mod. Phys. A **17**, 3689-3728 (2002) [arXiv:hep-ph/0206123 [hep-ph]]. https://doi.org/10.1142/S0217751X02011357

[7] Z. Komargodski, "Baryons as Quantum Hall Droplets," arXiv:1812.09253 [hep-th].

[8] E. Witten, "Theta dependence in the large N limit of four-dimensional gauge theories," Phys. Rev. Lett. **81**, 2862 (1998) [hep-th/9807109]. https://doi.org/10.1103/PhysRevLett.81.2862

[9] E. Witten, "Anti-de Sitter space, thermal phase transition, and confinement in gauge theories," Adv. Theor. Math. Phys. **2**, 505 (1998), [hep-th/9803131]. https://doi.org/10.4310/ATMP.1998.v2.n3.a3

[10] T. Sakai and S. Sugimoto, "Low energy hadron physics in holographic QCD," Prog. Theor. Phys. **113**, 843 (2005), [arXiv:1906.12117 [hep-th]]. https://doi.org/10.1143/PTP.113.843

[11] H. Hata, T. Sakai, S. Sugimoto and S. Yamato, "Baryons from instantons in holographic QCD," Prog. Theor. Phys. **117**, 1157 (2007) [hep-th/0701280 [hep-th]]. https://doi.org/10.1143/PTP.117.1157

[12] F. Bigazzi, A. L. Cotrone and A. Olzi, "Hall Droplet Sheets in Holographic QCD," JHEP **02**, 194 (2023), [arXiv:2211.05147 [hep-th]]. https://doi.org/10.1007/JHEP02(2023)194

[13] S. Dubovsky, A. Lawrence and M. M. Roberts, "Axion monodromy in a model of holographic gluodynamics," JHEP **1202**, 053 (2012) [arXiv:1105.3740 [hep-th]]. https://doi.org/10.1007/JHEP02(2012)053

[14] L. Bartolini, F. Bigazzi, S. Bolognesi, A. L. Cotrone and A. Manenti, "Theta dependence in Holographic QCD," JHEP **1702**, 029 (2017) [arXiv:1611.00048 [hep-th]]. https://doi.org/10.1007/JHEP02(2017)029

[15] R. Argurio, M. Bertolini, F. Bigazzi, A. L. Cotrone and P. Niro, "QCD domain walls, Chern-Simons theories and holography," JHEP **1809**, 090 (2018) [arXiv:1806.08292 [hep-th]]. https://doi.org/10.1007/JHEP09(2018)090

[16] O. Aharony, J. Sonnenschein and S. Yankielowicz, "A Holographic model of deconfinement and chiral symmetry restoration," Annals Phys. **322**, 1420 (2007) [hep-th/0604161]. https://doi.org/10.1016/j.aop.2006.11.002

[17] F. Bigazzi, A. L. Cotrone and A. Olzi, work in progress

[18] F. Bigazzi, A. Caddeo, A. L. Cotrone, P. Di Vecchia and A. Marzolla, "The Holographic QCD Axion," JHEP **12**, 056 (2019), [arXiv:1906.12117 [hep-th]]. https://doi.org/10.1007/JHEP12(2019)056

[19] F. Bigazzi, A. L. Cotrone and A. Olzi, "Axionic Strings, Domain Walls and Baryons," Phys. Rev. D **108** (2023) no.2, 2, [arXiv:2212.09783 [hep-ph]]. https://doi.org/10.1103/PhysRevD.108.026019

[20] Y. L. Ma, M. A. Nowak, M. Rho and I. Zahed, "Baryon as a Quantum Hall Droplet and the Cheshire Cat Principle," Phys. Rev. Lett. **123**, 172301 (2019) [arXiv:1907.00958 [hep-th]]. https://doi.org/10.1103/PhysRevLett.123.172301

[21] A. Karasik, "Skyrmions, Quantum Hall Droplets, and one current to rule them all," SciPost Phys. **9**, 008 (2020) [arXiv:2003.07893 [hep-th]]. https://doi.org/10.21468/SciPostPhys.9.1.008

[22] Y. L. Ma and M. Rho, "Dichotomy of Baryons as Quantum Hall Droplets and Skyrmions: Topological Structure of Dense Matter," Symmetry **13**, no.10, 1888 (2021) [arXiv:2009.09219 [nucl-th]]. https://doi.org/10.3390/sym13101888

[23] A. Karasik, "Vector dominance, one flavored baryons, and QCD domain walls from the "hidden" Wess-Zumino term," SciPost Phys. **10**, no.6, 138 (2021) [arXiv:2010.10544 [hep-th]]. https://doi.org/10.21468/SciPostPhys.10.6.138

[24] R. Kitano and R. Matsudo, "Vector mesons on the wall," JHEP **03**, 023 (2021) [arXiv:2011.14637 [hep-th]]. https://doi.org/10.1007/JHEP03(2021)023

[25] H. Nastase and J. Sonnenschein, "CSBIon – a charged soliton of the 3-dimensional CS + BI Abelian gauge theory," Phys. Rev. D **107** (2023) no.12, 125011 [arXiv:2210.06581 [hep-th]]. https://doi.org/10.1103/PhysRevD.107.125011

[26] M. Rho, "Probing Fractional Quantum Hall Sheets in Dense Baryonic Matter," [arXiv:2211.14890 [nucl-th]]

[27] Y. L. Ma and W. C. Yang, "Topology and Emergent Symmetries in Dense Compact Star Matter," Symmetry **15** (2023) no.3, 776 [arXiv:2301.02105 [nucl-th]]. https://doi.org/10.3390/sym15030776

[28] F. Bigazzi and F. Castellani, "Resonance contributions to nucleon spin structure in holographic QCD," JHEP **04** (2024), 037 [arXiv:2308.16833 [hep-ph]]. https://doi.org/10.1007/JHEP04(2024)037

[29] F. Lin and Y. L. Ma, "Baryons as vortexes on the η' domain wall," JHEP **05** (2024), 270 doi:10.1007/JHEP05(2024)270 [arXiv:2310.16438 [hep-th]]. https://doi.org/10.1007/JHEP05(2024)270

[30] M. Rho, "The Smile of Cheshire Cat At High Density," [arXiv:2408.00692 [nucl-th]]

EPJ Web of Conferences **314**, 00015 (2024) https://doi.org/10.1051/epjconf/202431400015
QCD@Work 2024

Hadronic contribution to the muon $g-2$: Theoretical challenges with the MUonE experiment

Fulvio Piccinini[1,*]

[1]INFN Sezione di Pavia, via A. Bassi 6, Pavia, Italy

Abstract. After briefly reviewing the present status on the theoretical predictions for the hadronic contribution to the muon anomalous magnetic moment, the MUonE approach is introduced, with a discussion on the challenges for the needed theoretical predictions.

1 Introduction

The muon magnetic moment anomaly $a_\mu = (g_\mu - 2)/2$ is a fundamental observable in particle physics, which allows to test the structure of the Standard Model (SM) as a local Quantum Field Theory with unprecedented accuracy [1]. Its present experimental value [2], known with a precision of 0.19 ppm, stems from the average of the final result of the E821 experiment at the Brookhaven National Laboratory [3] with the Run−1 and Run−2/3 measurements at the Fermilab Muon $g-2$ experiment [2, 4]. The SM prediction for a_μ, as compiled by the Muon $g-2$ Theory Initiative in 2020 [5], yields a discrepancy of the order of 5.0σ with the up-to-date experimental result, which would be an evidence of New Physics beyond the SM, provided all sources of systematic error are under control. The latter include also the uncertainty of the SM theoretical prediction for a_μ, which is dominated by non-perturbative contributions, such as the Leading Order Hadronic Vacuum Polarization (HVP,LO) contribution $a_\mu^{\rm HVP,LO}$ and the Hadronic Light-by-Light contribution $a_\mu^{\rm LxL}$, the latter being at present subleading with respect to the former [5]. Traditionally two independent methods are used to calculate $a_\mu^{\rm HVP,LO}$, the data-driven approach based on dispersive relations and ab-initio Lattice QCD simulations.

In the data-driven method, thanks to their analyticity and unitarity properties, loop integrals containing insertions of hadronic vacuum polarization in the photon propagators can be calculated as dispersion integrals over the cross section of a virtual photon decaying into hadrons [1, 5]. In particular, $a_\mu^{\rm HVP,LO}$ can be expressed as

$$a_\mu^{\rm HVP,LO} = \frac{\alpha^2}{3\pi^2}\int_{m_\pi^2}^{\infty} \frac{K(s)}{s} R(s)\, ds\,, \tag{1}$$

where $K(s)$ is a kernel function and $R(s)$ is the hadronic $R-$ratio defined as

$$R(s) = \frac{3s}{4\pi\alpha^2}\sigma^0(e^+e^- \to \text{hadrons}(+\gamma))\,. \tag{2}$$

*e-mail: fulvio.piccinini@pv.infn.it

The hadronic cross section $\sigma^0(s)$ in Eq. (2) should not include the effect of the vacuum polarization effect, to avoid double counting. In the integral of Eq. (1), in particular in the region below the charm and bottom resonances, where QCD perturbation theory becomes not reliable, the cross section entering $R(s)$ is taken from the measurements of the hadronic cross section at flavour factories, through a direct energy scan, where the collider center of mass (c.m.) energy is changed over an allowed energy window, or through the radiative return method, where the collider c.m. energy is fixed at the nominal value and the emission of an additional photon is exploited to explore a di-pion invariant mass window below the c.m. energy. The problem of calculating $a_\mu^{\rm HVP,LO}$ is shifted to the combination of several measurements corresponding to several hadronic final states in the integral of Eq. (1). Different approaches for the calculation of the time-like dispersion integral are available in the literature [6–8]. Even if the estimated precision of the predictions are at the few 0.1% level and in mutual agreement, the accuracy of this method is complicated by the existence of several resonances in the low-energy cross section. Furthermore, the CMD-3 collaboration has recently published a new measurement of the $e^+e^- \to \pi^+\pi^-$ cross section [9, 10], which disagrees with the results of previous experiments [11–13], pointing out some tension among different estimates of the pion form factor [1].

The method based on LQCD amounts to numerically calculate [1, 5]

$$a_\mu^{\rm HVP,LO} = \left(\frac{\alpha}{\pi}\right)^2 \int_0^\infty dQ^2\, f(Q^2) \left\{\Pi(Q^2) - \Pi(0)\right\} , \tag{3}$$

where the vacuum polarization function $\Pi(Q^2)$ is related to the Fourier transform of the electromagnetic current correlator

$$\Pi_{\mu,\nu}(Q) = \int d^4x\, e^{iQ\cdot x} \langle J_\mu(x) J_\nu(0)\rangle = (Q_\mu Q_\nu - \delta_{\mu\nu} Q^2)\Pi(Q^2) , \tag{4}$$

J_μ is the current in terms of quark fields, $J_\mu = \frac{2}{3}\bar{u}\gamma_\mu u - \frac{1}{3}\bar{d}\gamma_\mu d - \frac{1}{3}\bar{s}\gamma_\mu s + \cdots$, and $f(Q^2)$ is an appropriate kernel function. Different ways of performing the volume and momentum integration of the above Eqs. (3) and (4) with lattice regularization characterize different approaches to $a_\mu^{\rm HVP,LO}$ by various LQCD collaborations with corresponding results. Until recently the theoretical uncertainties associated to the LQCD approaches by different collaborations were estimated at the few % level, not competitive with the time-like dispersive approach. During 2020 the BMW collaboration published the first LQCD result for $a_\mu^{\rm HVP,LO}$ with estimated accuracy below the % level [14], with further improvement in [15], in strong tension with the data-driven determinations and in substantial agreement with the SM. All the present tensions among predictions based on different methods suggest possible inaccuracies in the SM prediction of a_μ, preventing a reliable comparison with the experimental value.

2 Calculating $a_\mu^{\rm HVP}$ with space-like data in the MUonE experiment

A novel and independent data-driven approach to evaluate $a_\mu^{\rm HVP}$ has been proposed in [16]. The method is based on the measurement of the hadronic contribution to the running of the QED coupling $\Delta\alpha(t)$ in the space-like region $t < 0$, where the LO HVP contribution is a smooth function [17] [2]:

$$a_\mu^{\rm HVP,LO} = \frac{\alpha}{\pi} \int_0^1 dx\,(1-x)\,\Delta\alpha_{\rm had}[t(x)] , \tag{5}$$

[1] We remark that the weight of the $e^+e^- \to \pi^+\pi^-$ process in the time-like dispersion integral is more than 70%.

[2] The kernel function $\frac{\alpha}{\pi}(1-x)$ refers to the LO HVP contribution. Recently, also the Next-to-Leading Order (NLO) [18, 19] and the Next-to-Next-to-Leading Order (NNLO) [18] expressions for the kernel function have been computed.

where $\Delta\alpha_{\rm had}(t)$ is the hadronic contribution to the running of the QED coupling, evaluated at the space-like squared four-momentum transfer:

$$t(x) = -\frac{x^2 m_\mu^2}{1-x} < 0\,. \tag{6}$$

The running QED coupling is expressed as:

$$\alpha(t) = \frac{\alpha(0)}{1-\Delta\alpha(t)}, \tag{7}$$

where $\alpha(0) = \alpha$ is the fine-structure constant, and

$$\Delta\alpha(t) = \Delta\alpha_{\rm lep}(t) + \Delta\alpha_{\rm had}(t) + \Delta\alpha_{\rm top}(t). \tag{8}$$

The hadronic contribution $\Delta\alpha_{\rm had}(t)$ can be extracted from data of a scattering experiment by subtracting from $\Delta\alpha(t)$ the purely leptonic part $\Delta\alpha_{\rm lep}(t)$, which can be calculated to very high precision in QED, and the top-quark contribution, which can be reliably calculated in perturbation theory.

Based on the method [16], the MUonE experiment has been proposed [20], with the aim of measuring $\Delta\alpha_{\rm had}(t)$ from μe scattering at low energy, through the CERN M2 muon beam with energy $E_\mu \sim 160$ GeV off atomic electrons of a light target. The integrand of Eq. (5) is shown as a function of x in the left panel of Fig. 1 while the comparison between hadronic and leptonic contribution is shown on the right panel. In principle, this method could reach a

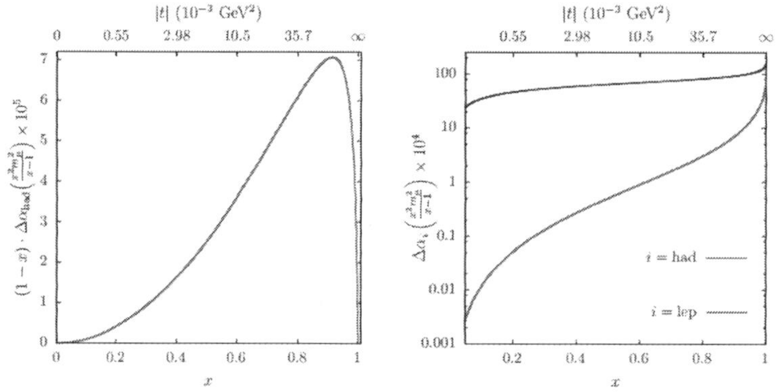

Figure 1. Left: the integrand $(1-x)\Delta\alpha_{\rm had}[t(x)] \times 10^5$ as a function of x and t. Right: $\Delta\alpha_{\rm had}[t(x)] \times 10^4$ (red) and, for comparison, $\Delta\alpha_{\rm lep}[t(x)] \times 10^4$ (blue), as a function of x and t (upper scale). From [20].

competitive precision below 0.5% on $a_\mu^{\rm HVP,LO}$, provided the systematic errors are kept under control. After the original proposal, a Letter-of-Intent [21] has been submitted to the CERN SPS Committee in 2019. A detailed description of the proposed apparatus can be found in [21, 22].

In a simplified approach at Leading Order (LO) accuracy, the MUonE experimental data on $\frac{d\sigma}{dt}$ can be related to $\Delta\alpha_{\rm had}(t)$ through the relation

$$R_{\rm had}^{LO}(t) = \frac{d\sigma^{LO}(\Delta\alpha_{\rm had}(t) \neq 0)}{d\sigma^{LO}(\Delta\alpha_{\rm had}(t) = 0)} \simeq 1 + 2\Delta\alpha_{\rm had}(t)\,, \tag{9}$$

where the numerator of the ratio is intended to be taken from data, while the denominator should rely on Monte Carlo simulations. In the range $0 < |t| < |t_{max}| = 0.175$ GeV^2, the hadronic contribution $\Delta\alpha_{\rm had}(t)$ has a tiny variation from 0 to about 10^{-3} at the peak of the integrand of Eq. (5). A competitive determination of $a_\mu^{\rm HVP,LO}$ requires a precision of $O(10^{-2})$ in the measurement of the hadronic running, which translates into an unprecedented precision of $O(10^{-5})$ in the shape of the differential cross section. Reaching this accuracy requires a huge statistics of data, in the order of few times 10^{12} events. More details on the MUonE data analysis strategies can be found in [21, 22]. Indipendent procedures to determine $a_\mu^{\rm HVP,LO}$ with MUonE data has been proposed in [23–25].

3 Theoretical calculations for elastic muon-electron scattering

The MUonE precision goal requires a theoretical computation of the differential cross section for $\mu^\pm e^- \to \mu^\pm e^-$ with an accuracy of 10 ppm on the shape of differential cross sections. As a consequence, QED corrections have to be included with the highest available precision. In fact, just moving from LO to NLO accuracy in QED, the exclusive description of the events receives important corrections. For instance, the correlation between muon and electron scattering angles of the elastic process is spoiled, as can be observed in Fig. 2. A comprehensive

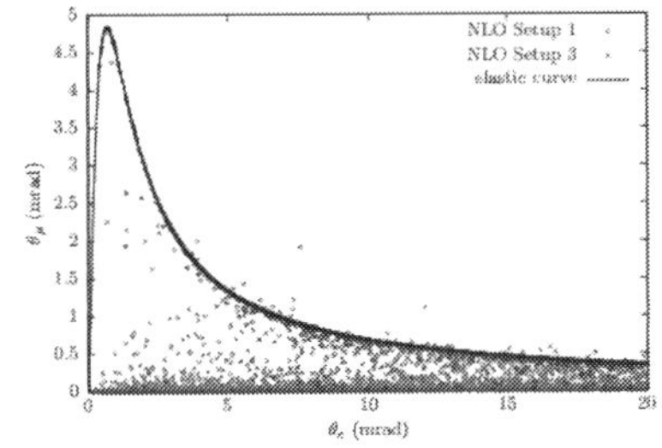

Figure 2. The correlation between the electron scattering angles θ_e and θ_μ for the process $\mu^+ e^- \to \mu^+ e^-$ at LO and NLO QED, for two different event selections. From [36].

analysis of the issues in the calculation of the radiative corrections to $\mu\, e$ scattering at MUonE has been presented in [26]. In order to match the extreme precision requirements of MUonE, the calculation must include QED radiative corrections at least at NNLO. These have been computed, with different approximations, by two groups with indipendent methods for dealing with the infrared singularities [27–33]. Also NNLO hadronic contributions have been computed in [34, 35]. The calculations have been implemented in two fully differential Monte Carlo (MC) generators, Mesmer [27, 28, 36] and McMule [29, 37], which are currently under development. The LO Z-exchange contribution is relevant at the 10^{-5} scale on distributions while the NLO weak corrections have been evaluated in [36] and found to be below the 10^{-6} scale. As a general rule, the NNLO corrections are of the order of 10^{-4}, so they are not enough in view of the final MUonE precision. In this respect, a consistent matching of resummed multi-photon effects (described by the Parton Shower and/or the YFS techniques) with fixed order calculations at NLO and at NNLO accuracy will be required. Another challenging approach is the calculation of the corrections at the next order in QED, i.e. with the so-called N3LO accuracy, along the electronic line. In fact, for a sufficiently inclusive event selection, the radiation from the electronic line is by far the dominant component at NLO and NNLO. Important results for the needed building blocks have been already achieved [38–41].

4 Theoretical calculations for backgrounds

In addition to the calculation of radiative corrections to the elastic process, the impact of the possible background processes must be precisely evaluated. In this regard, the MESMER event generator has been extended to include the simulation of $\mu^{\pm}e^{-} \to \mu^{\pm}e^{-}\ell^{+}\ell^{-}$ with $\ell = \{e, \mu\}$ [28] and $\mu^{\pm}e^{-} \to \mu^{\pm}e^{-}\pi^{0}$ [42]. While the latter process has been found to be completely negligible, the former processes can be considered as part of the NNLO corrections to the signal. The real emission of an additional $e^{+}e^{-}$ pair gives very large contributions, which can be mitigated by appropriate event selections, as shown in [28]. However, since the initial-state electrons are bound in a low-Z atomic target (beryllium-9 or carbon-12), the main source of experimental background is given by the muon-nucleus scattering with the emission of an addition leptonic pair. This process is particularly important because it can resemble an elastic event if one of the final leptons is not detected. Moreover, its cross section scales with the atomic number Z^2 while the muon-electron scattering scales with Z. Considering that the approximation contained in the GEANT4 toolkit [43] of muon scattering angle $\theta_{\mu} = 0$ is not acceptable for MUonE running conditions, a new LO exact calculation, in the approximation of infinite nuclear mass, has been carried out in [44] and implemented in the MC event generator MESMER. The results of the simulations show that, even considering appropriate event selections, this process is relevant on the 10^{-4} scale. In view of the final MUonE precision, NLO QED corrections to the process should be calculated, together with the contribution of the $e^{+}e^{-}$ pair production through the muon scattering with two nuclei.

5 MUonE and New Physics

Since MUonE aims at a measurement of the SM hadronic contribution to the running of the QED coupling, the analysis should be robust against possible New Physics contamination. In fact, given the extreme target precision, it could be possible that the present bounds on New Physics do not guarantee such a requirement. The issue has been studied using the Effective Field Theory approach for heavy new degrees of freedom and explicit models for hypothetical light weekly coupled mediators [45–47]. Considering all the available bounds, all the analysis conclude that any possible contamination from not yet excluded New Physics candidates is very unlikely to give contributions at the 10^{-5} scale to MUonE observables. Only in the case of a model of vector unparticles with broken scale invariance, the present constraints on the couplings with leptons could allow a contamination in the MUonE measurements.

In addition to the above studies, other authors investigated the MUonE potential in exploring light New Physics models through the $2 \to 3$ processes $\mu^{\pm}e \to \mu^{\pm}eX$ [49, 50] and $\mu^{\pm}N \to \mu^{\pm}NX$ [51], i.e. through the real excitation of the new degree of freedom, finding room for improvement on the existing bounds on couplings and masses. This is very interesting and it would not interfere with the measurement of $\Delta\alpha(t)$. However, in order to understand the real discovery potential, further developments on the detector configuration would be necessary.

6 Summary

In the present contribution, after a brief account of the present status on the theoretical predictions for the muon $g-2$, the space-like approach for the HVP calculation has been illustrated, with a discussion on the theoretical challenges and future developments required to successfully accomplish the MUonE physics program.

The author wishes to thank the QCD@Work Workshop Organizers for their kind invitation, support and warm hospitality. The pleasant and fruitful collaboration with the colleagues of the MUonE project is gratefully acknowledged. The work summarized in the present contribution is partially supported by the Italian Ministero dell'Università e Ricerca (MUR) and European Union - Next Generation EU through the research grant number 20225X52RA "MUS4GM2: Muon Scattering for $g-2$" under the program PRIN 2022.

References

[1] F. Jegerlehner, *The Anomalous Magnetic Moment of the Muon* (Springer Tracts Mod. Phys. **274**, Cham, Switzerland, 2017) pag. 1-693

[2] D. P. Aguillard, et al., Measurement of the Positive Muon Anomalous Magnetic Moment to 0.20 ppm, Phys. Rev. Lett. **131** (16) (2023) 161802. https://doi.org/10.1103/PhysRevLett.131.161802

[3] G. W. Bennett, et al., Final Report of the Muon E821 Anomalous Magnetic Moment Measurement at BNL, Phys. Rev. D 73 (2006) 072003. https://doi.org/10.1103/PhysRevD.73.072003

[4] B. Abi, et al., Measurement of the Positive Muon Anomalous Magnetic Moment to 0.46 ppm, Phys. Rev. Lett. **126** (14) (2021) 141801. https://doi.org/10.1103/PhysRevLett.126.141801

[5] T. Aoyama, et al., The anomalous magnetic moment of the muon in the Standard Model, Phys. Rept. **887** (2020) 1–166. https://doi.org/10.1016/j.physrep.2020.07.006

[6] A. Keshavarzi, D. Nomura, T. Teubner, $g-2$ of charged leptons, $\alpha(M_Z^2)$, and the hyperfine splitting of muonium, Phys. Rev. **D101** (1) (2020) 014029. http://arxiv.org/abs/1911.00367

[7] M. Davier, A. Hoecker, B. Malaescu, Z. Zhang, A new evaluation of the hadronic vacuum polarisation contributions to the muon anomalous magnetic moment and to $\boldsymbol{\alpha}(\mathbf{m_Z^2})$, Eur. Phys. J. **C80** (3) (2020) 241, [Erratum: Eur. Phys. J. **C80**, 410 (2020)]. http://arxiv.org/abs/1908.00921

[8] M. Benayoun, L. Delbuono, F. Jegerlehner, $BHLS_2$, a New Breaking of the HLS Model and its Phenomenology, Eur. Phys. J. **C80** (2) (2020) 81, [Erratum: Eur. Phys. J. **C80**, 244 (2020)]. http://arxiv.org/abs/1903.11034

[9] F. V. Ignatov, et al., Measurement of the $e^+e^- \to \pi^+\pi^-$ cross section from threshold to 1.2 GeV with the CMD-3 detector, Phys. Rev. **D109** (2024) 112002. https://doi.org/10.1103/PhysRevD.109.112002

[10] F. V. Ignatov, et al., Measurement of the pion formfactor with CMD-3 detector and its implication to the hadronic contribution to muon $(g-2)$, Phys. Rev. Lett. **132** (2024) 231903. https://doi.org/10.1103/PhysRevLett.132.231903

[11] J. P. Lees, et al., Precise Measurement of the $e^+e^- \to \pi^+\pi^-(\gamma)$ Cross Section with the Initial-State Radiation Method at BABAR, Phys. Rev. **D86** (2012) 032013. https://doi.org/10.1103/PhysRevD.86.032013

[12] M. Ablikim, et al., Measurement of the $e^+e^- \to \pi^+\pi^-$ cross section between 600 and 900 MeV using initial state radiation, Phys. Lett. **B753** (2016) 629–638, [Erratum: Phys. Lett. **B812**, 135982 (2021)]. https://doi.org/10.1016/j.physletb.2015.11.043

[13] A. Anastasi, et al., Combination of KLOE $\sigma(e^+e^- \to \pi^+\pi^-\gamma(\gamma))$ measurements and determination of $a_\mu^{\pi^+\pi^-}$ in the energy range $0.10 < s < 0.95$ GeV^2, JHEP **03** (2018) 173. https://doi.org/10.1007/JHEP03(2018)173

[14] S. Borsanyi, et al., Leading hadronic contribution to the muon magnetic moment from lattice QCD, Nature **593** (7857) (2021) 51–55. https://doi.org/10.1038/s41586-021-03418-1

[15] S. Boccaletti, et al., High precision calculation of the hadronic vacuum polarisation contribution to the muon anomaly, https://arxiv.org/pdf/2407.10913

[16] C. M. Carloni Calame, M. Passera, L. Trentadue, G. Venanzoni, A new approach to evaluate the leading hadronic corrections to the muon g-2, Phys. Lett. **B746** (2015) 325–329. https://doi.org/10.1016/j.physletb.2015.05.020

[17] B. Lautrup, A. Peterman, E. de Rafael, Recent developments in the comparison between theory and experiments in quantum electrodynamics, Phys. Rept. **3** (1972) 193–260. https://doi.org/10.1016/0370-1573(72)90011-7

[18] E. Balzani, et al., Hadronic vacuum polarization contributions to the muon $g-2$ in the space-like region, Phys. Lett. **B834** (2022) 137462. https://doi.org/10.1016/j.physletb.2022.137462

[19] A.V. Nesterenko, Timelike and spacelike kernel functions for the hadronic vacuum polarization contribution to the muon anomalous magnetic moment, J. Phys. G **49** (2022) 055001, https://doi.org/10.1088/1361-6471/ac5d0a; J. Phys. G **50** (2022) 029401, https://doi.org/10.1088/1361-6471/aca3c1

[20] G. Abbiendi, et al., Measuring the leading hadronic contribution to the muon $g-2$ via μe scattering, Eur. Phys. J. **C77** (3) (2017) 139. https://doi.org/10.1140/epjc/s10052-017-4633-z

[21] G. Abbiendi, et al., Letter of Intent: the MUonE project, Tech. Rep. CERN-SPSC-2019-026 (2019). https://cds.cern.ch/record/2677471

[22] G. Abbiendi, Status of the MUonE experiment, Phys. Scripta **97** (5) (2022) 054007. https://doi.org/10.1088/1402-4896/ac6297

[23] D. Greynat and E. de Rafael, Hadronic vacuum polarization and the MUonE proposal, JHEP **05** (2022) 084. https://doi.org/10.1007/JHEP05(2022)084

[24] F. Ignatov, R.N. Pilato, T. Teubner and G. Venanzoni, An alternative evaluation of the leading-order hadronic contribution to the muon $g-2$ with MUonE, Phys. Lett. **B848** (2024) 138344. https://doi.org/10.1016/j.physletb.2023.138344

[25] D. Boito, C.Y. London, P. Masjuan and C. Rojas, Model-independent extrapolation of MUonE data with Padé and D-Log approximants, Phys. Rev. **D110** (2024) 074012. https://doi.org/10.1103/PhysRevD.110.074012

[26] P. Banerjee, et al., Theory for muon-electron scattering @ 10 ppm: A report of the MUonE theory initiative, Eur. Phys. J. **C 80** (6) (2020) 591. https://doi.org/10.1140/epjc/s10052-020-8138-9

[27] C. M. Carloni Calame, M. Chiesa, S. M. Hasan, G. Montagna, O. Nicrosini, F. Piccinini, Towards muon-electron scattering at NNLO, JHEP **11** (2020) 028. https://doi.org/10.1007/JHEP11(2020)028

[28] E. Budassi, C. M. C. Calame, M. Chiesa, C. L. Del Pio, S. M. Hasan, G. Montagna, O. Nicrosini, F. Piccinini, NNLO virtual and real leptonic corrections to muon-electron scattering, JHEP **11** (2021). https://doi.org/10.1007/JHEP11(2021)098

[29] A. Broggio, et al., Muon-electron scattering at NNLO, JHEP **01** (2023) 112. https://doi.org/10.1007/JHEP01(2023)112

[30] P. Mastrolia, M. Passera, A. Primo, U. Schubert, Master integrals for the NNLO virtual corrections to μe scattering in QED: the planar graphs, JHEP **11** (2017) 198. https://doi.org/10.1007/JHEP11(2017)198

[31] S. Di Vita, S. Laporta, P. Mastrolia, A. Primo, U. Schubert, Master integrals for the NNLO virtual corrections to μe scattering in QED: the non-planar graphs, JHEP **09** (2018) 016. https://doi.org/10.1007/JHEP09(2018)016

[32] R. Bonciani, et al., Two-Loop Four-Fermion Scattering Amplitude in QED, Phys. Rev. Lett. **128** (2) (2022) 022002. https://doi.org/10.1103/PhysRevLett.128.022002

[33] T. Ahmed, et al., Two-loop vertices with vacuum polarization insertion, JHEP **01** (2024) 010. https://doi.org/10.1007/JHEP01(2024)010

[34] M. Fael, Hadronic corrections to μ-e scattering at NNLO with space-like data, JHEP **02** (2019) 027. https://doi.org/10.1007/JHEP02(2019)027

[35] M. Fael, M. Passera, Muon-Electron Scattering at Next-To-Next-To-Leading Order: The Hadronic Corrections, Phys. Rev. Lett. **122** (19) (2019) 192001. https://doi.org/10.1103/PhysRevLett.122.192001

[36] M. Alacevich, C. M. Carloni Calame, M. Chiesa, G. Montagna, O. Nicrosini, F. Piccinini, Muon-electron scattering at NLO, JHEP **02** (2019) 155. https://doi.org/10.1007/JHEP02(2019)155

[37] P. Banerjee, T. Engel, A. Signer, Y. Ulrich, QED at NNLO with McMule, SciPost Phys. **9** (2020) 027. https://doi.org/10.21468/SciPostPhys.9.2.027

[38] M. Fael, et al., Massive Vector Form Factors to Three Loops, Phys. Rev. Lett. **128** (17) (2022) 172003. https://doi.org/10.1103/PhysRevLett.128.172003

[39] S. Badger, J. Kryś, R. Moodie, S. Zoia, Lepton-pair scattering with an off-shell and an on-shell photon at two loops in massless QED, JHEP **11** (2023) 041. https://doi.org/10.1007/JHEP11(2023)041

[40] T. Engel, The LBK theorem to all orders, JHEP **07** (2023) 177. https://doi.org/10.1007/JHEP07(2023)177

[41] T. Engel, Multiple soft-photon emission at next-to-leading power to all orders JHEP **03** (2024) 004. https://doi.org/10.1007/JHEP03(2024)004

[42] E. Budassi, C. M. Carloni Calame, C. L. Del Pio, F. Piccinini, Single $\pi 0$ production in μe scattering at MUonE, Phys. Lett. **B829** (2022) 137138. https://doi.org/10.1016/j.physletb.2022.137138

[43] S. Agostinelli, et al., GEANT4–a simulation toolkit, Nucl. Instrum. Meth. **A506** (2003) 250–303. https://doi.org/10.1016/S0168-9002(03)01368-8

[44] G. Abbiendi, et al., Lepton pair production in muon-nucleus scattering, Phys. Lett. **B854** (2024) 138720. https://doi.org/10.1016/j.physletb.2024.138720

[45] A. Masiero, et al., New physics at the MUonE experiment at CERN, Phys. Rev. **D102** (7) (2020) 075013. https://doi.org/10.1103/PhysRevD.102.075013

[46] P. S. B. Dev, W. Rodejohann, X.-J. Xu, Y. Zhang, MUonE sensitivity to new physics explanations of the muon anomalous magnetic moment, JHEP **05** (2020) 053. https://doi.org/10.1007/JHEP05(2020)053

[47] O. Atkinson, M. Black, C. Englert, A. Lenz, A. Rusov, MUonE, muon $g-2$ and electroweak precision constraints within 2HDMs, Phys. Rev. **D106** (11) (2022) 115031. https://doi.org/10.1103/PhysRevD.106.115031

[48] D. N. Le, V. D. Le, D. T. Le, V. C. Le, Unparticle effects at the MUonE experiment, Eur. Phys. J. **C83** (11) (2023) 1037. https://doi.org/10.1140/epjc/s10052-023-12213-3

[49] K. Asai, K. Hamaguchi, N. Nagata, S.-Y. Tseng, J. Wada, Probing the Lμ-Lτ gauge boson at the MUonE experiment, Phys. Rev. **D106** (5) (2022) L051702. https://doi.org/10.1103/PhysRevD.106.L051702

[50] I. Galon, D. Shih, I. R. Wang, Dark photons and displaced vertices at the MUonE experiment, Phys. Rev. **D107** (9) (2023) 095003. https://doi.org/10.1103/PhysRevD.107.095003

[51] G. Grilli di Cortona, E. Nardi, Probing light mediators at the MUonE experiment, Phys. Rev. **D105** (11) (2022) L111701. https://doi.org/10.1103/PhysRevD.105.L111701

Effects of atomic electron momentum distribution on resonant dark sector production*

Fernando Arias-Aragón[1], *Luc* Darmè[2], *Giovanni* Grilli di Cortona[3,**], and *Enrico* Nardi[1,4]

[1]Istituto Nazionale di Fisica Nucleare, Laboratori Nazionali di Frascati, Frascati, 00044, Italy
[2]Université Claude Bernard Lyon 1, CNRS/IN2P3, Institut de Physique des 2 Infinis de Lyon, UMR 5822, F-69622, Villeurbanne, France
[3]Istituto Nazionale di Fisica Nucleare, Laboratori Nazionali del Gran Sasso, Assergi, 67100, L'Aquila (AQ), Italy
[4]Laboratory of High Energy and Computational Physic, HEPC-NICPB, Rävala 10, 10143, Tallin, Estonia

Abstract. Exploring resonant positron annihilation on atomic electrons offers a promising method for detecting light dark sector particles that interact with e^+e^- pairs. However, to accurately estimate production rates, a thorough understanding of atomic electron momentum distributions is essential. We introduce a comprehensive approach using the Compton profile of target materials to accurately incorporate electron velocity effects into resonant annihilation cross-sections. We show that taking into account the finite motion of atomic electrons not only can improve the search for new physics, but can also allow to enhance the accuracy of hadronic cross-section measurements.

1 Introduction

Compelling evidence for the existence of physics beyond the Standard Model is provided by established phenomena like neutrino masses, dark matter and the cosmological baryon asymmetry. New physics may manifest in a new sector with new particles and interactions. Intense positron beams impinging on fixed targets can be used to search for new light particles with feeble couplings to electrons and positrons. The power of this strategy relies on the fact that, if the condition for resonant e^+e^- annihilation into new states occurs, it would lead to a huge enhancement of the production rates [1, 2], a technique that is already being exploited by the experimental community [3, 4].

In experiments with thick fixed targets of large nuclear charge, a continuous scanning over the suitable centre-of-mass energy range can be achieved by keeping the beam energy fixed, as the positron will lose energy while passing through the target. On the other hand, in thin targets of low nuclear charge, the positron energy loss traversing the material is negligible and the beam energy must be tuned to span the desired range. In both cases, reliable estimates of resonant production rates and signal shapes can be obtained only after a careful characterization of the atomic electron momentum distribution.

*Presented by G. Grilli di Cortona at the QCD@Work: International Workshop on QCD, 18-21 June 2024, Trani, Italy

**e-mail: giovanni.grilli@lngs.infn.it

In this contribution, we discuss the impact of atomic electron momentum distribution on resonant dark sector production in fixed target experiments, presented in Ref. [5] (see also Ref [6]). Furthermore, we present a new method, discussed for the first time in Ref. [7], that can allow a measurement of the $e^+e^- \to$ hadrons cross-section leveraging the relativistic velocity of atomic electrons in targets with large atomic number Z.

2 Cross sections

In this section we discuss how to compute the cross section for the process $e^+e^- \to X$, where the particle X is resonantly produced, and for di-muon production. The latter process will allow us to obtain $\sigma_{\rm had} = \sigma(e^+e^- \to \text{hadrons}) = R(s)\,\sigma^{\rm th}_{\mu\mu}(s)$ via the experimentally measured R-ratio $R(s) = N_{\rm had}(s)/N_{\mu\mu}(s)$, where $N_{\mu\mu}$ denotes the number of di-muon events, while $N_{\rm had}$ is the number of hadronic events.

2.1 Resonant production

The differential cross section for positron annihilation off an atomic electron can be written as:

$$d\sigma_q = \frac{d^3p}{(2\pi)^3}\frac{d^3k_A}{(2\pi)^3}\frac{d^3k_B}{(2\pi)^3}\frac{(2\pi)^4\delta^{(4)}(\hat{k}_A+\hat{k}_B-\hat{p})}{2E_X 2E_{k_A} 2E_B\,|v_A - v_B|}\left|\phi_{A,q}(\boldsymbol{k}_A)\right|^2|\mathcal{M}_{\rm free}|^2\,|\phi_B(\boldsymbol{k}_B)|^2 \tag{1}$$

where q collectively labels the orbital quantum numbers, the subscripts A and B denote respectively electron and positron quantities, and $\phi(\boldsymbol{k})$, $\boldsymbol{k}$ and E denote their wave function, momentum and energy respectively. The term $\delta^{(4)}(\hat{k}_A+\hat{k}_B-\hat{p}) = \delta^{(3)}(\boldsymbol{k}_A+\boldsymbol{k}_B-\boldsymbol{p})\delta(E_A+E_B-E_X)$ enforces three-momentum and energy conservation, with $E_A = m_e$ the atomic electron energy neglecting its binding energy. Finally, p and E_X denote the momentum and energy of the final X particle, while $E_{k_A} = \sqrt{k_A^2+m_e^2}$. Taking the positrons in the beam as free particles with a well defined momentum $\boldsymbol{p}_B$, its wave function satisfies: $\int \frac{dk_B^3}{(2\pi)^3}|\phi_B(\boldsymbol{k}_B)|^2 = 1$, and $|\phi_B(\boldsymbol{k}_B)|^2 = (2\pi)^3\delta^{(3)}(\boldsymbol{p}_B - \boldsymbol{k}_B)$. On the other hand, atomic electrons are confined in space, implying a certain probability distribution associated with their momenta. The electron momentum density function $n(\boldsymbol{k}_A)$ is normalised to the atomic number Z of the target atoms as follows

$$n(\boldsymbol{k}_A) = \sum_q\left|\phi_q(\boldsymbol{k}_A)\right|^2,\quad \int\frac{d^3k_A}{(2\pi)^3}\,n(\boldsymbol{k}_A) = Z\,. \tag{2}$$

The electron momentum density distribution $n(\boldsymbol{k}_A)$ will be discussed in detail in Section 2.3. Neglecting the binding energy and assuming isotropy of the atomic electron momentum distribution, we can sum (1) over q, perform the integration over d^3p, and finally obtain

$$\frac{d\sigma}{dk_A} = \frac{|\mathcal{M}_{\rm free}|^2}{16\pi}\,\frac{k_A\,n(k_A)}{p_B|E_B k_A x_0(k_A) - E_{k_A}p_B|}. \tag{3}$$

Energy conservation implies that

$$x_0(k_A) = \frac{2E_AE_B + 2m_e^2 - m_X^2 - k_A^2}{2k_Ap_B},\qquad k_A^{\rm max,min} = \left|p_B \pm \sqrt{(E_A+E_B)^2 - m_X^2}\right|. \tag{4}$$

The energy distribution of positrons in the beam, described by a Gaussian $\mathcal{G}(E, E_B, \sigma_B)$ centered at E_B and with a standard deviation σ_B, must be taken into account while computing the cross-section

$$\sigma(E_B, \sigma_B) = \int dE\,\mathcal{G}(E, E_B, \sigma_B)\,\sigma(E), \tag{5}$$

where $\sigma(E)$ is the cross-section obtained after integrating Eq. (3).

2.2 Di-muon production

The differential cross-section for the annihilation of a positron with an atomic electron in a certain orbital n, ℓ with momentum-space wave-function $\phi_{n\ell}(\boldsymbol{k}_A)$ is

$$d\sigma = \frac{d^3p_1 d^3p_2 d^3k_A}{(2\pi)^5 16 E_1 E_2} \frac{|\phi_{nl}(\boldsymbol{k}_A)|^2\, |\mathcal{M}_{\rm free}|^2}{|E_B k_A^z - E_{k_A} p_B|} \delta^4(\hat{k}_A + \hat{p}_B - \hat{p}_f). \tag{6}$$

In (6), $\mathcal{M}_{\rm free}$ is the matrix element, and $\delta^{(4)}(\hat{k}_A + \hat{p}_B - \hat{p}_f) = \delta^{(3)}(\boldsymbol{k}_A + \boldsymbol{p}_B - \boldsymbol{p}_1 - \boldsymbol{p}_2)\delta(E_A + E_B - E_1 - E_2)$ denotes the three-momentum and energy conservation. We have neglected the electron binding energy in the energy conservation condition by setting $E_A = m_e$, justified by the separation of scales with the centre-of-mass (c.m.) energy in the process.

Performing all the integration, see Ref. [7] for more details, we finally obtain:

$$\frac{d^2\sigma}{ds\, dc_{\theta_2}} = \int_{m_\mu}^{\infty} dE_2 \int_0^{\infty} dk_A \frac{|\mathcal{M}_{\rm free}|^2}{32\pi^2} \frac{|\phi_{nl}(k_A)|^2}{16\pi^3} \frac{(2\pi - \arccos d)\,\Pi\left(\frac{d}{2}\right)}{p_B |E_B k_A c_{\theta_A} - E_{k_A} p_B|\, s_{\theta_2} s_{\theta_A} \sqrt{1-d^2}}, \tag{7}$$

with

$$\begin{aligned} d &= \frac{a - \mathcal{E}^2}{b}, \\ a &= p_2^2 + m_\mu^2 + k_A^2 + p_B^2 + 2k_A p_B c_{\theta_A} - 2p_2 c_{\theta_2}(p_B + k_A c_{\theta_A}), \qquad b = 2k_A p_2 s_{\theta_2} s_{\theta_A}, \\ \mathcal{E} &= E_A + E_B - E_2. \end{aligned} \tag{8}$$

In (7) we assumed an isotropic distribution for the electron momentum, and used spherical coordinates $d^3k_A = k_A^2 dk_A dc_{\theta_A} d\varphi_A$ and $d^3p_2 = p_2^2 dp_2\, dc_{\theta_2} d\varphi_2$, where $c_{\theta_A} = \cos\theta_A$ and $c_{\theta_2} = \cos\theta_2$. Moreover the Π function restricts the integration to values for which $|d| < 1$ and enforces energy conservation. Furthermore, $c_{\theta_A} = (2m_e^2 + 2E_A E_B - s - k_A^2)/(2p_B k_A)$.

2.3 Atomic electron momentum distribution

A fundamental aspect in the cross-section calculation is the electron momentum distribution $n(k_A)$. This quantity can be directly extracted from measurements of the Compton profile (CP) [8]. The electron momentum density from spherically averaged Compton profiles is defined as [9]

$$J(p) = \frac{1}{2}\int_{|p|}^{\infty} \rho(k) k dk, \tag{9}$$

with $\rho(k)$ the electron momentum distribution, and it is normalised as $\int_{-\infty}^{\infty} J(p)dp = Z$. In our notation, this implies that $\rho(k) = \frac{n(k)}{2\pi^2}$ and the electron momentum density distribution is

$$n(k) = -\frac{(2\pi)^2}{k}\frac{dJ(k)}{dk}. \tag{10}$$

In Fig. 1 we show the comparison between cross sections for the resonant process $e^+e^- \to X$ evaluated as a function of the beam energy, where X is a vector particle with mass m_X and interaction with electrons described by $\mathcal{L}_X \subset g_V X_\mu \bar{e}\gamma^\mu e$. We discuss the results for two targets: diamond, as it is used by the PADME Collaboration in the search for the X_{17} boson [10–12], and tungsten, as an example of a large-Z material. The results for the diamond target are given in the left panel of Fig. 1, where the blue curve is obtained under the assumption of atomic electrons at rest, so that the broadening of the curve is entirely due

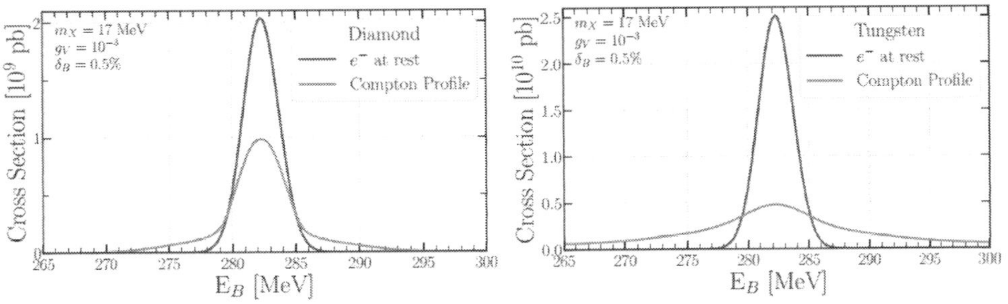

Figure 1. Resonant production cross section for a new vector boson with mass $m_X = 17$ MeV and coupling $g_V = 10^{-3}$, including the effects of a 0.5% beam energy spread, for a diamond (left) and a tungsten (right) target. The blue curve assumes electrons at rest, while the green one accounts for the atomic electron momentum distribution via the CP.

to the beam energy spread. The green curve, on the other hand, is obtained using the CP from Ref. [9]. This last curve is more spread due to the atomic electron motion. The particular shape of the green curve is due to the fact that the four atomic electrons belonging to the valence shells contribute to the central peak, while the two core electrons contribute to the broader tails.

The right panel shows the same result for tungsten, where the larger momentum of core electrons leads to a more significant energy spread. As a consequence, in both cases the smearing of the resonance has an important effect on searches for resonant peaks at fixed target experiments.

3 Atoms as electron accelerators

The motion of atomic electrons has a strong impact on the production of light new bosons via electron-positron resonant annihilation. In the following, we will focus on searches for light dark sector particles, in particular the X_{17} [10–15]. Finally, we propose to measure the cross-section for $e^+e^- \to$ hadrons at experiments where positron beams scatter on fixed targets [7], taking advantage of the atomic electron momentum distribution to scan the relevant energy range.

3.1 Impact on new physics searches

Resonantly produced vector bosons, a possible explanation of the ATOMKI anomaly [13–15], are affected by the target electron motion. Figure 2 shows the sensitivity of the PADME experiment to the X_{17} particle mass and coupling, assuming a 12 energy bin scan in the range $E_B = [265, 297]$ MeV, using $6 \cdot 10^{11}$ total positron on a 100 μm diamond target and with a beam energy spread of 0.5%.[1] The gray areas are excluded by several experiments [16–20]. The shaded violet regions show the 1σ and 2σ range for $m_X \sim 16.98 \pm 0.21$ MeV, obtained by combining the statistical uncertainties from the ATOMKI measurements [13–15] and adding in quadrature a systematic error of 0.20 MeV. The result obtained by assuming electrons at rest is depicted by the blue dashed line [12]. Finally, the projected sensitivity including the electron velocity effects via the CP is shown by the orange solid line. These last two curves were obtained by assuming $\sim 7.5 \times 10^4$ background events from Bhabha scattering and

[1]For the actual strategy employed by the PADME Collaboration, see [4].

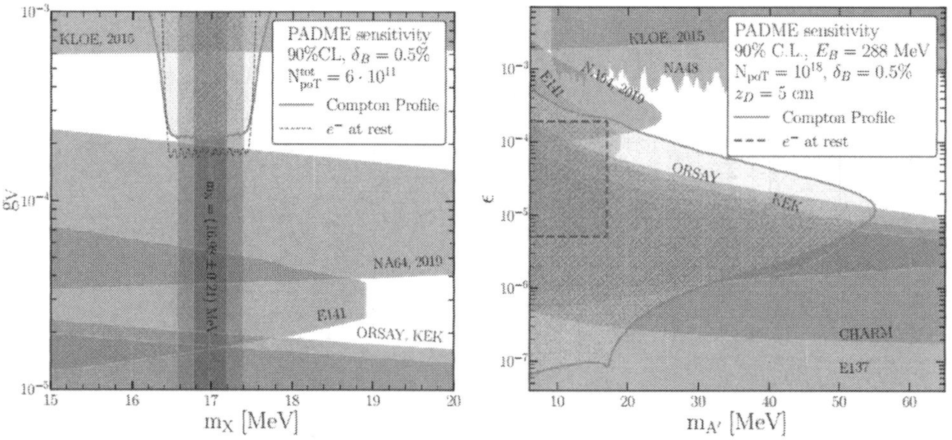

Figure 2. Projected sensitivity of the PADME run-III (left) on g_V as a function of the X_{17} mass and on the dark photon kinetic mixing ϵ as a function of the dark photon mass (right). Figures from [5].

neglecting the $\gamma\gamma$ background. In this analysis, systematic uncertainties have been neglected. As a result, correctly accounting for the atomic electron momentum distribution leads to a certain loss in sensitivity because the signal gets distributed over a broader range of energies, reducing the signal-to-noise ratio.

Taking advantage of the atomic electron motion gives also an opportunity to probe larger masses. In fact, the atomic electron motion can increase the centre of mass energy of the collision. The right panel of Figure 2 shows the exclusion limits for dark photon searches where the dark photon decays only to an electron-positron pair with coupling $\epsilon = g_V/e$. The gray regions represent the exclusion limits from several experiments [16, 17, 19–24]. The orange shaded area shows the 2.7 event limit (corresponding to 90% C.L.) that can be reached at an experiment with a positron beam with $E_B = 288$ MeV scattering on a 5 cm-thick tungsten target, assuming 10^{18} positrons on target and no background. The larger mass reach with respect to the blue region obtained assuming target electron at rest is due to the contribution of atomic electrons with large momenta.

3.2 A proposal for measuring the hadronic cross section

The large momenta of the core electron of high-Z materials can be used to precisely measure the hadronic cross section. A suitable positron beam is the one from the CEBAF facility at Jefferson Lab (JLab). The JLab injector could produce $1-5\,\mu$A unpolarized positron beams accelerated up to energies of 12 GeV, with a negligible energy spread. One year of data taking could produce up to 10^{21} positrons on target. This could be enough to measure a large number of di-muon events, as shown by the blue line of Figure 3 left, by scattering 10^{21} positrons of $E_B = 12$ GeV on a $500\,\mu$m uranium target.[2] This result is compared to the $\mu\mu\gamma$ events detected by KLOE [25]. The Figure shows that JLAB has the potential to collect more statistics than KLOE even at c.m. energies of order 1 GeV.

Furthermore, at CERN positron beams of energy between 100 and 200 GeV are available. It is, therefore, natural to ask how an increase in energy would change this measurement.

[2]We chose uranium as a target since it has the largest momentum spread in its core electrons. Other heavy targets, like tungsten, would give a weaker but comparable result.

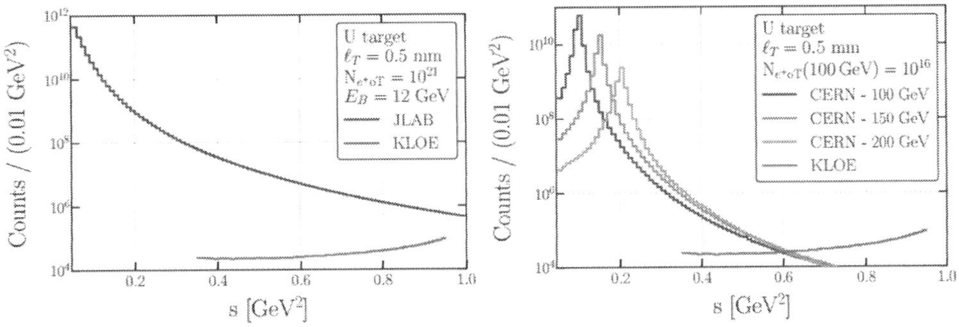

Figure 3. Left: number of $\mu\mu$ events produced at JLAB (blue) with a 12 GeV beam energy, compared to the di-muon events measured in KLOE [25]. Right: number of $\mu\mu$ events produced at CERN beam line (blue, green and orange) with a 100, 150 and 200 GeV beam energy, compared to the di-muon events measured in KLOE [25]. Figures from [7].

Those beams, however, may reach a statistics of $\sim 6.5 \cdot 10^{12}\ e^+$oT/yr for a 100 GeV beam. This is not sufficient to allow for a useful measurement, as shown in the right panel of Figure 3. We explore nevertheless the energy dependence of the measurement by assuming an (unrealistic) $N_{e^+\text{oT}} \sim 10^{16}$ for a beam energy $E_B = 100\,\text{GeV}$ and a $500\,\mu$m uranium target. We account for the scaling of the beam intensity at $E_B > 100\,\text{GeV}$ by using the following parametrization $N_{e^+\text{oT}}(p') = N_{e^+\text{oT}}(p)e^{-\frac{B}{p_0}(p'-p)}$, where p_0 is the energy of the primary protons (400 GeV), $B = 10$, and $N_{e^+\text{oT}}(100\,\text{GeV}) = 10^{16}$ [26]. These results are shown in the right panel of Fig. 3 for $E_B = 100, 150, 200$ GeV (blue, green and orange histogram, respectively). The red histogram shows the $\mu\mu\gamma$ events from KLOE. Above $s \sim 0.6$ GeV^2 an increase in the beam energy does not lead to an improvement of the statistics: the decrease in beam intensity offsets the advantage of increasing the beam energy. Therefore, considering the larger hadron contamination at higher energy, lower beam energies are preferable.

4 Conclusions

We showed the impact of atomic electron momentum distribution in new physics searches at fixed target experiments. In particular, we showed that the high momentum of core electron in large-Z materials can improve the mass reach in models of dark photons with couplings only to e^+e^- pairs. Finally, we proposed a novel strategy to measure the hadronic cross section. This could play a crucial role in solving the issues in the determinations of the hadron vacuum polarization contribution to the muon anomalous magnetic moment extracted from $e^+e^- \to$ hadrons data, and could rule out possible new physics explanations of the disagreement [27, 28].

G.G.d.C received partial financial support by the INFN "Iniziativa specifica" PADME@LNF. The work of E.N. was supported by the Estonian Research Council grant PRG1884.

References

[1] E. Nardi, Dark Photon production in positron beam dump experiments via resonant annihilation, in *New directions in Dark Matter and Neutrino Physics*, Perimeter Institute, July 20-22, 2017, (PI Video Library: http://pirsa.org/17070015) (2017)

[2] E. Nardi, C.D.R. Carvajal, A. Ghoshal, D. Meloni, M. Raggi, Resonant production of dark photons in positron beam dump experiments, Phys. Rev. D **97**, 095004 (2018), 1802.04756. 10.1103/PhysRevD.97.095004

[3] Y.M. Andreev et al., Improved exclusion limit for light dark matter from e+e- annihilation in NA64, Phys. Rev. D **104**, L091701 (2021), 2108.04195. 10.1103/PhysRevD.104.L091701

[4] PADME Collaboration, in preparation (2024)

[5] F. Arias-Aragón, L. Darmé, G. Grilli di Cortona, E. Nardi, Production of Dark Sector Particles via Resonant Positron Annihilation on Atomic Electrons, Phys. Rev. Lett. **132**, 261801 (2024), 2403.15387. 10.1103/PhysRevLett.132.261801

[6] R. Plestid, M.B. Wise, Atomic binding corrections for high energy fixed target experiments (2024), 2403.12184

[7] F. Arias-Aragón, L. Darmé, G. Grilli di Cortona, E. Nardi, Atoms as electron accelerators for measuring the $e^+e^- \to$ hadrons cross section (2024), 2407.15941

[8] M.J. Cooper, Compton scattering and electron momentum determination, Reports on Progress in Physics **48**, 415 (1985). 10.1088/0034-4885/48/4/001

[9] J.C. Aguiar, C.R. Quevedo, J.M. Gomez, H.O. Di Rocco, Theoretical Compton profile of diamond, boron nitride and carbon nitride, Physica B Condensed Matter **521**, 361 (2017). 10.1016/j.physb.2017.07.016

[10] M. Raggi, V. Kozhuharov, Proposal to Search for a Dark Photon in Positron on Target Collisions at DAΦNE Linac, Adv. High Energy Phys. **2014**, 959802 (2014), 1403.3041. 10.1155/2014/959802

[11] M. Raggi, V. Kozhuharov, P. Valente, The PADME experiment at LNF, EPJ Web Conf. **96**, 01025 (2015), 1501.01867. 10.1051/epjconf/20159601025

[12] L. Darmé, M. Mancini, E. Nardi, M. Raggi, Resonant search for the X17 boson at PADME, Phys. Rev. D **106**, 115036 (2022), 2209.09261. 10.1103/PhysRevD.106.115036

[13] A.J. Krasznahorkay et al., Observation of Anomalous Internal Pair Creation in Be8 : A Possible Indication of a Light, Neutral Boson, Phys. Rev. Lett. **116**, 042501 (2016), 1504.01527. 10.1103/PhysRevLett.116.042501

[14] A.J. Krasznahorkay et al., New results on the ^{8}Be anomaly, J. Phys. Conf. Ser. **1056**, 012028 (2018). 10.1088/1742-6596/1056/1/012028

[15] A.J. Krasznahorkay, M. Csatlós, L. Csige, J. Gulyás, A. Krasznahorkay, B.M. Nyakó, I. Rajta, J. Timár, I. Vajda, N.J. Sas, New anomaly observed in He4 supports the existence of the hypothetical X17 particle, Phys. Rev. C **104**, 044003 (2021), 2104.10075. 10.1103/PhysRevC.104.044003

[16] M. Davier, H. Nguyen Ngoc, An Unambiguous Search for a Light Higgs Boson, Phys. Lett. B **229**, 150 (1989). 10.1016/0370-2693(89)90174-3

[17] A. Konaka et al., Search for Neutral Particles in Electron Beam Dump Experiment, Phys. Rev. Lett. **57**, 659 (1986). 10.1103/PhysRevLett.57.659

[18] Y.M. Andreev et al. (NA64), Search for pseudoscalar bosons decaying into e^+e^- pairs in the NA64 experiment at the CERN SPS, Phys. Rev. D **104**, L111102 (2021), 2104.13342. 10.1103/PhysRevD.104.L111102

[19] E.M. Riordan et al., A Search for Short Lived Axions in an Electron Beam Dump Experiment, Phys. Rev. Lett. **59**, 755 (1987). 10.1103/PhysRevLett.59.755

[20] A. Anastasi et al., Limit on the production of a low-mass vector boson in $e^+e^- \to U\gamma$, $U \to e^+e^-$ with the KLOE experiment, Phys. Lett. B **750**, 633 (2015), 1509.00740. 10.1016/j.physletb.2015.10.003

[21] Y.D. Tsai, P. deNiverville, M.X. Liu, Dark Photon and Muon $g-2$ Inspired Inelastic Dark Matter Models at the High-Energy Intensity Frontier, Phys. Rev. Lett. **126**, 181801 (2021), 1908.07525. 10.1103/PhysRevLett.126.181801

[22] S. Andreas, C. Niebuhr, A. Ringwald, New Limits on Hidden Photons from Past Electron Beam Dumps, Phys. Rev. D **86**, 095019 (2012), 1209.6083. 10.1103/PhysRevD.86.095019

[23] L. Marsicano, M. Battaglieri, M. Bondi', C.D.R. Carvajal, A. Celentano, M. De Napoli, R. De Vita, E. Nardi, M. Raggi, P. Valente, Dark photon production through positron annihilation in beam-dump experiments, Phys. Rev. D **98**, 015031 (2018), 1802.03794. 10.1103/PhysRevD.98.015031

[24] E. Depero et al. (NA64), Hunting down the X17 boson at the CERN SPS, Eur. Phys. J. C **80**, 1159 (2020), 2009.02756. 10.1140/epjc/s10052-020-08725-x

[25] D. Babusci et al. (KLOE), Precision measurement of $\sigma(e^+e^- \to \pi^+\pi^-\gamma)/\sigma(e^+e^- \to \mu^+\mu^-\gamma)$ and determination of the $\pi^+\pi^-$ contribution to the muon anomaly with the KLOE detector, Phys. Lett. B **720**, 336 (2013), 1212.4524. 10.1016/j.physletb.2013.02.029

[26] L. Gatignon, Design and Tuning of Secondary Beamlines in the CERN North and East Areas (2020). 10.17181/CERN.T6FT.6UDG

[27] L. Darmé, G. Grilli di Cortona, E. Nardi, The muon $g-2$ anomaly confronts new physics in e and μ final states scattering, JHEP **06**, 122 (2022), 2112.09139. 10.1007/JHEP06(2022)122

[28] L. Darmé, G. Grilli di Cortona, E. Nardi, Indirect new physics effects on σhad confront the (g-2)μ window discrepancies and the CMD-3 result, Phys. Rev. D **108**, 095056 (2023), 2212.03877. 10.1103/PhysRevD.108.095056

EPJ Web of Conferences **314**, 00017 (2024) https://doi.org/10.1051/epjconf/202431400017
QCD@Work 2024

Electroweak phase transition in a vector dark matter scenario

Nico Benincasa[1,*], *Luigi* Delle Rose[2,3,**], *Luca* Panizzi[2,3,***], *Maimoona* Razzaq[2,3,****], and *Savio* Urzetta[2,†]

[1]School of Physics, University of Electronic Science and Technology of China, 611731 Chengdu, China
[2]Dipartimento di Fisica, Università della Calabria, Arcavacata di Rende, I-87036, Cosenza, Italy
[3]INFN, Gruppo Collegato di Cosenza, Arcavacata di Rende, I-87036, Cosenza, Italy

Abstract. This study explores the parameter space of a minimal extension of the Standard Model with a non-abelian $SU(2)$ group, in which the gauge bosons are stable and acquire mass through a mechanism of spontaneous symmetry breaking involving a new scalar doublet which interacts with the Higgs boson through a quartic coupling. The exploration aims to assess whether it is possible to obtain a first-order phase transition while ensuring that the gauge bosons are viable dark matter candidates. Theoretical, astrophysical and collider bounds are considered. The results are then tested against the sensitivity of future experiments for the detection of gravitational wave signals.

1 Introduction

Cosmological and astrophysical observations strongly indicate that baryonic matter is not the primary component of matter in the Universe [1]. Non-baryonic matter manifests mainly through gravitational effects, since is electrically and colour neutral, and is therefore referred to as dark matter (DM). Many theories have been put forward to extend the Standard Model (SM) with new DM particles. If a DM state interacts with SM fields (weakly, as it has so far escaped detection), it is in principle possible to pose bounds on its mass and on the size of the interactions. DM models can be constrained via direct and indirect detection experiments, and at particle accelerators such as the Large Hadron Collider. The relative importance of such constraints depends on the hypotheses about the interactions of DM with SM particles. Models where the DM interacts exclusively through the Higgs boson are in general less constrained due to the uncertainties in the measurement of Higgs couplings. Since recently, gravitational waves (GWs) offer a novel and complementary approach to exploring these models, especially those that involve additional scalar fields capable of triggering first-order phase transitions, thus leading to GW emissions. The main focus of our work is to explore one of such models, where the DM is represented by the gauge bosons of a new non-abelian gauge group, which acquire mass through a mechanism of spontaneous symmetry breaking in the dark sector driven by a new scalar field.

*e-mail: nico.alexis.benincasa@ut.ee
**e-mail: luigi.dellerose@unical.it
***e-mail: luca.panizzi@unical.it
****poster author, e-mail: maimoona.razzaq@unical.it
†e-mail: sav.urze98@gmail.com

2 The model

We explore an extension of the SM that includes a new $SU(2)_D$ gauge symmetry, under which all SM particles are singlets. In this framework, we introduce a scalar doublet that breaks the $SU(2)_D$ symmetry in the dark sector through a Higgs mechanism. The custodial $SO(3)$ symmetry left in the dark sector after spontaneous symmetry breaking ensures that three gauge bosons remain stable and mass-degenerate, with a common mass of $m_{V_D} = g_D v_D/2$, where g_D is the $SU(2)_D$ gauge coupling and v_D is the vacuum expectation value of the dark scalar doublet. Because the gauge bosons are triplet under $SO(3)$ and all other particles are singlets, this custodial symmetry prohibits the decay of the these gauge bosons [2]. The Lagrangian of the model is given by

$$\mathcal{L} = \mathcal{L}_{SM} - \frac{1}{4} F_{D\mu\nu} F_D^{\mu\nu} + (D_\mu \Phi_D)^\dagger (D^\mu \Phi_D) + \mu_D^2 (\Phi_D^\dagger \Phi_D) - \lambda_D (\Phi_D^\dagger \Phi_D)^2 - \lambda_{HD} (\Phi_D^\dagger \Phi_D)(\Phi^\dagger \Phi), \quad (1)$$

where $\mathcal{L}_{SM}$ is the SM Lagrangian, $F_D^{\mu\nu}$ is the field strength associated to $SU(2)_D$, Φ_D is the dark scalar doublet, and Φ is the SM Higgs doublet (see also [3–5]). After the Electroweak and dark symmetry breakings, we can write the scalar doublets in the unitary gauge as

$$\Phi = \frac{1}{\sqrt{2}} \begin{pmatrix} 0 \\ v + h_1 \end{pmatrix}, \quad \Phi_D = \frac{1}{\sqrt{2}} \begin{pmatrix} 0 \\ v_D + h_2 \end{pmatrix},$$

where v and v_D are the vacuum expectation values of the Electroweak and SU(2)$_D$ symmetries, respectively, and h_1, h_2 are real scalar fields in the interaction eigenbasis. The portal coupling λ_{HD} between the SM and the dark sector induces a mixing between the two scalar fields h_1 and h_2. The free parameters of the model are four, and they can be chosen to be the gauge coupling g_D, the gauge boson mass m_{V_D}, the mass of the new scalar m_{H_D} and the mixing angle θ_D between the two real scalars in the mass eigenbasis, h and H_D.

3 Constraints

The parameter space of the model is subject to several constraints, ensuring theoretical consistency and agreement with experimental data:

- **Theoretical Constraints:** Perturbative unitarity and boundedness from below are applied to ensure the theoretical consistency of the model.
- **Collider Constraints:** Bounds are imposed based on direct searches for new scalar particles and on the compatibility with the measured Higgs couplings at the LHC [6].
- **Electroweak Precision Tests:** Electroweak precision observables S, T and U [7, 8] are used to parameterize the effects of new physics on Electroweak observables.
- **Dark Matter Relic Density:** The relic density of DM, as measured by Planck [9], imposes stringent constraints on the parameters. The DM relic density, obtained through the freeze-out mechanism, can be approximated by:

$$\Omega_{DM} \approx 0.1 \frac{x_f}{\sqrt{g_*(m_{V_D})}} \frac{10^{-8}\,\text{GeV}^{-2}}{\langle \sigma v \rangle} \approx 0.12$$

  where x_f is the freeze-out parameter, $g_*(m_{V_D})$ represents the effective number of relativistic degrees of freedom at temperature $T = m_{V_D}$ and $\langle \sigma v \rangle$ is the thermally averaged annihilation cross-section. There are several annihilation channels at tree-level that influence the relic density of vector DM during freeze-out, shown in fig. 1. Notably, the annihilation channel on the right exhibits a unique characteristic, as it involves one DM candidate in the final state, a process that is not possible in models based on Z_2 symmetry [2].

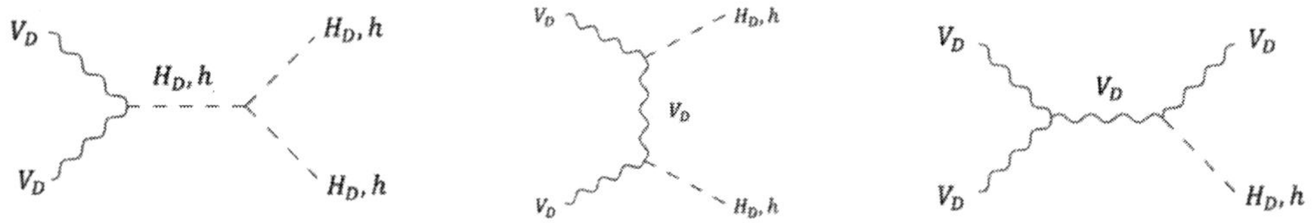

Figure 1. Topologies describing DM annihilation processes.

- **Direct and Indirect Detection:** Constraints from direct detection experiments (e.g., XENON1T, LUX-Zepelin) [10–12] and indirect detection experiments (e.g., HESS, FERMI LAT) [13, 14] are applied to further narrow down the parameter space.

4 Gravitational waves from first-order phase transitions

In a first-order phase transition, the system tunnels from a metastable to the ground state through bubble nucleation. In the early universe the bubbles expand, collide and eventually convert energy into a background of GWs. The power spectrum of these can be entirely characterized by the following phase transition parameters, which can be, in turn, computed from the Lagrangian of the model:

- α: the strength of the phase transition, defined as

$$\alpha \equiv \frac{\Delta\epsilon}{\rho_{\text{rad}}}\Big|_{T=T_n}, \quad \Delta\epsilon \equiv \epsilon\Big|_{\text{false vacuum}} - \epsilon\Big|_{\text{true vacuum}} \tag{2}$$

  where $\Delta\epsilon$ is the vacuum energy released during the phase transitions and $\rho_{\text{rad}} = \frac{\pi^2}{30} g_* T^4$ [15] and T_n is the nucleation temperature.

- β/H: the inverse of the transition time duration defined as follows [16]:

$$\frac{\beta}{H} = T \frac{d(S/T)}{dT}\Big|_{T=T_n} \tag{3}$$

  where S represents the 3-dimensional Euclidean action. A smaller value of β/H typically corresponds to a stronger phase transition, which in turn leads to a more pronounced GW signal.

- T_p: the percolation temperature is the temperature at which bubbles of true vacuum expand enough and merge with one another to form a continuous phase across space such that the false vacuum fraction is ~ 0.71. In absence of a significant amount of supercooling, as in our scenario, $T_p \simeq T_n$, where T_n is defined as the temperature at which $\Gamma/H^4 \simeq 1$, with Γ the nucleation rate.

- v_w: the bubble wall velocity refers to the speed at which the true vacuum bubble expands through the surrounding plasma, see [17–19] for a recent discussion.

5 Discussion

The parameter space of the model, sifted with both the theoretical and experimental constraints discussed in in section 3, and projected onto the plane (m_{V_D}, m_{H_D}) is shown in fig. 2.

The colour code emphasises the interplay between the predicted relic density and the phase transition strength, here quantified by the ratio v_n/T_n, with $v_n^2 = (v^2 + v_D^2)|_{T=T_n}$ evaluated at the nucleation temperature. Specifically, the gray points correspond to regions where the relic density is underabundant within 5σ of the measured Planck value, *i.e.* $\Omega h^2 \leq 0.125$, and all observational constraints (astrophysical and collider) are satisfied. Among such points, those with blue colour indicate regions where the relic density matches the observed value within 3σ, i.e., $\Omega h^2 = 0.120 \pm 0.003$ [9], while cyan points mark regions where a strong phase transition occurs, with $v_n/T_n \geq 1$. The red stars highlight points where the relic density is within the observed range and a strong phase transition is present. In this figure, the presence of two

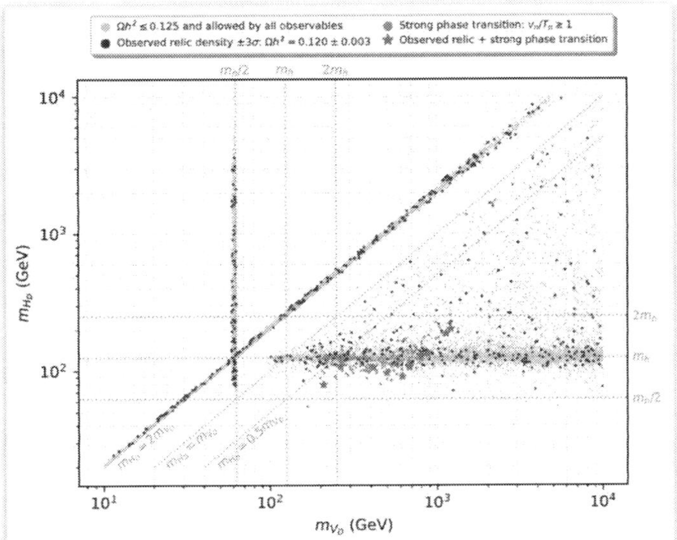

Figure 2. Points in the (m_V, m_{H_D}) projection of the parameter space which satisfy experimental and theoretical bounds. The gray points corresponds to underabundant scenarios with relic density within 5σ from the Planck measurement. The blue points reproduce the correct relic density within 3σ. Points in cyan corresponds to a strong first-order phase transition. Red stars correspond to points with a strong phase transition and the correct relic density.

resonant annihilation processes is clearly evident and manifests in the clustering of points along the vertical and oblique directions. When $m_{H_D} = 2m_{V_D}$, annihilation proceeds through the H_D resonance, while for $m_h = 2m_{V_D}$, it proceeds through the h resonance. The lower triangular region, below $m_{V_D} = m_{H_D}$ and above $m_{H_D} = m_h/2$, is a non-resonant region. The region with a higher density of points, where $m_{H_D} \approx m_h$ and $m_{V_D} > m_h$, is less constrained by collider observables.

To further explore the connection between DM and GWs, we analyze the GW signal produced during phase transitions. The stochastic background of GWs depends on three main mechanisms: the propagation of sound waves, bubble collisions and turbulence effects in the plasma. Therefore, the GW signal is expressed as [23]

$$h^2\Omega_{\rm GW} \simeq h^2\Omega_{\rm col} + h^2\Omega_{\rm sw} + h^2\Omega_{\rm turb}. \tag{4}$$

In fig. 3, the GW signal $h^2\Omega^{\rm peak}$ is plotted as a function of the peak frequency $f^{\rm peak}$. The color coding in the figure corresponds to the strength of the phase transition and the relic density of

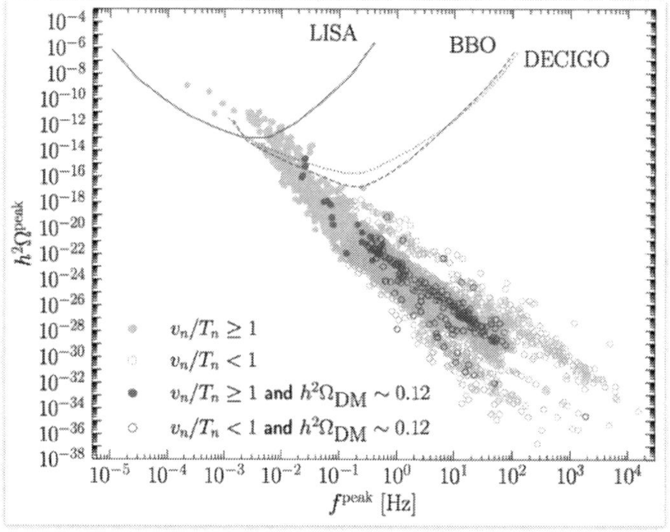

Figure 3. GW signal $h^2\Omega^{\rm peak}$ as a function of peak frequency $f^{\rm peak}$, with color coding representing the strength of the phase transition and dark matter relic density. Sensitivity curves of future GW detectors are included [20–22].

DM. Additionally, the sensitivity curves of some future GW detectors are included to show the expected detection capabilities for these signals.

6 Conclusion

We have studied a spin-one DM model based on a $SU(2)_D$ gauge group in a dark sector in which a custodial symmetry ensures the DM stability. The dark gauge bosons acquire masses through the $SU(2)_D$ symmetry breaking triggered by a dark scalar doublet. This symmetry breaking resulted in a phase transition in the dark sector. We analyzed the nature of the phase transition and characterised its property. If a first-order phase transition was found, we examined the corresponding GW signal and assessed its detectability by future GW observatories. While the forthcoming LISA experiment [20] could detect GW signals associated with sufficiently strong first-order phase transitions in this model, a much larger sensitivity reach is necessary to probe the region of the parameter space in which the DM candidate could comply with all current constraints while reproducing, at the same time, the correct relic abundance.

References

[1] G. Bertone, D. Hooper, J. Silk, Physics reports **405**, 279 (2005)
[2] T. Hambye, JHEP **01**, 028 (2009), `0811.0172`
[3] A. Belyaev, A. Deandrea, S. Moretti, L. Panizzi, D.A. Ross, N. Thongyoi (2022), `2204.03510`
[4] N. Benincasa, L. Delle Rose, L. Panizzi, M. Razzaq, S. Urzetta (2024), `in preparation`
[5] N. Baouche, A. Ahriche, G. Faisel, S. Nasri, Physical Review D **104**, 075022 (2021)

[6] H. Bahl, T. Biekötter, S. Heinemeyer, C. Li, S. Paasch, G. Weiglein, J. Wittbrodt, Computer Physics Communications **291**, 108803 (2023)

[7] M.E. Peskin, T. Takeuchi, Phys. Rev. D **46**, 381 (1992)

[8] M.E. Peskin, T. Takeuchi, Phys. Rev. Lett. **65**, 964 (1990)

[9] N. Aghanim, Y. Akrami, M. Ashdown, J. Aumont, C. Baccigalupi, M. Ballardini, A.J. Banday, R. Barreiro, N. Bartolo, S. Basak et al., Astronomy & Astrophysics **641**, A6 (2020)

[10] E. Aprile, J. Aalbers, F. Agostini, M. Alfonsi, F. Amaro, M. Anthony, F. Arneodo, P. Barrow, L. Baudis, B. Bauermeister et al., Physical review letters **119**, 181301 (2017)

[11] E. Aprile, J. Aalbers, F. Agostini, M. Alfonsi, L. Althueser, F. Amaro, M. Anthony, F. Arneodo, L. Baudis, B. Bauermeister et al., Physical review letters **121**, 111302 (2018)

[12] D. Akerib, C. Akerlof, D.Y. Akimov, A. Alquahtani, S. Alsum, T. Anderson, N. Angelides, H. Araújo, A. Arbuckle, J. Armstrong et al., Nuclear Instruments and Methods in Physics Research Section A: Accelerators, Spectrometers, Detectors and Associated Equipment **953**, 163047 (2020)

[13] M. Ackermann et al. (Fermi-LAT), Phys. Rev. Lett. **115**, 231301 (2015), 1503.02641

[14] H. Abdalla, F. Aharonian, F.A. Benkhali, E. Angüner, C. Armand, H. Ashkar, M. Backes, V. Baghmanyan, V.B. Martins, R. Batzofin et al., Physical Review Letters **129**, 111101 (2022)

[15] J.R. Espinosa, T. Konstandin, J.M. No, G. Servant, JCAP **06**, 028 (2010), 1004.4187

[16] C. Grojean, G. Servant, Phys. Rev. D **75**, 043507 (2007), hep-ph/0607107

[17] S. De Curtis, L.D. Rose, A. Guiggiani, A.G. Muyor, G. Panico, JHEP **03**, 163 (2022), 2201.08220

[18] S. De Curtis, L. Delle Rose, A. Guiggiani, A. Gil Muyor, G. Panico, JHEP **05**, 194 (2023), 2303.05846

[19] S. De Curtis, L. Delle Rose, A. Guiggiani, A. Gil Muyor, G. Panico, JHEP **05**, 009 (2024), 2401.13522

[20] P. Amaro-Seoane et al. (LISA) (2017), 1702.00786

[21] V. Corbin, N.J. Cornish, Class. Quant. Grav. **23**, 2435 (2006), gr-qc/0512039

[22] N. Seto, S. Kawamura, T. Nakamura, Phys. Rev. Lett. **87**, 221103 (2001), astro-ph/0108011

[23] C. Caprini, M. Hindmarsh, S. Huber, T. Konstandin, J. Kozaczuk, G. Nardini, J.M. No, A. Petiteau, P. Schwaller, G. Servant et al., Journal of cosmology and astroparticle physics **2016**, 001 (2016)

EPJ Web of Conferences **314**, 00018 (2024) https://doi.org/10.1051/epjconf/202431400018
QCD@Work 2024

A holographic approach to heavy non $q\bar{q}$-states using configurational entropy

Miguel Angel Martin Contreras[1,*] and *Alfredo* Vega[2,**]

[1]School of Nuclear Science and Technology, University of South China, Hengyang, China. No 28, West Changsheng Road, Hengyang City, Hunan Province, China
[2]Instituto de Física y Astronomía, Universidad de Valparaíso, A. Gran Bretaña 1111, Valparaíso, Chile

Abstract. This work uses the connection between hadronic stability and configurational entropy to explore hadronic structures written in terms of non-quadratic dilaton $(\kappa\, z)^{2-\alpha}$. These hadronic structures are described using the relation between the parameters κ and α with the constituent mass. We test Z_c, Z_b, and ψ as non-$q\bar{q}$ states defined as hybrid meson, hadroquarkonium, hadronic molecule, or diquark-antidiquark pair. We find that, holographically speaking, Z_c and Z_B are better described as a hybrid meson and ψ as hadrocharmonium.

1 Introduction

One of the most famous and successful bottom-up models is the so-called *softwall* (SWM) [1]. In this model, confinement is induced by slightly breaking the AdS conformal symmetry by adding a dilaton field $\Phi(z) = (\kappa\, z)^2$ that sets the masses of hadrons living at the boundary. This set of masses is organized in linear Regge trajectories. However, the quadratic nature of the SWM dilaton is tied up to the constituent mass inside hadrons. We will use this idea to describe non-$q\bar{q}$ states based on the constituents and their structures inside hadrons.

Non-$q\bar{q}$ states or exotic hadrons are all the hadronic states with quantum numbers not allowed by the usual constituent quark model. They have been observed in hadronic facilities since the early 2000s. However, much discussion remains about their inner structure and constituent (or cluster) interaction. On the holographic description, we have works written in the context of dilaton-based models [2–4].

2 Hadrons in AdS/QCD

In AdS/CFT correspondence, hadrons are non-perturbative boundary operators dual to bulk fields. For a given hadronic boundary operator defined as $O = f(q, \bar{q}, G_{\mu\nu}, D_\mu)$, the scaling dimension $\dim O = \Delta + L + \gamma$ establishes the hadronic identity in the bulk AdS through the bulk mass M_5 [5], where Δ is the energy dimension of the pure constituent operator, L is the angular momentum and γ is an anomalous dimension. This relation is a consequence of the so-called field/operator duality, which tells you how the bulk fields match the boundary information [6]. For massive bulk p-forms in AdS_5 written as $A_p(z, q) = A_p(q)\, \psi(z, q)$, this duality ensures that $\psi(z, q)|_{z\to 0} = C\, z^{\Delta - p}$.

*e-mail: miguelangel.martin@usc.edu.cn
**e-mail: alfredo.vega@uv.cl

Notice that in the most general case for the bulk mass, we have $M_5 = M_5(\Delta, L, \gamma)$. This fact implies that *a hadron in holographic models is a bag of constituents characterized by the bulk mass* M_5. However, this picture has no information about the inner hadronic structure.

Along with this definition, there is the emergence of confinement in these models. In pure AdS space, bulk modes are not normalizable because of field/operator duality: localized normalizable states at the boundary become diffuse (non-localized) and non-normalizable. However, we can avoid this problem by introducing energy scales that fix hadronic masses. One of the possible scenarios to do this is the so-called *bottom-up perspective.*

3 Bottom-up approach in the static dilaton context

The emergence of bounded states of hadrons is a consequence of color confinement. This observation is realized when we break the conformal symmetry slightly in the bulk by introducing an energy scale that will fix the hadron mass [1, 7]. This is the idea behind the so-called bottom-up models. From a wider perspective, *bottom-up holography describes boundary physics with adequate bulk fields inspired by boundary phenomenology.*

Concerning the hadronic spectrum, we will consider the *softwall model prescription*, where confinement is achieved by placing a dilaton field $\Phi(z)$ in the AdS bulk. In pure AdS, free modes are not normalizable (in the sense of the Breitenlohner-Freedman bound [8]) and non-localizable. By placing a dilaton with an energy scale, the states become normalizable, i.e., the spectrum is now bounded.

For the sake of clarity, let us focus on bulk vector fields dual to vector non-$q\bar{q}$. Keep in mind that these ideas can be extended to higher $p-$forms. For this bulk vector field, minimally coupled to a static dilaton $\Phi(z)$, we will define the following action density:

$$I_{\text{hadrons}} = -\frac{1}{2}\int d^5x\,\sqrt{-g}\,e^{-\Phi(z)}\left[\frac{1}{2\,g_5^2}F_{mn}\,F^{mn} + M_5^2\,A_m\,A^m\right], \tag{1}$$

where g_5 is a coupling constant that fixes action units. It is defined from the OPE of the large N QCD two-point function [9], and $M_5^2\,R^2 = (\Delta + L - 1)(\Delta + L - 3)$. Notice that we are considering a massive vector field since for multiquark states, the bulk mass is different from zero. Also, the presence of the bulk mass does not affect the boundary gauge invariance since we can always choose the gauge $A_z = 0$ that preserves the boundary structure [10]. This action will be living in the Poincare Patch defined as:

$$dS^2 = \frac{R^2}{z^2}\left[dz^2 + \eta_{\mu\nu}\,dx^\mu\,dx^\nu\right], \tag{2}$$

with R the AdS radius, z the holographic coordinate, and $\eta_{\mu\nu}$ the Minkowski metric with negative signature.

Once we set the framework, we perform the Fourier transformation of the bulk field A_m and split it into the boundary source and the bulk mode: $A_m(z, q) = \psi(z, q)\,A_m(q)$. Then, we impose the static gauge $A_z = 0$. These definitions allow us to write down the equations of motion for the bulk mode in the Sturm-Liouville form, whose equation is given by:

$$\partial_z\left[e^{-B(z)}\,\partial_z\,\psi(z, q)\right] + (-q^2)\,e^{-B(z)}\,\psi(z, q) - \frac{M_5^2\,R^2}{z^2}\,e^{-B(z)}\,\psi(z, q) = 0. \tag{3}$$

The function B is written regarding the warp factor and the dilaton. In the most general case, it is written as $B(z) = \Phi(z) + \beta\,\log(R/z)$, where $\beta = -(3 - 2\,p)$ represents the spin of the hadron. For vector hadrons ($p = 1$), we have $\beta = -1$. This function B is useful to define the effective holographic potential. To do so, we perform a *Bogoliubov transformation* $\psi(z, q) =$

$e^{\Phi(z)/2}\phi(z,q)$ which allows us to write the bulk equations of motion into a Schrödinger-like one for the field $\phi(z,q)$:

$$-\phi''(z,q) + V(z)\,\phi(z) = (-q^2)\,\phi(z,q), \tag{4}$$

with the following definition:

$$V(z) = \frac{B'(z)^2}{4} - \frac{B''(z)}{2} + \frac{M_5^2\,R^2}{z^2}. \tag{5}$$

If we impose the on-shell mass condition on the equations above, i.e. $-q^2 = M_n^2$, the Regge trajectories emerge as the holographic potential $V(z)$ eigenvalues. We will consider the static dilaton exposed in [11], where the authors proposed a new dilaton to address the isovector mesons spectra. This dilaton is defined as a deformation of the standard softwall model:

$$\Phi(z) = (\kappa\, z)^{2-\alpha}\,, \tag{6}$$

where κ is an energy scale that fixes the hadron masses, and α is a deformation that accounts for quark content. Both parameters are flavor-dependent. However, if we have a fit for different states ranging from light to heavy quarks, we can define a set of calibration curves. These curves allow us to infer, from phenomenological bounds, the contribution of the structure to the holographic model. For further details, see [4, 11]. The next section will apply this machinery to the non-$q\bar{q}$ states.

4 Non-$q\bar{q}$ states

The relation between bulk mass and hadronic identity is not univocally determined in this context of hadrons considered as a bag of constituents. For example, for 1-form bulk fields, dual to vector hadrons on the boundary, we observe that $\Delta = 6$ could be associated with multiquark states with four constituent quarks or hybrid structures with three constituent gluons. The very same picture emerges for other bulk fields as scalar or fermionic.

A possible form to circumvent this issue was exposed in [4, 11] in the context of non-linear Regge trajectories and models based on dilaton fields. The parameter set (κ, α) for the isovector meson family (ρ, ω, ϕ,j/ψ, Υ) establishes a running with a constituent mass *threshold* $\bar{m}$:

$$\alpha(\bar{m}) = a_\alpha - b_\alpha e^{-c_\alpha \bar{m}^2} \tag{7}$$

$$\kappa(\bar{m}) = a_\kappa - b_\kappa e^{-c_\kappa \bar{m}^2}, \tag{8}$$

where $\{a_i, b_i, c_i\}$ with $i = \kappa, \alpha$, are fitting constants determined from isovector meson spectrum modeling [11]. This threshold mass is understood as a *stoichiometric* relation describing what kind of constituents there are in the hadron and in what proportion they contribute to the hadron structure. For example, for a $q\bar{q}$ state, it is straightforward to infer that the mass threshold should be the arithmetic mean between the masses, that is, $\bar{m} = \frac{m_1+m_2}{2}$.

For other more general and exotic structures, we can extend the arithmetic mean idea to the general stoichiometric relation given by the following expressions:

$$\bar{m}_{eh} = \sum_{i=1}^{N}(P_i^{\text{quark}}\,\bar{m}_{q_i} + P_i^{\text{Gluon}}\,m_{G_i} + P_i^{\text{meson}}\,m_{\text{meson}_i}), \tag{9}$$

Table 1. Constituent masses (in GeV) used in the numerical fit. See [12, 13].

Const	Mass	Const	Mass	Const	Mass	Const	Mass
q_u	0.360	q_d	0.340	q_c	1.550	q_b	4.730
Gluon	0.700	J/ψ	3.097	$\rho(770)$	0.77526	Υ	9.460

along with the constraint condition:

$$\sum_{i=1}^{N}(P_i^{\text{quark}} + P_i^{\text{Gluon}} + P_i^{\text{meson}}) = 1. \tag{10}$$

In the expressions above, we model an exotic hadron, with mass m_{eh}, having a structure given by N constituents (quarks, gluons, or mesons). The contribution to the exotic structure is measured by the normalized weights $P_i^{\text{constituent}}$. Thus, the problem of describing exotic mesons in dilaton-based models is reduced to how we determine the set of weights $P_i^{\text{constituents}}$. We will consider heavy exotic states that can be described as diquark clusters, hadroquarkonium, and hadronic molecules. The constituent weights are defined by phenomenological bounds read from the exotic hadron theory [14, 15], which encloses the parameter space in the numerical fitting of exotic states. We will test these ideas with the spectrum of exotic hadrons composed of Z_c, ψ, and Z_b states.

Table 2. Calibration curves and threshold masses for exotic hadrons in the non-quadratic softwall model.

Calibration Curves for $\bar{m}$		
$\kappa(\bar{m}) = 15.2085 - 14.808\, e^{-0.0524\,\bar{m}^2}$		
$\alpha(\bar{m}) = 0.8454 - 0.8455\, e^{-0.4233\,\bar{m}^2}$		
Stoichiometric relations for exotic hadrons		
Structure	**$\bar{m}$ (GeV)**	Δ
Diquark for Z_c	$m_{4q} = m_c = 1.55$	6
Diquark for Z_b	$m_{4q} = m_b = 4.730$	6
Hadroquarkonium for ψ	$\bar{m}_{HQc} = \frac{1}{2}m_{J/\psi} + \frac{1}{4}(\bar{m}_u + \bar{m}_d) = 1.717$	6
Hadronic Molecule for ψ	$\bar{m}_{HM,\psi} = \frac{1}{3}m_{J/\psi} + \frac{2}{3}m_\rho = 1.549$	6
Hadronic Molecule for Z_c	$\bar{m}_{HM,Z_c} = 0.283\, m_{J/\psi} + 0.717\, m_\rho = 1.4323$	6
Hadronic Molecule for Z_b	$\bar{m}_{HM,Z_B} = 0.458\, m_{\Upsilon(1S)} + 0.542\, m_\rho = 4.753$	6
Hybrid meson for Z_c	$\bar{m}_{Z_c} = 0.98\, m_c + 0.2\, m_G = 1.533$	5 (One gluon)
Hybrid meson for Z_b	$\bar{m}_{Z_b} = 0.99\, m_b + 0.01\, m_G = 4.6897$	7 (Two gluons)

Holographically, the exotic spectrum depends entirely on the dilaton field and geometry. The hadronic identity is defined by the bulk mass $M_5(\Delta)$. All of this information is condensed into the holographic potential for vector hadronic states, defined as:

$$V_{\text{non-}q\bar{q}}(z, \kappa, \Delta) = V_{q\bar{q}}(z, \kappa, \alpha) + \frac{M_5^2(\Delta)\, R^2}{z^2}, \tag{11}$$

where $V_{q\bar{q}}(z, \kappa, \alpha)$ is the holographic potential used to describe isovector meson states in the context of holographic nonlinear Regge trajectories [11]. The parameter set is summarized in table 1. The calibration curves and the threshold mass for each exotic structure is written in table 2. Table 3 exposes the non-$q\bar{q}$ spectrum fit results. Experimental masses are read from PDG [12].

Table 3. Holographic mass spectrum for the non-$q\bar{q}$ states candidates according to their threshold mass $\bar{m}$. The parameters κ and α are read from the calibration curves, and the stoichiometric relations are exposed in table 2. Experimental data come from [12].

Z_c states with $I^G(J^{CP}) = 1^+(1^{+-})$			
Non-$q\bar{q}$ state	$Z_c(3900)$	$Z_c(4200)$)	$Z_c(4430)$
Exp. Masses (MeV)	3887.1 ± 3.6	4296^{+35}_{-32}	4478^{+13}_{-18}
Diquark-Antidiquark (MeV) $\bar{m}_{4Q} = 1550$ MeV $\kappa = 2151$ MeV and $\alpha = 0.5387$	4004.8 (3.0%)	4384.9 (2.1%)	4706.6 (5.1%)
Hadronic Molecule (MeV) $\bar{m}_{HM} = 1432.3$ MeV $\kappa = 1907$ MeV and $\alpha = 0.4887$	3817.1 (1.8%)	4214.7 (1.9%)	4552.1 (1.6%)
Hybrid Meson (MeV) $\bar{m}_{Z_c} = 1533$ MeV $\kappa = 2114.8$ MeV and $\alpha = 0.5317$	3721.9 (4.2%)	4156.4 (3.2%)	4513.2 (0.8%)
ψ states with $0^+(1^{--})$			
Non-$q\bar{q}$ state	$\psi(4230)$	$\psi(4360)$)	$\psi(4660)$
Exp. Masses (MeV)	4222.5 ± 2.4	4374 ± 7	4630 ± 6
Hadrocharmonium (MeV) $\bar{m}_{HQc} = 1717$ MeV $\kappa = 2522.6$ MeV and $\alpha = 0.6036$	4222.5 (0.2%)	4577.4 (4.6%)	4871.8 (5.2%)
Hadronic Molecule (MeV) $\bar{m}_{HM} = 1549.1$ MeV $\kappa = 2149.2$ MeV and $\alpha = 0.5384$	4003.5 (5.2%)	4383.7 (0.2%)	4705.6 (1.6%)
Z_b states with $I^G(J^{CP}) = 1^+(1^{+-})$			
Non-$q\bar{q}$ state	$Z_b(10610)$	$Z_b(10650)$)	—
Exp. Masses (MeV)	10609 ± 6	10652.2 ± 1.5	—
Diquark-Antidiquark (MeV) $\bar{m}_{HM} = 4753$ MeV $\kappa = 11269.5$ MeV and $\alpha = 0.8633$	10224.5 (3.6%)	10517.6 (1.3%)	—
Hybrid Meson (MeV) $\bar{m}_{Zb} = 4689.7$ MeV $\kappa = 11102.3$ MeV and $\alpha = 0.8633$	10257.9 (3.3%)	10512.5 (1.3%)	—

5 Configurational entropy and stability

Configurational entropy [16] is associated with the information content in solutions to a given set of equations of motion and accounts for the microstate configurations emerging in a given macrostate at a fixed temperature. Thus, higher possible microstate configurations imply a higher values of configurational entropy (CE), inferring that the system could be unstable. In information theory, CE is connected to the information content in such EOM solutions. The information content is measured in terms of *complexity and localization*. In other words, highly localized solutions in space will lead to lower values of CE. Thus, ground states should be more stable than excited ones [17].

In the case of AdS/QCD, CE can be used to test hadronic stability [4]. The key point is how locality is a synonym of confinement. From the NRQCD potential model, a hadron should have a *well-localized* wave function. A particular set of constituents living inside hadrons, with defined structures, will have an associated CE. Preferred structures would be those with smaller CE. This localized information is encoded into the bulk dual modes. Thus, studying the CE of bulk modes will give us an insight into which structure is more stable.

Mesons in bottom-up models appear from the field/operator duality. The key point is the dimension of boundary hadronic operators, accounting for constituents nor the hadron inner structure. This constituent information is encoded in the bulk mass. Any information about the hadronic inner structure is not present *ab initio*. This map between boundary and bulk physics also implies that normalizable bulk modes are dual to mesons at the boundary. From the AdS/CFT point of view, this claim clashes with field/operator duality since *spatially localized wave functions at the boundary should be delocalized at the bulk*. This is the case for $\mathcal{N} = 4$ SYM theory dual to a Type IIB supergravity, where bulk states are unbounded [18]. However, long ago, we learned that one of the consequences of releasing the bulk conformal invariance is precisely this reinterpretation of the operator/field duality where extended and localized objects at the boundary are dual to localized objects at the bulk [19]. This locality in the bulk objects seems natural in non-conformal theories [20]. This idea supports computing hadronic properties (mass spectrum, form factors, decay constants, thermal densities, etc.) in bottom-up models. Therefore, in models where non-conformality exists, *boundary local operators have local dual bulk objects*. This claim directly connects to the configurational entropy. Therefore, we can say *locality at the bulk (involving lower configurational entropy) implies stability at the boundary*.

Let us move to the holographic calculation of the configurational entropy. We will follow the prescription exposed in [21, 22]. The CE algorithm can be summarized as follows. Start by defining the on-shell energy-momentum tensor as

$$T_{mn} = \frac{2}{\sqrt{-g}} \frac{\partial \left[\sqrt{-g}\, \mathcal{L}_{\text{Hadron}} \right]}{\partial\, g^{mn}}. \tag{12}$$

Then, extract the T_{00} component that defines the energy density for the bulk modes. In our particular case of dilaton-base models, the energy density $\rho(z)$ acquires the form:

$$\rho(z) \equiv T_{00} = \frac{e^{-B(z)}}{2} \left(\frac{z}{R} \right)^3 \left\{ \left[\frac{1}{g_5^2} \left(M_n^2\, \psi_n^2 + \psi_n'^2 \right) - \frac{M_5^2\, R^2}{z^2} \psi_n^2 \right] \right\} \Omega. \tag{13}$$

Then, Fourier-transform the density to compute the *modal fraction* as:

$$f(k) = \frac{|\bar{\rho}(k)|^2}{\int dk |\bar{\rho}(k)|^2}. \tag{14}$$

The next step is to normalize this modal fraction with the its maximum value, i.e., $\tilde{f}(k) = f(k) / f(k)_{\text{Max}}$. Now, compute *differential configurational entropy* using the Shannon-like expression:

$$S_{DCE} = - \int dk\, \tilde{f}(k) \log \tilde{f}(k). \tag{15}$$

We computed the DCE for each non-$q\bar{q}$ possible structure proposed in this work. The numerical results are depicted in figure 1. According to the claim that less CE implies a more stable structure, we can conclude that Z_c and Z_b could be better described as *hybrid mesons*, and ψ (or Y) states fit better with *hadroquarkonium proposal*.

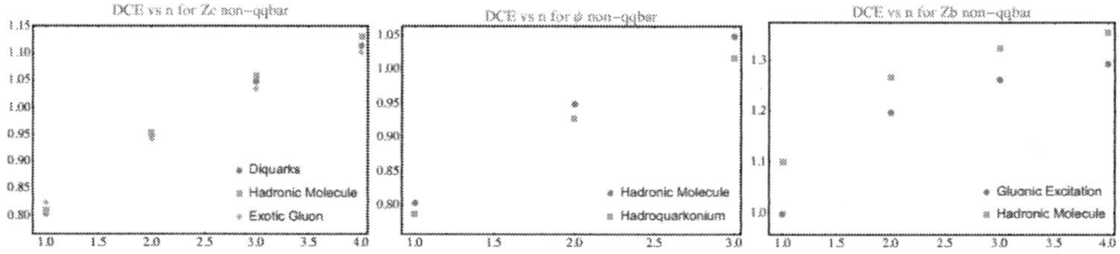

Figure 1. Differential configurational entropy (DCE), in natural entropy units, as a function of excitation number for non-$q\bar{q}$ candidates. From DCE grounds, those structures with less entropy are the most stable and, therefore, preferred.

6 Conclusions

This manuscript has discussed two main ideas: configurational entropy to measure stability and the implication that this measure can predict the constituent arrangement in nature from holographic grounds. These ideas are the cornerstone of our research and are crucial in advancing our understanding of theoretical physics. Recalling to hadronic stability is equivalent to addressing the wave function locality. This discussion stretches the hypothesis that normalizable modes living in the AdS bulk are dual to meson modes. This claim is supported by slightly breaking the bulk conformal invariance, inducing bounded KK-towers dual to hadronic Regge trajectories. In bottom-up models, we achieve this by introducing a dilaton field. Thus, well-localized (bounded) bulk modes imply localized and confined objects at the boundary. In this holographic context, we found that for the heavy non-$q\bar{q}$ states analyzed, the Z_c and Z_b states should be cataloged as *hybrid mesons*, and the ψ states as *hadrocharmonium*.

M. A. Martin Contreras would like to acknowledge the financial support provided by the National Natural Science Foundation of China (NSFC) under Grant No 12350410371.

References

[1] A. Karch, E. Katz, D. T. Son and M. A. Stephanov, *Linear confinement and AdS/QCD*, Phys. Rev. D **74**, 015005 (2006). https://doi.org/10.1103/PhysRevD.74.015005

[2] G. Karapetyan and R. da Rocha, *Configurational entropy of heavy-quark QCD exotica*, Eur. Phys. J. Plus **136** (2021) No.10, 993 (2021). https://doi.org/10.1140/epjp/s13360-021-01942-7

[3] P. Colangelo and F. Loparco, *Configurational Entropy can disentangle conventional hadrons from exotica*, Phys. Lett. B **788**, 500-504 (2019). https://doi.org/10.1016/j.physletb.2018.11.053

[4] M. A. Martin Contreras and A. Vega, *Holographic stability for non-$q\bar{q}$ candidates*, Phys. Rev. D **108** No.12, 126024 (2023). https://doi.org/10.1103/PhysRevD.108.126024

[5] O. Aharony, S. S. Gubser, J. M. Maldacena, H. Ooguri and Y. Oz, *Large N field theories, string theory and gravity*, Phys. Rept. **323**, 183-386 (2000). https://doi.org/10.1016/S0370-1573(99)00083-6

[6] J. Polchinski and M. J. Strassler, *Deep inelastic scattering and gauge / string duality*, JHEP **05**, 012 (2003). https://iopscience.iop.org/article/10.1088/1126-6708/2003/05/012

[7] H. Boschi-Filho and N. R. F. Braga, *QCD / string holographic mapping and glueball mass spectrum*, Eur. Phys. J. C **32**, 529-533 (2004). https://doi.org/10.1140/epjc/s2003-01526-4

[8] P. Breitenlohner and D. Z. Freedman, *Stability in Gauged Extended Supergravity*, Annals Phys. **144**, 249 (1982). https://doi.org/10.1016/0003-4916(82)90116-6

[9] J. Erlich, E. Katz, D. T. Son and M. A. Stephanov, *QCD and a holographic model of hadrons*, Phys. Rev. Lett. **95**, 261602 (2005). https://doi.org/10.1103/PhysRevLett.95.261602

[10] N. R. F. Braga, M. A. Martin Contreras and S. Diles, *Decay constants in soft wall AdS/QCD revisited*, Phys. Lett. B **763**, 203-207 (2016). https://doi.org/10.1016/j.physletb.2016.10.046

[11] M. A. Martin Contreras and A. Vega, *Nonlinear Regge trajectories with AdS/QCD*, Phys. Rev. D **102** No.4, 046007 (2020). https://doi.org/10.1103/PhysRevD.102.046007

[12] S. Navas *et al.* [Particle Data Group], *Review of particle physics*, Phys. Rev. D **110**, No.3, 030001 (2024). https://doi.org/10.1103/PhysRevD.110.030001

[13] W. S. Hou, et al, *Glueball states in a constituent gluon model*, Phys. Rev. D **64**, 014028 (2001). https://doi.org/10.1103/PhysRevD.64.014028

[14] N. Brambilla, S. Eidelman, C. Hanhart, A. Nefediev, C. P. Shen, C. E. Thomas, A. Vairo and C. Z. Yuan, *The XYZ states: experimental and theoretical status and perspectives*, Phys. Rept. **873**, 1-154 (2020). https://doi.org/10.1016/j.physrep.2020.05.001

[15] R. F. Lebed, R. E. Mitchell and E. S. Swanson, *Heavy-Quark QCD Exotica*, Prog. Part. Nucl. Phys. **93**, 143-194 (2017). https://doi.org/10.1016/j.ppnp.2016.11.003

[16] M. Gleiser and N. Stamatopoulos, *Information Content of Spontaneous Symmetry Breaking*, Phys. Rev. D **86**, 045004 (2012). https://doi.org/10.1103/PhysRevD.86.045004

[17] M. Gleiser and D. Sowinski,*Information-Entropic Stability Bound for Compact Objects: Application to Q-Balls and the Chandrasekhar Limit of Polytropes*, Phys. Lett. B **727**, 272-275 (2013). https://doi.org/10.1016/j.physletb.2013.10.005

[18] A. Hamilton, D. N. Kabat, G. Lifschytz and D. A. Lowe, *Holographic representation of local bulk operators*, Phys. Rev. D **74**, 066009 (2006). https://doi.org/10.1103/PhysRevD.74.066009

[19] K. H. Rehren, *Local quantum observables in the anti-de Sitter conformal QFT correspondence*, Phys. Lett. B **493**, 383-388 (2000). https://doi.org/10.1016/S0370-2693(00)01168-0

[20] I. Bena,*On the construction of local fields in the bulk of AdS(5) and other spaces*, Phys. Rev. D **62**, 066007 (2000). https://doi.org/10.1103/PhysRevD.62.066007

[21] A. E. Bernardini and R. da Rocha, *Entropic information of dynamical AdS/QCD holographic models*, Phys. Lett. B **762**, 107-115 (2016). https://doi.org/10.1016/j.physletb.2016.09.023

[22] M. A. Martin Contreras, A. Vega and S. Diles, *Isospectrality and configurational entropy as testing tools for bottom-up AdS/QCD*, Phys. Lett. B **854**, 138723 (2024). https://doi.org/10.1016/j.physletb.2024.138723

EPJ Web of Conferences **314**, 00019 (2024) https://doi.org/10.1051/epjconf/202431400019
QCD@Work 2024

Top-mass determination from leptonic final states

Gennaro Corcella[1,*]

[1]INFN, Laboratori Nazionali di Frascati, Via E. Fermi 54, I-00044 Frascati (RM), Italy

Abstract. The top-quark mass is a fundamental parameter of the Standard Model, as it plays a crucial role in the electroweak precision tests, stability of the vacuum and inflation. I review the method and the main results contained in a recent ATLAS analysis which measures the top mass by using the invariant mass of the leptons coming from W and B-hadron decays. The extracted top mass turns out to be the most precise single measurement by ATLAS.

1 Introduction

The mass of the top quark (m_t) is a fundamental quantity of the Standard Model of strong and electroweak interactions, which enters in the electroweak precision tests and constrained the value of the Higgs boson mass even before its discovery. Furthermore, it is of paramount importance in inflation and in the determination of the stability of the universe.

Since the top quark is the only quark which decays before hadronization, the interpretation of the top-mass measurements has been the subject of a long-standing debate, since the most accurate measurements have been obtained by using Monte Carlo event generators, which are not exact QCD calculationss. Much work has been carried out in addressing the uncertainty/discrepancy of the extracted mass when expressed in terms of well-defined field theory definitions, such as the pole mass (see, e.g., [1–3] for reviews). It is well understood that the mass obtained from direct measurements relying on top decays ($t \to bW$) must be close to the pole mass, up to corrections which depend on the top width or non-pertubative effects, such as colour reconnection or underlying event. Having in mind this discussion on the nature of the measured top mass, this contribution deals with a recent top-mass measurement carried out at the LHC by the ATLAS Collaboration [4] at a centre-of-mass energy $\sqrt{s}$ = 13 TeV and integrated luminosity $\mathcal{L}$ = 36 fb^{-1}. This analysis uses leptonic final states in top decays and the so-called soft-muon tagging (SMT), i.e., it exploits the invariant mass $m_{\ell\mu}$, where ℓ is a charged lepton from the W and μ comes from the decay of a b-flavoured hadron. The advantage of this method is that, as it is based on fully-leptonic final states, it exhibits a mild sensitivity to jet calibration and to the $t\bar{t}$ production phase; on the other hand, one should expect a major dependence on b-quark fragmentation, as it relies on the description of the hadronization of a b into a B-hadron, eventually decaying into a muon. Nevertheless, as will be detailed, such an analyis yields the most precise single m_t measurement by ATLAS.

In the following section, I will give some insight on the analysis [4], paying special attention to the Monte Carlo modelling and uncertainties in the top mass measurement. I will finally make some concluding remarks.

*e-mail: gennaro.corcella@lnf.infn.it

2 Measuring the top mass using the soft-muon tagging

The analysis presented hereafter makes use of data collected in 2015 and 2016 by the ATLAS Collaboration at the LHC, namely centre-of-mass energy $\sqrt{s} = 13$ TeV and integrated luminosity $\mathcal{L} = 36.1$ fb^{-1}. The selected $t\bar{t}$ events feature lepton+jets final states, i.e. one W decays leptonically ($W \to \ell\nu$) and the other one into jets ($W \to jj'$). At parton level, one has a final state $b\ell\nu\bar{b}jj'$, i.e. four jets, one charged lepton and missing energy due to the neutrino. Furthermore, as pointed out in the introduction, at least one of the b-flavoured jets in $t \to bW$ is required to yield a muon μ, coming from the semileptonic decay of a B-hadron. The invariant mass $m_{\ell\mu}$ is then employed to extract m_t.

Final-state jets are clustered according to the anti-k_T algorithm [5] with a radius parameter $R = 0.4$ and must have rapidity $|\eta_j| < 2.5$ and transverse momentum $p_{T,j} > 25$ GeV. As for primary leptons, i.e. those coming from W decays, the corresponding cuts are $p_{T,\ell} > 27$ GeV and $|\eta_\ell| < 2.5$. One also sets an invariant opening angle between jets and primary leptons $\Delta R_{j\ell} > 0.4$, where $\Delta R = \sqrt{(\Delta\eta)^2 + (\Delta\phi)^2}$, ϕ being the azimuthal angle. In the selected sample each event must have at least one SMT, namely a b-jet where the B-hadron decays into a soft muon. The cuts applied on such a muon are $p_{T,\mu} > 8$ GeV, $|\eta_\mu| < 2.5$ and an angular distance $\Delta R_{j\mu} < 0.4$ from its own jet candidate. Furthermore, one requires $\Delta R_{\ell\mu} < 2$ between primary leptons and soft muons, in such a way to enhance the probability that both leptons come from the same top and minimize the fraction of events where they instead originate from two different top quarks.

The simulation of the $t\bar{t}$ samples is carried out by using the *hvq* setup of POWHEG [6] with the NLO NNPDF3.0 parton distribution functions [7]. Off-shell and width effects are included in the MADSPIN approximation, based on Ref. [8], while showers, underlying event and hadronization are modelled by the PYTHIA 8.2 code [9]. HERWIG 7.1.3 [22] is instead used for the sake of comparison and estimating the Monte Carlo systematics. Bottom- and charm-hadron decays and mixing are handled by the EVTGEN 1.2.0 program [10], with the production fractions and branching ratios rescaled to agree with the PDG values [12]. In the *hvq* POWHEG framework, NLO QCD corrections are applied to $t\bar{t}$ production, while gluon radiation in top decays is completely handled by the PYTHIA shower. Matrix-element corrections to parton showers in top decays, namely NLO tree-level corrections $t \to bWg$, are included along the lines of [11]. The $h_{\rm damp}$ parameter in POWHEG, ruling the transverse momentum of the first emission, is set to $h_{\rm damp} = 1.5\, m_t$; moreover, the total $t\bar{t}$ cross section is rescaled to the NNLO value computed by the Top++ code [13], i.e. $\sigma_{t\bar{t}} \simeq (832^{+46}_{-51})$ pb. Backgrounds to $t\bar{t}$ production are simulated by SHERPA 2.2.1 [14] or POWHEG itself: the details of the background simulation are discussed in [4] and are not reported here for the sake of brevity.

The specific PYTHIA tuning employed in this analysis deserves some thorough discussion. The ATLAS Collaboration released the A14 tuning [15], based on LEP and Tevatron data, as well as on a few ATLAS measurements at $\sqrt{s} = 7$ TeV, in order to account for effects which are relevant at hadron colliders, such as underlying event, colour reconnection or multiple interactions. However, since this top-mass determination crucially depends on the fragmentation of b quarks into B-hadrons, the ATLAS Collaboration decided to retune the PYTHIA b-fragmentation parameters, for the sake of a reliable measurement. In PYTHIA, b-quark hadronization is described according to the so-called Lund-Bowler non-perturbative fragmentation function [16, 17]:

$$f(z) = \frac{1}{z^{1+br_b m_b^2}}(1-z)^a \exp\left(-bm_T^2/z\right). \qquad (1)$$

In Eq. (1) m_b is the b-quark mass, m_T the B-hadron transverse mass, i.e. $m_T = \sqrt{p_T^2 + m_B^2}$, z the hadron longitudinal energy fraction with respect to the parent quark. a, b and r_b are instead free parameters which must be tuned to experimental data. However, while a and b are sensitive to light-hadron production, r_b is specific to b-quark fragmentation. Therefore, in [4] a and b are kept to their default values, while r_b, whose default is $r_b = 0.85$ in A14, is instead retuned to data from LEP (ALEPH [18], DELPHI [19] and OPAL [20] experiments) and SLD [21] on B-hadron production in e^+e^- annihilation at the Z pole [1]. The details of the fit are described in [4] and lead to the value $r_b = 1.05 \pm 0.02$: the obtained setting is referred to as A14-r_b. The comparison of the tuned PYTHIA simulation with the e^+e^- data is presented in Fig. 2, which also contains the predictions of the HERWIG 7.1.3 event generator [22]: the improvement of the A14-r_b setup with respect to the default one is well visible. Before presenting the results on the top-mass extraction, it should be pointed out that, in principle, using in pp collisions a Monte Carlo fit to e^+e^- data may be arguable, since in a hadron environment one has effects like initial-state radiation, underlying event or colour reconnection between initial and final state, which do not have a counterpart in lepton colliders. From this point of view, a fit to precise pp data would be preferable. However, the analysis setup of [4] indeed minimizes such effects, as one just deals with leptonic final states, and the standalone A14 fit includes information coming from LHC data. Furthermore, in the narrow-width approximation for the top quark, shower and hadronization of b quarks in $t \to bW$ are expected to be similar to $Z \to b\bar{b}$, and hence a tuning of r_b to LEP and SLD is expected to be reliable even for b-quark fragmentation in top decays.

The top mass is then extracted by comparing the POWHEG+PYTHIA prediction, based on the A14-r_b tuning, with the data on the $m_{\ell\mu}$ invariant mass, as displayed in Fig. 2, for distributions obtained after the fit. In [4], checks are carried out fitting separately the electron and muon channels, as well as the samples where the leptons from W and B have the same (SS) or opposite (OS) signs. The fit yields the following result:

$$m_t = 174.41 \pm 0.39\ (\text{stat.}) \pm 0.66\ (\text{syst.}) \pm 0.25\ (\text{recoil}). \qquad (2)$$

While for a discussion on the determination of the systematic and statistical errors one can refer to [4], the so-called recoil uncertainty quoted in (2) deserves further comments. In fact, in the treatment of the PYTHIA parton shower off the b quark in $t \to bW$, one has the option to change the gluon recoil scheme from recoiling against the b (often labelled as RTB), which is the default setting, to recoiling against the W (RTW) or the top (RTT). The RTW and RTT options typically lead to wide-angle radiation in top decay and deposits of energy out of the clustered b-jet, while the RTB one is more consistent with the shower collinear approximation. As a matter of fact, in [4] the change of the recoil modelling is assumed to be another source of uncertainty. Some hints on the reliability of the various options can be obtained by comparing the PYTHIA results for the B-hadron energy fraction in top decays, defined in an equivalent way to x_B in Fig. 1, with the NLO+NLL resummed calculation of Ref. [23], which uses the Kartvelishvili non-perturbative fragmentation function [24], tuned to ALEPH data [18] to model hadronization. Ref. [23] also investigates bottom-quark fragmentation in top decays in Mellin moment space, including the non-perturbative information by means of a fit to DELPHI data in N-space [19]. Overall, it is found that the RTB setting data gives the best comparison with [23], while the RTT and RTW options lead to a shift in the extracted mass of about 0.25 GeV. Therefore, Ref. [4] quotes a recoil uncertainty of 0.25 GeV, as in Eq. (2), in the measured top mass with the SMT method.

[1] b-quark fragmentation in Refs. [18–21] is described in terms of the energy fraction $x_B = 2p_B p_Z/m_Z^2$, p_B and p_Z being the four-momenta of the B-hadron and of the Z boson, respectively.

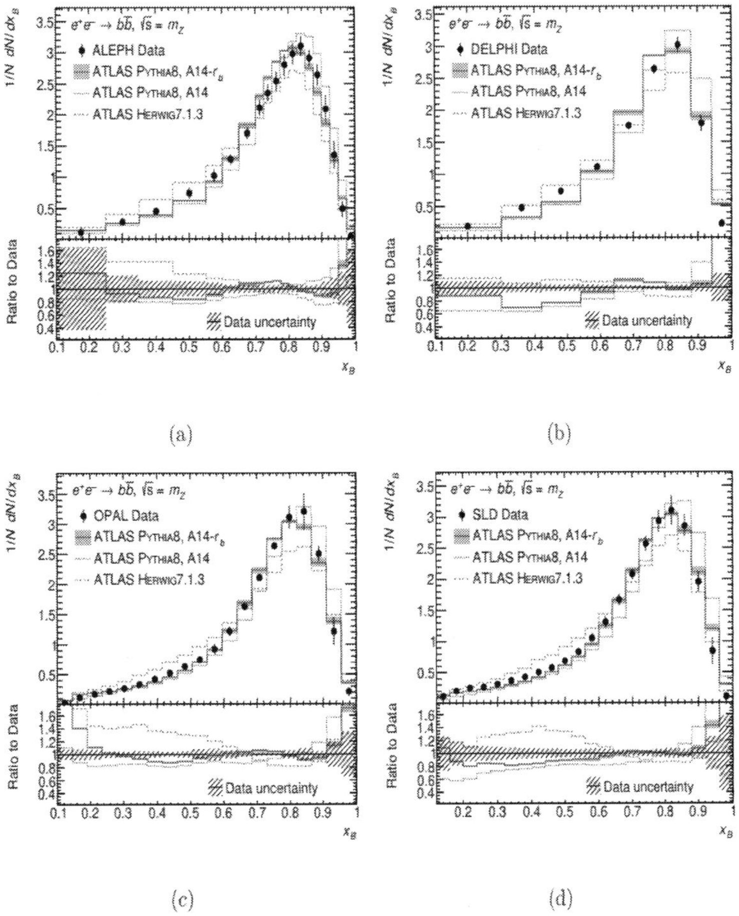

Figure 1. Comparison of the x_B data from LEP and SLD at the Z pole with PYTHIA and HERWIG simulations. As for PYTHIA, both A14 and A14-r_b setups are employed. Figures taken from [4].

3 Conclusions

I reviewed a recent analysis undertaken by the ATLAS Collaboration on the top-mass determination by using the invariant mass of a lepton and a soft muon, coming from a W and a B-hadron in top decays, respectively. This measurements fully relies on leptonic final states in top decays, which minimizes uncertainties due to the $t\bar{t}$ production stage and

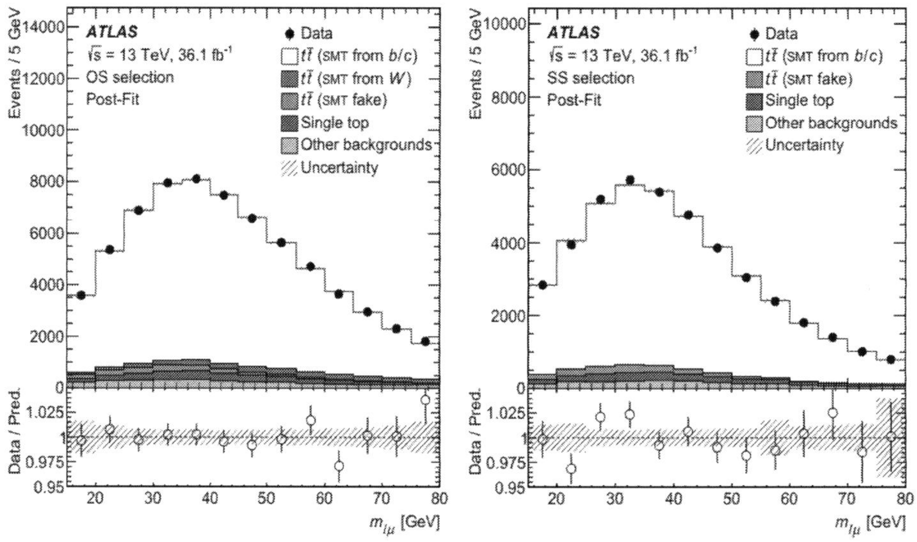

Figure 2. Data on the invariant mass $m_{\ell\mu}$ compared with tuned POWHEG+PYTHIA for $t\bar{t}$ signal and background as discussed in [4]. The OS and SS labels refer to whether the lepton from the W and the soft muon from the B-hadron have opposite (OS) or same (SS) sign. The uncertainty band includes statistical and systematic uncertainties. Figures taken from [4].

jet energy scale and calibration. For the sake of a reliable measurement, the bottom-quark fragmentation parameter r_b in the PYTHIA event generator was retuned to LEP and SLD data. As a matter of fact, for the time being, this top-mass extraction turns out to be the most precise single measurement from the reconstruction of the top-decay products, i.e. m_t = 174.41 ± 0.39 (stat.) ± 0.66 (syst.) ± 0.25 (recoil). While the determination of the systematic and statistical uncertainties is rather standard, particular care was taken about the assessment of the one due to the choice of the PYTHIA recoil scheme for gluon radiation in top decays. For this purpose, the comparison with a resummed calculation for b-quark fragmentation in top decays turned out to be very useful to address the best option for the recoil scheme and give a meaningful estimate of the uncertainty. Overall, the presented top-mass measurement is in agreement with other determination and, thanks to the negligible error due to the jet activity, it is expected to play a relevant role in the future combinations.

The presented work was carried out during the association of the author with the ATLAS Collaboration as an ACE (Analysis Consultant and Expert). The author wishes to warmly thank Lucio Cerrito, who proposed the association with the Rome Tor Vergata ATLAS group, as well as the other members of the analysis team Veronique Boisvert, Umberto De Santis, Francesco Giuli, Michele Pinamonti and Marco Vanadia.

References

[1] G. Corcella, The top-quark mass: challenges in definition and determination. Front. in Phys. **7**, 54 (2019). https://doi.org/10.3389/fphy.2019.00054

[2] P. Nason, The Top Quark Mass at the LHC. Frascati Phys. Ser. **65**, (2017) 65-70

[3] A.H. Hoang, What is the Top Quark Mass? Ann. Rev. Nucl. Part. Sci. **70**, 225 (2020). https://doi.org/10.1146/annurev-nucl-101918-023530

[4] G. Corcella and ATLAS Collaboration, Measurement of the top-quark mass using a leptonic invariant mass in pp collisions at $\sqrt{s}$ = 13 TeV with the ATLAS detector. JHEP **06**, 019 (2023). https://doi.org/10.1007/JHEP06(2023)019

[5] M. Cacciari, G.P. Salam and G. Soyez, The anti-k_t jet clustering algorithm. JHEP **04**, 023 (2008). https://doi.org/10.1088/1126-6708/2008/04/063

[6] S. Alioli, P. Nason, C. Oleari and E. Re, A general framework for implementing NLO calculations in shower Monte Carlo programs: the POWHEG BOX. JHEP **06**, 043 (2010). https://doi.org/10.1007/JHEP06(2010)043

[7] NNPDF Collaboration, Parton distributions for the LHC Run II. JHEP **04**, 040 (2015). https://doi.org/10.1007/JHEP04(2015)040

[8] S. Frixione, E. Laenen, P. Motylinski and B.R. Webber, Angular correlations of lepton pairs from vector boson and top quark decays in Monte Carlo simulations.JHEP **04**, 081 (2007). https://doi.org/10.1088/1126-6708/2007/04/081

[9] T. Sjöstrand, S. Mrenna and P.Z. Skands, A Brief Introduction to PYTHIA 8.1 Comput. Phys. Commun. **178**, 852 (2008). https://doi.org/10.1016/j.cpc.2008.01.036

[10] D.J. Lange, The EvtGen particle decay simulation package Nucl. Instrum. Meth. A **462**, 152 (2001). https://doi.org/10.1016/S0168-9002(01)00089-4

[11] E. Norrbin and T. Sjöstrand, QCD radiation off heavy particles. Nucl. Phys. B **603**, 297 (2001). https://doi.org/10.1016/S0550-3213(01)00099-2

[12] Particle Data Group Collaboration, Review of Particle Physics. Phys. Rev. D **98**, 030001 (2018). https://doi.org/10.1103/PhysRevD.98.030001

[13] M. Czakon and A. Mitov, Top++: A Program for the Calculation of the Top-Pair Cross-Section at Hadron Colliders. Comput. Phys. Commun. **185**, 2930 (2014). https://doi.org/10.1016/j.cpc.2014.06.021

[14] Sherpa Collaboration, Event Generation with Sherpa 2.2. SciPost Phys. **7**, 034 (2019). https://doi.org/10.21468/SciPostPhys.7.3.034

[15] ATLAS Collaboration, ATLAS Pythia 8 tunes to 7 TeV data. ATL-PHYS-PUB-2014-021

[16] B. Andersson, G. Gustafson, G. Ingelman and T. Sjöstrand, Parton Fragmentation and String Dynamics. Phys. Rept. **97**, 31 (1983). https://doi.org/10.1016/0370-1573(83)90080-7

[17] M.G. Bowler, e^+e^- production of heavy quarks in the string model. Z. Phys. C **11**, 169 (1981). https://doi.org/10.1007/BF01574001

[18] ALEPH Collaboration, Study of the fragmentation of b quarks into B mesons at the Z peak. Phys. Lett. B **512**, 30 (2001). https://doi.org/10.1016/S0370-2693(01)00690-6

[19] DELPHI Collaboration, A study of the b-quark fragmentation function with the DELPHI detector at LEP I and an averaged distribution obtained at the Z Pole. Eur. Phys. J. C **71**, 1557 (2011). https://doi.org/10.1140/epjc/s10052-011-1557-x

[20] OPAL Collaboration, Inclusive analysis of the b quark fragmentation function in Z decays at LEP. Eur. Phys. J. C **29**, 463 (2003). https://doi.org/10.1140/epjc/s2003-01229-x

[21] SLD Collaboration, Precise measurement of the b quark fragmentation function in Z^0 boson decays. Phys. Rev. Lett. **84**, 4300 (2000). https://doi.org/10.1103/PhysRevLett.84.4300

[22] J. Bellm et al., Herwig 7.0/Herwig++ 3.0 release note. Eur. Phys. J. C **76**, 196 (2016). https://doi.org/10.1140/epjc/s10052-016-4018-8

[23] M. Cacciari, G. Corcella and A. Mitov, Soft gluon resummation for bottom fragmentation in top quark decay. JHEP **12**, 015 (2002). https://doi.org/10.1088/1126-6708/2002/12/015

[24] V.G. Kartvelishvili, A.K. Likhoded and V.A. Petrov, On the Fragmentation Functions of Heavy Quarks Into Hadrons. Phys. Lett. B **78**, 615 (1978). https://doi.org/10.1016/0370-2693(78)90653-6

EPJ Web of Conferences **314**, 00020 (2024) https://doi.org/10.1051/epjconf/202431400020
QCD@Work 2024

Rapidity-only evolution of TMDs

Giovanni Antonio Chirilli[1,*]

[1]Department of Mathematics and Physics, University of Salento, and Sezione di Lecce, Istituto Nazionale di Fisica Nucleare, Via Arnesano, Lecce, I-73100, Italy

Abstract. The most known scheme to regulates the rapidity/UV divergences of the Transverse Momentum Distribution operators due to the infinite light-like gauge links are the Collis Soper Sterman (CSS) formalism or the Soft Collinear Effective Theory (SCET) formalism. An alternative procedure is provided by the scheme used in the small-x_B physics. The corresponding evolution equations differ already in leading order. Because of the future Electron-Ion Collider experiment, which will probe the TMDs at values of the Bjorken x in the region between small-x_B to $x_B \sim 1$, the different formalisms need to be reconciled. I will discuss the conformal properties of TMD operators and present the result of the conformal rapidity evolution of TMD operators in the Sudakov region. In particular, I will present the calculation of the scale of the coupling constant obtained using the BLM procedure.

1 Introduction

Recent developments in the study of transverse-momentum dependent parton distributions (TMDs) [1–3] have greatly expanded their use in processes such as semi-inclusive deep inelastic scattering (SIDIS) and particle production in hadron-hadron collisions. TMDs, which are defined as quark or gluon matrix elements accompanied by light-like gauge links, are subject to rapidity divergences due to the nature of these links. To deal with this, various regularization schemes have been proposed. The most common approaches involve the CSS or SCET formalisms, but alternative methods from small-x physics have also gained attention. However, discrepancies between these methods, even at leading order, pose theoretical challenges that need to be resolved, especially in view of the upcoming Electron-Ion Collider (EIC) experiments, which will probe a wide range of Bjorken-x values, from small-x_B to values closer to 1.

A key strategy to tackle these discrepancies is to derive leading-order evolution equations from conformal considerations. Since perturbative QCD (pQCD) is conformally invariant at leading order, one might be able to derive evolution equations without an explicit running coupling at this order. Given that TMD operators include light-like Wilson lines, their behavior should, in principle, follow the transformations of the conformal subgroup that preserves this light-like direction. However, regularizing rapidity divergences remains an obstacle, as no regularization method currently preserves full conformal invariance. A conformal regulator for leading-order calculations could still be a valuable tool, with higher-order calculations

*e-mail: giovanni.chirilli@le.infn.it

accounting for the breaking of conformal symmetry due to the QCD running coupling. Insights from $\mathcal{N} = 4$ Super-Yang-Mills (SYM) theory, which retains conformal symmetry, could provide a useful framework, as results in this theory often match the most complex parts of pQCD calculations. This approach has been applied to the rapidity evolution of color dipoles, leading to conformal invariance in the leading-order evolution of the Balitsky-Kovchegov (BK) equation, and offers a promising direction for similar studies in TMD evolution.

Th focus is the extension of the TMD formalism to small-x_B processes, a region that is becoming increasingly important for future experimental studies like those at the EIC. Traditionally, TMDs have been used at larger Bjorken-x, but expanding the formalism to the small-x_B regime has led to the development of new evolution equations based on rapidity cutoffs inspired by small-x_B physics [5, 6, 10, 11]. These new equations can smoothly interpolate between linear evolution at moderate x_B and the nonlinear dynamics at small x_B, capturing important transitions between different evolution regimes such as DGLAP, BFKL, and Sudakov forms. Despite this success, the evolution equations in the intermediate region are complex and present practical difficulties, such as the indeterminate argument of the coupling constant at leading order, which can only be determined through next-to-leading order calculations. This issue has been addressed through the Brodsky-Lepage-Mackenzie (BLM) [9] approach, which has been successfully applied to similar small-x evolution studies in color dipoles.

In this context, the results presented here aim to explore these theoretical issues and propose pathways to construct a unified framework for TMD evolution that is applicable across a broad range of kinematic regimes, from small to moderate x_B, in preparation for the future data from the EIC and other experiments.

2 Rapidity factorization for particle production in hadron-hadron collisions

As a working example of the of the rapidity-only formalism, let us consider the Drell-Yan (DY) hadronic tensor $W_{\mu\nu}(q)$, which in the TMD-factorization scheme [7] can be written as

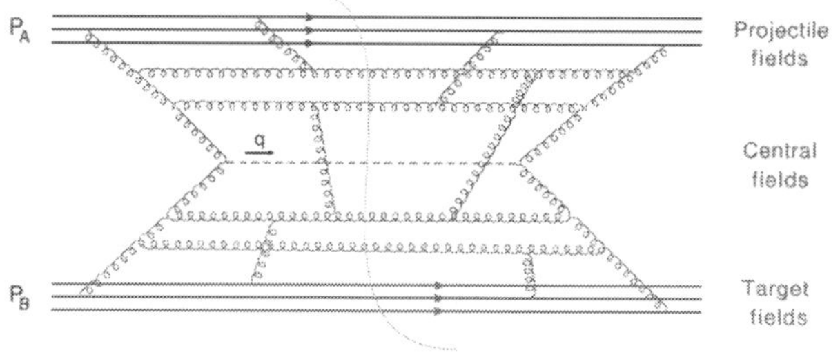

Figure 1. Diagrammatic representation of rapidity factorization in particle production in hadron-hadron collisions.

$$W_{\mu\nu}(q) = \sum_{\text{flavors}} e_f^2 \int d^2k_\perp \mathcal{D}^{(i)}_{f/A}(x_A, k_\perp)\mathcal{D}^{(i)}_{f/B}(x_B, q_\perp - k_\perp)C_{\mu\nu}(q, k_\perp)$$
$$+ \text{ power corrections } + \text{ Y} - \text{terms} \tag{1}$$

Here $\mathcal{D}_{f/A}(x_A, k_\perp)$ is the TMD density of a quark f in hadron A with fraction of momentum x_A and transverse momentum $k_\perp$, $\mathcal{D}_{f/B}(x_B, q_\perp - k_\perp)$ is a similar quantity for hadron B, and coefficient functions $C_i(q, k)$ are determined by the cross section $\sigma(f\bar{f} \to \mu^+\mu^-)$ of production of DY pair of invariant mass q^2 in the scattering of two quarks.

The TMD densities $\mathcal{D}_{f/A}(x_A, k_\perp)$ and $\mathcal{D}_{f/B}(x_B, k_\perp)$ are defined by quark-antiquark operators with gauge links going to $-\infty$. For example, the TMD f_1 responsible for the total DY cross section for unpolarized hadrons is defined by

$$f_1^f(x_B, k_\perp) = \frac{1}{16\pi^3}\int dz_+ d^2 z_\perp \, e^{-ix_B z^+ \sqrt{\frac{s}{2}} + i(k,z)_\perp} \langle p_N|\bar{\psi}_f(z_+, z_\perp)[z, z - \infty n]\not{n}\psi_f(0)|p_N\rangle \quad (2)$$

where $|p_N\rangle$ is an unpolarized nucleon with momentum $p_N \simeq p_N^-$ and $n = (\frac{1}{\sqrt{2}}, 0, 0, \frac{1}{\sqrt{2}})$ is a lightlike vector in the "+" direction (almost) collinear to vector p_A. For a straight-line gauge link, known as Wilson line, connecting points x and y we use the notation

$$[x, y] \equiv \mathrm{P}e^{ig\int du\, (x-y)^\mu A_\mu(ux+(1-u)y)} \quad (3)$$

Here, we will consider the rapidity-only evolution of the quark operator

$$\bar{\psi}(x^+, x_\perp)[x, x \pm \infty n]\,[\pm\infty n + x_\perp, \pm\infty n + y_\perp]\Gamma[\pm\infty n + y, y]\psi(y^+, y_\perp) \quad (4)$$

with Γ is one of the matrices $\gamma^-, \gamma^-\gamma_5, \gamma^-\gamma_\perp$ so we single out "good" projections in the light-cone language.

A diagrammatic representation of the factorization in rapidity is given in Fig. 1. Here the fields are cut with respect to the light-cone component of their momenta. The gluon fields, which are the dominant degree of freedom at this regime, are factorized in target fields (in red) with light-cone component $|k^+| < \sigma_b$, projectile fields (in green) with light-cone component $|k^-| < \sigma_a$, and central fields (in blue) for the coefficient functions. Here, σ_a and σ_b are two parameters which cut the gluon longitudinal momenta to distinguish them in target, projectile and central fields.

In section 4 we will present the evolution of the TMDs operator arising either from the target fields or the projectile ones with respect to the rapidity cut-off σ.

3 Conformal invariance of TMD Operators

Before delving into the calculation of the evolution equation, let us present the TMDs conformal group. The algebra of the full conformal group $SO(4, 2)$ consists of 16 generators: four operators P^μ, six $M^{\mu\nu}$, four special conformal generators K^μ, and dilatation operator D. One can easily check that in the leading order the following 11 operators act on TMDs covariantly

$$P^i, P^-, M^{12}, M^{-i}, D, K^i, K^-, M^{-+} \quad (5)$$

while the action of operators P^+, M^{+i}, and K^+ do not preserve the form of the TMD operator. The action of the generators (5) on the TMD operator is the same as the action on the field F^{-i} without gauge link attachments. The corresponding group consists of transformations which leave the hyperplane $z^- = 0$ and vector n invariant. Those include shifts in transverse and "+" directions, rotations in the transverse plane, Lorentz rotations/boosts created by M^{-i}, dilatations, and special conformal transformations

$$z'_\mu = \frac{z_\mu - a_\mu z^2}{1 - 2a \cdot z + a^2 z^2} \quad (6)$$

with $a = (a^+, 0, a_\perp)$. In terms of "embedding formalism" [8] defined in 6-dim space, this subgroup is isomorphic to "Poincare + dilatations" group of the 4-dim subspace orthogonal to our physical light-like "+" and "-" directions. In ref. [10], we have shown that the LO evolution equation in the Sudakov approximation is invariant under the TMDs group represented by the generators (5) provided that the cut-off is chosen in the appropriate way.

4 Rapidity-only evolution equation with running coupling

The evolution equation of the quark TMD (4) is obtained calculating the diagrams in Fig.2 with rapidity cut-off defined as

$$A^\sigma_\mu(x) = \int \frac{d^4k}{16\pi^4}\,\theta(\sigma\varrho - |k^+|)e^{-ik\cdot x}A_\mu(k) \tag{7}$$

However, instead of a rigid one imposed by θ-function, it turned out that it is more convenient

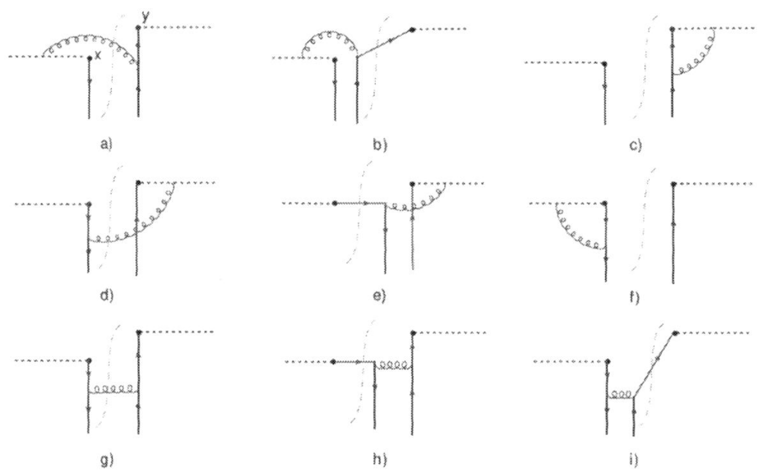

Figure 2. One-loop diagrams for TMD operator (4) in the background quark field. The dashed lines denote gauge links.

to use smooth cutoff in $|k^+|$. This can be encoded in coordinate space by defining the rapidity-regularized operators as

$$\bar\psi^\sigma(x^+, x_\perp) \equiv \bar\psi(x^+, x_\perp, -\frac{1}{\rho\sigma})[x^+, -\infty]_x, \quad \psi^\sigma(y^+, y_\perp) \equiv [-\infty, y^+]_y\psi(y^+, y_\perp, -\frac{1}{\rho\sigma}) \tag{8}$$

We will consider the evolution equation in the Sudakov region $\sigma x_B s \gg k_\perp^2 \sim q_\perp^2$, where s is the Mandelstam variable, $q_\perp^2$ is the transverse momenta of the produced lepton-pair or Higgs boson in DY process (or of the produced hadron in SIDIS), while $k_\perp^2$ is the typical transverse momenta of the parton participating in the interaction.

The diagrams contributing the LO evolution equation are given in Fig. 2. The calculation can be performed directly in coordinate space [10], or in momentum space and then perform

Fourier transform in coordinate space [11]. The result in coordinate space is

$$\Big(\sigma\frac{d}{d\sigma}+\sigma'\frac{d}{d\sigma'}\Big)\bar\psi^{\sigma'}(x^+,x_\perp)\Gamma\psi^{\sigma}(y^+,y_\perp)$$
$$=\frac{\alpha_s}{4\pi^2}c_F\int dz^+\Big\{\Big[i\frac{\ln\varrho(-x^++z^++i\epsilon)-\ln\frac{\sigma b_\perp^2 s}{4}e^\gamma}{-x^++z^++i\epsilon}+\text{c.c.}\Big]\bar\psi^{\sigma'}(z^+,x_\perp)\Gamma\psi^{\sigma}(y^+,y_\perp)$$
$$+\Big[i\frac{\ln\varrho(-y^++w^++i\epsilon)-\ln\frac{\sigma' b_\perp^2 s}{4}e^\gamma}{-y^++w^++i\epsilon}+\text{c.c.}\Big]\bar\psi^{\sigma'}(x^+,x_\perp)\Gamma\psi^{\sigma}(y^+,y_\perp)\Big\} \tag{9}$$

The solution of this equation has the form

$$\bar\psi^{\sigma'}(x^+,x_\perp)\Gamma\psi^{\sigma}(y^+,y_\perp) = e^{-\frac{\alpha_s c_F}{4\pi}\left(\ln\frac{\sigma'}{\sigma'_0}\ln\sigma'\sigma'_0+\ln\frac{\sigma}{\sigma_0}\ln\sigma\sigma_0\right)}$$
$$\times\int dz^+\Big[\frac{i\Gamma(1-\frac{\alpha_s c_F}{2\pi}\ln\frac{\sigma'}{\sigma'_0})}{(z^+-x^++i\epsilon)^{1-\frac{\alpha_s c_F}{2\pi}\ln\frac{\sigma'}{\sigma'_0}}}+\text{c.c.}\Big]\int dw^+\Big[\frac{i\Gamma(1-\frac{\alpha_s c_F}{2\pi}\ln\frac{\sigma}{\sigma_0})}{(w^+-y^++i\epsilon)^{1-\frac{\alpha_s c_F}{2\pi}\ln\frac{\sigma}{\sigma_0}}}+\text{c.c.}\Big]$$
$$\times\frac{1}{4\pi^2}(b_\perp^2 e^\gamma\sqrt{s/8})^{-\frac{\alpha_s c_F}{2\pi}\left(\ln\frac{\sigma'}{\sigma'_0}+\ln\frac{\sigma}{\sigma_0}\right)}\bar\psi^{\sigma'_0}(z^+,x_\perp)\Gamma\psi^{\sigma_0}(w^+,y_\perp) \tag{10}$$

The evolution equation (10) is invariant under the conformal sub-group represented by the generators (5) if we take $\sigma=\sigma'=\frac{\varsigma\sqrt2}{\varrho|\Delta_\perp|}$ where ς is an evolution parameter.

In the LO, the rapidity evolution equation has the argument of the coupling constant left undetermined. To include it, we adopted the BLM method: calculate the contribution of the first quark loop to our TMD evolution and promote $-\frac{1}{6\pi}n_f$ to full $b_0=\frac{11}{12\pi}N_c-\frac{1}{6\pi}n_f$. To this end, each gluon propagator in diagrams in Fig. 2 should be replaced by a one-loop correction, i.e.

$$\frac{1}{p^2+i\epsilon}\ \to\ \frac{1}{p^2+i\epsilon}\Big(1+b_0\alpha_s(\mu)\ln\frac{\tilde\mu^2}{-p^2-i\epsilon}\Big)$$
$$\frac{1}{p^2-i\epsilon}\ \to\ \frac{1}{p^2-i\epsilon}\Big(1+b_0\alpha_s(\mu)\ln\frac{\tilde\mu^2}{-p^2+i\epsilon}\Big)$$
$$2\pi\delta(p^2)\theta(p_0)\ \to\ \frac{i\theta(p_0)}{p^2+i\epsilon}\Big(1+b_0\alpha_s(\mu)\ln\frac{\tilde\mu^2}{-p^2-i\epsilon}\Big)-\frac{i\theta(p_0)}{p^2-i\epsilon}\Big(1+b_0\alpha_s(\mu)\ln\frac{\tilde\mu^2}{-p^2+i\epsilon}\Big) \tag{11}$$

where $\tilde\mu^2\equiv\bar\mu_{\rm MS}^2e^{5/3}$.

The evolution equation for the TMD operator with running coupling takes the form

$$\Big(\sigma\frac{d}{d\sigma}+\sigma'\frac{d}{d\sigma'}\Big)\bar\psi^{\sigma'}(\beta'_B,x_\perp)\Gamma\psi^{\sigma}(\beta_B,y_\perp)=-\frac{c_F}{2\pi}\bar\psi^{\sigma'}(\beta'_B,x_\perp)\Gamma\psi^{\sigma}(\beta_B,y_\perp)$$
$$\times\Big[\alpha_s(\mu_{\sigma'})\ln\Big(-\frac{i}{4}(\beta'_B+i\epsilon)\sigma' s b_\perp^2e^\gamma\Big)+\alpha_s(\mu_\sigma)\ln\Big(-\frac{i}{4}(\beta_B+i\epsilon)\sigma s b_\perp^2e^\gamma\Big)\Big] \tag{12}$$

where $b_\perp\equiv\Delta_\perp=(x-y)_\perp$ We see that in the Sudakov region we can define TMD operator (4) with two independent "left" and "right" cutoffs σ and σ' defined in Eqs. (8) and the evolutions with respect to those cutoffs are independent [except for $b_\perp=(x-y)_\perp$]. The solution of Eq. (12) has the form

$$\bar\psi^{\sigma'}(\beta'_B,x_\perp)\Gamma\psi^{\sigma}(\beta_B,y_\perp) = e^{-\frac{2c_F}{\pi b_0^2}\left[\left(\frac{1}{\alpha_s(\tilde b_\perp^{-1})}-\frac{b_0}{2}\ln[-i(\tau'_B+i\epsilon)]\right)\ln\frac{\alpha_s(\mu_{\sigma'})}{\alpha_s(\mu_{\sigma'_0})}+\frac{1}{\alpha_s(\mu_{\sigma'})}-\frac{1}{\alpha_s(\mu_{\sigma'_0})}\right]}$$
$$\times e^{-\frac{2c_F}{\pi b_0^2}\left[\left(\frac{1}{\alpha_s(\tilde b_\perp^{-1})}-\frac{b_0}{2}\ln[-i(\tau_B+i\epsilon)]\right)\ln\frac{\alpha_s(\mu_\sigma)}{\alpha_s(\mu_{\sigma_0})}+\frac{1}{\alpha_s(\mu_\sigma)}-\frac{1}{\alpha_s(\mu_{\sigma_0})}\right]}\bar\psi^{\sigma'_0}(\beta'_B,x_\perp)\Gamma\psi^{\sigma_0}(\beta_B,y_\perp) \tag{13}$$

where $\tilde b_\perp^2=\frac{b_\perp^2}{2}e^{\gamma/2}$ and $\tau_B=\frac{\beta_B}{|\beta_B|}$, $\tau'_B=\frac{\beta'_B}{|\beta'_B|}$, $\mu_\sigma^2\equiv b_\perp^{-1}\sqrt{\sigma|\beta_B|s}$, $\mu_{\sigma'}^2\equiv b_\perp^{-1}\sqrt{\sigma|\beta'_B|s}$. As at leading order, the structure of the Sudakov evolution (13) resembles two independent exponential factors which describe two independent evolutions of operators (8). However, this property should not be expected beyond the Sudakov region.

5 Conclusions

We described the rapidity evolution of quark (and gluon) TMDs using small-x_B methods. By employing a rapidity-only cutoff, we simplified the evolution to depend on a single parameter – rapidity. However, as highlighted in the introduction, the argument of the coupling constant in such evolution is not determined at leading order. To address this, we applied the BLM scale-setting procedure to both quark and gluon TMDs, leading to the result that the effective BLM scale for Sudakov evolution is halfway between the transverse momentum and the longitudinal energy of the TMD in logarithmic scale. This approach mirrors previous successful applications in NLO BK evolution [12].

Despite using small-x_B methods, our findings are valid for any value of x_B, provided certain kinematic conditions are met. The distinction between small and moderate x_B only emerges at the endpoint of evolution, where different evolution regimes, such as DGLAP for moderate x_B and BFKL/BK for small-x_B, should be applied.

We also observed the importance of choosing a smooth rapidity cutoff to preserve the analytic properties of Feynman diagrams and avoid infrared (IR) divergences, a consideration that distinguishes TMD evolution from dipole evolution. Finally, future work will focus on extending this framework to TMD factorization and calculating cross sections for processes like Higgs production and Drell-Yan at $q_\perp \sim$ a few GeV, as well as comparing our results with those from the CSS method at higher orders. This ongoing research aims to provide a more comprehensive understanding of TMD evolution across different kinematic regimes.

References

[1] J. C. Collins and D. E. Soper, Nucl. Phys. B **194**, 445-492 (1982) doi:10.1016/0550-3213(82)90021-9

[2] J. C. Collins, D. E. Soper and G. F. Sterman, Nucl. Phys. B **250**, 199-224 (1985) doi:10.1016/0550-3213(85)90479-1

[3] X. d. Ji, J. p. Ma and F. Yuan, Phys. Rev. D **71**, 034005 (2005) doi:10.1103/PhysRevD.71.034005 [arXiv:hep-ph/0404183 [hep-ph]]

[4] M. G. Echevarria, A. Idilbi and I. Scimemi, JHEP **07**, 002 (2012) doi:10.1007/JHEP07(2012)002 [arXiv:1111.4996 [hep-ph]]

[5] I. Balitsky and A. Tarasov, JHEP **10**, 017 (2015) doi:10.1007/JHEP10(2015)017 [arXiv:1505.02151 [hep-ph]]

[6] I. Balitsky and A. Tarasov, JHEP **06**, 164 (2016) doi:10.1007/JHEP06(2016)164 [arXiv:1603.06548 [hep-ph]]

[7] J. Collins, Camb. Monogr. Part. Phys. Nucl. Phys. Cosmol. **32**, 1-624 (2011)

[8] S. Rychkov, doi:10.1007/978-3-319-43626-5 [arXiv:1601.05000 [hep-th]]

[9] S. J. Brodsky, G. P. Lepage and P. B. Mackenzie, Phys. Rev. D **28**, 228 (1983) doi:10.1103/PhysRevD.28.228

[10] I. Balitsky and G. A. Chirilli, Phys. Rev. D **100**, no.5, 051504 (2019) doi:10.1103/PhysRevD.100.051504 [arXiv:1905.09144 [hep-ph]]

[11] I. Balitsky and G. A. Chirilli, Phys. Rev. D **106**, no.3, 034007 (2022) doi:10.1103/PhysRevD.106.034007 [arXiv:2205.03119 [hep-ph]]

[12] I. Balitsky and G. A. Chirilli, Phys. Rev. D **77**, 014019 (2008) doi:10.1103/PhysRevD.77.014019 [arXiv:0710.4330 [hep-ph]]

EPJ Web of Conferences **314**, 00021 (2024) https://doi.org/10.1051/epjconf/202431400021
QCD@Work 2024

General mass treatment for Z boson production in association with a heavy quark at hadron colliders

M. Guzzi[1,*], *P.* Nadolsky[2,3,**], *L.* Reina[4,***], *D.* Wackeroth[5,****], and *K.* Xie[2,†]

[1]Department of Physics, Kennesaw State University, 370 Paulding Avenue, Kennesaw, GA 30144, USA
[2]Department of Physics and Astronomy, Michigan State University, East Lansing, MI 48824, USA
[3]Department of Physics, Southern Methodist University, Dallas, TX 75275-0181, USA
[4]Physics Department, Florida State University, Tallahassee, FL 32306-4350, USA
[5]Department of Physics, University at Buffalo, The State University of New York, Buffalo, New York 14260-1500, USA

Abstract. We present the application of the ACOT and S-ACOT general mass variable flavor number schemes to proton-proton collisions with particular attention to the production of final states with at least one heavy quark. *Subtraction* and *residual* heavy-quark parton distribution functions are introduced to facilitate the implementation of this scheme at higher orders in perturbative QCD. The calculation of Z-boson hadronic production with at least one b jet beyond the lowest order in QCD is considered for illustration purposes.

1 Introduction

Modern global QCD analyses aimed at determining the proton's parton distribution functions (PDFs) and the substantial influx of high-precision measurements from the Large Hadron Collider (LHC), necessitate precise and accurate theoretical predictions for standard-candle processes in perturbation theory. The magnitude of higher-order radiative corrections in perturbative QCD calculations competes with multiple factors. Heavy-quark (HQ) contributions can play a significant role among those, and their dynamics can influence the cross-section calculations for a wide range of crucial processes at current and future colliders. Global PDF analyses at next-to-next-to-leading order (NNLO) and beyond (see refs. [1, 2]) extend on a wide range of collision energies and are sensitive to mass effects such as phase-space (PS) suppression, and large radiative corrections to collinear $Q\bar{Q}$ production. The magnitude of these effects is comparable to NNLO, and next-to-next-to-next-to-leading order (N^{3}LO) corrections. With the recent progress in theory calculations beyond NNLO in the QCD strong coupling (e.g., N^{3}LO DIS [3–9], DIS with massive quarks [10], N^{3}LO Drell-Yan [11–15], N^{3}LO Higgs production [16–20], and jet production in DIS [21, 22]), HQ contributions must be consistently considered in the cross section calculation for particle reactions to advance the precision frontier of the LHC Run-3 and beyond. That is, all the fitted cross sections should

*e-mail: mguzzi@kennesaw.edu
**e-mail: nadolsky@smu.edu
***e-mail: reina@hep.fsu.edu
****e-mail: dw24@buffalo.edu
†e-mail: xiekepi1@msu.edu

be computed in a factorization scheme able to interpolate between calculations in which the mass is fully retained in the appropriate kinematic regimes (Fixed Flavor Number (FFN) scheme), and calculations where the number of active flavors N_f increases with the typical energy scale and where mass effects may be approximated when the HQ mass is negligible compared to the other scales (e.g., Zero-Mass (ZM) variable flavor number (VFN) scheme). This matching must guarantee the continuity of the theoretical predictions and avoid double counting of scattering contributions. These interpolating schemes are generically referred to as General Mass Variable Flavor Number (GMVFN) schemes. In this work, we apply the Aivazis-Collins-Olness-Tung (ACOT) GMVFN scheme [23, 24], and its simplified version (S-ACOT) [25–27] to the production of a Z boson with at least one heavy quark in proton-proton collisions [28]. Various versions of the ACOT/S-ACOT schemes have been used to effectively describe heavy-flavor dynamics in DIS. For instance, the S-ACOT-χ at NNLO for neutral [27] and charged currents [29], and the H-VFN scheme [30] have been applied. Similar schemes have also been used in inclusive charm/bottom production in proton-proton collisions, such as the S-ACOT-Massive-Phase-Space(-MPS) scheme [31–33], and in D-meson hadroproduction, like the S-ACOT-m_T [34]. The concept of *subtraction* and *residual* HQ PDFs is introduced to facilitate the practical implementation of the ACOT scheme for $Z + b$ production. These are expressed as convolutions between PDFs and universal operator matrix elements (OMEs) which represent the transition from a massless parton to a HQ and include the HQ mass dependence. To allow for fast calculations, subtraction PDFs are pre-calculated and provided to external users as part of the CT18 [35] family of PDFs with grids in the LHAPDF6 format [36] and are publicly available through the HEPForge repository [37]. We illustrate how subtraction PDFs are incorporated in the calculation of the hadronic production of a Z-boson in association with at least one b-quark jet beyond the lowest-order in QCD, using theory predictions for the differential cross section from previous work [38–40].

2 GMVFN hadronic cross section for $Z + b$ at NLO in QCD

The general structure of a GMVFN scheme hadronic cross section calculation can schematically be represented by the combination of three terms

$$d\sigma_{\rm GMVFN} = d\sigma_{\rm FC} + d\sigma_{\rm FE} - d\sigma_{\rm Sub} \qquad (1)$$

where $d\sigma_{\rm FC}$ represents the Flavor Creation (FC) contributions which contain HQs only in the final state and are computed with $m_Q \neq 0$, $d\sigma_{\rm FE}$ represents the Flavor Excitation (FE) terms which are HQ-initiated processes where the large collinear logarithms have been conveniently resummed into a HQ PDF, and $d\sigma_{\rm Sub}$ is the subtraction which incorporates terms that remove the overlapping part between the FC and FE terms to avoid double counting. The implementation of Eq. (1) in the ACOT scheme for $Z+b$ production in proton-proton collisions at NLO in QCD requires that the FC contributions represented by the subprocess $i + j \to Z + b\bar{b} + X$ with $ij = \{gg, q\bar{q}, qg, \bar{q}g\}$, where X is any additional radiated light parton, are matched to the FE terms $i + j \to Z + b + X$ where $ij = \{Qg, \bar{Q}g, qQ, \bar{q}Q, \bar{Q}q\}$ at $O(\alpha_s^3)$. The partonic channels contributing to $d\sigma^{\rm NLO}_{\rm GMVFN}$ can explicitly be given in terms of PDF *subtractions* as follows (see Ref. [28] for more details on the general framework)

$$\begin{aligned} d\sigma^{\rm NLO}_{\rm FC} &= f_g \rhd \left[a_s^2\, d\widehat{\sigma}^{(2)}_{gg\to ZQ\bar{Q}} + a_s^3\, d\widehat{\sigma}^{(3)}_{gg\to ZQ\bar{Q}(g)} \right] \lhd f_g \\ &+ \sum_{i=q,\bar{q}} f_i \rhd \left[a_s^2\, d\widehat{\sigma}^{(2)}_{q\bar{q}\to ZQ\bar{Q}} + a_s^3\, d\widehat{\sigma}^{(3)}_{q\bar{q}\to ZQ\bar{Q}(g)} \right] \lhd f_i \\ &+ a_s^3 \sum_{i=q,\bar{q}} \left[f_g \rhd d\widehat{\sigma}^{(3)}_{gq\to ZQ\bar{Q}(q)} \lhd f_i + f_i \rhd d\widehat{\sigma}^{(3)}_{qg\to ZQ\bar{Q}(q)} \lhd f_g \right]; \end{aligned} \qquad (2)$$

$$\begin{aligned} d\sigma_{\rm FE}^{\rm NLO} &= f_g \triangleright \left[a_s\, d\widehat{\sigma}^{(1)}_{gQ\to ZQ} + a_s^2\, d\widehat{\sigma}^{(2)}_{gQ\to ZQ(g)}\right] \triangleleft f_Q \\ &+ a_s^2 \sum_{i=q,\bar{q}} f_i \triangleright d\widehat{\sigma}^{(2)}_{qQ\to ZQ(q)} \triangleleft f_Q + (\text{exch.}); \end{aligned} \tag{3}$$

$$\begin{aligned} d\sigma_{\rm sub}^{\rm NLO} &= a_s\, f_g \triangleright d\widehat{\sigma}^{(1)}_{gQ\to ZQ} \triangleleft \widetilde{f}_Q^{(\rm NLO)} + a_s^2\, f_g \triangleright d\widehat{\sigma}^{(2)}_{gQ\to ZQ(g)} \triangleleft \widetilde{f}_Q^{(1)} \\ &+ a_s^2 \sum_{i=q,\bar{q}} f_i \triangleright d\widehat{\sigma}^{(2)}_{qQ\to ZQ(q)} \triangleleft \widetilde{f}_Q^{(1)} + (\text{exch.}), \end{aligned} \tag{4}$$

where $a_s \equiv \alpha_s(\mu, N_f)/(4\pi)$ and Q is the b-quark. The caret symbol "$\widehat{\ }$" in Eqns. (2-4) denotes infrared-safe contributions (i.e., dimensional regularization is applied for initial-state massless partons and the associated $1/\epsilon$ singularities are subtracted). Besides the HQ mass and other kinematic variables, $d\widehat{\sigma}_{ij}$ depends on the two longitudinal momentum fractions (e.g., x_A and x_B) of the initial-state partons. Therefore, to make the implementation more transparent, we introduce the triangular arrows $\triangleright$ and $\triangleleft$ to define convolution products with two variables where the triangular arrows point to the corresponding integration variable that is integrated in the convolution

$$[f \triangleright H](x_A, x_B) \equiv \int_{x_A}^1 \frac{d\xi_A}{\xi_A} f(\xi_A) H(\widehat{x}_A, x_B); \qquad [H \triangleleft f](x_A, x_B) \equiv \int_{x_B}^1 \frac{d\xi_B}{\xi_B} H(x_A, \widehat{x}_B) f(\xi_B).$$

With one variable, we then have

$$\int_x^1 \frac{d\xi}{\xi} f(\xi) g\left(\frac{x}{\xi}\right) = [f \triangleright g](x) = [g \triangleleft f](x). \tag{5}$$

Finally, the *subtraction* HQ PDF $\widetilde{f}_Q$ in Eq. (4), in its first two orders in a_s, is defined as [28, 31, 32]

$$\widetilde{f}_Q^{(1)} = a_s [A_{Qg}^{(1)} \triangleleft g], \; \widetilde{f}_Q^{(2)} = a_s^2 \sum_{i=g,q,\bar{q}} [A_{Qi}^{(2)} \triangleleft f_i], \tag{6}$$

and the NLO subtraction PDF is[1]

$$\widetilde{f}_Q^{(\rm NLO)}(x,\mu) \equiv \widetilde{f}_Q^{(1)} + \widetilde{f}_Q^{(2)} = a_s [A_{Qg}^{(1)} \triangleleft g](x,\mu) + a_s^2 \sum_{i=g,q,\bar{q}} [A_{Qi}^{(2)} \triangleleft f_i](x,\mu), \tag{7}$$

where the $A_{ij}^{(n)}$ $(n = 1, 2, \dots)$ are perturbative coefficients obtained from the calculation of operator matrix elements (OMEs) that emerge from mass factorization [41–43]. The difference $d\sigma_{\rm FE} - d\sigma_{\rm sub}$ can equivalently be reorganized in terms of HQ PDF *residuals* as follows

$$\begin{aligned} d\sigma_{\rm GMVFN}^{\rm NLO} &= d\sigma_{\rm FC}^{\rm NLO} + a_s\, f_g \triangleright \left[d\widehat{\sigma}^{(1)}_{gQ\to ZQ}\right] \triangleleft \delta f_Q^{(\rm NLO)} \\ &+ a_s^2\, f_g \triangleright \left[d\widehat{\sigma}^{(2)}_{gQ\to ZQ(g)}\right] \triangleleft \delta f_Q^{(1)} + a_s^2 \sum_{i=q,\bar{q}} f_i \triangleright \left[d\widehat{\sigma}^{(2)}_{qQ\to ZQ(q)}\right] \triangleleft \delta f_Q^{(1)} + (\text{exch.}). \end{aligned} \tag{8}$$

The quark-antiquark annihilation channel $q\bar{q} \to ZQ\bar{Q}$ does not contain FE terms and therefore does not involve subtraction contributions.

3 Results

The NLO ACOT theory prediction for the invariant mass distribution of the $Z+b$ system $M_{Z,b}$ is illustrated in Fig. 1 where shown are also the scale uncertainty and the interplay between

[1]Here "NLO" refers to the a_s order of the subtraction coefficients.

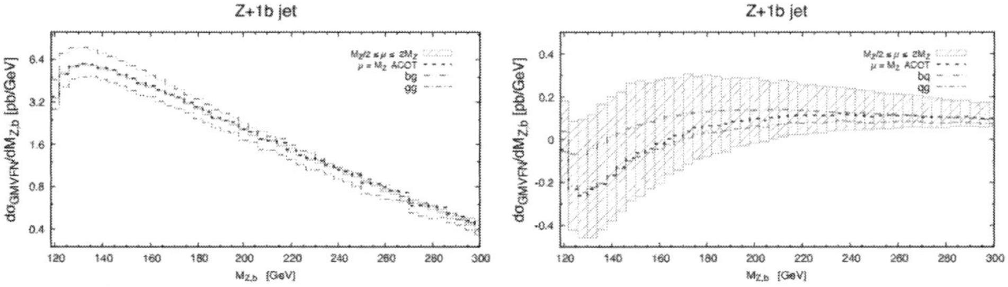

Figure 1. $M_{Z,b}$ differential distribution with scale dependence $M_Z/2 < \mu < 2M_Z$ at order $O(\alpha_s^3)$ for the ACOT GMVFN scheme with FC and FE terms in the gg(left) and qg(right) channels. The CT18NLO PDFs are used.

FC and FE contributions in the gg and qg channels. A more detailed discussion with results for other distributions is in Ref. [28]. The ACOT-$gg(qg)$ central prediction is represented by a black dashed line, while its scale dependence $M_Z/2 \leq \mu \leq 2M_Z$ is represented by a light-blue band. The FC and FE terms are represented by a red dot-dashed and a blue dot-dot-dashed line, respectively. We observe that the convergence appears to be faster in the qg channel, where the FE cross section is largely cancelled by the subtraction terms at $M_{Z,b} \approx 120$ GeV in contrast to the gg channel, where the matching of terms seems to be more sensitive to phase-space integration and applied cuts. In Fig. 2 we show a comparison between the ACOT and S-ACOT theory predictions within their corresponding scale uncertainty bands (light-blue for ACOT and light-red for S-ACOT). $\delta[\%]$ in the lower inset, represents the percent difference between S-ACOT and ACOT relative to ACOT, within the scale uncertainty of the ACOT prediction. Differences are in general around 2-3% and smaller, but they can be larger at higher values of the hard scale of the process. However, all differences are well within the ACOT scale dependence.

4 Conclusions

We extended the framework of the ACOT/S-ACOT GMVFN scheme, based on QCD factorization to the case of proton-proton collisions where at least one heavy quark is produced. Such a GMVFN scheme is applied to Z boson production in association with a least one b-quark jet at $O(\alpha_s^3)$ (NLO) in QCD as an illustrative case, and results are presented for the invariant mass distributions of the $Z+b$ system at the LHC with a collision energy of $\sqrt{s} = 13$ TeV. The practical realization of the theory calculation for the $Z + b$ cross section is facilitated by the introduction of subtraction and residual HQ PDFs. Tabulated grids for CT18 NLO and CT18 NNLO PDF ensembles, in which the b-quark PDFs are replaced by either the residual or subtraction b-quark PDFs, are provided in the LHAPDF6 format [36] and distributed through a repository at HEPForge [37]. These grids are process-independent and can be applied to construct ACOT-like GMVFN theory predictions for other processes.

The work of L.R. is supported in part by the U.S. Department of Energy under grant DE- SC0010102. M.G. is partially supported by the National Science Foundation under Grants No. PHY-2112025 and No. PHY-2412071. P.M.N. was partially supported by the U.S. Department of Energy under Grant No. DE-SC0010129. The work of D.W. is supported in part by the National Science Foundation under Grants No. PHY- 2014021 and No. PHY-2310363. K.X. is supported by the U.S. National Science Foundation under Grants No. PHY-2310291 and PHY-2310497.

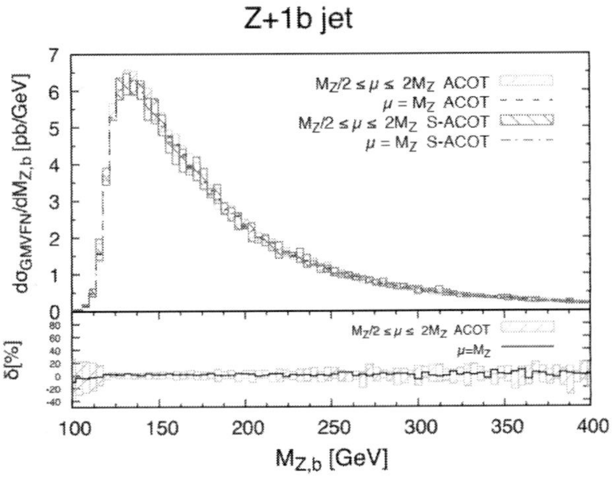

Figure 2. $M_{Z,b}$ differential distribution with scale dependence $M_Z/2 < \mu < 2M_Z$ at order $O(\alpha_s^3)$ for the ACOT and S-ACOT GMVFN schemes. The CT18NLO PDFs are used.

References

[1] J. McGowan, T. Cridge, L.A. Harland-Lang, R.S. Thorne, Approximate N^{3}LO parton distribution functions with theoretical uncertainties: MSHT20aN^{3}LO PDFs, Eur. Phys. J. C **83**, 185 (2023), [Erratum: Eur.Phys.J.C 83, 302 (2023)], 2207.04739. 10.1140/epjc/s10052-023-11236-0

[2] R.D. Ball et al. (NNPDF), The path to N^{3}LO parton distributions, Eur. Phys. J. C **84**, 659 (2024), 2402.18635. 10.1140/epjc/s10052-024-12891-7

[3] J.A.M. Vermaseren, A. Vogt, S. Moch, The Third-order QCD corrections to deep-inelastic scattering by photon exchange, Nucl. Phys. B **724**, 3 (2005), hep-ph/0504242. 10.1016/j.nuclphysb.2005.06.020

[4] S. Moch, J.A.M. Vermaseren, A. Vogt, The Longitudinal structure function at the third order, Phys. Lett. B **606**, 123 (2005), hep-ph/0411112. 10.1016/j.physletb.2004.11.063

[5] S. Moch, M. Rogal, A. Vogt, Differences between charged-current coefficient functions, Nucl. Phys. B **790**, 317 (2008), 0708.3731. 10.1016/j.nuclphysb.2007.09.022

[6] S. Moch, J.A.M. Vermaseren, A. Vogt, Third-order QCD corrections to the charged-current structure function F(3), Nucl. Phys. B **813**, 220 (2009), 0812.4168. 10.1016/j.nuclphysb.2009.01.001

[7] H. Kawamura, N.A. Lo Presti, S. Moch, A. Vogt, On the next-to-next-to-leading order QCD corrections to heavy-quark production in deep-inelastic scattering, Nucl. Phys. B **864**, 399 (2012), 1205.5727. 10.1016/j.nuclphysb.2012.07.001

[8] J. Davies, A. Vogt, S. Moch, J.A.M. Vermaseren, Non-singlet coefficient functions for charged-current deep-inelastic scattering to the third order in QCD, PoS **DIS2016**, 059 (2016), 1606.08907. 10.22323/1.265.0059

[9] J. Blümlein, P. Marquard, C. Schneider, K. Schönwald, The massless three-loop Wilson coefficients for the deep-inelastic structure functions F_2, F_L, xF_3 and g_1, JHEP **11**, 156 (2022), 2208.14325. 10.1007/JHEP11(2022)156

[10] J. Ablinger, A. Behring, J. Blümlein, A. De Freitas, A. von Manteuffel, C. Schneider, K. Schoenwald, The three-loop single-mass heavy flavor corrections to deep-inelastic scattering (2024), 2407.02006

[11] J. Baglio, C. Duhr, B. Mistlberger, R. Szafron, Inclusive production cross sections at N^3LO, JHEP **12**, 066 (2022), 2209.06138. 10.1007/JHEP12(2022)066

[12] C. Duhr, F. Dulat, B. Mistlberger, Charged current Drell-Yan production at N^3LO, JHEP **11**, 143 (2020), 2007.13313. 10.1007/JHEP11(2020)143

[13] C. Duhr, B. Mistlberger, Lepton-pair production at hadron colliders at N^3LO in QCD, JHEP **03**, 116 (2022), 2111.10379. 10.1007/JHEP03(2022)116

[14] X. Chen, T. Gehrmann, N. Glover, A. Huss, T.Z. Yang, H.X. Zhu, Dilepton Rapidity Distribution in Drell-Yan Production to Third Order in QCD, Phys. Rev. Lett. **128**, 052001 (2022), 2107.09085. 10.1103/PhysRevLett.128.052001

[15] X. Chen, T. Gehrmann, N. Glover, A. Huss, T.Z. Yang, H.X. Zhu, Transverse mass distribution and charge asymmetry in W boson production to third order in QCD, Phys. Lett. B **840**, 137876 (2023), 2205.11426. 10.1016/j.physletb.2023.137876

[16] C. Anastasiou, C. Duhr, F. Dulat, F. Herzog, B. Mistlberger, Higgs Boson Gluon-Fusion Production in QCD at Three Loops, Phys. Rev. Lett. **114**, 212001 (2015), 1503.06056. 10.1103/PhysRevLett.114.212001

[17] B. Mistlberger, Higgs boson production at hadron colliders at N^3LO in QCD, JHEP **05**, 028 (2018), 1802.00833. 10.1007/JHEP05(2018)028

[18] F. Dulat, B. Mistlberger, A. Pelloni, Precision predictions at N^3LO for the Higgs boson rapidity distribution at the LHC, Phys. Rev. D **99**, 034004 (2019), 1810.09462. 10.1103/PhysRevD.99.034004

[19] F.A. Dreyer, A. Karlberg, Vector-Boson Fusion Higgs Production at Three Loops in QCD, Phys. Rev. Lett. **117**, 072001 (2016), 1606.00840. 10.1103/PhysRevLett.117.072001

[20] L. Cieri, X. Chen, T. Gehrmann, E.W.N. Glover, A. Huss, Higgs boson production at the LHC using the q_T subtraction formalism at N^3LO QCD, JHEP **02**, 096 (2019), 1807.11501. 10.1007/JHEP02(2019)096

[21] T. Gehrmann, A. Huss, J. Niehues, A. Vogt, D.M. Walker, Jet production in charged-current deep-inelastic scattering to third order in QCD, Phys. Lett. B **792**, 182 (2019), 1812.06104. 10.1016/j.physletb.2019.03.003

[22] J. Currie, T. Gehrmann, E.W.N. Glover, A. Huss, J. Niehues, A. Vogt, N^3LO corrections to jet production in deep inelastic scattering using the Projection-to-Born method, JHEP **05**, 209 (2018), 1803.09973. 10.1007/JHEP05(2018)209

[23] M. Aivazis, F.I. Olness, W.K. Tung, Leptoproduction of heavy quarks. 1. General formalism and kinematics of charged current and neutral current production processes, Phys. Rev. D **50**, 3085 (1994), hep-ph/9312318. 10.1103/PhysRevD.50.3085

[24] M. Aivazis, J.C. Collins, F.I. Olness, W.K. Tung, Leptoproduction of heavy quarks. 2. A Unified QCD formulation of charged and neutral current processes from fixed target to collider energies, Phys. Rev. D **50**, 3102 (1994), hep-ph/9312319. 10.1103/PhysRevD.50.3102

[25] W.K. Tung, S. Kretzer, C. Schmidt, Open heavy flavor production in QCD: Conceptual framework and implementation issues, J. Phys. G **28**, 983 (2002), hep-ph/0110247. 10.1088/0954-3899/28/5/321

[26] M. Kramer, 1, F.I. Olness, D.E. Soper, Treatment of heavy quarks in deeply inelastic scattering, Phys. Rev. **D62**, 096007 (2000), hep-ph/0003035. 10.1103/PhysRevD.62.096007

[27] M. Guzzi, P.M. Nadolsky, H.L. Lai, C.P. Yuan, General-Mass Treatment for Deep Inelastic Scattering at Two-Loop Accuracy, Phys. Rev. D **86**, 053005 (2012), 1108.5112. 10.1103/PhysRevD.86.053005

[28] M. Guzzi, P. Nadolsky, L. Reina, D. Wackeroth, K. Xie, A general mass variable flavor number scheme for Z boson production in association with a heavy quark at hadron colliders (2024), 2410.03876

[29] J. Gao, T.J. Hobbs, P.M. Nadolsky, C. Sun, C.P. Yuan, General heavy-flavor mass scheme for charged-current DIS at NNLO and beyond, Phys. Rev. D **105**, L011503 (2022), 2107.00460. 10.1103/PhysRevD.105.L011503

[30] A. Kusina, F. Olness, I. Schienbein, T. Jezo, K. Kovarik, T. Stavreva, J. Yu, Hybrid scheme for heavy flavors: Merging the fixed flavor number scheme and variable flavor number scheme, Phys. Rev. D **88**, 074032 (2013), 1306.6553. 10.1103/PhysRevD.88.074032

[31] K. Xie, Ph.D. thesis, Southern Methodist U. (2019)

[32] K. Xie, J.M. Campbell, P.M. Nadolsky, A general-mass scheme for prompt charm production at hadron colliders, in *28th International Workshop on Deep Inelastic Scattering and Related Subjects* (2021), 2108.03741

[33] K. Xie, M. Guzzi, P. Nadolsky, Probing heavy-flavor parton distribution functions at hadron colliders, in *Snowmass 2021* (2022), 2203.06207

[34] I. Helenius, H. Paukkunen, Revisiting the D-meson hadroproduction in general-mass variable flavour number scheme, JHEP **05**, 196 (2018), 1804.03557. 10.1007/JHEP05(2018)196

[35] T.J. Hou et al., New CTEQ global analysis of quantum chromodynamics with high-precision data from the LHC, Phys. Rev. D **103**, 014013 (2021), 1912.10053. 10.1103/PhysRevD.103.014013

[36] A. Buckley, J. Ferrando, S. Lloyd, K. Nordström, B. Page, M. Rüfenacht, M. Schönherr, G. Watt, LHAPDF6: parton density access in the LHC precision era, Eur. Phys. J. C **75**, 132 (2015), 1412.7420. 10.1140/epjc/s10052-015-3318-8

[37] K. Xie et al., The Simplified ACOT scheme with Massive Phase Space, https://sacotmps.hepforge.org/

[38] F. Febres Cordero, L. Reina, D. Wackeroth, NLO QCD corrections to $Zb\bar{b}$ production with massive bottom quarks at the Fermilab Tevatron, Phys. Rev. D **78**, 074014 (2008), 0806.0808. 10.1103/PhysRevD.78.074014

[39] F. Febres Cordero, L. Reina, D. Wackeroth, W- and Z-boson production with a massive bottom-quark pair at the Large Hadron Collider, Phys. Rev. D **80**, 034015 (2009), 0906.1923. 10.1103/PhysRevD.80.034015

[40] D. Figueroa, S. Honeywell, S. Quackenbush, L. Reina, C. Reuschle, D. Wackeroth, Electroweak and QCD corrections to Z-boson production with one b jet in a massive five-flavor scheme, Phys. Rev. D **98**, 093002 (2018), 1805.01353. 10.1103/PhysRevD.98.093002

[41] E. Witten, Heavy Quark Contributions to Deep Inelastic Scattering, Nucl. Phys. B **104**, 445 (1976). 10.1016/0550-3213(76)90111-5

[42] M. Buza, Y. Matiounine, J. Smith, R. Migneron, W.L. van Neerven, Heavy quark coefficient functions at asymptotic values $Q^2 \gg m^2$, Nucl. Phys. B **472**, 611 (1996), hep-ph/9601302. 10.1016/0550-3213(96)00228-3

[43] M. Buza, Y. Matiounine, J. Smith, W. van Neerven, Charm electroproduction viewed in the variable flavor number scheme versus fixed order perturbation theory, Eur. Phys. J. C **1**, 301 (1998), hep-ph/9612398. 10.1007/BF01245820

EPJ Web of Conferences **314**, 00022 (2024) https://doi.org/10.1051/epjconf/202431400022
QCD@Work 2024

CMS experimental results

Maitreyee Mukherjee[1,*], on behalf of the CMS Collaboration

[1]Panjab University, Chandigarh, India

Abstract. The main goal of QCD studies is to improve our detailed description of the Standard Model physics. Studies of QCD are key to understand production of all BSM signals and backgrounds at the LHC. QCD measurements are important for data interpretation, background studies and precision studies. This study is focussed on the recent interesting results from the CMS Experiment at CERN.

1 Introduction

LHC provides experimental access to large phase space, accurate results from various final state, studies with PDFs, etc. These help to perform many important and interesting QCD studies experimentally. In this article, the recent results from CMS Experiment [1] have been presented. The results are mainly related to Higgs Physics, Exotica, Top Quark, Standard Model Physics and B-Physics.

2 Recent results related to Higgs Boson

2.1 Higgs Boson mass and width

Higgs Boson mass (m_H) is one of the most important free parameters of SM. Other Higgs Boson properties depend on mass. With $H \rightarrow ZZ^* \rightarrow 4l$ decay, the value reads, $m_H = 125.08 \pm 0.12(\pm 0.10\ (stat) \pm 0.07\ (syst))\ GeV$ using $Run\ 1 + 2$ results [2]. Higgs Boson width is measured in $H \rightarrow ZZ$ channel, instead of direct measurement. Full Run 2 data has been used. The on-shell and off-shell productions have been compared in different decay channels. The measured value is, $\Gamma_H = 2.9^{+1.9}_{-1.4}$ MeV. Observed (expected) upper and lower 95% CL is, $0.6 < \Gamma_H < 7.0\ (0.1 < \Gamma_H H < 10.5)$ MeV. The measured Higgs widths are compatible with the SM prediction [3].

With the channel $H \rightarrow \gamma\gamma$ the value of mass becomes, $m_H = 125.38 \pm 0.14\ [\pm 0.11(stat) \pm 0.08(syst)]$ GeV The 1σ and 2σ CL intervals are determined considering the asymptotic approximation is valid.

2.2 Search for rare decays of the Higgs boson into a photon and meson

This process is sensitive to light quark Yukawa coupling, may become significant in BSM scenarios. Higgs boson decays into a photon and a $\rho^0(770)$, $\phi\ or\ K^{*0}(892)$ meson are

*e-mail: maitreyee.mukherjee@cern.ch

searched by using proton-proton collision data collected by the CMS experiment at the LHC at $\sqrt{s} = 13\ TeV$. Events are selected in which the mesons decay into pairs of charged pions, pairs of charged kaons, and charged kaon-pion pairs respectively. $B\,(H \to \phi\gamma) \sim 10^{-6}$, $B\,(H \to \rho^0\gamma) \sim 10^{-5}$ and $B\,(H \to K^{*0}\gamma) \sim 10^{-19}$, the decays are powerful probes of possible anomalous flavour-changing Higgs boson couplings. This analysis uses single photon and meson/VBF jet trigger for partial Run-2 data (ggF and VBF modes) and lepton triggers for full Run-2 (VH production). Analysis is done by the reconstruction of photon and di-track invariant mass, mass windows applied to di-track system to mitigate meson misidentification. Signal selection has been improved using multivariate classifier.

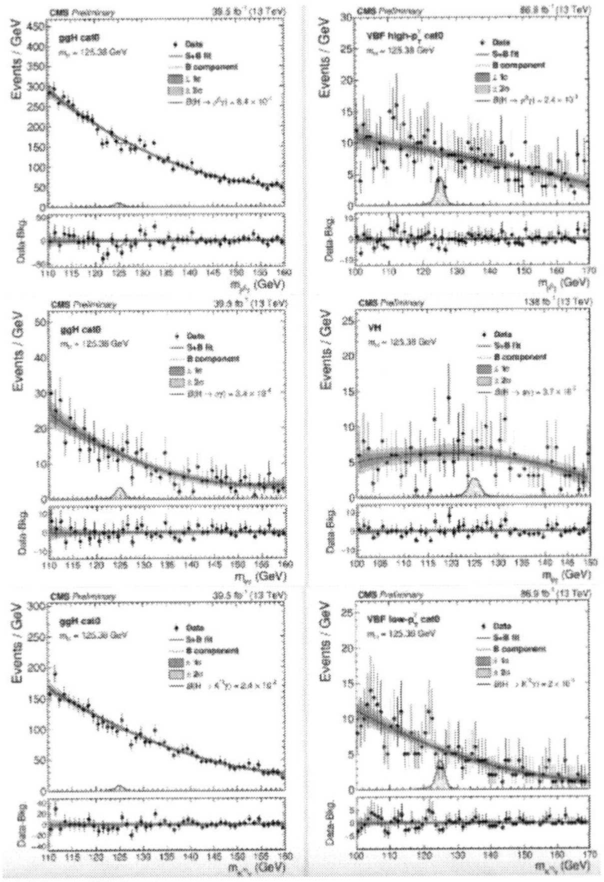

Figure 1. Post-fit m_{M_γ} distributions in data and the background model for ggH cat0 (upper left) and VBF high-$p_T^\gamma/cat0$ (upper right) of the $H \to \rho^0\gamma$ search, ggH cat0 (middle left) and the VH category (middle right) of the $H \to \phi\gamma$ search, ggH cat0 (lower left) and the VBF low-$p_T^\gamma/cat0$ (lower right) of the $H \to K^*(892)^0\gamma$ search. The signal simulation is shown normalized at a branching fraction corresponding to the expected upper limit [4]. Signal and background fit components, as well as 1- and 2-σ fit uncertainty bands are also shown.

Upper limits at the 95% confidence level on the Higgs boson branching fractions constitute the most stringent experimental limits to date [4], as shown in the Fig. 1.

The interesting results of a model-independent search for the pair production of new bosons decaying into muons in proton-proton collisions at $\sqrt{s} = 13TeV$ has been discussed in Ref [5].

3 Recent results related to Exotica

3.1 Search for low-mass long-lived particles decaying to displaced jets in proton-proton collisions at $\sqrt{s} = 13.6$ TeV

A search for low-mass long-lived particles (LLPs) decaying into jets has been performed using proton-proton collision data corresponding to an integrated luminosity of 34.7 fb^{-1}, collected at a center-of-mass energy of 13.6 TeV in 2022 [6]. Novel techniques in trigger, reconstruction, and machine learning were developed for and employed in this search, leading to significant improvements over existing results. The observed yields are consistent with the background predictions. The best limits to date are set for LLPs with masses between 15 and 55 GeV and with proper decay lengths smaller than $\sim 1\ m$. The search provides the first exclusions of hadonically decaying tau leptons with decay lengths smaller than $\sim 1\ m$. The branching fraction limits outperform previous results by a factor of up to 10 (8) as shown in Fig. 2.

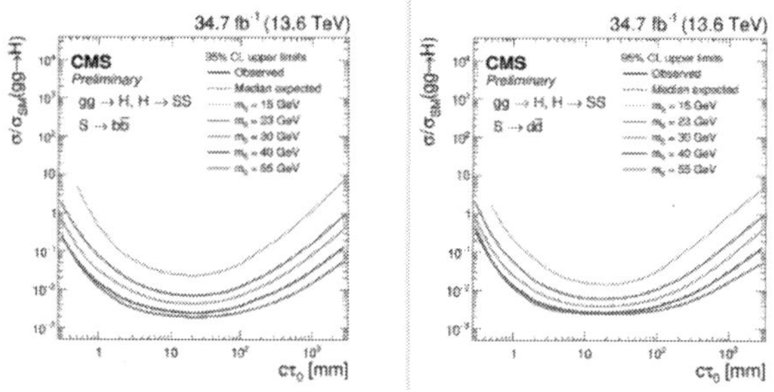

Figure 2. The 95% CL upper limits on the branching fraction B ($H \to SS$) for $S \to b\bar{b}$ (upper left) and $S \to d\bar{d}$ (upper right) for different LLP masses m_s and proper decay lengths $c\tau_0$ [6].

3.2 Search for new physics with emerging jets in proton-proton collisions at $\sqrt{s} = 13$ TeV

A search for emerging jet signatures arising from a strongly interacting dark sector produced in hadron collisions has been presented, using data corresponding to a total integrated luminosity of 138 fb^{-1} at $\sqrt{s} = 13\ TeV$ [7]. The signal model contains a family of dark quarks that couples to the SM quarks via a scalar mediator X_{dark}. Dark pions (π_{dark}) with a significant lifetime are produced by the hadronization of the dark quarks and then decay to SM particles at vertices displaced from the original interaction point. As the scalar mediator is assumed to be produced in pairs, with each decaying to a SM quark and a dark quark, the signature of this process is two SM jets plus two jets of particles with the constituents emerging from displaced vertices. Both unflavored and flavor-aligned couplings between the SM quarks and the dark quarks are examined in the search. The selection criteria were optimized for each class of couplings and as a function of the mediator particle mass, the dark pion mass, and the dark pion lifetime, using both a traditional cut-based selection and a graph neural

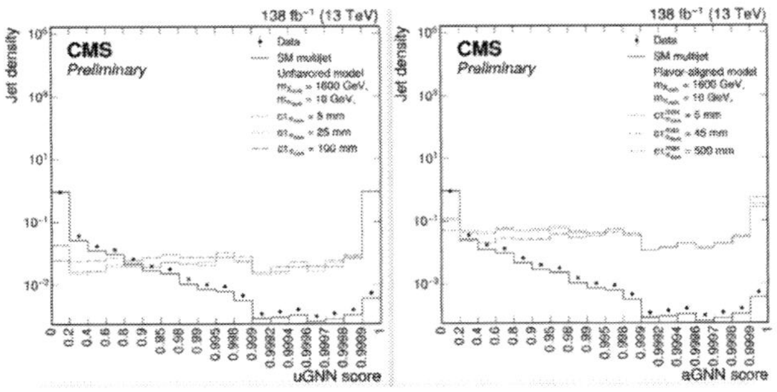

Figure 3. Distributions of the GNN output score for jets obtained from SM multijet simulation events (dark gray line), signal jets from the unflavored model (colored lines, left), and signal jets from the flavor-aligned models (colored lines, right). Results of the uGNN (aGNN) are shown on the left (right). The performance of the GNN varies only slightly over a wide range of the $c\tau$ of the dark mesons. The distributions are normalized to unit area [7].

network. The observed 95% confidence level exclusion limits agree with the expected limits. For the unflavored model, dark mediator masses up to $m_{X_{dark}} < 1950$ GeV are excluded for $c_{\tau_{\pi_{dark}}} \sim 100$ mm and $m_{\pi_{dark}} \sim 10$ GeV, while the flavor-aligned model result excludes $m_{X_{dark}} < 1850$ GeV at $c_{\tau_{\pi_{dark}}} \sim 500$ mm for $m_{\pi_{dark}} \sim 10$ GeV. This result as presented in Fig. 3, greatly surpasses the previous search for emerging jets in the unflavored scenario, increasing the experimental limit of the dark mediator particle by $\sim$ 500 GeV to set the most stringent limits to date, and provides the first direct exclusion of the flavor-aligned scenario.

4 Recent results related to top quarks

4.1 Observation of quantum entanglement in top quark pair production in proton-proton collisions at $\sqrt{s} = 13$ TeV

It is fascinating to see how the largest accelerator on Earth can also provide tools to explore fundamental quantum mechanics effects. A measurement of the extent of entanglement in top quark-antiquark ($t\bar{t}$) events produced in proton-proton collisions at a center-of-mass energy of 13 TeV is performed with the data recorded by the CMS experiment in 2016, and corresponding to an integrated luminosity of 36.3 fb^{-1} [8]. This measurement exploits the spin correlation variable D, which at the $t\bar{t}$ production threshold, and in absence of BSM contributions, provides access to the full spin correlation information. We directly measure entanglement at the parton level. Also, our analysis is the first to consider non-relativistic bound-state effects in the production threshold by including the ground state of toponium, η^t. The D variable represents an entanglement proxy, where less than -$\frac{1}{3}$ signals the presence of entanglement (Fig. 4). This proxy is measured using events containing two oppositely charged electrons or muons produced in pp collisions at a center-of-mass energy of 13 TeV. A combined signal model of ($t\bar{t} + \eta^t$) has been utilized in the measurement. A binned likelihood fit has been used to extract the entanglement proxy, directly from the distribution of cos ϕ, where ϕ is the angle between the two charged decay leptons in their respective parent top quark rest frames. This result has an observed (expected) significance of 5.1 (4.7) σ, corresponding to the observation of top quark entanglement. The measured value of D is in good agreement with the MC modelling in this phase space when including the expected η^t bound state contribution.

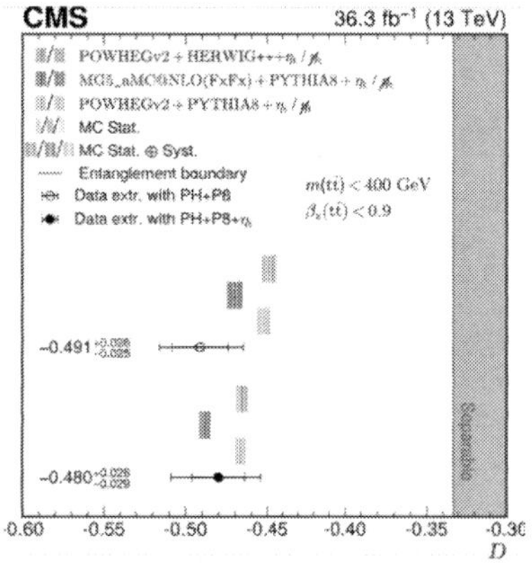

Figure 4. Measurement of the entanglement proxy D in data (black filled or open point) compared with MC predictions including (solid line) or not including (dashed line) contributions from the η_t−state. The legend denotes MC predictions without the toponium state with a slash. The boundary for entanglement is indicated by the shaded region at $D = -\frac{1}{3}$ [8].

4.2 Measurement of inclusive and differential cross sections for single top quark production in association with a W boson in proton-proton collisions at $\sqrt{s} = 13$ TeV

The first measurement of the inclusive and normalised differential cross sections for the production of single top quarks in association with a W boson in proton-proton collisions at a centre-of-mass energy of 13.6 TeV are presented. This is the first single top measurement at Run 3 from CMS. This is a challenging measurement as it is a rare process. The data used

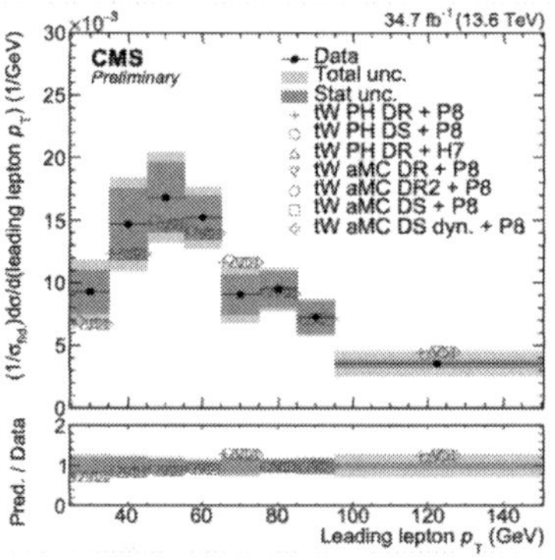

Figure 5. Normalised fiducial differential tW production cross section as function of the p_T of the leading lepton. Predictions from $POWHEG(PH) + PYTHIA8(P8)DR$ and $DS, POWHEG + HERWIG7(H7)DR, MadGraph-5\_aMC@NLO(aMC) + PYTHIA8DR, DR2, DS$, and DS with a dynamic factor are also shown. The grey band represents the statistical uncertainty and the orange band the total uncertainty. In the lower panel, the ratio of the prediction to the data is shown [9].

were recorded with the CMS detector at the LHC during 2022, and correspond to an integrated luminosity of 34.7 fb^{-1} [9]. The analyzed events contain one muon and one electron in the final state. For the inclusive measurement, multivariate discriminants are used exploiting the kinematic properties of the events to separate the signal from the dominant top quark antiquark production background. A cross section of $84.1 \pm 2.1(stat)^{+9.8}_{-10.2}(syst) \pm 3.3(lumi)$ pb is obtained, as shown in the Fig. 5, consistent with the predictions of the standard model. A fiducial region is defined according to the detector acceptance for performing the differential measurements. The resulting differential distributions are unfolded to particle level and show good agreement with the predictions at next-to-leading order in perturbative QCD.

5 Recent results related to the standard model physics

5.1 Measurement of the Drell-Yan forward-backward asymmetry and of the effective leptonic weak mixing angle using proton-proton collisions at 13 TeV

The electroweak mixing angle, $sin^2\theta_W = 1 - \frac{m_W^2}{m_Z^2}$, relates the masses of the W and Z bosons and is a key parameter in the standard model (SM) of particle physics. The forward-backward asymmetry in Drell–Yan production and the effective leptonic electroweak mixing angle are measured using a sample of proton-proton collisions at $\sqrt{s} = 13$ TeV collected by the CMS experiment and corresponding to an integrated luminosity of 137 fb^{-1} [10]. The measurement uses both dimuon and dielectron events, and is performed as a function of the dilepton mass and rapidity. Using the CT18Z set of parton distribution functions (PDF), we obtain,

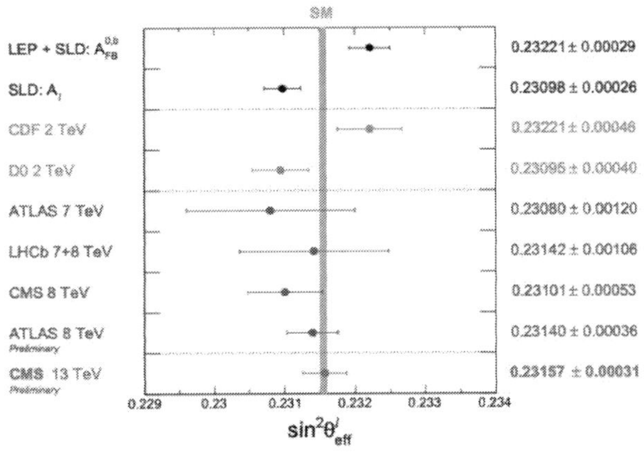

Figure 6. Comparison of the $sin^2\theta^l_{eff}$ values measured in this analysis with previous measurements and the SM prediction [10].

$sin^2\theta^l_{eff} = 0.23157 \pm 0.00010\ (stat) \pm 0.00015\ (syst) \pm 0.00009\ (theo) \pm 0.00027\ (PDF)$, the total uncertainty being 0.00031. The measured value agrees with the standard model prediction. The total uncertainty varies between 0.00024 and 0.00035, depending on the PDF set. With significantly smaller uncertainty than that of previous CMS results, increase in the size of the data samples, improved analysis techniques and the inclusion of central-forward dielectrons, this is the most precise measurement of the effective weak mixing angle at a hadron collider, with a precision comparable to the results obtained at LEP and SLD, as presented in the Fig. 6.

5.2 Strong coupling constant (α_s) in CMS at 13 TeV

Strong coupling constant (α_s) is the only free parameter except quark masses. In CMS, values have been measured in a wide spectrum of analyses at 13 TeV. For NNLO prediction, fit uncertainty has become dominant. Uncertainties are cancelled out for ratios of cross sections. Simultaneous measurements (dijet + $t\bar{t}$) can improve the results. NNLO predictions missing for some measurements like jet substructure. Running of α_s is consistent with the Renormalization Group Equations over 3 orders of magnitude (Refer to Fig. 7).

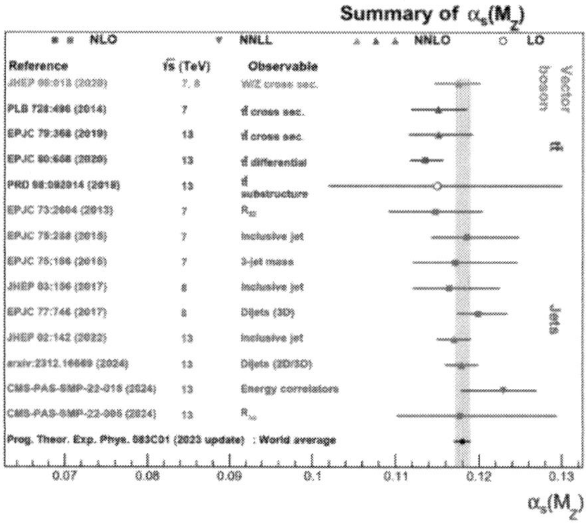

Figure 7. Summary of the strong coupling constant (α_s) measurements for 13 TeV.

6 Recent results related to the B-physics

6.1 Measurement of time-dependent CP violation in $B_s^0 \rightarrow J/\psi\phi$ (1020) decays with the CMS detector

A study of CP violation in $B_s^0 \rightarrow J/\psi\phi(1020)$ decays is presented, based on the data collected by the CMS experiment at the LHC. The measurement is based on a data sample of roughly 491 000 reconstructed signal events corresponding to an integrated luminosity of 96.5 fb^{-1}, collected in proton-proton collisions at 13 TeV in 2017-2018 [11]. A time- and flavor-dependent angular analysis of the $\mu^+\mu^-K^+K^-$ final state is performed to extract the values of the CP-violating phase ϕ_s, the decay width difference ($\Delta\Gamma_s$), and absolute mass difference (Δm_s) between the light and heavy B_s^0 mass eigenstates, their average decay width Γ_s, and the direct CP violation parameter $|\lambda|$. The analysis utilizes a pioneering inclusive flavor tagging framework based on state-of-the-art machine learning techniques.

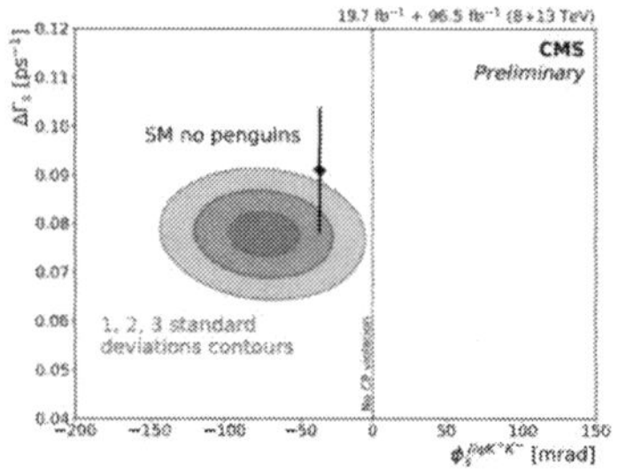

Figure 8. The two-dimensional one, two, and three standard deviations contours in the $\phi_s - \Delta\Gamma_s$ plane for the combined results. The contours take into account both statistical and systematic uncertainties. The SM prediction is represented by the black rectangle, with the central value indicated with the black diamond [11].

CMS has used a graph neural network (cutting-edge AI algorithm for accurate flavour tagging) to separate mesons from anti-mesons. In combination with similar CMS results from Run 1, CMS reports a ϕ_s value deviating from zero with a significance of 3.2 standard deviations. This result provides the first evidence for CP violation in these B_s^0 decays (as presented in the Fig. 8), a milestone in our CP violation program.

References

[1] CMS Collaboration, "The CMS Experiment at the CERN LHC". JINST **3** S08004 (2008). 10.1088/1748-0221/3/08/S08004

[2] CMS Collaboration, "A measurement of the Higgs boson mass in the diphoton decay channel". PLB **805** 135425 (2020). https://doi.org/10.1016/j.physletb.2020.135425

[3] CMS Collaboration, "Measurement of the Higgs boson mass and width using the four leptons final state". **CMS-PAS-HIG-21-019** (2023). https://cds.cern.ch/record/2871702?ln=en

[4] CMS Collaboration, "Search for rare decays of the Higgs boson into a photon and a ρ, ϕ or K^{*0} meson". **CMS-PAG-HIG-23-005** (2024). https://cds.cern.ch/record/2894541

[5] CMS Collaboration, "Model-independent search for pair production of new bosons decaying into muons in proton-proton collisions at $\sqrt{s}$ = 13 TeV". **CMS-PAS-HIG-21-004** (2024) https://cds.cern.ch/record/2893561

[6] CMS Collaboration, "Search for low-mass long-lived particles decaying to displaced jets in proton-proton collisions at $\sqrt{s}$ = 13.6 TeV". **CMS-PAS-EXO-23-013** (2024). https://cds.cern.ch/record/2893044

[7] CMS Collaboration, "Search for new physics with emerging jets in proton-proton collisions at $\sqrt{s}$ = 13 TeV". **CMS-PAS-EXO-22-015** (2024). https://cds.cern.ch/record/2889672

[8] CMS Collaboration, "Observation of quantum entanglement in top quark pair production in proton-proton collisions at $\sqrt{s}$ = 13 TeV". **CMS-TOP-23-001** (2024). https://doi.org/10.48550/arXiv.2406.03976

[9] CMS Collaboration, "Measurement of inclusive and differential cross sections for single top quark production in association with a W boson in proton-proton collisions at $\sqrt{s}$ = 13.6 TeV". **CMS-PAS-TOP-23-008** (2024). https://doi.org/10.48550/arXiv.2409.06444

[10] CMS Collaboration, "Measurement of the Drell-Yan forward-backward asymmetry and of the effective leptonic weak mixing angle using proton-proton collisions at 13 TeV". **CMS-PAS-SMP-22-010** (2024). https://cds.cern.ch/record/2893842?ln=en

[11] CMS Collaboration, "Measurement of time-dependent CP violation in $B_s^0 \to J/\psi\phi(1020)$ decays with the CMS detector". **CMS-PAS-BPH-23-004** (2024). https://cds.cern.ch/record/2894821?ln=en

EPJ Web of Conferences **314**, 00023 (2024) https://doi.org/10.1051/epjconf/202431400023
QCD@Work 2024

Event shape distribution and the NLP corrections

Neelima Agarwal[1,*], *Melissa van* Beekveld[2,**], *Eric* Laenen[3,4,5,***], *Shubham* Mishra[6,7,****], *Ayan* Mukhopadhyay[6,†], and *Anurag* Tripathi[6,‡]

[1]Department of Physics, Chaitanya Bharathi Institute of Technology, Gandipet, Hyderabad, Telangana State 500075, India
[2]Rudolf Peierls Centre for Theoretical Physics, Clarendon Laboratory, Parks Road, University of Oxford, Oxford OX1 3PU, UK
[3]Institute of Physics, University of Amsterdam, Science Park 904, 1098 XH Amsterdam, The Netherlands
[4]Nikhef, Theory Group, Science Park 105, 1098 XG, Amsterdam, The Netherlands
[5]Institute for Theoretical Physics, Utrecht University
[6]Department of Physics, Indian Institute of Technology Hyderabad, Kandi, Sangareddy, Telangana State 502285, India
[7]Department of Physics, Bharat Institute of Engineering and Technology, Mangalpally, Ibrahimpatnam, Hyderabad, Telangana State, 501510, India

Abstract. For the event shapes thrust and c-parameter at next-to-leading order, we examine the origins of next-to-leading power corrections. We track down the origin of these factors for each shape of the event and compare our findings with a recent method that makes use of the eikonal approximation and momentum changes derived from the Low-Burnett-Kroll-Del Duca theorem. We do an analytical and numerical analysis of the differences. Both precise and approximate findings are described in terms of elliptic integrals for the c-parameter; yet, it exhibits patterns that are similar to those observed in the thrust results near the elastic limit.

1 Introduction

In perturbative QCD, precise cross-section estimates must be obtained at the same rate as collider-physics observations get more precise. Two contrasting approaches are used in this endeavor. Initially, precise higher-order computations in the coupling α_s are performed, and pertinent techniques are developed. Second, under particular kinematic limits, when certain classes of logarithmic terms become strengthened, all-order results are achieved. There are numerous methods that can be used to resum these logarithmic terms to any order, and their variety keeps growing. This holds true in particular for the near-elastic zone, also known as the threshold region, where the phase space of emitted particles is bounded. In such a

*e-mail: dragarwalphysics@gmail.com
**e-mail: melissa.vanbeekveld@physics.ox.ac.uk
***e-mail: eric.laenen@nikhef.nl
****e-mail: shubhamhere82@gmail.com
†e-mail: ayanmukhopadhyay5@gmail.com
‡e-mail: tripathi@phy.iith.ac.in

situation, the cancellation of infrared singularities, guaranteed by the KLN theorem [1, 2] leaves large logarithmic remainders at any order in perturbation theory. To be more specific, if ξ is a dimensionless kinematic variable, such that $\xi \to 0$ towards the elastic region, the corresponding differential cross-section has the generic form

$$\frac{d\sigma}{d\xi} = \sum_{n=0}^{\infty}\left(\frac{\alpha_s}{\pi}\right)^n \left[\sum_{m=0}^{2n-1} c_{nm}^{\mathrm{LP}}\left(\frac{\log^m \xi}{\xi}\right)_+ + c_n^{(\delta)}\delta(\xi) + \sum_{m=0}^{2n-1} c_{nm}^{\mathrm{NLP}} \log^m \xi + \ldots\right]. \tag{1}$$

The first term on the right in the above equation is well-known to originate from soft and/or collinear radiation and, together with the second term, makes up the *leading power* (LP) terms. Much is known about LP terms to arbitrary order, and there have been numerous approaches towards their resummation and power correction studies. The last term on the right corresponds to terms that are *next-to-leading power* (NLP). These are produced via collinear gluon, soft quark/anti-quark, and soft gluon emissions (next to). Even if they are supressed by a power of ξ, the growth of these logarithmic terms towards threshold makes them numerically relevant for the precision studies. The exact arrangement of these NLP terms to all orders and arbitrary logarithmic accuracy is yet unclear, in contrast to the LP terms. In this work [3], we focus on such NLP terms for two event shapes in e^+e^- collisions: thrust [4, 5] and the C-parameter [6]. The calculation at NLO for spherocity can be found in [7]. These observables are interesting in this regard because in contrast to most of the previous studies in direct QCD, all QCD effects reside in the final state, and because their definition involves special phase space constraints that were not considered so far. For these observables, our aim is to trace the origin of the NLP terms near the elastic limit, to examine to what extent there is a common pattern of NLP terms, and assess their size. Here we examine to what extent NLP terms for two event shapes can be predicted using the kinematical shift method, as well as the soft quark emission approximation [8, 9].

The leading logarithmic terms can alternatively be computed using the approach of shifted kinematics [10], in this approach the matrix element that can capture terms up to NLP at NLO is given by the formula

$$\overline{\sum}|\mathcal{M}|^2 = g_s^2 N_c(N_c^2-1)\frac{2p_1\cdot p_2}{(p_1\cdot p_3)(p_2\cdot p_3)}|\mathcal{M}_0(p_1-\delta p_1, p_2-\delta p_2)|^2\,, \tag{2}$$

where $|\mathcal{M}_0(p_1,p_2)|^2$ is the matrix element squared at the leading order (LO), and $\overline{\sum}$ denotes the sum (average) over the final (initial) state spins and colours, p_3 is the momentum of the emitted radiation, and p_1, p_2 are the momenta of the particles already present at the Born level. The shifts in the momenta are given by

$$\delta p_1^\mu = -\frac{1}{2}\left(\frac{p_2\cdot p_3}{p_1\cdot p_2}p_1^\mu - \frac{p_1\cdot p_3}{p_1\cdot p_2}p_2^\mu + p_3^\mu\right),\ \delta p_2^\mu = -\frac{1}{2}\left(\frac{p_1\cdot p_3}{p_1\cdot p_2}p_2^\mu - \frac{p_2\cdot p_3}{p_1\cdot p_2}p_1^\mu + p_3^\mu\right). \tag{3}$$

Expressions (2) and (3) yield the dominant NLP contributions to the NLO matrix element.

2 Event shape distribution

In this work the process taken under consideration, and shown in figure below is

$$e^+(p_b) + e^-(p_a) \to \gamma^*(q) \to q(p_1) + \overline{q}(p_2) + g(p_3)\,. \tag{4}$$

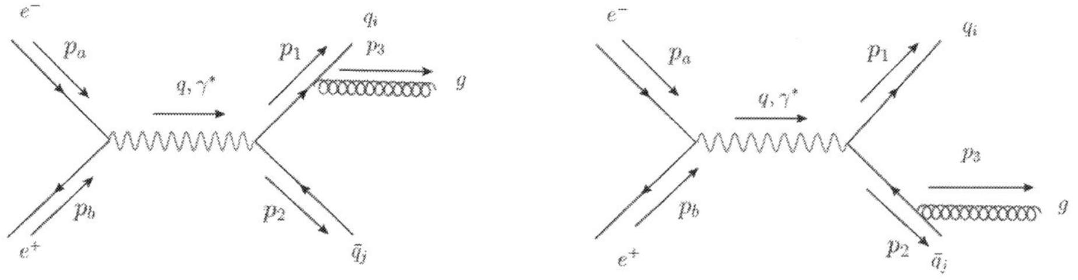

Figure 1. Feynman diagrams for the real emission of a gluon.

The event shape distribution at NLO has the form

$$\frac{d\sigma}{dX} = \frac{1}{2s}\int d\Phi_3 \overline{\sum}|\mathcal{M}(x_1,x_2)|^2\delta(X - X(x_1,x_2,x_3)), \tag{5}$$

where s is the center of mass energy squared, and $d\Phi_3$ is the three-body phase space factor. The scaleless energy fraction variables x_i are defined for each parton as

$$x_i = \frac{2E_i}{Q}, \tag{6}$$

where E_i is the energy of ith parton, and Q is the centre of mass energy. The energy fraction variables satisfy the constraint $x_1 + x_2 + x_3 = 2$. The matrix element squared for the process in eq. (4) is

$$\overline{\sum}|\mathcal{M}(x_1,x_2)|^2 = 8(e^2e_q)^2g_s^2C_FN_c\frac{1}{3Q^2}\left[\frac{x_1^2+x_2^2}{(1-x_1)(1-x_2)}\right], \tag{7}$$

where $\alpha = \left(e^2/4\pi\right)$, e_q is the charge of quarks in the unit of fundamental electric charge e, $\alpha_s = g_s^2/4\pi$, $C_F = (N_c^2-1)/2N_c$, and N_c is the number of quark colours. The two event shape variables that we are going to study in this work are thrust and C-parameter. Thrust is defined as

$$T = \max_{\hat{\mathbf{n}}}\frac{\sum_i|\mathbf{p_i}\cdot\hat{\mathbf{n}}|}{\sum_i E_i}. \tag{8}$$

For a final state with three massless particles, eq. (8) takes the simple form

$$T = \max\{x_1,x_2,x_3\}. \tag{9}$$

The C-paramter is given by

$$C = 3 - \frac{3}{2}\sum_{i,j}\frac{\left(p_i\cdot p_j\right)^2}{(p_i\cdot q)\left(p_j\cdot q\right)}. \tag{10}$$

For our process, we define a rescaled c-parameter, the expression for which is

$$c \equiv \frac{C}{6} = \frac{(1-x_1)(1-x_2)(1-x_3)}{(2-x_1-x_2)(2-x_2-x_3)(2-x_1-x_3)}. \tag{11}$$

We are going to use the above expressions of thrust and c-parameter[1] (rescaled c parameter instead of C-parameter) to compute their respective event shape distribution in upcoming sections.

2.1 Thrust distribution

To obtain the thrust distribution one integrates over the x_1, x_2 variables in eq. (5) with the appropriate limits:

$$\frac{1}{\sigma_0(s)}\frac{d\sigma}{dT} = \frac{2\alpha_s}{3\pi}\int_0^1 dx_1 \int_0^{1-x_1} dx_2 \,\frac{x_1^2+x_2^2}{(1-x_1)(1-x_2)} \times \delta(T-\max(x_1,x_2,x_3)) \tag{12}$$

where $x_3 = 2 - x_1 - x_2$, and $\sigma_0(s)$ is the LO cross-section. For our purposes, we categorize the contributions from three different regions of the phase space integration in eq. (12). These regions are defined by which of x_1, x_2, or x_3 is the largest, and we refer to them as region-I, region-II, and region-III, respectively. The contribution from region I is then

$$\frac{1}{\sigma_0(s)}\frac{d\sigma}{dT}\bigg|_{\mathrm{I}} = \frac{2\alpha_s}{3\pi}\int_0^1 dx_2 \int_0^{1-x_2} dx_1 \,\frac{x_1^2+x_2^2}{(1-x_1)(1-x_2)}\,\delta(T-x_1)\,. \tag{13}$$

Instead of T, we shall mostly use the variable $\tau = 1 - T$, which vanishes in the zero-radius dijet limit. Upon integrating and expanding around $\tau = 0$

$$\frac{1}{\sigma_0(s)}\frac{d\sigma}{d\tau}\bigg|_{\mathrm{I}} = \frac{2\alpha_s}{3\pi}\Big(\frac{-3-4\log\tau}{2\tau} + 2\log\tau + \frac{3\tau}{2} - \tau\log\tau + O(\tau^2)\Big)\,. \tag{14}$$

Region-I captures the leading log (LL) at LP and NLP both. In region II, due to the symmetry under the interchange of $x_1 \leftrightarrow x_2$, the contribution from this region is identical to eq. (14). Thus

$$\frac{1}{\sigma_0(s)}\frac{d\sigma}{d\tau}\bigg|_{\mathrm{II}} = \frac{2\alpha_s}{3\pi}\Big(\frac{-3-4\log\tau}{2\tau} + 2\log\tau + \frac{3\tau}{2} - \tau\log\tau + O(\tau^2)\Big)\,. \tag{15}$$

In region III, the thrust axis is aligned with the momentum of the gluon. The contribution is given by

$$\frac{1}{\sigma_0(s)}\frac{d\sigma}{dT}\bigg|_{\mathrm{III}} = \frac{2\alpha_s}{3\pi}\int_0^1 dx_2 \int_0^{1-x_2} dx_1 \,\frac{x_1^2+x_2^2}{(1-x_1)(1-x_2)}\,\delta(T+x_1+x_2-2)\,. \tag{16}$$

Integrating and expanding in τ gives

$$\frac{1}{\sigma_0(s)}\frac{d\sigma}{d\tau}\bigg|_{\mathrm{III}} = \frac{2\alpha_s}{3\pi}\Big(-2-2\log\tau + (2-2\log\tau)\tau + O(\tau^2)\Big)\,. \tag{17}$$

The region-III captures only the LL at NLP, and no contributions from LP gets captured here. Combining the contributions from these three regions we finally get

$$\frac{1}{\sigma_0(s)}\frac{d\sigma}{d\tau}\bigg|_{\mathrm{NLO}} = \frac{2\alpha_s}{3\pi}\Big(\frac{-3-4\log\tau}{\tau} - 2 + 2\log\tau + (5\tau-4\log\tau)\tau + O(\tau^2)\Big)\,. \tag{18}$$

[1] $c = C/6$

Note that in eq. (18) at NLP all three regions produce LL terms, but a partial cancellation takes place when combined.

The expression of matrix element squared under shifted formalism for our process derived using eq. (2) reads as

$$\overline{\sum}|\mathcal{M}_{\rm shift}(p_1,p_2)|^2 = 8(e^2e_q)^2N_cg_s^2C_F\,\frac{1}{3Q^2}\left(\frac{2x_1+2x_2-2}{(1-x_1)(1-x_2)}\right). \tag{19}$$

The result of thrust distribution in region-I using fomalism of shifts results in

$$\frac{1}{\sigma_0(s)}\frac{d\sigma}{d\tau}\bigg|_{\rm I} = \frac{2\alpha_s}{3\pi}\left(\frac{-2-2\log\tau}{\tau}+2+2\log\tau+O(\tau^2)\right). \tag{20}$$

When we compare the above expression to eq. (14), we see that the LLs at both the LP and NLP are correctly captured, while the NLL terms at LP are only partially reproduced. Similarly the contribution from region-III for small values of τ is given as

$$\frac{1}{\sigma_0(s)}\frac{d\sigma}{d\tau}\bigg|_{\rm III} = \frac{2\alpha_s}{3\pi}\Big(-4\tau\log\tau+O(\tau^2)\Big). \tag{21}$$

This is as expected, as the hard gluon/soft quark region is not part of the shifted kinematics method. Combining contributions from all three regions the thrust distribution from shifted kinematics formalism at NLO reads or

$$\frac{1}{\sigma_0(s)}\frac{d\sigma}{d\tau}\bigg|_{\rm shift} = \frac{2\alpha_s}{3\pi}\bigg[\frac{-4-4\log\tau}{\tau}+4+4\log\tau-(4\log\tau)\tau+O(\tau^2)\bigg]. \tag{22}$$

The LL at LP are captured correctly, however LL at NLP are only captured partially.

In order to find the remaining LLs we use the soft quark approximation, the expression of matrix element squared in this approximation is

$$\overline{\sum}|\mathcal{M}_{\rm rem}(x_1,x_2)|^2 = 8(e^2e_q)^2N_cC_Fg_s^2\frac{1}{3Q^2}\left(\frac{1-x_2}{1-x_1}+\frac{1-x_1}{1-x_2}\right). \tag{23}$$

Thrust distribution computed from soft quark approximation results in

$$\frac{1}{\sigma_0(s)}\frac{d\sigma}{d\tau}\bigg|_{\rm rem} = \frac{2\alpha_s}{3\pi}\left(\frac{1}{\tau}-6-2\log\tau+5\tau\right). \tag{24}$$

Thus we notice that the soft quark approximation captures the remaining logarithmic terms at the LP and NLP for thrust distribution. The above expression when combined with shifted kinematics result in eq. (22) produces all the terms of thrust distribution given in eq. (18).

2.2 C-parameter distribution

The c-parameter distribution for the process under consideration is given by the expression

$$\frac{1}{\sigma_0(s)}\frac{d\sigma}{dc} = \frac{1}{2s}\int d\Phi_3\,\overline{\sum}|\mathcal{M}(x_1,x_2)|^2\,\delta(c(y,z)-c). \tag{25}$$

where the expression of matrix element squared and c-parameter in terms of new set of variables y and z is given by

$$\overline{\sum} |\mathcal{M}(y,z)|^2 = 8(e^2 e_q)^2 N_c C_F g_s^2 \frac{1}{3Q^2}\left(\frac{2+y(y-2yz(1-z)-2)}{y^2 z(1-z)}\right), \tag{26}$$

and

$$c(y,z) = \frac{(1-y)(1-z)yz}{(1-y(1-z))(1-yz)}. \tag{27}$$

The new variables are defined as

$$\begin{aligned} y &= 2 - x_1 - x_2\,, \\ z &= \frac{1-x_2}{y}\,, \end{aligned} \tag{28}$$

The expression for c-parameter distribution as given in eq. (25) upon substituting the above expressions of matrix element squared and c-parameter reads as

$$\left.\frac{1}{\sigma_0(s)}\frac{d\sigma}{dc}\right|_{\text{NLO}} = \frac{2\alpha_s}{3\pi}\int_0^1 dy \int_0^1 dz \frac{2(y(z-1)+1)^2(yz-1)^2(y(2y(z-1)z+y-2)+2)}{(y-1)^2 y^2 (z-1)z(2z-1)} \times\Big(\delta(z-z_1)+\delta(z-z_2)\Big). \tag{29}$$

After the z integration the limits of y change to (y_1, y_2), see below in eq. (31). We have now

$$\left.\frac{1}{\sigma_0(s)}\frac{\mathrm{d}\sigma}{\mathrm{d}c}\right|_{\text{NLO}} = \frac{2\alpha_s}{3\pi}\int_{y_1}^{y_2} dy \frac{2(1-y)(y\left(c(y-2)^2+(y-3)y+4\right)-2)}{c(cy+y-1)\sqrt{y(cy+y-1)\left(c(y-2)^2+(y-1)y\right)}}. \tag{30}$$

$$y_1 = \frac{1+4c-\sqrt{1-8c}}{2(1+c)}, \qquad y_2 = \frac{1+4c+\sqrt{1-8c}}{2(1+c)}. \tag{31}$$

The integration over y in eq. (30) produces incomplete elliptic integrals of three types, each with somewhat involved arguments and coefficients. After conversion to their so-called complete counterparts and carefully collecting their coefficients, the final expression can be organized in a compact form as follows

$$\left.\frac{1}{\sigma_0(s)}\frac{d\sigma}{dc}\right|_{\text{NLO}} = \frac{2\alpha_s}{3\pi}\Big(e(c)\,\mathrm{E}[m_1(c)] + p(c)\,\Pi[n_1(c), m_1(c)] + k(c)\,\mathrm{K}[m_1(c)]\Big), \tag{32}$$

where E, Π, and K are the standard complete elliptic integrals of the first, second, and third kind. The arguments of these elliptic integrals in eq. (32) have the following form

$$n_1(c) = \frac{2\sqrt{1-8c}}{1+\sqrt{1-8c}-4c}, m_1(c) = \frac{2\sqrt{1-8c}}{1+\sqrt{1-8c}-4c-8c^2}. \tag{33}$$

The arguments have monotonic behavior. The elliptic integrals in eq. (32) have the following asymptotic behavior as $c \to 0$

$$\begin{aligned} \mathrm{E}[m_1(c)] &= 1 + O(c^3 \log c)\,, \\ \Pi[n_1(c), m_1(c)] &= -\frac{\log c}{8c^2} + \frac{6\log c - 3}{8c} + \frac{15 - 8\log c}{16} + \frac{9\log c + 10}{3}c + O(c^2)\,, \\ \mathrm{K}[m_1(c)] &= -\frac{3\log c}{2} + O(c^3 \log c)\,. \end{aligned} \tag{34}$$

The expression for the coefficients appearing in eq. (32) around $c = 0$ are

$$\begin{aligned} e(c) &= -\frac{3}{c} + 9 + 12c + 36c^2 + O(c^3)\,, \\ p(c) &= 32c + 112c^2 + O(c^3)\,, \\ k(c) &= 4 - 20c + 48c^2 + O(c^3)\,. \end{aligned} \tag{35}$$

From eqs. (33), (34) and (35) the c-parameter distribution at NLO for small-c reads,

$$\frac{1}{\sigma_0(s)}\frac{d\sigma}{dc}\bigg|_{\rm NLO} = \frac{2\alpha_s}{3\pi}\bigg(\frac{-3-4\log c}{c} - 3 + 4\log c + O(c)\bigg). \tag{36}$$

We next compute the c-parameter distribution using the shifted kinematics method, and again assess to what extent LP and NLP terms in the exact NLO calculation are reproduced. The approximation is The shifted matrix element in eq. (19), when written in terms of the transformed variables (y, z), takes the form

$$\sum|\mathcal{M}_{\rm shift}(y,z)|^2 = 8(e^2e_q)^2N_cC_Fg_s^2\frac{1}{3Q^2}\left(\frac{2(y-1)}{y^2(z-1)z}\right). \tag{37}$$

After the z integration we have

$$\frac{1}{\sigma_0(s)}\frac{d\sigma}{dc}\bigg|_{\rm shift} = \frac{2\alpha_s}{3\pi}\int_{y_1}^{y_2} dy\frac{4(y-1)^2}{c\sqrt{y\,(y+cy-1)\,(c(y-2)^2+(y-1)y)}}\,. \tag{38}$$

The above integral again results in to elliptic integrals as given in eq. (32). The small-c behavior of coefficients of elliptic integrals obtained in this case are

$$\begin{aligned} e_s(c) &= -\frac{4}{c} + 16 - 4c + 72c^2 + O(c^3)\,, \\ p_s(c) &= 32c + 128c^2 + O(c^3)\,, \\ k_s(c) &= -8c + O(c^3)\,. \end{aligned} \tag{39}$$

Using these results, the expression of c-paramter distribution at NLO up to NLP is given by

$$\frac{1}{\sigma_0(s)}\frac{d\sigma}{dc}\bigg|_{\rm shift} = \frac{2\alpha_s}{3\pi}\bigg(\frac{-4-4\log c}{c} + 4 + 8\log c + O(c)\bigg). \tag{40}$$

On the other hand the expression obtained from soft quark approximation reads as

$$\frac{1}{\sigma_0(s)}\frac{d\sigma}{dc}\bigg|_{\rm rem} = \frac{2\alpha_s}{3\pi}\bigg(\frac{1}{c} - 7 - 4\log c + O(c)\bigg). \tag{41}$$

Thus we notice from the two expressions above that the shifted approach along with the soft quark approximation reproduces the LL up to NLP at NLO accurately.

3 Conclusions

Fixed-order studies in perturbative QCD face challenges from IR singularities, which persist even after cancellation, leaving logarithmic corrections behind. These logarithms, categorized as LL and NLL at different powers, including LP and NLP, must be resummed to derive meaningful results from these studies. We calculated these logarithms with the primary aim of determining the LL at NLP for the event shape distributions of thrust and c-parameter. We found that combining the shifted approach with the soft quark approximation effectively captured the LL up to NLP. This method can be applied to other event shape variables and processes to evaluate the accuracy of the shifted approach and to identify the significant LL at NLP for resummation purposes.

References

[1] T. Kinoshita, A. Ukawa, Mass Singularities of Feynman Amplitudes, Lect. Notes Phys. **39**, 55 (1975). 10.1007/BFb0013300

[2] T.D. Lee, M. Nauenberg, Degenerate Systems and Mass Singularities, Phys. Rev. **133**, B1549 (1964). 10.1103/PhysRev.133.B1549

[3] N. Agarwal, M. van Beekveld, E. Laenen, S. Mishra, A. Mukhopadhyay, A. Tripathi, Next-to-leading power corrections to event-shape variables, Pramana **98**, 60 (2024), 2306.17601. 10.1007/s12043-024-02743-0

[4] E. Farhi, A QCD Test for Jets, Phys. Rev. Lett. **39**, 1587 (1977). 10.1103/PhysRevLett.39.1587

[5] S. Catani, G. Turnock, B.R. Webber, L. Trentadue, Thrust distribution in e+ e- annihilation, Phys. Lett. B **263**, 491 (1991). 10.1016/0370-2693(91)90494-B

[6] J.F. Donoghue, F.E. Low, S.Y. Pi, Tensor Analysis of Hadronic Jets in Quantum Chromodynamics, Phys. Rev. D **20**, 2759 (1979). 10.1103/PhysRevD.20.2759

[7] S. Mishra, Next-to-leading power corrections to spherocity distribution at NLO in QCD (2024), 2403.16449

[8] M. van Beekveld, W. Beenakker, E. Laenen, C.D. White, Next-to-leading power threshold effects for inclusive and exclusive processes with final state jets, JHEP **03**, 106 (2020), 1905.08741. 10.1007/JHEP03(2020)106

[9] M. van Beekveld, W. Beenakker, E. Laenen, A. Misra, C.D. White, The role of soft quarks in next-to-leading power threshold effects, PoS **RADCOR2019**, 053 (2019). 10.22323/1.375.0053

[10] V. Del Duca, E. Laenen, L. Magnea, L. Vernazza, C.D. White, Universality of next-to-leading power threshold effects for colourless final states in hadronic collisions, JHEP **11**, 057 (2017), 1706.04018. 10.1007/JHEP11(2017)057

Cwebs in multiparton scattering amplitude: Structures at four loops

Neelima Agarwal[1], *Sourav* Pal[2], *Aditya* Srivastav[3,*], and *Anurag* Tripathi[3]

[1]Chaitanya Bharathi Institute of Technology, Hyderabad, India
[2]Physical Research Laboratory, Ahmedabad, India
[3]Indian Institute of Technology Hyderabad, India

Abstract. Soft function exponentiates in terms of the soft anomalous dimension Γ_S; the Feynman diagrams contributing to it are called Cwebs. The colour and kinematics of a Cweb mix via a web mixing matrix – calculation of web mixing matrices at higher loop orders is a nontrivial task using replica trick. Here I discuss a recent development of an efficient algorithm to find the Cwebs that are present at any loop order, and the result of mixing matrices, and exponentiated colour factors associated with Boomerang Cweb at four loops connecting three and four Wilson lines.

1 Introduction

The study of infrared (IR) singularities in non-abelian gauge theories is essential for understanding high-energy scattering processes in collider physics. These universal singularities cancel when real emissions are added to virtual corrections but produce large logarithms in specific regions. Subtraction schemes address these issues, while the QCD factorization theorem isolates IR divergences via soft functions constructed from Wilson lines. These renormalized functions follow a renormalization group equation, with solutions involving the soft anomalous dimension.

In processes with multipartons, the soft anomalous dimension becomes a matrix, a central focus of the research discussed in this article. Calculations of the anomalous dimension have been carried out to one, two, and even three loops, with ongoing efforts to extend these to four loops. The article explores an alternative method for determining the soft function's exponentiation as diagrammatic exponentiation, which introduces the concept of Cwebs [1, 2]. The logarithm of the Soft function is a sum over all the Cwebs at each perturbative order:

$$\mathcal{S} = \exp\left[\sum_{W}\sum_{d,d'\in W} K(d)\, R_W(d,d')\, C(d')\right]. \tag{1}$$

The d here denotes a diagram in a Cweb W and its corresponding kinematic and colour factor are denoted by $K(d)$ and $C(d)$. The action of web mixing matrix R_W on the colour of a diagram $C(d)$ generates its exponentiated colour factor $\widetilde{C}(d)$,

$$\widetilde{C}(d) = \sum_{d'\in W} R_W(d,d')\, C(d'). \tag{2}$$

*e-mail: shrivastavadi333@gmail.com

Cwebs are sets of Feynman diagrams whose contributions are summed over gluon attachments on Wilson lines, with recent extensions introducing Boomerang Cwebs [3, 4] for massive particle scattering and at five loops in massless case [5].

In this paper, the results on the mixing matrices for Boomerang Cwebs, which describe the interactions of multiple massive Wilson lines, are provided. Modifications to existing algorithms for enumerating Cwebs have been made and implemented in an updated Mathematica code, `CwebGen 2.0`, which is used to perform the calculations presented in this study. The structure of the paper includes a review of Cweb properties, a description of the modified algorithm, and explicit results for Boomerang Cwebs at four loops. These findings contribute to a deeper understanding of the soft anomalous dimension.

2 Cwebs and web mixing matrices

For the convenience of the reader, we collect here some of the definitions that are used in this article.

Definitions

Web: A set of diagrams closed under shuffles of the gluon attachments — originating from different gluon propagators or three-point or four-point gluon vertices — on each Wilson line.

Cweb: A set of diagrams closed under shuffles of the gluon attachments — originating from different gluon correlators

Multi-pronged-boomerang correlator: An m-point gluon correlator ($m \geq 2$) whose all ends attach to one Wilson line. The contributions of these for massless Wilson lines vanish identically.

Boomerang Cweb: A Cweb that contains at least one multi-pronged-boomerang correlator with $m = 2$ and none with $m > 2$.

Weight factor (s-factor): The weight factor $s(d)$ for a diagram d is defined as the number of ways in which the gluon correlators can be *sequentially* shrunk to their common origin.

Column weight vector: We can construct a column weight vector out of the s-factors for a Cweb with n diagrams as

$$S = \{s(d_1), s(d_2), \ldots, s(d_n)\}. \tag{3}$$

$\mathbf{W_n^{(c_2,\ldots,c_p)}(k_1,\ldots,k_n)}$: A Cweb constructed out of c_m m-point connected gluon correlators ($m = 2, \ldots, p$), where k_i denotes number of attachments on i^{th} Wilson line.

Note that the perturbative expansion for an m-point connected gluon correlator starts at $O(g^{m-2})$, while each attachment to a Wilson line carries a further power of g, the perturbative expansion for a Cweb can be written as

$$W_n^{(c_2,\ldots,c_p)}(k_1,\ldots,k_n) = g^{\sum_{i=1}^n k_i + \sum_{r=2}^p c_r(r-2)} \sum_{j=0}^{\infty} W_{n,j}^{(c_2,\ldots,c_p)}(k_1,\ldots,k_n)\, g^{2j}, \tag{4}$$

which defines the perturbative coefficients $W_{n,j}^{(c_2,\ldots,c_p)}(k_1,\ldots,k_n)$. Here the perturbative order of this Cweb is $g^{\sum_{i=1}^n k_i + \sum_{r=2}^p c_r(r-2)}$.

Cwebs are the proper building blocks of the logarithm of the Soft function — they are useful in the organization and counting of diagrammatic contributions at higher perturbative orders [1, 2]. The contribution of colour and kinematic factors to a web W can be arranged in a more transparent manner if we diagonalize the mixing matrix R:

$$W = \left(K^T Y^{-1}\right) YRY^{-1} (YC) = \sum_{j=1}^{r} \left(K^T Y^{-1}\right)_j (YC)_j \, , \tag{5}$$

where Y is the diagonalizing matrix and $YRY^{-1} \equiv \mathcal{D}$ is the diagonal matrix that we get; furthermore we have arranged $\mathcal{D}_{jj} = 1$ for $1 \leq j \leq r$. $(YC)_j$ are also referred to as exponentiated colour factors and the corresponding kinematic factors are $\left(K^T Y^{-1}\right)_j$. In this article we will present the exponentiated colour factors in terms of $(YC)_j$.

Properties of mixing matrices

1. *Idempotence*: A Cweb mixing matrix is idempotent, that is, $R^2 = R$
2. *Zero row sum rule*: The entries of R obey the zero row sum rule $\sum_{d'} R(d, d') = 0$.
3. *Column sum rule*: Along with these general properties, the mixing matrices also obey a conjectured column-sum rule of the form
$$\sum_d s(d) R(d, d') = 0 \, . \tag{6}$$
4. *Uniqueness*: For a given column weight vector $S = \{s(d_1), s(d_2), \ldots, s(d_k)\}$ with all $s(d_i) \neq 0$, the mixing matrix is unique.

In section 5, we will verify these properties for four loop Boomerang Cwebs. In the next section, we describe a recursive algorithm that generates Cwebs present at $O(g^{2l+2})$ using Cwebs at $O(g^{2l})$.

3 An improved algorithm to generate Cwebs

We find that it is sufficient to use Cwebs at l loops connecting m Wilson lines to generate all Cwebs at $l + 1$ loops connecting the same number of Wilson lines. Based on these facts, an improved and efficient algorithm to generate Cwebs recursively is given below:

1. To generate Cwebs at $l + 1$ loops connecting at most $l + 1$ Wilson lines starting from a Cweb at l loops connecting m $(1 \leq m \leq l + 1)$ lines
   (a) Connect any two existing Wilson lines by introducing a two-point gluon correlator (for massive Wilson lines a two point correlator can also be attached to a single line),
   (b) Connect any existing k-point gluon correlator to an existing Wilson line.
2. To generate Cwebs at $l + 1$ loops connecting the highest number $(l + 2)$ of Wilson lines, the following steps need to be applied to Cwebs at l loops connecting the highest number $(l + 1)$ of lines allowed at this order:
   (a) Connect a new Wilson line to any of the existing lines by introducing a two-point gluon correlator.
   (b) Connect any existing k-point gluon correlator to a new Wilson line.
3. Discard the duplicate Cwebs.

4 A brief description of `CwebGen 2.0`

We present below the algorithm of current version of the in-house Mathematica code `CwebGen 2.0`, that incorporates replica trick algorithm and is a significantly improved version of the codes that were used in [1, 2]. This code has been used in this work to obtain the Cweb mixing matrices.

- The code begins by generating the colour of one of the Cweb diagrams; however, it is different from the usual colour assignment as we also assign replica variables $\{i, j, k, \ldots\}$ to each of the gluon correlators that are present in the diagram.
- A subroutine then shuffles the attachments on each of the Wilson lines belonging to different correlators, and correspondingly different replica variables, to generate all the diagrams of the Cweb.
- Next hierarchies between the replica variables are generated. For example, for a Cweb with two correlators, with replica variables i and j, the code generates hierarchies $\{h\} = \{i = j, i > j, i < j\}$. Then number of distinct replica variables N_r for each hierarchy is calculated, for example, for hierarchy $i = j$, $N_r = 1$, whereas, for $i > j$, and $i < j$, $N_r = 2$. Using this, $M_{N_r}(h)$ is obtained.

  Note that previous versions of the code could generate all the possible hierarchies only if the number of replica variables was less than or equal to 4. `CwebGen 2.0` can generate the hierarchies for any number of replica variables.
- Another subroutine then generates replica ordered colour factors $\mathbf{R}[C(d)|h]$ for each of the hierarchies for every diagram of the Cweb. A diagonalizing matrix Y_W is constructed to diagonalize R_W using its right eigenvectors. Y_W then acts on the column vector containing the colour factors of each of the diagrams of the Cweb, and then gives the corresponding independent exponentiate colour factors.

The algorithm and its implementation in two older versions of the same code can be found in [1, 2]. To make a comparison of the present version with the previous versions, we have calculated the mixing matrix of $\mathrm{W}_5^{(4)}(1, 1, 1, 1, 4)$ appearing at four loops [1]. The largest dimension mixing matrix is 24×24 and it took 7 days in the first version [1], upon subsequent improvement it took 6.4 hours [2]. The same calculation by `CwebGen 2.0` takes only 1.54 seconds. Thus, the latest version of code is almost 15000 times faster as compared to the previous versions.

5 Boomerang Cwebs at Four loops

Using `CwebGen 2.0` and discarding the duplicates we get 8 and 20 Boomerang Cwebs connecting four and three Wilson lines respectively. In [3], we had obtained the diagonal blocks of mixing matrices for all the 8 Boomerang Cwebs that are present at four loops using the formalism introduced in [8]. Of course, we could not obtain the complete mixing matrices and exponentiated colour factors in that work which is the subject matter of this paper.

Before going into the details of the calculation, reader is advised to review the concepts used here such as Normal ordering, Fused-Web, and Uniqueness theorem, which were introduced in [8]. We now present the results for one Cweb connecting four Wilson lines. The calculations for all such Cwebs is provided in [4].

$W_4^{(1,0,1)}(3, 1, 1, 1)$: *a four-line Boomerang Cweb*

Boomerang Cweb $\mathrm{W}_4^{(1,0,1)}(3, 1, 1, 1)$, shown in fig (1), connects four Wilson lines with one multi-pronged-boomerang correlator (m=2) and a four point gluon correlator. This Cweb

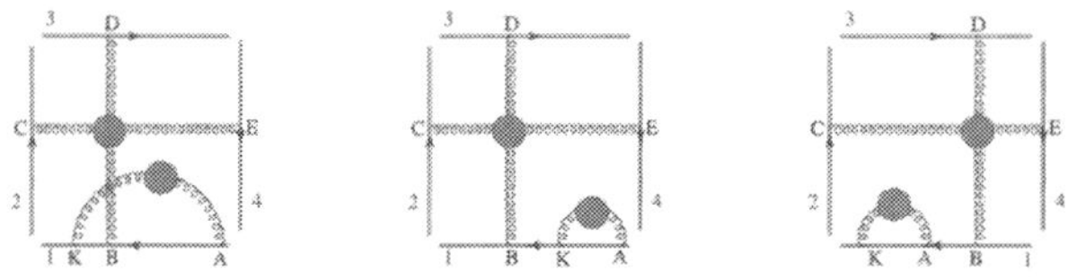

Figure 1. Diagrams of Boomerang Cweb $W_4^{(1,0,1)}(3,1,1,1)$.

has three possible shuffles on Wilson line 1. The shuffle of gluon attachments from the two correlators generates three diagrams as shown in fig. (1); thus the order of mixing matrix for this Cweb is three. To proceed we choose the order of the diagrams as in the fig. (1). This order is labeled using the order of attachments on line 1: $C_1 = \{ABK\}$, $C_2 = \{AKB\}$ and $C_3 = \{BAK\}$. The mixing matrix R and the diagonalizing matrix Y has been calculated using `CwebGen 2.0` and we get,

$$R = \frac{1}{2}\begin{pmatrix} 2 & -1 & -1 \\ 0 & 1 & -1 \\ 0 & -1 & 1 \end{pmatrix}, \qquad Y = \begin{pmatrix} 1 & -1 & 0 \\ 0 & -1 & 1 \\ 0 & 1 & 1 \end{pmatrix}. \tag{7}$$

This matches with the result obtained in [3].

The column weight vector S for this Cweb can be calculated by determining the s-factors for each diagram. The s-factor for C_1 is zero as one can not shrink any of the two correlators to the origin of Wilson lines. In the diagram C_2 one can shrink the multi-pronged-boomerang correlator first then the four point correlator, thus there is only one way to shrink all the correlators to origin, making $s = 1$. The s-factor for C_3 is also one except the fact that here the multi-pronged-boomerang correlator has to be shrunk after the four point correlator. Thus the column weight vector becomes $S = \{0, 1, 1\}$.

The matrix R satisfies all the known properties of mixing matrix: it is idempotent, $R^2 = R$, satisfies the zero row sum rule as in each row of matrix sum of entries is zero, and, the conjectured column sum rule. The rank of this mixing matrix is two as there are only two independent rows in R. We obtain the independent exponentiated colour factors given as,

$$(YC)_1 = 0, \qquad (YC)_2 = -i f^{abn} f^{bch} f^{deh}\, \mathbf{T}_1^a \mathbf{T}_1^n \mathbf{T}_2^c \mathbf{T}_3^d \mathbf{T}_4^e. \tag{8}$$

The vanishing of one of the independent ECFs can be understood in terms of $\widetilde{C}$. Applying R on the C we get,

$$\widetilde{C}_1 = \frac{1}{2}(2C_1 - C_2 - C_3), \qquad \widetilde{C}_2 = \frac{1}{2}(C_2 - C_3), \qquad \widetilde{C}_3 = \frac{1}{2}(C_3 - C_2). \tag{9}$$

Recall that when a Cweb is Normal Ordered, the block D gives mixing between the reducible diagrams. All these diagrams contain Boomerang that give C_A factors, and thus their ECFs vanish.

6 Conclusions

In this article we study Boomerang Cwebs that connect three and four Wilson lines at four loops which are important ingredients in the studies of scattering involving massive particles.

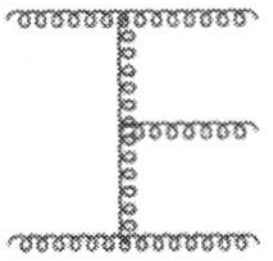

Figure 2. Exponentiated colour factor obtained from all the Boomerang Cwebs.

To enumerate Cwebs at a given perturbative order, we have introduced an improved version of the algorithm that was developed in [1]. We presented some details of the current version of the in-house Mathematica code `CwebGen 2.0`, that incorporates this new algorithm, and the replica trick, and it is a significantly faster version of the codes that were used in [1, 2].

We have computed the mixing matrices, the diagonalizing matrices and the exponentiated colour factors for all these Cwebs. We have verified that our results match with the predictions of the diagonal blocks presented in [3]. We found that the general structure of exponentiated colour factor of all the twenty eight Boomerang Cwebs, shown in fig. (2), is same as the general structure of ECFs for the Cwebs without the multi-pronged-boomerang correlators at four loops [1, 2]. This is an artifact of the Cwebs with multi-pronged-boomerang correlator, for which quadratic Casimir C_A (the colour of a Boomerang) is absent from all the exponentiated colour factors. This is in agreement with the calculations of Boomerang Cwebs at three loops [9]. We found that the ECFs corresponding to the D block of mixing matrices vanish which is consistent with [9].

References

[1] N. Agarwal, A. Danish, L. Magnea, S. Pal, A. Tripathi, Multiparton webs beyond three loops, JHEP **05**, 128 (2020), `2003.09714`. 10.1007/JHEP05(2020)128

[2] N. Agarwal, L. Magnea, S. Pal, A. Tripathi, Cwebs beyond three loops in multiparton amplitudes, JHEP **03**, 188 (2021), `2102.03598`. 10.1007/JHEP03(2021)188

[3] N. Agarwal, S. Pal, A. Srivastav, A. Tripathi, Deciphering colour building blocks of massive multiparton amplitudes at 4-loops and beyond, JHEP **02**, 258 (2023), `2212.06610`. 10.1007/JHEP02(2023)258

[4] N. Agarwal, S. Pal, A. Srivastav, A. Tripathi, Correlator webs of massive multiparton amplitudes at four loops: A study of boomerang webs, Phys. Rev. D **109**, 094038 (2024), `2307.15924`. 10.1103/PhysRevD.109.094038

[5] S. Mishra, S. Pal, A. Srivastav, A. Tripathi, Multiparton Cwebs at five loops (2023), `2305.17452`

[6] N. Agarwal, L. Magnea, C. Signorile-Signorile, A. Tripathi, The infrared structure of perturbative gauge theories, Phys. Rept. **994**, 1 (2023), `2112.07099`. 10.1016/j.physrep.2022.10.001

[7] E. Gardi, E. Laenen, G. Stavenga, C.D. White, Webs in multiparton scattering using the replica trick, JHEP **11**, 155 (2010), `1008.0098`. 10.1007/JHEP11(2010)155

[8] N. Agarwal, S. Pal, A. Srivastav, A. Tripathi, Building blocks of Cwebs in multiparton scattering amplitudes, JHEP **06**, 020 (2022), `2204.05936`. 10.1007/JHEP06(2022)020

[9] E. Gardi, M. Harley, R. Lodin, M. Palusa, J.M. Smillie, C.D. White, S. Yeomans, Boomerang webs up to three-loop order, JHEP **12**, 018 (2021), `2110.01685`. 10.1007/JHEP12(2021)018

EPJ Web of Conferences **314**, 00025 (2024) https://doi.org/10.1051/epjconf/202431400025
QCD@Work 2024

Topology of the large-N expansion in SU(N) Yang-Mills theory and spin-statistics theorem

Marco Bochicchio[1,*], *Mauro* Papinutto[1,2,**], and *Francesco* Scardino[1,2,***]

[1]Physics Department, INFN Roma1, Piazzale A. Moro 2, Roma, I-00185, Italy
[2]Physics Department, "Sapienza" University of Rome and INFN Roma1, Piazzale Aldo Moro 2, I-00185 Roma, Italy

Abstract. Recently, we computed the generating functional of Euclidean asymptotic correlators at short-distance of single-trace twist-2 operators in large-N SU(N) Yang-Mills (YM) theory to the leading-nonplanar order. Remarkably, it has the structure of the logarithm of a functional determinant, but with the sign opposite to the one arising from the spin-statistics theorem for the glueballs. To solve the sign puzzle, we reconsider the proof that in 't Hooft large-N expansion of YM theory the leading-nonplanar contribution to the generating functional consists of the sum over punctures of n-punctured tori. We discover that for twist-2 operators it contains – in addition to the n-punctured tori – the normalization of tori with $1 \leq p \leq n$ pinches and $n-p$ punctures. Once the existence of the new sector is taken into account, the violation of the spin-statistics theorem disappears. Besides, the new sector contributes trivially to the nonperturbative S matrix because – for example – the n-pinched torus represents nonperturbatively a loop of n glueball propagators with no external leg. This opens the way for an exact solution limited to the new sector that may be solvable thanks to the vanishing S matrix.

It has been known for more than forty years that U(N) Yang-Mills (YM) theory admits 't Hooft large-N topological expansion [1] for the n-point connected correlators of gauge-invariant single-trace operators. In the present paper [1] we work out the SU(N) case, almost literally summarizing [2]. In 't Hooft large-N expansion, the Feynman diagrams in double-line representation [1] contributing to the above correlators – after a suitable gluing of reversely oriented lines – are topologically classified [1, 3] by the sum on the genus g of n-punctured closed Riemann surfaces, each topology being weighted by N^{χ}, with $\chi = 2-2g-n$ the Euler characteristic of the Riemann surface. 't Hooft expansion extends to large-N QCD [1, 3] – and generally to large-N gauge theories – by eventually including punctured Riemann surfaces with boundaries. It exactly matches the topology and weights [1, 3] of a string theory [4], with closed-string coupling $g_s = \frac{1}{N}$. This matching has been advocated for the existence [1, 3, 5] of a nonperturbative string solution of large-N YM theory, QCD and, more generally, gauge theories, where – in the asymptotically free case – the dimensionful parameter of the string theory, i.e. the string tension, is identified [6] with the square of the renormalization-group (RG) invariant scale Λ_{RG} of the gauge theory. In the supposed string solution of large-N QCD glueballs are excitations of closed strings, mesons of open strings, and the above

*e-mail: marco.bochicchio@roma1.infn.it
**e-mail: mauro.papinutto@roma1.infn.it
***e-mail: francesco.scardino@roma1.infn.it

[1]M. Bochicchio invited talk at the conference QCD@work2024.

Riemann surfaces arise as their world-sheets – the new ingredient of the canonical string solution [6] being the existence of a conformal field theory living on the string world-sheets that is employed to compute the S-matrix amplitudes [4] and possibly the correlators [7] of the gauge theory. Actually, the existence of the supposed canonical string solution of (YM theory) QCD is constrained [6] by the large-N nonperturbative (non-)renormalization properties [8] of $\Lambda_{(YM)QCD}$. By assuming 't Hooft topological expansion, the generating functional $\mathcal{W}^E[J_O] = \log \mathcal{Z}^E[J_O]$ of Euclidean connected correlators of single-trace operators O in SU(N) YM theory, with

$$\mathcal{Z}^E[J_O] = \frac{1}{\mathcal{Z}^E} \int \mathcal{D}A \, e^{-S_{YM}+\sum_s \int J_{O_s} O_s} , \tag{1}$$

reads $\mathcal{W}^E[J_O] = \mathcal{W}^E_{\text{sphere}}[J_O] + \mathcal{W}^E_{\text{torus}}[J_O] + \cdots$ (Fig. 1). Nonperturbatively, $\mathcal{W}^E_{\text{sphere}}[J_O]$,

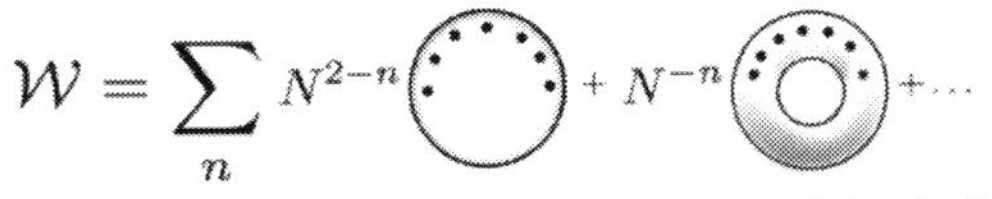

Figure 1. 't Hooft topological expansion of the generating functional that includes n-punctured spheres and tori.

which perturbatively is the ('t Hooft-)planar contribution [1], is a sum of tree diagrams involving glueball propagators and vertices, while $\mathcal{W}^E_{\text{torus}}[J_O]$, which perturbatively is the leading-non('t Hooft-)planar contribution, is a sum of glueball one-loop diagrams. Nonperturbatively, $\mathcal{W}^E_{\text{torus}}[J_O]$ should have the structure of the logarithm of a functional determinant [9]. Indeed, in the yet-to-come nonperturbative solution of large-N YM theory, the very same correlators should be computed by the correlators of a glueball field Φ with an infinite number of components, the corresponding generating functional being schematically [9]

$$\mathcal{Z}^E_{\text{glueball}}[J] = \frac{1}{\mathcal{Z}^E_{\text{glueball}}} \int \mathcal{D}\Phi \, e^{-S_{\text{glueball}}(\Phi)+\int \Phi *_1 J} , \tag{2}$$

with $S_{\text{glueball}}(\Phi) = \frac{1}{2}\int \Phi *_2 (-\Delta + M^2)\Phi + \frac{1}{N}\frac{1}{3}\Phi *_3 \Phi *_3 \Phi + \cdots$, where $*_2$ and $*_1$ are fixed below, the ellipses and $*_3$ respectively stand for n-glueball vertices with $n > 3$ and some presently unknown operation on the glueball fields. Hence, nonperturbatively the connected generating functional $\mathcal{W}^E_{\text{glueball}}[J] = \log \mathcal{Z}^E_{\text{glueball}}[J]$ reads to one loop of glueballs [9]

$$\mathcal{W}^E_{\text{glueball}}[J] = -S_{\text{glueball}}(\Phi_J) + \int \Phi_J *_1 J + \cdots - \frac{1}{2}\log \text{Det}\left(*_2(-\Delta + M^2) + \frac{1}{N} *_3 \Phi_J *_3 + \cdots \right), \tag{3}$$

with Φ_J determined by $\left.\frac{\delta S_{\text{glueball}}}{\delta \Phi}\right|_{\Phi_J} = *_1 J$. The minus sign in front of $\log \text{Det}$ in $\mathcal{W}^E_{\text{glueball}}[J]$ arises from the spin-statistics theorem [10], since all the gauge-invariant glueball interpolating fields have integer spin, and thus the glueballs should be bosons. The dictionary between $\mathcal{W}^E[J_O]$ and $\mathcal{W}^E_{\text{glueball}}[J]$ arises from matching the corresponding spectral representations – as a sum of free propagators with residues R_{sm} – for the 2-point correlators [11] of O_s at $N = \infty$ that, by fixing $*_2$ according to the canonical normalization of the glueball kinetic term, uniquely determines the coupling of J to the tower of glueball fields $\Phi *_1 J = \sum_{sm} \Phi_{sm} \sqrt{R_{sm}} J_s$. Until recently nothing has been quantitatively known on $\mathcal{W}^E[J_O]$ and $\mathcal{W}^E_{\text{glueball}}[J]$. Actually, since the early days of large-N QCD, it has been qualitatively known that asymptotic freedom (AF) applies to large-N correlators, and specifically to the nonperturbative 2-point correlators [12, 13]. Accordingly, the ultraviolet (UV), i.e. short-distance, constraints on the large-N spectral representation of 2-point correlators implied by AF were quantitatively investigated for multiplicatively renormalizable operators [14] and twist-2 operators that mix by renormalization [15]. More recently, we computed the UV asymptotics of the generating functional $\mathcal{W}^E[J_O]$ of Euclidean correlators of twist-2 operators in large-N YM theory [16] that follows from RG, Callan-Symanzik equation and AF. For simplicity, we report just a particular case of our calculation that involves balanced collinear twist-2 operators [16, 17] with even spin and maximal-spin component in the p_+ direction in

Minkowskian space-time. In the light-cone gauge $A_+ = 0$ they read [16, 17]

$$\mathbb{O}_s = \frac{1}{2N}\bar{A}^a(x)\mathcal{Y}^{\frac{5}{2}}_{s-2}(\overrightarrow{\partial}_+, \overleftarrow{\partial}_+)A^a(x), \qquad \mathcal{Y}^{\frac{5}{2}}_{s-2}(\overrightarrow{\partial}_+, \overleftarrow{\partial}_+) = \overleftarrow{\partial}_+(i\overrightarrow{\partial}_+ + i\overleftarrow{\partial}_+)^{s-2}C^{\frac{5}{2}}_{s-2}\left(\frac{\overrightarrow{\partial}_+-\overleftarrow{\partial}_+}{\overrightarrow{\partial}_++\overleftarrow{\partial}_+}\right)\overrightarrow{\partial}_+, \tag{4}$$

where $s = 2, 4, 6, \ldots$, the sum over repeated color indices, $a = 1, 2, \ldots N^2 - 1$, is understood, and $C^{\frac{5}{2}}_{s-2}$ are Gegenbauer polynomials. Our calculation was performed in three steps that we briefly recall. First, we computed to the lowest perturbative order – directly from its functional-integral definition as a Gaussian integral

$$\mathcal{Z}_{\text{conf}}[J_{\mathbb{O}}] = \frac{1}{Z}\int \mathcal{D}A\mathcal{D}\bar{A}\, e^{\int -i\bar{A}^a\Box A^a + \sum_s J_{\mathbb{O}_s}\mathbb{O}_s\, d^4x} \tag{5}$$

the generating functional of the connected conformal correlators in Minkowskian space-time and – by analytical continuation – the corresponding object in Euclidean space-time [16], respectively

$$\mathcal{W}_{\text{conf}}[J_{\mathbb{O}}] = -(N^2-1)\log\operatorname{Det}\left(\mathbb{I} + \tfrac{1}{2}i\Box^{-1}\tfrac{J_{\mathbb{O}_s}}{N}\otimes\mathcal{Y}^{\frac{5}{2}}_{s-2}\right), \qquad \mathcal{W}^E_{\text{conf}}[J_{\mathbb{O}^E}] = -(N^2-1)\log\operatorname{Det}\left(\mathbb{I} + \tfrac{1}{2}\Delta^{-1}\tfrac{J_{\mathbb{O}^E_s}}{N}\otimes\mathcal{Y}^{E\frac{5}{2}}_{s-2}\right), \tag{6}$$

where $\mathbb{I}$ is the identity in space-time, the sum over repeated s indices is understood and $\mathcal{Y}^{E\frac{5}{2}}_{s-2}$ is obtained from $\mathcal{Y}^{\frac{5}{2}}_{s-2}$ by the substitution $\partial_+ \to i\partial_z$ [16, 17]. Second, we worked out the UV asymptotics, as all the coordinates are uniformly rescaled by a factor $\lambda \to 0$, of renormalized correlators of collinear twist-2 operators that mix by renormalization with derivatives along the maximal-spin direction of collinear twist-2 operators of lower spin and same canonical dimensions [16]

$$G^{(n)}_{k_1\ldots k_n}(\lambda x_1, \ldots, \lambda x_n; \mu, g(\mu)) = \sum_{j_1\ldots j_n} Z_{k_1 j_1}(\lambda)\ldots Z_{k_n j_n}(\lambda)\,\lambda^{-\sum_{i=1}^n(s_{j_i}+2)}G^{(n)}_{j_1\ldots j_n}(x_1, \ldots, x_n; \mu, g(\tfrac{\mu}{\lambda})), \tag{7}$$

reducing the above mixing to the multiplicatively renormalizable case [16]

$$G^{(n)}_{j_1\ldots j_n}(\lambda x_1, \ldots, \lambda x_n; \mu, g(\mu)) = Z_{\mathbb{O}_{j_1}}(\lambda)\ldots Z_{\mathbb{O}_{j_n}}(\lambda)\,\lambda^{-\sum_{i=1}^n(s_{j_i}+2)}G^{(n)}_{j_1\ldots j_n}(x_1, \ldots, x_n; \mu, g(\tfrac{\mu}{\lambda})), \tag{8}$$

thanks to the inductive construction to all orders of perturbation theory [16, 18] of a finite change of renormalization scheme for the renormalized operators, where $Z_{sj}(\lambda) = \delta_{sj}Z_{\mathbb{O}_s}(\lambda) = \delta_{sj}\left(\frac{g(\mu)}{g(\frac{\mu}{\lambda})}\right)^{\frac{\gamma_{0\mathbb{O}_s}}{\beta_0}}$ is diagonal and one-loop exact [18], and independent of N [16], while in any scheme

$$G^{(n)}_{j_1\ldots j_n}(x_1, \ldots, x_n; \mu, g(\tfrac{\mu}{\lambda})) \sim G^{(n)}_{\text{conf}\, j_1\ldots j_n}(x_1, \ldots, x_n) + g^2(\tfrac{\mu}{\lambda})\,\mathcal{G}^{(n,2)}_{j_1\ldots j_n}(x_1, \ldots, x_n; \mu) + \cdots. \tag{9}$$

Consequently [16], since $g(\frac{\mu}{\lambda}) \to 0$ as $\lambda \to 0$ by AF,

$$G^{(n)}_{j_1\ldots j_n}(\lambda x_1, \ldots, \lambda x_n; \mu, g(\mu)) \sim \frac{Z_{\mathbb{O}_{j_1}}(\lambda)\ldots Z_{\mathbb{O}_{j_n}}(\lambda)}{\lambda^{\sum_{i=1}^n(s_{j_i}+2)}}G^{(n)}_{\text{conf}\, j_1\ldots j_n}(x_1, \ldots, x_n). \tag{10}$$

Third, by Eq. (10) we lifted $\mathcal{W}^E_{\text{conf}}[J_{\mathbb{O}^E}]$ to the generating functional of RG-improved UV asymptotic correlators $\mathcal{W}^E_{\text{asym}}[J_{\mathbb{O}^E}, \lambda]$ that inherits the very same structure of the logarithm of a functional determinant

$$\mathcal{W}^E_{\text{asym}}[J_{\mathbb{O}^E}, \lambda] = -(N^2-1)\log\operatorname{Det}\left(\mathbb{I} + \frac{1}{2}\frac{Z_{\mathbb{O}_s}(\lambda)}{\lambda^{s+2}}\Delta^{-1}\frac{J_{\mathbb{O}^E_s}}{N}\otimes\mathcal{Y}^{E\frac{5}{2}}_{s-2}\right). \tag{11}$$

The conformal generating functional admits the large-N expansion that trivially follows from Eq. (6), $\mathcal{W}_{\text{conf}}[J_{\mathbb{O}^E}] = \mathcal{W}_{\text{conf sphere}}[J_{\mathbb{O}^E}] + \mathcal{W}_{\text{conf torus}}[J_{\mathbb{O}^E}]$, and the asymptotic generating functional as well $\mathcal{W}^E_{\text{asym}}[J_{\mathbb{O}^E}, \lambda] = \mathcal{W}^E_{\text{asym sphere}}[J_{\mathbb{O}^E}, \lambda] + \mathcal{W}^E_{\text{asym torus}}[J_{\mathbb{O}^E}, \lambda]$, with

$$\mathcal{W}^E_{\text{asym torus}}[J_{\mathbb{O}^E}, \lambda] = +\log\operatorname{Det}\left(\mathbb{I} + \frac{1}{2}\frac{Z_{\mathbb{O}_s}(\lambda)}{\lambda^{s+2}}\Delta^{-1}\frac{J_{\mathbb{O}^E_s}}{N}\otimes\mathcal{Y}^{E\frac{5}{2}}_{s-2}\right). \tag{12}$$

Remarkably, the above equation reproduces the log Det structure of the glueball one-loop generating functional in Eq. (3), which it should be UV asymptotic to thanks to the AF, but with the sign opposite to the one arising from the spin-statistics theorem for the glueballs. Yet, we verify the correctness of the above sign, which is inherited from Eqs. (6) and (11), according to the sign arising from the spin-statistics theorem for gluons in one-loop perturbation theory, and the fact that in the SU(N) theory there are exactly $N^2 - 1$ gluons. The aim of the present paper is to solve the sign puzzle and discuss the implications of its solution. Essentially, there are two possible way-outs: Either the spin-statistics theorem is violated or 't Hooft topological expansion (Fig. 1) and the corresponding effective action in Eq. (3) need considerable refinements for the correlators of twist-2 operators. As we shall show momentarily, only the second alternative applies in large-N YM theory.

In fact, the spin-statistics theorem applies in full generality only to theories with a finite number of local fields [19–22]. If the number of fields is infinite, there exist rigorous counterexamples [19–21] to the spin-statistics theorem. They involve massive infinite-dimensional representations of the Lorentz group that are the relevant ones in YM theory because of its mass gap. Yet, they have infinite mass degeneracy, since they decompose [20, 21], according to Wigner theorem [23], into the sum of irreducible representations of the Poincaré group corresponding to an infinite number of particles of any spin, all having the same mass. In large-N YM theory this would imply the existence of vertical Regge trajectories that is hardly acceptable both theoretically and numerically because of contrary evidence from analysis [24] of lattice calculations.

Hence, it remains the second alternative. In order to understand why new topologies arise in the generating functional of correlators of twist-2 operators, we have reconsidered the proof of 't Hooft topological expansion in the U(N) versus SU(N) YM theory. The delicate point of the proof in the SU(N) theory is that – contrary to the original U(N) case [1] – the gluon propagator has a leading and subleading $\frac{1}{N}$ contribution in its double-line representation [25, 26]: $\langle A^i_{\ j} A^k_{\ l}\rangle \propto I^{ik}_{jl} - P^{ik}_{jl}$, with $A^i_{\ j} = A^a(\lambda^a)^i_{\ j}$, λ^a in the fundamental representation, $\mathrm{Tr}(\lambda^a\lambda^b) = \frac{1}{2}\delta^{ab}$, $[\lambda^a, \lambda^b] = if^{abc}\lambda^c$, $I^{ik}_{jl} = \delta^i_{\ l}\delta^k_{\ j}$ the components of the identity I in the u(N) Lie algebra, where the product of two matrices, B and C, is defined by $(BC)^{im}_{jn} = B^{ik}_{jl}C^{lm}_{kn}$, with the color trace $\mathrm{Tr}\, B = B^{ij}_{ji}$, $P^{ik}_{jl} = \frac{1}{N}\delta^i_{\ j}\delta^k_{\ l}$ the components of the u(1) projector P, and $i, l, j, k = 1, \ldots, N$. Yet, the subleading propagator, if it is attached to the vertices of the action, $V_3 \propto \frac{g}{\sqrt{N}} f^{abc}$ and $V_4 \propto \frac{g^2}{N} f^{abe} f^{ecd}$, does not contribute [25, 26]. This is rather obvious [26], since the subleading propagator is a u(1) contribution, while the action vertices are purely nonabelian. Therefore, in the SU(N) theory the topology of connected vacuum diagrams containing vertex insertions matches 't Hooft topological expansion with weights N^χ, where the gluon propagator is just the leading one [25, 26]. At this point, generalizing the structure of V_3 and V_4, we extend the above argument to connected correlators of single-trace operators provided that the corresponding local vertices are proportional to a matrix product [26] of $(T^a)^b_c = -if^{abc}$, $(T^{a_2} \ldots T^{a_{n-1}})^{a_1}_{a_n} = 2\,\mathrm{Tr}(\lambda^{a_1}[\lambda^{a_2}, \ldots [\lambda^{a_{n-1}}, \lambda^{a_n}] \ldots])$ that requires at least 3-gluon operators in perturbation theory. Indeed, the above vertices in double line have no $\frac{1}{N}$ correction [26] and their contraction with the subleading propagator vanishes because they involve nested commutators. Therefore, there is a vast – but quite specific – class of single-trace operators which the 't Hooft topological expansion applies to in SU(N) YM theory. Yet, we point out that the above proof does not apply to 2-gluon operators, and specifically to twist-2 operators. Indeed, in this case – even in the adjoint representation – their local vertex involves δ^{ab}, as opposed to f^{abc}, and contains a $\frac{1}{N}$ correction (Fig. 2a), exactly as the gluon propagator does. In fact, it has been suggested to consider the U(N) [1, 5, 25] – rather than the SU(N) – theory, where perturbatively no subleading correction both to the gluon propagator and (bare) 2-gluon vertices occur – though gauge-invariant single-trace twist-2 operators have not been specifically discussed. Nevertheless, this seemingly easy way-out is pointless for asymptotically free theories that allow us to resum the (renormalized) RG-improved generating functional into the logarithm of a functional determinant

$$\mathcal{W}^E_{U(N)\mathrm{asym}}[J_\mathbb{O}, \lambda] = -(N^2-1)\log \mathrm{Det}\left(\mathbb{I} + \frac{1}{2}\frac{Z_{\mathbb{O}_s}(\lambda)}{\lambda^{s+2}}\Delta^{-1}\frac{J_{\mathbb{O}^E_s}}{N} \otimes \mathcal{Y}^{E\frac{5}{2}}_{s-2}\right)$$
$$- \log \mathrm{Det}\left(\mathbb{I} + \frac{1}{2}\frac{1}{\lambda^{s+2}}\Delta^{-1}\frac{J_{\mathbb{O}^E_s}}{N} \otimes \mathcal{Y}^{E\frac{5}{2}}_{s-2}\right), \tag{13}$$

while the U(1) contribution in the second line decouples and is exactly conformal and not only asymptotic because the U(1) theory is free. Indeed, asymptotically in the UV the above result is dominated by only the SU(N) contribution for the correlators of all the twist-2 operators but the stress-energy tensor, since for them $\gamma_{0\mathbb{O}_s}$ is positive [16, 27]. Hence, the sign problem for the U(N) theory is exactly the same as for the SU(N) theory. To solve the sign problem in the SU(N) theory, we rewrite Eq. (6) identically as

$$\mathcal{W}^E_{\text{conf}}[J_{\mathbb{O}^E}] = -\log \text{Det}\left(\mathcal{I} + \frac{1}{2}(I-P)\Delta^{-1}\frac{J_{\mathbb{O}^E_s}}{N} \otimes \mathcal{Y}^{E\frac{5}{2}}_{s-2}\right)$$
$$= -\log \text{Det}\left[\left(\mathcal{I} + \frac{1}{2}\Delta^{-1} I\frac{J_{\mathbb{O}^E_s}}{N} \otimes \mathcal{Y}^{E\frac{5}{2}}_{s-2}\right)\left(\mathcal{I} - \frac{1}{2}\left(\mathcal{I} + \frac{1}{2}\Delta^{-1} I\frac{J_{\mathbb{O}^E_s}}{N} \otimes \mathcal{Y}^{E\frac{5}{2}}_{s-2}\right)^{-1}\Delta^{-1} P\frac{J_{\mathbb{O}^E_s}}{N} \otimes \mathcal{Y}^{E\frac{5}{2}}_{s-2}\right)\right], \quad (14)$$

where the term containing $I-P$ involves the SU(N) propagator, $\mathcal{I}$ is the identity in both color and space-time and the first equality above follows by noticing that the color trace Tr in the loop expansion of Eq. (14) containing the insertion of n sources $J_{\mathbb{O}^E_s}$ produces the overall factor $\text{Tr}(I-P)^n = \text{Tr}(I-P) = N^2-1$ that occurs in Eq. (6). For keeping 't Hooft double-line representation, which only involves the leading propagator, also beyond the planar limit of the SU(N) theory, we transfer in Eq. (14) the $\frac{1}{N}$ dependence from the propagator to the local vertex for 2-gluon operators (Fig. 2a) so that

$$\mathcal{W}^E_{\text{conf}}[J_{\mathbb{O}^E}] = -\log \text{Det}\left(\mathcal{I} + \frac{1}{2}\Delta^{-1} I\frac{J_{\mathbb{O}^E_s}}{N} \otimes \mathcal{Y}^{E\frac{5}{2}}_{s-2}\right)$$
$$-\log \text{Det}\left(\mathcal{I} - \frac{1}{2}\left(\mathcal{I} + \frac{1}{2}\Delta^{-1} I\frac{J_{\mathbb{O}^E_s}}{N} \otimes \mathcal{Y}^{E\frac{5}{2}}_{s-2}\right)^{-1}\Delta^{-1} P\frac{J_{\mathbb{O}^E_s}}{N} \otimes \mathcal{Y}^{E\frac{5}{2}}_{s-2}\right). \quad (15)$$

The first log Det above is the ('t Hooft-)planar contribution – involving the leading local vertex (Fig. 2a) – and the second log Det the leading-nonplanar one – involving at least one subleading local vertex (Fig. 2a) carrying the factor P. The latter gives rise to new topologies. For example, for 2-point correlators (Fig. 2b), by the standard 't Hooft gluing of reversely oriented lines, the first diagram in Fig. 2b – that

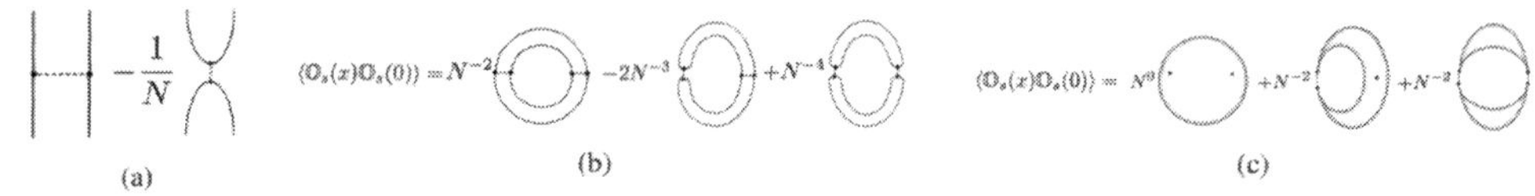

Figure 2. (a) Local vertex for 2-gluon operators in double-line. The dotted lines identify the punctures. (b) 2-point correlators of twist-2 operators to the leading perturbative order, with weights as in Fig. 3a. (c) Topology and weights of diagrams in Fig. 2b.

is the planar contribution – leads to a 2-punctured sphere and the remaining ones – by representing a 2-punctured disk as an infinite strip and gluing the opposite edges of the strip to get an infinite cylinder, i.e. a 2-punctured sphere – to possibly disconnected punctured spheres with at least two punctures pairwise identified (Fig. 2c). By generalizing the new diagrams in Fig. 2b, the leading-nonplanar contributions to the second log Det in Eq. (15) involve the new topological sector (Fig. 3a). After the

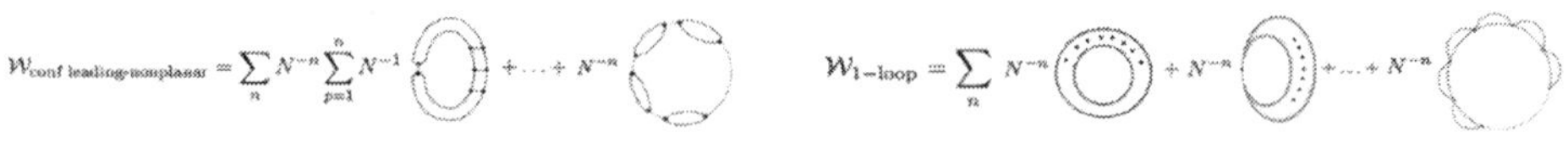

Figure 3. (a) The refined perturbative expansion of the conformal leading-nonplanar generating functional, with n pairs of punctures and p insertions of P, which give rise to the weights N^{-p} compensated for by the color traces. (b) After gluing reversely oriented lines, the new topologies arise from the normalization, obtained by cutting the pinches, of p-pinched tori with $n-p$ punctures.

RG improvement, the leading-nonplanar asymptotic generating functional reads

$$\mathcal{W}^E_{\text{asym nonplanar}}[J_{\mathbb{O}^E}, \lambda] = -\log \text{Det}\left(\mathcal{I} - \frac{1}{2}\left(\mathcal{I} + \frac{1}{2}\frac{Z_{\mathbb{O}_s}(\lambda)}{\lambda^{s+2}}\Delta^{-1} I\frac{J_{\mathbb{O}^E_s}}{N} \otimes \mathcal{Y}^{E\frac{5}{2}}_{s-2}\right)^{-1}\frac{Z_{\mathbb{O}_s}(\lambda)}{\lambda^{s+2}}\Delta^{-1} P\frac{J_{\mathbb{O}^E_s}}{N} \otimes \mathcal{Y}^{E\frac{5}{2}}_{s-2}\right)$$
$$= +\log \text{Det}\left(\mathbb{I} + \frac{1}{2}\frac{Z_{\mathbb{O}_s}(\lambda)}{\lambda^{s+2}}\Delta^{-1}\frac{J_{\mathbb{O}^E_s}}{N} \otimes \mathcal{Y}^{E\frac{5}{2}}_{s-2}\right). \quad (16)$$

Remarkably, now the overall sign of the first log Det in Eq. (16) is consistent with the bosonic statistics for the glueballs, but at the price of introducing a refined topological expansion where, by generalizing the new topologies in Fig. 2c for each n, in addition to the n-punctured tori, punctured spheres that are the normalization [28, 29] of p-pinched and $n-p$-punctured tori (Fig. 3b) arise, with $1 \leq p \leq n$. Besides, the punctured smooth tori are suppressed in perturbation theory with respect to the pinched ones, since the smooth tori inevitably involve the action vertices that carry powers of 't Hooft gauge coupling. Yet, in the RG-improved generating functional the smooth tori are essential to provide the renormalization factors in Eq. (16) due to the anomalous dimensions also for the new topologies that do not couple perturbatively to the action vertices V_3 and V_4, but mix with the punctured smooth tori by perturbative renormalization, since they have the same weights.

We provide now a nonperturbative interpretation of our refined topological expansion for twist-2 operators – that, incidentally, nonperturbatively explains the above mixing – in terms of the effective theory of glueballs that is closely related – though not exactly coincident – to the dual-graph representation of Riemann surfaces [28, 29] recalled below. It has been known for more than forty years that, in the effective theory of glueballs, punctured spheres correspond to glueball tree graphs [11, 12]. Specifically, the sphere with two punctures corresponds to an infinite sum of glueball propagators [11], while the sphere with three punctures corresponds in Minkowskian space-time to vertices that involve sums of three or two glueball poles [12] (Fig. 4a). For simplicity, in the following we skip n-glueball vertices with $n > 3$ as in Eq. (3). Incidentally, we point out that the last graph in Fig. 4a contributes zero to the S matrix because of the missing glueball external leg. Analogously, the 2-punctured torus may contribute, in addition to glueball one-loop two-leg graphs, also graphs with only one or zero external glueball legs (Fig. 4b) and, similarly, our new topologies may contribute the graphs in Fig. 4c. Incidentally, the partial matching of the graphs in Fig. 4c with the ones in Fig. 4b nonperturbatively explains the aforementioned mixing by renormalization. Analogous statements hold for the graphs in Figs. 3b and 5. Both graphs in Fig. 4c contribute zero to the S matrix as well. More generally, all the

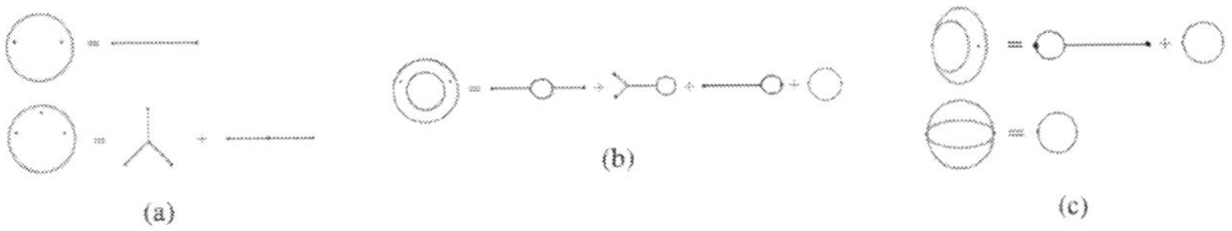

Figure 4. (a) Glueball tree graphs dual to punctured spheres in the effective theory of glueballs. (b) Glueball one-loop graphs dual to the 2-punctured torus. (c) Glueball one-loop graphs dual to pinched tori.

new topologies contribute zero to the S matrix, since they miss at least one external glueball leg, as for the last two graphs in Fig. 5. It is an open problem – though – whether an effective action of the kind in Eq. (3) exists that also nonperturbatively captures the contribution of the entire new topological sector for the correlators. However, independently of the existence of the above effective action, for the graphs

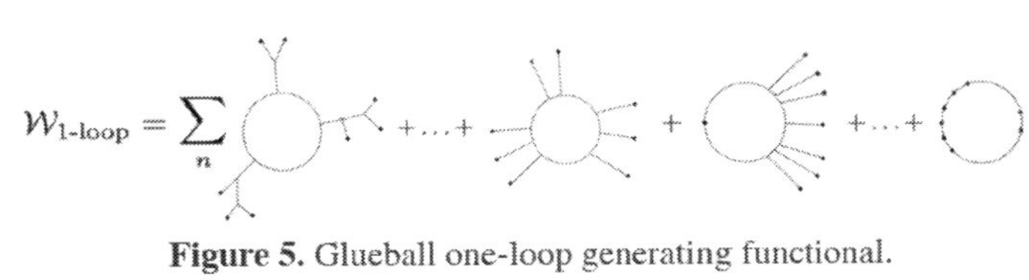

Figure 5. Glueball one-loop generating functional.

that only contain bivalent vertices, i.e. the maximally pinched ones in Fig. 3b, which are in one-to-one correspondence with diagrams only involving the subleading vertex and no leading vertex in Eq. (16), we verify the spin-statistics theorem directly by means of our asymptotic computation

$$\mathcal{W}^{E}_{\text{asym maximally pinched}}[J_{\mathbb{O}^E}, \lambda] = -\log \operatorname{Det}\left(\mathcal{I} - \frac{1}{2}\frac{Z_{\mathbb{O}_s}(\lambda)}{\lambda^{s+2}}\Delta^{-1} P \frac{J_{\mathbb{O}^E_s}}{N} \otimes \mathcal{Y}^{E\frac{s}{2}}_{s-2}\right)$$

$$= -\log \operatorname{Det}\left(\mathbb{I} - \frac{1}{2}\frac{Z_{\mathbb{O}_s}(\lambda)}{\lambda^{s+2}}\Delta^{-1} \frac{J_{\mathbb{O}^E_s}}{N} \otimes \mathcal{Y}^{E\frac{s}{2}}_{s-2}\right) \tag{17}$$

thanks to the minus sign in the r.h.s. above. Finally, the occurrence of pinched tori with punctures resembles Deligne-Mumford (DM) compactification [28, 29] of the moduli space of punctured closed Riemann surfaces that arises in canonical string theories [29] because of their underlying conformal structure [29]. Yet, in general our pinched tori do not occur in the DM compactification of n-punctured tori, whose components only involve punctured closed Riemann surfaces with $\chi < 0$ [28] and, specifically, spheres with at least three punctures [28, 29] – because in the DM compactification punctures and pinches never collide [28, 29], according to the conformal structure of canonical string theories [29] – while the components of our pinched tori may contain 2-punctured spheres with $\chi = 0$. Indeed, contrary to the DM compactification [28], the dual graphs to our pinched tori may contain bivalent vertices as in the last graphs in Figs. 4c and 5. As a consequence, no canonical closed-string theory that admits the DM compactification may contain the new topological sector.

By summarizing, in large-N SU(N) YM theory a new topological sector exists that refines 't Hooft topological expansion for the correlators of twist-2 operators, both perturbatively and nonperturbatively. The new topologies solve the sign puzzle, specifically in the maximally pinched sector. They also dominate the UV asymptotics of the correlators of twist-2 operators, but contribute zero to the nonperturbative S matrix, since nonperturbatively they consist of tori with at least one pinch, corresponding to glueball one-loop diagrams with at least one missing glueball external leg. As in general the new topologies do not arise from the DM compactification of the moduli space of punctured tori, no canonical string theory admitting it may exist for the correlators of large-N YM theory in the new topological sector, but it may exist for the S-matrix amplitudes. Finally, the existence of the new topological sector – specifically, the maximally pinched one – that contributes zero to the S matrix, but nontrivially to the correlators, opens the way for a nonperturbative solution limited to the new sector by a topological field/string theory – noncanonical in the sense of the present paper – along the lines foreseen in [9], by reinterpreting the coupling to D-branes in [9] as the generating functional of correlators in the new sector [2], rather than of the collinear S-matrix amplitudes.

References

[1] G. 't Hooft, A Planar Diagram Theory for Strong Interactions, Nucl. Phys. B **72**, 461 (1974). 10.1016/0550-3213(74)90154-0

[2] M. Bochicchio, M. Papinutto, F. Scardino, On the structure of the large-N expansion in SU(N) Yang-Mills theory (2024), 2401.09312

[3] G. Veneziano, Some Aspects of a Unified Approach to Gauge, Dual and Gribov Theories, Nucl. Phys. B **117**, 519 (1976). 10.1016/0550-3213(76)90412-0

[4] G. Veneziano, An Introduction to Dual Models of Strong Interactions and Their Physical Motivations, Phys. Rept. **9**, 199 (1974). 10.1016/0370-1573(74)90027-1

[5] O. Aharony, S.S. Gubser, J.M. Maldacena, H. Ooguri, Y. Oz, Large N field theories, string theory and gravity, Phys. Rept. **323**, 183 (2000), hep-th/9905111. 10.1016/S0370-1573(99)00083-6

[6] M. Bochicchio, Renormalization in large-N QCD is incompatible with open/closed string duality, Phys. Lett. B **783**, 341 (2018), 1703.10176. 10.1016/j.physletb.2018.06.072

[7] S.S. Gubser, I.R. Klebanov, A.M. Polyakov, Gauge theory correlators from noncritical string theory, Phys. Lett. B **428**, 105 (1998), hep-th/9802109. 10.1016/S0370-2693(98)00377-3

[8] M. Bochicchio, The large-N Yang-Mills S-matrix is ultraviolet finite, but the large-N QCD S-matrix is only renormalizable, Phys. Rev. D **95**, 054010 (2017), 1701.07833. 10.1103/PhysRevD.95.054010

[2] M. Bochicchio to appear on arXiv.

[9] M. Bochicchio, An asymptotic solution of Large-N QCD, for the glueball and meson spectrum and the collinear S-matrix, AIP Conf. Proc. **1735**, 030004 (2016). 10.1063/1.4949387

[10] R.F. Streater, A.S. Wightman, PCT, spin and statistics, and all that, Vol. 52 (Princeton University Press, 2000)

[11] A.A. Migdal, Multicolor QCD as Dual Resonance Theory, Annals Phys. **109**, 365 (1977). 10.1016/0003-4916(77)90181-6

[12] E. Witten, Baryons in the 1/N Expansion, Nucl. Phys. B **160**, 57 (1979). 10.1016/0550-3213(79)90232-3

[13] A.M. Polyakov, Gauge Fields and Strings, Vol. 3 of *Contemporary concepts in physics* (Taylor & Francis, 1987), ISBN 9783718603930

[14] M. Bochicchio, Glueball and meson propagators of any spin in large-N QCD, Nucl. Phys. B **875**, 621 (2013), `1305.0273`. 10.1016/j.nuclphysb.2013.07.023

[15] M. Bochicchio, Higher-Spin Currents, Operator Mixing and UV Asymptotics in Large-*N* QCD-like Theories, Universe **9**, 57 (2023). 10.3390/universe9020057

[16] M. Bochicchio, M. Papinutto, F. Scardino, UV asymptotics of n-point correlators of twist-2 operators in SU(N) Yang-Mills theory, Phys. Rev. D **108**, 054023 (2023), `2208.14382`. 10.1103/PhysRevD.108.054023

[17] M. Bochicchio, M. Papinutto, F. Scardino, n-point correlators of twist-2 operators in SU(N) Yang-Mills theory to the lowest perturbative order, JHEP **08**, 142 (2021), `2104.13163`. 10.1007/JHEP08(2021)142

[18] M. Bochicchio, On the geometry of operator mixing in massless QCD-like theories, Eur. Phys. J. C **81**, 749 (2021), `2103.15527`. 10.1140/epjc/s10052-021-09543-5

[19] Harish-Chandra, Infinite irreducible representations of the Lorentz group, Proceedings of the Royal Society of London. Series A. Mathematical and Physical Sciences **189**, 372 (1947)

[20] G. Feldman, P.T. Matthews, Multimass fields, spin, and statistics, Phys. Rev. **154**, 1241 (1967). 10.1103/PhysRev.154.1241

[21] R.F. Streater, Local fields with the wrong connection between spin and statistics, Communications in Mathematical Physics **5**, 88 (1967). 10.1007/BF01646839

[22] R. Casalbuoni, Majorana and the Infinite Component Wave Equations, PoS **EMC2006**, 004 (2006), `hep-th/0610252`. 10.22323/1.037.0004

[23] E. Wigner, On unitary representations of the inhomogeneous Lorentz group, Annals of mathematics pp. 149–204 (1939)

[24] M. Bochicchio, Glueball and Meson Spectrum in Large-N QCD, Few Body Syst. **57**, 455 (2016). 10.1007/s00601-016-1100-6

[25] M. Marino, Instantons and large N: an introduction to non-perturbative methods in quantum field theory (Cambridge University Press, 2015)

[26] F. Maltoni, K. Paul, T. Stelzer, S. Willenbrock, Color Flow Decomposition of QCD Amplitudes, Phys. Rev. D **67**, 014026 (2003), `hep-ph/0209271`. 10.1103/PhysRevD.67.014026

[27] A.V. Belitsky, D. Mueller, Broken conformal invariance and spectrum of anomalous dimensions in QCD, Nucl. Phys. B **537**, 397 (1999), `hep-ph/9804379`. 10.1016/S0550-3213(98)00677-4

[28] M. Chan, Moduli spaces of curves: Classical and tropical, Notices of the American Mathematical Society, (2021). https://doi.org/10.1090/noti2360

[29] E. Witten, Notes On Super Riemann Surfaces And Their Moduli, Pure Appl. Math. Quart. **15**, 57 (2019), `1209.2459`. 10.4310/PAMQ.2019.v15.n1.a2

Quantum anomalies and parity-odd CFT correlators for chiral states of matter*

Claudio Corianò[1,2,3,**] and *Stefano* Lionetti[1,2,***]

[1]Dipartimento di Matematica e Fisica, Università del Salento and INFN Sezione di Lecce, Via Arnesano 73100 Lecce, Italy
[2]National Center for HPC, Big Data and Quantum Computing, Via Magnanelli 2, 40033 Casalecchio di Reno, Italy
[3]Institute of Nanotechnology, National Research Council (CNR-NANOTEC), Lecce 73100, Italy

Abstract. Chiral currents influence the parity-odd sector of CFT correlators in momentum space, playing a crucial role in the evolution of the quark-gluon plasma in the early universe. We demonstrate that these parity-odd interactions, which couple quarks and gluons to gravitons, can be fully determined in terms of their anomaly content by solving the conformal constraints in momentum space. This process involves a single nonlocal, massless axion-like interaction in the longitudinal channel, which remains protected against thermal and finite density effects.

1 Introduction

Interest in studying anomalies in chiral matter, particularly the Adler-Bell-Jackiw (ABJ) anomaly, has grown significantly over the past two decades. This growth spans both condensed matter theory and high-energy physics, focusing on the study of matter under high densities and its behavior in transport. Experimental work, such as heavy ion collision studies, has observed and confirmed anomalous behavior in matter when exposed to strong fields and finite density chiral backgrounds. The chiral interaction indicates that for massless fermions interacting with electric and magnetic fields $(\vec{E}, \vec{B})$, the chiral fermion number N_5 is not conserved. Instead, it is altered due to the anomaly, expressed as

$$\frac{dN_5}{dt} = \frac{e^2}{2\pi^2} \int d^3x \, \mathbf{E} \cdot \mathbf{B}. \tag{1}$$

Axial-vector currents associated with global anomalous $U(1)$ symmetries can couple to gluons through quark loops, and to gravitons via both quark and gluon loops, facilitated by chiral anomaly interactions. These interactions are crucial in shaping the dynamics of the quark-gluon plasma and become particularly significant in chirally asymmetric environments, where they can intensify chiral asymmetries, especially in the early universe.

*Presented by S. Lionetti at QCD@Work, 18 - 21 June 2024, Trani, Italy

**e-mail: claudio.coriano@le.infn.it
***e-mail: stefano.lionetti@unisalento.it

The phenomenon can be integrated into the magnetohydrodynamic (MHD) equations governing the early universe plasma, leading to the generation of a chiral magnetic cascade. A prominent example of a related phenomenon is the chiral magnetic effect (CME) in the quark-gluon plasma. The CME is characterized by the generation of an electric charge separation along an external magnetic field, driven by an imbalance in chirality caused by the anomaly. In this context, the chiral current is introduced as

$$\vec{J}_5 = \frac{e^2 \mu_5}{2\pi^2} \vec{B}, \tag{2}$$

where $\mu_5 = \mu_L - \mu_R$ represents the chiral chemical potential, defined by the difference between the chemical potentials of left-handed (L) and right-handed (R) Weyl fermions in the initial state, and $\vec{B}$ is the magnetic field. The current $\vec{J}_5$ acts as an external source that triggers an anomalous response at the quantum level, leading to the generation of an electric field $\vec{E}$, which is aligned with the magnetic field.
This macroscopic phenomenon induces a collective motion within the Dirac sea. The topological nature of the imbalance ($\mu_5 \neq 0$) imparts a distinctive quality to the CME current, ensuring its non-dissipative behavior even in the presence of radiative corrections. The presence of a topological protection of the anomaly has been explictly shown in [1, 2].

1.1 Connection with axion physics

The analysis of parity-odd interactions in the conformal setting may involve pseudoscalar operators, naturally coupled to pseudoscalar fields in a conformal phase of the early universe. Since their introduction as a possible solution of the strong CP-problem in QCD within the Peccei-Quinn theory, pseudoscalar fields, particularly those coupled to gravity, can leave distinctive signatures in the stochastic gravitational wave background. Their interaction can be investigated in conformal field theory in momentum space (CFT$_p$). For axion-like fields, their coupling to gauge fields can induce spin-1 helicity asymmetries before any spontaneous symmetry breaking occurs. A correlator like $\langle TTO \rangle$, where O is a pseudoscalar coupled to an axion-like field ϕ, is directly linked to a gravitational anomaly. The local effective action in the infrared, when a conformal symmetry-breaking scale Λ is present, takes the form

$$\mathcal{L}_{axion} \supset \frac{\phi}{\Lambda} R_{\mu\nu} \tilde{R}^{\mu\nu}, \tag{3}$$

where $R_{\mu\nu}$ is the Ricci tensor. This correlator describes the interaction between the pseudoscalar and two gravitational waves at a semiclassical level in an effective vertex in the forms of local realization. This interaction is analogous to the familiar $(\phi/f)F\tilde{F}$ coupling in axion physics, which is associated with the $\langle JJO \rangle$ correlator.

2 Anomalies and CFT$_p$: Non-Lagrangian formulations

Of all the anomalies, the chiral and conformal ones are among the most discussed. The violation of the chiral symmetry at the quantum level manifests as a breakdown in the conservation of the axial current $J_5^\mu = \bar{\psi}\gamma^\mu\gamma_5\psi$

$$\nabla_\mu J_5^\mu = a_1\, \varepsilon^{\mu\nu\rho\sigma} F_{\mu\nu} F_{\rho\sigma} + a_2\, \varepsilon^{\mu\nu\rho\sigma} R^\alpha{}_{\beta\mu\nu} R^\beta{}_{\alpha\rho\sigma} \tag{4}$$

where a_1 and a_2 are numerical coefficients, $F_{\mu\nu}$ is the electromagnetic tensor and $R^\alpha{}_{\beta\mu\nu}$ is the Riemann tensor. Initially discovered in the context of high-energy physics, the chiral

anomaly's significance extends far beyond. Its implications span from cosmology [3] to condensed matter physics [4], with phenomena such as the quantum Hall effect, the chiral magnetic effect and applications to topological insulators.
On the other hand, the violation of conformal symmetry at the quantum level is reflected by the energy-momentum tensor's failure to be traceless

$$g_{\mu\nu}\langle T^{\mu\nu}\rangle = b_1 E_4 + b_2 C^{\mu\nu\rho\sigma}C_{\mu\nu\rho\sigma} + b_3\nabla^2 R + b_4 F^{\mu\nu}F_{\mu\nu} + f_1\varepsilon^{\mu\nu\rho\sigma}R_{\alpha\beta\mu\nu}R^{\alpha\beta}_{\rho\sigma} + f_2\varepsilon^{\mu\nu\rho\sigma}F_{\mu\nu}F_{\rho\sigma} \quad (5)$$

where $C^{\mu\nu\rho\sigma}$ is the Weyl tensor, E_4 is the Gauss-Bonnet term and b_i and f_i are numerical coefficients. The last two terms on the right-hand side of the equation are parity-odd and have been the center of recent discussions (see for example [5–9]). The anomalous laws described above translate into Ward identities that dictate the behaviour of various three-point correlation functions. We are going to present a short overview on parity-odd correlation functions which are affected by such anomalies in the conformal limit.
The idea of using conformal Ward identities (CWIs) to determine the structure of three-point functions in momentum space was presented for the first time independently in [10] and [11], the second of which outlines a method that includes the tensor case. Investigations into parity-odd CFT correlators have only recently emerged [12–17].
CFT$_p$, especially with broken parity, finds applications in cosmology, condensed matter physics, holography and the conformal bootstrap program. Indeed, the recent discovery of applications to topological insulators and Weyl semimetals has sparked new interest in the topic [18].

3 Decomposition of the correlators

The chiral and conformal Ward identities describe the behavior of the longitudinal and trace part of correlators, respectively. It is therefore useful to decompose a current J^μ and the energy-momentum tensor $T^{\mu\nu}$ in terms of their transverse-traceless part and longitudinal-trace one (also called "local")

$$J^\mu(p) = j^\mu(p) + j^\mu_{loc}(p), \qquad T^{\mu\nu}(p) = t^{\mu\nu}(p) + t^{\mu\nu}_{loc}(p) \quad (6)$$

where

$$\begin{aligned} j^\mu(p) &= \pi^\mu_\alpha(p)\,J^\alpha(p), & j^\mu_{loc}(p) &= \frac{p^\mu p_\alpha}{p^2}\,J^\alpha(p), \\ t^{\mu\nu}(p) &= \Pi^{\mu\nu}_{\alpha\beta}(p)\,T^{\alpha\beta}(p), & t^{\mu\nu}_{loc}(p) &= \Sigma^{\mu\nu}_{\alpha\beta}(p)\,T^{\alpha\beta}(p), \end{aligned} \quad (7)$$

having introduced the transverse (π), transverse-traceless (Π) and longitudinal-trace (Σ) projectors, given respectively by

$$\begin{aligned} \pi^\mu_\alpha &= \delta^\mu_\alpha - \frac{p^\mu p_\alpha}{p^2}, \qquad \Pi^{\mu\nu}_{\alpha\beta} = \frac{1}{2}\left(\pi^\mu_\alpha\pi^\nu_\beta + \pi^\mu_\beta\pi^\nu_\alpha\right) - \frac{1}{d-1}\pi^{\mu\nu}\pi_{\alpha\beta}, \\ \Sigma^{\mu\nu}_{\alpha\beta} &= \frac{p_\beta}{p^2}\Big[2\delta^{(\nu}_\alpha p^{\mu)} - \frac{p_\alpha}{(d-1)}\left(\delta^{\mu\nu} + (d-2)\frac{p^\mu p^\nu}{p^2}\right)\Big] + \frac{\pi^{\mu\nu}(p)}{(d-1)}\delta_{\alpha\beta}. \end{aligned} \quad (8)$$

As an example, we now consider the $\langle JJJ_5\rangle$ correlator, built with two vector currents J and one axial current J_5. The vector currents are conserved, while the axial one is not. Therefore, the conservation Ward identities are given by

$$\begin{aligned} p_{1\mu_1}\langle J^{\mu_1}(p_1)J^{\mu_2}(p_2)J_5^{\mu_3}(p_3)\rangle &= 0, \qquad p_{2\mu_2}\langle J^{\mu_1}(p_1)J^{\mu_2}(p_2)J_5^{\mu_3}(p_3)\rangle = 0 \\ p_{3\mu_3}\langle J^{\mu_1}(p_1)J^{\mu_2}(p_2)J_5^{\mu_3}(p_3)\rangle &= -8\,a_1\,i\,\varepsilon^{p_1p_2\mu_1\mu_2} \end{aligned} \quad (9)$$

where the last equation follows directly from the chiral anomaly formula (4). Due to the equations above, the longitudinal components of the vector currents vanish and the correlator consists only of a transverse part and an anomalous longitudinal part in J_5. Using the projectors mentioned above, we can express the general form for this correlator as

$$\begin{aligned}\langle J^{\mu_1}(p_1)J^{\mu_2}(p_2)J_5^{\mu_3}(p_3)\rangle &= \langle j^{\mu_1}(p_1)j^{\mu_2}(p_2)j_{5\,loc}^{\mu_3}(p_3)\rangle + \langle j^{\mu_1}(p_1)j^{\mu_2}(p_2)j_5^{\mu_3}(p_3)\rangle \\ &= -8\,i\,a_1\,\frac{p_3^{\mu_3}}{p_3^2}\varepsilon^{p_1p_2\mu_1\mu_2} + \pi^{\mu_1}_{\alpha_1}(p_1)\pi^{\mu_2}_{\alpha_2}(p_2)\pi^{\mu_3}_{\alpha_3}(p_3)\,X^{\alpha_1\alpha_2\alpha_3}\end{aligned} \tag{10}$$

where the transverse-traceless component is parametrized by an undetermined tensor $X^{\alpha_1\alpha_2\alpha_3}$. In general, by imposing (anomalous) Ward identities such as those given in equations (4) and (5), we can fully determine the longitudinal-trace part of a correlator. Specifically, correlators affected by an anomaly exhibit a pole in the anomalous momentum within the longitudinal-trace part. On the other hand, the transverse-traceless component of a correlator is not directly constrained by these Ward identities. In order to fix such component, we need to impose the conformal constraints following from the invariance under dilatations and special conformal transformations. In the next section, we will briefly outline the procedure.

4 Conformal Ward identities

The conformal constraints consist of differential equations that a correlator needs to satisfy. By solving such equations, we can determine the most general expression of a correlator in CFT$_p$.

Consider an arbitrary n-point correlator $\langle\ldots\rangle$ with the overall delta function for momentum conservation removed. In momentum space, the invariance of a correlator under dilatations is reflected in the following first-order differential equation

$$\left[\sum_{j=1}^{n}\Delta_j-(n-1)d-\sum_{j=1}^{n-1}p_j^\alpha\frac{\partial}{\partial p_j^\alpha}\right]\langle\ldots\rangle=0 \tag{11}$$

while the invariance under special conformal transformations leads to a second-order differential equation

$$0=\mathcal{K}^\kappa\,\langle\ldots\rangle \tag{12}$$

where

$$\mathcal{K}^\kappa \equiv \sum_{j=1}^{n-1}\left(2\left(\Delta_j-d\right)\frac{\partial}{\partial p_j^\kappa}-2p_j^\alpha\frac{\partial}{\partial p_j^\alpha}\frac{\partial}{\partial p_j^\kappa}+\left(p_j\right)_\kappa\frac{\partial}{\partial p_j^\alpha}\frac{\partial}{\partial p_{j\alpha}}\right)+\mathcal{K}^\kappa_{spin} \tag{13}$$

The expression of the operator $\mathcal{K}^\kappa_{spin}$ depends on the tensorial structure of the correlator [11]. The general solution of these equations can be expressed in terms of integrals involving a product of three Bessel functions ('3K integrals')

$$I_{\alpha\{\beta_1\beta_2\beta_3\}}(p_1,p_2,p_3)=\int dx\,x^\alpha\prod_{j=1}^{3}p_j^{\beta_j}K_{\beta_j}\left(p_jx\right) \tag{14}$$

where K_ν is a modified Bessel function of the second kind

$$\begin{aligned}K_\nu(x)&=\frac{\pi}{2}\frac{I_{-\nu}(x)-I_\nu(x)}{\sin(\nu\pi)}, \qquad \nu\notin\mathbb{Z}\\ I_\nu(x)&=\left(\frac{x}{2}\right)^\nu\sum_{k=0}^{\infty}\frac{1}{\Gamma(k+1)\Gamma(\nu+1+k)}\left(\frac{x}{2}\right)^{2k}\end{aligned} \tag{15}$$

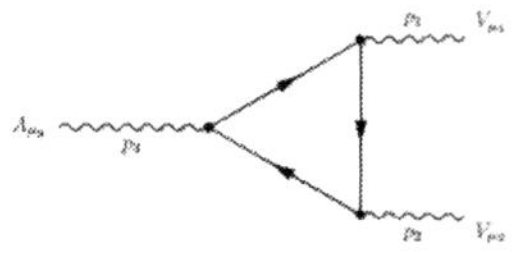

Figure 1. Feynman diagram for the chiral anomaly interaction $\langle JJJ_5\rangle$.

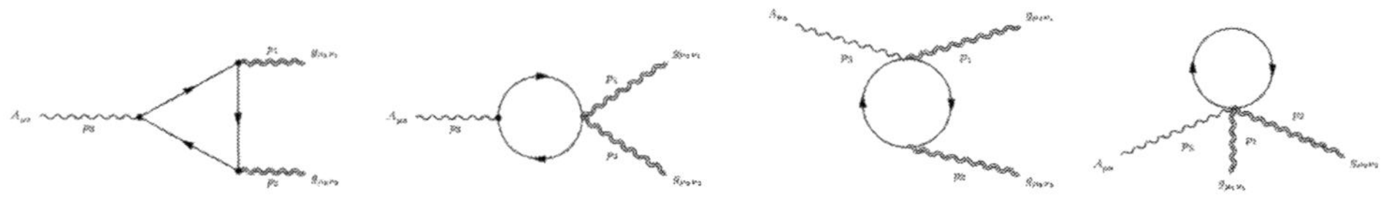

Figure 2. Topologies of Feynman diagrams contributing to the gravitational anomaly interaction $\langle TTJ_5\rangle$.

Details can be found in [11], including the connection to ordinary Feynman integrals for free field theory realizations. The analysis of the parity-odd sector in the presence of chiral and conformal anomalies is discussed in [14–17].

5 Solution of the CWIs: The correlators $\langle JJJ_5\rangle$, $\langle TTJ_5\rangle$ and $\langle J_5J_5J_5\rangle$

We now focus on the correlators that are affected by the chiral anomaly given in eq. (4). The first term on the righ-hand side of this equation can be analyzed by looking at the $\langle JJJ_5\rangle$. By imposing the conformal Ward identities described in the previous section, we can completely fix the correlator in the following form

$$\begin{aligned}\langle J^{\mu_1}(p_1)\, J^{\mu_2}(p_2)\, J_5^{\mu_3}(p_3)\rangle &= -8\, i\, a_1\, \frac{p_3^{\mu_3}}{p_3^2}\varepsilon^{p_1p_2\mu_1\mu_2} + \\ &\pi^{\mu_1}_{\alpha_1}(p_1)\pi^{\mu_2}_{\alpha_2}(p_2)\pi^{\mu_3}_{\alpha_3}(p_3)\; 8\, i\, a_1 \left[I_{3\{1,0,1\}}\, p_2^2\, \varepsilon^{p_1\alpha_1\alpha_2\alpha_3} - I_{3\{0,1,1\}}\, p_1^2\, \varepsilon^{p_2\alpha_1\alpha_2\alpha_3}\right]\end{aligned} \tag{16}$$

On the other hand, if we want examine the effect of the second term in eq. (4) we need to consider the $\langle TTJ_5\rangle$ correlator. By solving the conformal Ward identities, we can determine its general expression

$$\begin{aligned}\left\langle T^{\mu_1\nu_1}T^{\mu_2\nu_2}J_5^{\mu_3}\right\rangle &= 4ia_2\frac{p_3^{\mu_3}}{p_3^2}(p_1\cdot p_2)\left\{\left[\varepsilon^{\nu_1\nu_2p_1p_2}\left(g^{\mu_1\mu_2} - \frac{p_1^{\mu_2}p_2^{\mu_1}}{p_1\cdot p_2}\right) + (\mu_1\leftrightarrow\nu_1)\right] + (\mu_2\leftrightarrow\nu_2)\right\} + \\ &\Pi^{\mu_1\nu_1}_{\alpha_1\beta_1}(p_1)\,\Pi^{\mu_2\nu_2}_{\alpha_2\beta_2}(p_2)\,\pi^{\mu_3}_{\alpha_3}(p_3)\left[A_1\varepsilon^{p_1\alpha_1\alpha_2\alpha_3}p_2^{\beta_1}p_3^{\beta_2} + A_2\varepsilon^{p_1\alpha_1\alpha_2\alpha_3}\delta^{\beta_1\beta_2} + \left(\{p_1,\alpha_1,\beta_1\}\leftrightarrow\{p_2,\alpha_2,\beta_2\}\right)\right]\end{aligned} \tag{17}$$

with

$$\begin{aligned}A_1 &= -4\, i\, a_2\, p_2^2\, I_{5\{2,1,1\}}\\ A_2 &= -8\, i\, a_2\, p_2^2\left(p_3^2\, I_{4\{2,1,0\}} - 1\right).\end{aligned} \tag{18}$$

Let us now comment on the equations (16) and (17).

- Both formulas include, in their first row, an anomalous term in the longitudinal sector of the correlator, characterized by an anomalous pole $1/p_3^2$. In their second row, the formulas feature the transverse-traceless component, which is fixed in terms of 3K integrals by imposing conformal invariance.
- At first glance, one might assume that the chiral anomaly only affects the longitudal component of correlators, expecially when looking at equations like (9). However, this assumption does not hold true in a CFT, where the transverse component can indirectly depend on the anomaly content. This occurs because the conformal equations couple the longitudinal and transverse sectors, thereby leading to the appearance of the anomalous coefficients a_1 and a_2 in the transverse-traceless component as well.
- The correlators are fully determined by the anomalous coefficient a_1 or a_2, with no additional degrees of freedom. This is unusual for parity even 3-point correlators in a CFT which typically depend on multiple unfixed constants. Here, the parity-odd correlators are generated entirely by an anomaly and they vanish if the anomaly is absent ($a_1 = 0, a_2 = 0$).
- An alternative approach for studying this topic is given by the perturbative realization of CFT correlators which allows us to handle more simplified expressions proceeding with analyzing an ordinary Feynman expansion (see Fig. 1 and 2). The perturbative analysis serves as an important independent check for the non-perturbative conformal methods described in this paper and it also provides information on the emergence of the conformal and chiral anomalies. As demonstrated in [14, 16], the equations (16) and (17) are in perfect agreement with the perturbative results.

One can repeat the calculations for different correlators. In the case of the $\langle J_5 J_5 J_5\rangle$, where all the currents are anomalous, solving the conformal Ward identities yields the following result

$$\left\langle J_5^{\mu_1} J_5^{\mu_2} J_5^{\mu_3}\right\rangle = \frac{1}{3}\Big[\left\langle J_5^{\mu_1} J^{\mu_2} J^{\mu_3}\right\rangle + \left\langle J^{\mu_1} J_5^{\mu_2} J^{\mu_3}\right\rangle + \left\langle J^{\mu_1} J^{\mu_2} J_5^{\mu_3}\right\rangle\Big] \tag{19}$$

where the explicit form of the terms on the right-hand side can be obtained by appropriately permuting the indices and momenta in equation (16).
The methodology described in this paper can also be applied to correlators affected by trace anomalies. Such phenomena are discussed in [15].

5.1 Conclusions and outlook

In recent years, it has become increasingly clear that chiral anomalies—resulting from the breaking of classical global symmetries by quantum corrections—play a pivotal role in the dynamics of fundamental interactions. This influence extends not only to the vacuum case ($\mu = 0$), but also to scenarios involving chiral chemical potentials, as demonstrated in the context of the chiral magnetic effect. The potential for experimental verification of these interactions, which could reveal the nature of the pseudoparticles emerging from virtual corrections, as predicted by recent and past analyses, should be taken seriously at the experimental level. We have briefly outlined how conformal symmetry and a nonlocal interaction in a longitudinal channel determine the interactions underlying chiral and gravitational anomalies in gauge theories.

This work is partially funded by the European Union, Next Generation EU, PNRR project "National Centre for HPC, Big Data and Quantum Computing", project code CN00000013; by INFN, inziativa specifica *QG-sky* and by the grant PRIN 2022BP52A MUR "The Holographic Universe for all Lambdas" Lecce-Naples.

References

[1] C. Corianò, M. Cretì, S. Lionetti, R. Tommasi, Axionlike quasiparticles and topological states of matter: Finite density corrections of the chiral anomaly vertex, Phys. Rev. D **110**, 025014 (2024), 2402.03151. 10.1103/PhysRevD.110.025014

[2] C. Corianò, M. Cretì, S. Lionetti, R. Tommasi, Gravitational chiral anomaly at finite temperature and density, Phys. Rev. D **110**, 025008 (2024), 2404.06272. 10.1103/PhysRevD.110.025008

[3] K. Kamada, N. Yamamoto, D.L. Yang, Chiral effects in astrophysics and cosmology, Prog. Part. Nucl. Phys. **129**, 104016 (2023), 2207.09184. 10.1016/j.ppnp.2022.104016

[4] M.N. Chernodub, Y. Ferreiros, A.G. Grushin, K. Landsteiner, M.A.H. Vozmediano, Thermal transport, geometry, and anomalies, Phys. Rept. **977**, 1 (2022), 2110.05471. 10.1016/j.physrep.2022.06.002

[5] L. Bonora, Elusive anomalies, EPL **139**, 44001 (2022), 2207.03279. 10.1209/0295-5075/ac83e9

[6] L. Bonora, Perturbative and Non-Pertrubative Trace Anomalies, Symmetry **13** no.7, 1292 (2021) 2107.07918 10.3390/sym13071292

[7] S. Abdallah, S.A. Franchino-Viñas, M. Fröb, Trace anomalies for Weyl fermions: too odd to be true? (2023), 2304.08939 10.1088/1742-6596/2531/1/012004

[8] F. Bastianelli and L. Chiese, Chiral fermions, dimensional regularization, and the trace anomaly, Nucl. Phys. B **983**, 115914 (2022), 2203.11668 10.1016/j.nuclphysb.2022.115914

[9] R. Larue, J. Quevillon and R. Zwicky, Gravity-gauge anomaly constraints on the energy-momentum tensor, JHEP **05**, 307 (2024) 2312.13222 10.1007/JHEP05(2024)307

[10] C. Corianò, L. Delle Rose, E. Mottola, M. Serino, Solving the Conformal Constraints for Scalar Operators in Momentum Space and the Evaluation of Feynman's Master Integrals, JHEP **1307**, 011 (2013), 1304.6944. 10.1007/JHEP07(2013)011

[11] A. Bzowski, P. McFadden, K. Skenderis, Implications of conformal invariance in momentum space, JHEP **03**, 111 (2014), 1304.7760. 10.1007/JHEP03(2014)111

[12] S. Jain, R.R. John, Relation between parity-even and parity-odd CFT correlation functions in three dimensions, JHEP **12**, 067 (2021), 2107.00695. 10.1007/JHEP12(2021)067

[13] E.I. Buchbinder, B.J. Stone, Three-point functions of conserved currents in 4D CFT: General formalism for arbitrary spins, Phys. Rev. D **108**, 086017 (2023), 2307.11435. 10.1103/PhysRevD.108.086017

[14] C. Corianò, S. Lionetti, M.M. Maglio, Parity-odd 3-point functions from CFT in momentum space and the chiral anomaly, Eur. Phys. J. C **83**, 502 (2023), 2303.10710. 10.1140/epjc/s10052-023-11661-1

[15] C. Corianò, S. Lionetti, M.M. Maglio, CFT correlators and CP-violating trace anomalies, Eur. Phys. J. C **83**, 839 (2023), 2307.03038. 10.1140/epjc/s10052-023-11984-z

[16] C. Corianò, S. Lionetti, M.M. Maglio, Parity-violating CFT and the gravitational chiral anomaly, Phys. Rev. D **109**, 045004 (2024), 2309.05374. 10.1103/PhysRevD.109.045004

[17] C. Corianó, S. Lionetti, CFT Constraints on Parity-odd Interactions with Axions and Dilatons (2024), 2408.02580

[18] C. Corianò, M. Cretì, S. Lionetti, D. Melle, R. Tommasi, Axion-like Interactions and CFT in Topological Matter, Anomaly Sum Rules and the Faraday Effect, Adv. Phys. Res. (2024), 2403.15641. 10.1002/apxr.202400043

EPJ Web of Conferences **314**, 00027 (2024) https://doi.org/10.1051/epjconf/202431400027
QCD@Work 2024

Renormalized Entanglement Entropy and general entropic c−function

Mitsutoshi Fujita[1,*], *Song* He[2,3,**], and *Yuan* Sun[4,***] *Jun* Zhang[5,****]

[1]School of Nuclear Science and Technology, University of South China, Hengyang 421001, China
[2]Center for Theoretical Physics and College of Physics, Jilin University, Changchun 130012, People's Republic of China
[3]Max Planck Institute for Gravitational Physics (Albert Einstein Institute), Am Mühlenberg 1, 14476 Golm, Germany
[4]School of Physics and Electronics, Central South University, Changsha 418003, China
[5]Department of Physics and Astronomy, University of Alabama, 514 University Boulevard, Tuscaloosa, AL, 35487, USA

Abstract. Holographic entanglement entropy (EE) and its renormalized form in AdS solitons with gauge potential are computed across a range of dimensions. The renormalized EE is identified as a cutoff-independent universal component of EE. By taking into account Kaluza-Klein reduction and the constraints of the low-energy limit, the $(d-1)$-dimensional renormalized EE is deduced from the odd-dimensional counterpart. The region associated with the shrinking circle, which is examined at large values of l is indicative of this relationship. Transitions of the minimal surface are from a disk-shaped to a cylindrical configuration as l increases. A quantum phase transition occurs at a critical subregion size, with the renormalized EE exhibiting non-monotonic behavior in the vicinity of this size. In all dimensions, massive modes become decoupled at the low energy limit, whereas degrees of freedom that include Wilson lines influence the behavior at smaller energy scales.

1 Introduction

Quantum entanglement entropy is a pivotal concept in quantum mechanics, offering valuable insights into the entanglement phenomenon between distant parts of a quantum system. By examining the entanglement among different elements, EE gives a metric for the extent of information. Its implications involve quantum information theory, black hole physics, and condensed matter physics. Specifically, the EE of a subsystem A reveals the level of entanglement within a given quantum field theory, as referenced in the literature [1, 2]. The geometric aspect of field theories is captured as an area law, which resembles black hole entropy.

A motivation of this research is to analyze quantum entanglement of quantum field theory (QFT) with negative energy. $d = 4$ $\mathcal{N} = 4$ SYM theory on S^1 has negative energy. Because scalar and fermions become massive in addition to S^1 component of the gauge field, the IR

*e-mail: fujitamitsutoshi@usc.edu.cn
**e-mail: hesong@jlu.edu.cn
***e-mail: sunyuan@csu.edu.cn
****e-mail: jzhang163@crimson.ua.edu

behavior of this theory is dominated by $d = 3$ pure Yang-Mills theory [3]. A feature of this theory is confinement and a mass gap. The AdS soliton is dual to confining theory via the AdS/CFT. The mass of the AdS soliton gives negative values, which corresponds to the negative Casimir energy of gauge theory on the boundary $S^1 \times R^{1,2}$. Thus, the AdS soliton is stabler than other background satisfying the same asymptotic boundary conditions.

Degrees of freedom can be also computed from EE. For example, the coefficient of the logarithmic term gives the A-type anomaly for a spherical entangling surface in conformal field theory (CFT) [4]. Renormalized EE with a spherical entangling surface generally becomes central charge in CFT [5]. In the context of 4*d* CFT, it becomes the A-type anomaly, which obeys the C-theorem: the renormalized EE decreases in the infrared limit. The entropic C function from the EE with the striped subsystem also becomes degrees of freedom and the central charge of CFT [6]. What is then the difference between two entropy? The trace anomaly part is different. Renormalized EE only contains either A-type or B-type anomaly, while entropic c function contains both A-type and B-type anomalies due to the coexistence of these two anomalies.

In this study, we delve into the B-type anomaly of QFT dual to the AdS soliton with the gauge potential by using renormalized EE. Our findings reveal that the renormalized EE exhibits non-monotonic behavior relative to the length of the interval. There is the signified occurance of a quantum phase transition between disk and cylinder topologies. Notably, this phase transition occurs not at free theory but at QFT in the large N limit. We show that renormalized EE is a probe of alternations in the operator mass and the gauge potential. Consequently, the renormalized EE diminishes as the operator mass increases, showing the decoupling of massive degrees of freedom.

In section 2, the UV structure of the entanglement entropy is analyzed. In section 3, the renoramlized EE is analyzed in 4-dimensions. In section 4, holographic stress tensors in the AdS soliton with gauge potential is reviewed. In section 5, renormalized EE of QFT dual to AdS solitons with gauge potential is analyzed. The quantum phase transition of renormalized EE is captured.

2 The UV structure of the entanglement entropy

The ultraviolet (UV) structure of EE encompasses the most signifinant divergent component, which is proportional to the area of the entangling surface. We define s_{UV} as the UV structure of EE with the entangling surface $S^1 \times S^{d-3}$, whereas $s^{(0)}_{UV}$ is the UV structure of EE with the entangling surface S^{d-2}. These two structures are interconnected by a specific relationship [7]:

$$s_{UV} = \frac{l_\phi}{r} s^{(0)}_{UV}, \tag{1}$$

where l_ϕ is the periodicity of S^1 and r is the radius of a sphere.

The renormalized entanglement entropy is obtained by a combination of differentiation. The renormalized entanglement entropy is finite and independent of the UV cut-off.

$$\begin{aligned} s_{ren} &= \frac{1}{r} f_d(r\partial_r) r s_{EE} = L_d(r\partial_r) s_{EE} \\ &= \begin{cases} \dfrac{1}{(d-2)!!} r\partial_r (r\partial_r - 2)\ldots(r\partial_r - (d-3)) s_{EE}, & d = \text{odd}, \\ \dfrac{1}{(d-2)!!} (r\partial_r + 1)(r\partial_r - 1)\ldots(r\partial_r - (d-3)) s_{EE}, & d = \text{even}, \end{cases} \end{aligned} \tag{2}$$

Note that the first line is equal to $d-1$ dimensional renormalized entanglement entropy with the entangling surface S^{d-2} as follows [5]:

$$f_d(r\partial_r)s_{EE}^{(0)} = \begin{cases} \frac{1}{(d-2)!!}(r\partial_r - 1)(r\partial_r - 3)\ldots(r\partial_r - (d-2))s_{EE}^{(0)}, & d = \text{odd}, \\ \frac{1}{(d-2)!!}r\partial_r(r\partial_r - 2)\ldots(r\partial_r - (d-2))s_{EE}^{(0)}, & d = \text{even}. \end{cases} \tag{3}$$

We focus on odd dimensional results. According to (2), the renormalized entanglement entropy for $d=3$ and $d=5$, becomes

$$s_{ren} = r\partial_r s_{EE} \quad \text{for } d = 3, \tag{4}$$

$$s_{ren} = \frac{1}{3}r\partial_r(r\partial_r - 2)s_{EE} \quad \text{for } d = 5. \tag{5}$$

OUr analysis reveals that through a Kaluza-Klein reduction along the S^1 direction, the renormalized entanglement entropy effectively embodies $d-1$ dimensional one in the infrared limit. The first formula resembles the form of the 2-dimensional entropic c function, which is known to be non-negative and exhibits monotonic decrease in systems respecting Lorentz symmetry. In our study, we demonstrate that (4) in $d=3$ adheres to these properties for QFT dual to the deformed AdS soliton. Conversely, the behavior of the $4d$ renormalized entanglement entropy exhibits non-monotonic characteristics.

3 Renormalized entanglement entropy of 4d QFT

In this section, we examine the entanglement entropy and renormalized entanglement entropy across various scenarios, including a free scalar, Dirac fermion, and a 4-dimensional CFT. The trace anomaly in general CFT is linked to the logarithmic term inherent to EE.

The length scale r_1 of the subregion is related to rescaling of the metric:

$$\begin{aligned} r_1\partial_{r_1}s_{EE} =& \lim_{n\to 1}(-2\partial_n \int d^{d+1}x g_{\mu\nu}(x)\frac{\delta}{\delta g_{\mu\nu}(x)}[w - nw|_{n=1}]) \\ & \frac{1}{2\pi}\lim_{n\to 1}\partial_n\Big(\langle\int d^{d+1}x\sqrt{g}T_\mu{}^\mu(x)\rangle_{M_n} - n\langle\int d^{d+1}x\sqrt{g}T_\mu{}^\mu(x)\rangle_{M_1}\Big), \end{aligned} \tag{6}$$

where w is the effective action in the manifold with conical singularities M_n. EE has been replaced with a function of the partition function in M_n. The formula (6) relates EE to the trace anomaly.

To compute EE, we focus on a subsystem σ which is a cylindrical configuration where one direction is S^1. This subsystem aligns with the one in QFT corresponding to $d+1$-dimensional AdS solitons. By integrating the formula (6) and applying the trace anomaly formula, the log term of the entanglement entropy [8, 9] becomes $S_{EE} = s_A \log\left(\epsilon/R_1\right)$, where s has expression with respect to extrinsic curvatures [4]

$$s_A = \frac{a}{180}\int_{\partial\sigma} d^2x\sqrt{h}E_2 + \frac{c}{120}\int_{\partial\sigma} d^2x\sqrt{h}I_2, \tag{7}$$

where E_2 is the Euler density and I_2 is a Weyl invariant. Compared with the normalization of , $a = 360a_4$ and $c = 120c_4$. For a real scalar, $c = 1$ and for Dirac fermion, $c = 6$. Coefficients agree with trace anomalies.

The renormalized entanglement entropy for cylinder type topology in (2) becomes

$$S_{ren} = \frac{1}{2}(r\partial_r + 1)(r\partial_r - 1)s_{EE} = s_A = \frac{cl_\phi}{240r}. \tag{8}$$

Thus, the renormalized EE is consistent with the coefficient s_A. The renormalized entanglement entropy for a spherical entangling surface (3) becomes

$$S_{ren} = \frac{1}{2}r\partial_r(r\partial_r - 2)s_{EE} = s_A = \frac{a}{90}. \tag{9}$$

According to [5], the renormalized EE is consistent with s once again.

4 The AdS soliton with the gauge potential

The AdS soliton arises from performing a double Wick rotation of the AdS black hole. It corresponds to QFT with anti-periodic boundary conditions on fermions. The mass of the AdS soliton equates to the negative energy of the corresponding dual QFT [11]. This methodology has been applied for the Reissner Nordstrom AdS black hole as reported in reference [12]. An analytic continuation of the metric across both temporal and spatial dimensions is performed. The metric of the AdS soliton with a background gauge field becomes [13]

$$\begin{aligned} ds^2_{d+1} =& \frac{L^2}{z^2}\Big(\frac{dz^2}{h(z)} + h(z)d\phi^2 - dt^2 + dR^2 + R^2 d\Omega_{d-3}\Big), \\ A_\phi =& a^{(0)}_\phi\Big(1 - \Big(\frac{z}{z_+}\Big)^{d-2}\Big), \end{aligned} \tag{10}$$

where

$$h(z) = 1 - \Big(1 + \tilde{\epsilon} z_+^2 a_\phi^2\Big)\Big(\frac{z}{z_+}\Big)^d + \tilde{\epsilon} z_+^2 a_\phi^2 \Big(\frac{z}{z_+}\Big)^{2(d-1)}. \tag{11}$$

Here, we set $\tilde{\epsilon} = -1$ [1], The gauge field becomes as the source field of the conserved current and gives a non-zero VEV for the current $\langle J_\phi \rangle$. This gauge field becomes a Wilson line, changing the boundary condition (twisted boundary condition), by considering a gauge transformation. The Wilson line approaches to zero at the tip $z = z_+$ and the gauge connection is regular there. The radial coordinate is restricted to $z \leq z_+$.

Due to the regularity at the tip of the cigar, ϕ must have the periodicity $\phi \to \phi + 1/m_0$, where the Kaluza-Klein mass is defined as

$$m_0 = \frac{1}{4\pi z_+}\Big(d + (d-2)\frac{a_\phi^2}{z_+^2}\Big). \tag{12}$$

When one rewrites (12) with respect to m_0 and a_ϕ, there is a bound between two. The solution exists only if

$$a_\phi \leq \frac{2\pi m_0}{\sqrt{d(d-2)}} = a_c. \tag{13}$$

According to [10], the boundary energy becomes

$$M = -\frac{V_{d-2}}{m_0}\frac{L^{d-1}\bar{a}_\phi}{2\kappa^2 z_+^d}. \tag{14}$$

The boundary energy can become negative or positive depending on Wilson lines (gauge potential),.

$$\begin{cases} M < 0 & z_+ a_\phi < 1 \\ M > 0 & z_+ a_\phi > 1. \end{cases} \tag{15}$$

For $a_\phi = 0$, it becomes Casimier energy of dual $4d$ Super Yang-Mills theory [11]. Casimir energy is different among periodic, anti-periodic boundary conditions, and twisted boundary conditions.

[1] $\tilde{\epsilon} = 1$ for the Reissner Nordstrom AdS black hole.

5 The holographic entanglement entropy

Figure 1. The minimal surface for the small (left) and large (right) subregion in the gravitational background. Since the subregion is a spherical region, the minimal surface becomes a disk (cylinder) type of the small (large) subregion.

In this section, we compute the holographic entanglement entropy. We choose the entangling surface $S^1 \times S^{d-3}$: $R = r$ and $0 \leq \phi \leq l_\phi$ at a fixed time instance in the background (10). $R = R(z)$ is utilized as an embedding scalar. Note that this choice of entangling surface differs from those scenarios without the compactification of the ϕ direction. The surface action becomes

$$A = \int d^{d-1}x\mathcal{L} = \Omega_{d-3} l_\phi L^{d-1} \int dz \frac{R^{d-3}}{z^{d-1}} \sqrt{1 + h\dot{R}^2}, \tag{16}$$

where $\mathcal{L} = \sqrt{\det g_{ind}}$ and g_{ind} is the induced metric. The holographic entanglement entropy is given by

$$S^g_{EE} = \frac{2\pi}{\kappa^2} A \tag{17}$$

where A is minimized. Recall that the ratio L^{d-1}/κ^2 is dimensionless. We solve the EOM derived from (16). For this purpose, boundary conditions are imposed on $R(z)$. We consider two kinds of boundary conditions. The first one is a Disk type boundary conditions: $R(z_t) = 0$ and $R'(z_t) = \infty$. The second one is a cylinder type boundary condition: $z_t = z_+$. The minimal surface of each a type is written in Fig. 1 taken from [14].

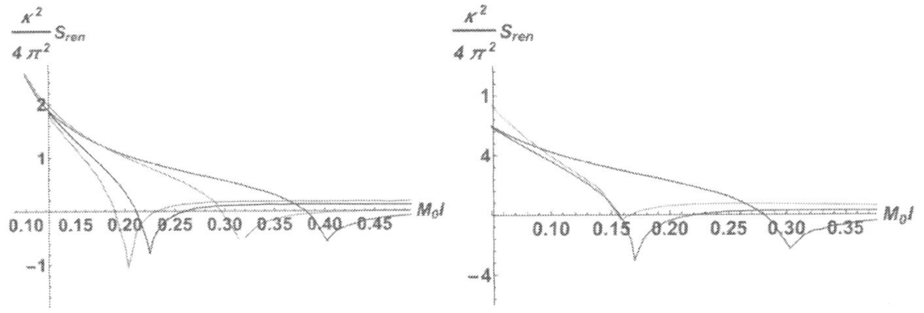

Figure 2. S_{ren} in 4 dimensions. (Left) S_{ren} for $d = 4$ is plotted for $a_\phi = \frac{i}{2}, 0, \frac{2}{3}, \frac{1}{\sqrt{2}}$ from the left to the right. The quantum phase transition occurs for $m_0 r_c = 0.21, 0.23, 0.32, 0.4$. Renormalized EE non-monotonically behaves around critical lengths. (Right) Renormalized EE for $m_0 = 3/5, 2/5, 1/\pi$ from the left to the right. The quantum phase transition occurs when $m_0 r_c = 0.24, 0.27, 0.4$. It implies that massive modes decouple others as soon as possible. The final state becomes a product state.

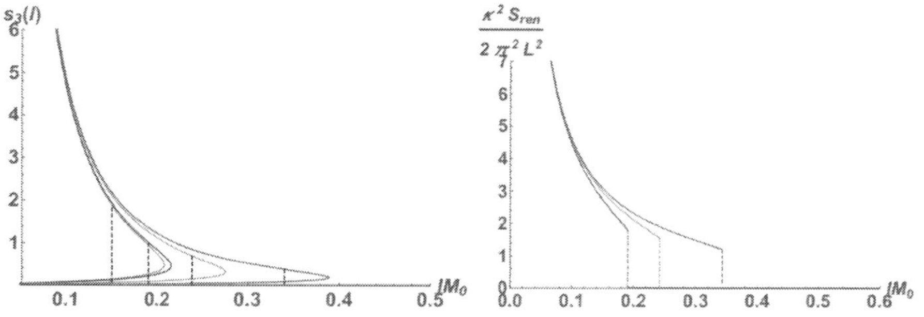

Figure 3. Left: $s_3(r) = dA_s/(3l_\phi dr)$ is plotted as a function of rm_0 for $a_\phi = i/2, 0, 1, 2/\sqrt{3}$ from the left to the right. The phase transition occurs at critical lengths $r_c = 0.15, 0.19, 0.24, 0.34$. $s_3(r)$ has multi-values. The upper curve agrees with the strong subadditivity. Right: S_{ren} for $d = 3$ for $a_\phi = 0, 1, 2/\sqrt{3}$ from the left to the right. S_{ren} monotonically decreases as a function of rm_0 and jumps to zero at a large distance. It implies that theory is a product state there.

For $4d$, renormalized EE is given by the formula (2), in which the differentiation is applied to s^g_{EE}. Renormalized EE is indicative of effective degrees of freedom from EE at energy scale $E \sim 1/r$. Renormalized EE is dipicted as a function of rm_0 ($r \equiv l$) in Fig. 2 sourced from [14]. Renormalized EE non-monotonically behaves near the critical length, which is analogous to a behavior of GPPZ flow. Because degrees of freedom with Wilson lines remain at large a_ϕ and high energy, the renormalized EE slowly decreases until the critical length. For low energies, the renoramlized EE fails to discern the effective degrees of freedom and almost stabilizes to a constant value. It implies that Wilson lines make particles of dual QFT light. The right figure shows that massive modes decouple others soon.

For $3d$, renormalized EE is plotted as a function of rm_0 in Fig. 3 taken from [14]. Renormalized EE is positive and monotonically decreasing because massive degrees of freedom decouple others in the low energy limit. This monotonic behavior arises because of Lorentz symmetry and strong subadditivity. The quantum phase transition occurs at critical lengths. Figures are consistent with the entropic c function in $R^{1,1}$. In the left Figure, $s_3(r)$ is ds_{EE}/dr. Figures also reflect that the minimal surface is concave.

6 Discussion

Renormalized EE exhibits a non-monotonic trend around a critical length, while it partially satisfies C-theorem because $S_{ren}(r \to 0) \geq S_{ren}(r \to \infty)$. S^{ren} in the small length limit is larger than S^{ren} in the large length limit. Only for $d = 3$, REE decreases and becomes positive because of its form $S_{ren} = r\partial r s^{EE}$ constrained by Lorentz invariance and strong subadditivity. [2] Renormalized EE also counts effective DOF of Wilson lines along the S^1 direction, capturing quantum phase transition.

Except for $d = 3$, qualitative behavior of renormalized EE is analogous to the generalized entropic C-function. Both quantities count effective DOF of Wilson lines and detect quantum phase transition. For $d = 3$, renormalized EE is different from the generalized entropic C-function by a product of r: $C^g = rS^{d=3}_{ren}$. Actually, the non-monotonic behavior of the generalized entropic C-function can be attributed to the interplay between r and $\partial s_{EE}/\partial r$.

[2]We would like to thank T. Takayanagi for pointing out this.

References

[1] C. Holzhey, F. Larsen and F. Wilczek, "Geometric and renormalized entropy in conformal field theory," Nucl. Phys. B **424**, 443 (1994) [arXiv:hep-th/9403108]; P. Calabrese and J. L. Cardy, "Entanglement entropy and quantum field theory," J. Stat. Mech. **0406**, P002 (2004) [arXiv:hep-th/0405152]

[2] P. Calabrese and J. Cardy, "Entanglement entropy and conformal field theory," J. Phys. A **42** (2009) 504005 [arXiv:0905.4013 [cond-mat.stat-mech]]; H. Casini and M. Huerta, "Entanglement entropy in free quantum field theory," J. Phys. A **42** (2009) 504007 [arXiv:0905.2562 [hep-th]]

[3] E. Witten, "Anti-de Sitter space, thermal phase transition, and confinement in gauge theories," Adv. Theor. Math. Phys. **2**, 505-532 (1998) [arXiv:hep-th/9803131 [hep-th]]

[4] S. N. Solodukhin, "Entanglement entropy, conformal invariance and extrinsic geometry," Phys. Lett. B **665**, 305-309 (2008) [arXiv:0802.3117 [hep-th]]

[5] H. Liu and M. Mezei, "A Refinement of entanglement entropy and the number of degrees of freedom," JHEP **04** (2013), 162 [arXiv:1202.2070 [hep-th]]

[6] T. Nishioka and T. Takayanagi, "AdS Bubbles, Entropy and Closed String Tachyons," JHEP **0701**, 090 (2007) [hep-th/0611035]

[7] M. Ishihara, F. L. Lin and B. Ning, "Refined Holographic Entanglement Entropy for the AdS Solitons and AdS black Holes," Nucl. Phys. B **872**, 392-426 (2013) [arXiv:1203.6153 [hep-th]]

[8] H. Casini and M. Huerta, "Entanglement entropy in free quantum field theory," J. Phys. A **42**, 504007 (2009) doi:10.1088/1751-8113/42/50/504007 [arXiv:0905.2562 [hep-th]]

[9] M. Huerta, "Numerical Determination of the Entanglement Entropy for Free Fields in the Cylinder," Phys. Lett. B **710**, 691-696 (2012) [arXiv:1112.1277 [hep-th]]

[10] V. Balasubramanian and P. Kraus, "A Stress tensor for Anti-de Sitter gravity," Commun. Math. Phys. **208**, 413 (1999) [hep-th/9902121]

[11] G. T. Horowitz and R. C. Myers, "The AdS / CFT correspondence and a new positive energy conjecture for general relativity," Phys. Rev. D **59**, 026005 (1998) [hep-th/9808079]

[12] S. A. Hartnoll, "Lectures on holographic methods for condensed matter physics," Class. Quant. Grav. **26**, 224002 (2009) [arXiv:0903.3246 [hep-th]]

[13] M. Fujita, S. He and Y. Sun, "Thermodynamical property of entanglement entropy and deconfinement phase transition," Phys. Rev. D **102** (2020) no.12, 126019 [arXiv:2005.01048 [hep-th]]

[14] M. Fujita, S. He, Y. Sun and J. Zhang, "Holographic renormalized entanglement and entropic c-function," JHEP **01**, 079 (2024) [arXiv:2309.03491 [hep-th]]

EPJ Web of Conferences **314**, 00028 (2024) https://doi.org/10.1051/epjconf/202431400028
QCD@Work 2024

Symplectic quantization: A new deterministic approach to the dynamics of quantum fields inspired by statistical mechanics*

Martina Giachello[1,2,**] and *Giacomo* Gradenigo[1,2,***]

[1]Gran Sasso Science Institute, Viale F. Crispi 7, 67100 L'Aquila, Italy
[2]INFN-Laboratori Nazionali del Gran Sasso, Via G. Acitelli 22, 67100 Assergi (AQ), Italy

Abstract. In this contribution we summarize the main features of a new algorithm (already presented in [1]) to sample numerically on the lattice the quantum fluctuations of fields by means of a deterministic pseudo-Hamiltonian dynamics in an enlarge space of variables. The main goal is to provide a numerical tool which is well defined in Minkowski space. The proposed approach introduces an additional time variable that plays the role of a true physical parameter that controls the deterministic dynamics. The sampling of quantum fluctuations is guaranteed by the presence of new additional conjugated momenta, which represent the rate of variation of ordinary fields with respect to the newly added time variable. From the pseudo-Hamiltonian dynamics one is then able, assuming ergodicity, to retrieve the Feynman path integral as the Fourier transform of a pseudo-microcanonical partition function.

1 Introduction

Since its introduction by Kenneth Wilson [2], lattice field theory has undergone significant advancements [3, 4] as a powerful tool for tackling non-perturbative problems in quantum field theory. This is particularly true for the study of strong interactions, including challenges like the calculation of hadronic masses [5] and the analysis of heavy-ion collisions [6]. However, despite these impressive accomplishments, a fundamental limitation remains: any numerical method on the lattice relies on an importance sampling protocol which is defined only for Euclidean field theory. Yet, while Wick rotation makes numerical approaches feasible, it simultaneously imposes a main limitation: it does not allow for a proper representation of processes that inherently depend on the causal structure of space-time, such as those occurring on the light cone, which is not defined in the Euclidean framework. The probability distribution $\exp(-S_E[\phi]/\hbar)$, to which is mapped in the Euclidean framework the complex phase factor $\exp(iS[\phi]/\hbar)$, essentially acts as an "equilibrium" measure for quantum fluctuations, enabling highly precise reproduction only of stable/equilibrium bound states [5]. However, it fails in capturing metastable resonances with brief lifetimes, such as tetraquark or pentaquark states [7, 8], or the dynamics of relativistic scattering processes, which involve different numbers of degrees of freedom between $\langle \text{IN}|$ and $|\text{OUT}\rangle$ states. For these reasons, we find it crucial to explore numerical tests of any new formulations of quantum field theory that can

*Presented by M.Giachello at QCD@Work, 18 - 21 June 2024, Trani, Italy

**e-mail: martina.giachello@gssi.it
***e-mail: giacomo.gradenigo@gssi.it

both define and investigate the dynamics of quantum fluctuations directly within Minkowski space-time. A promising idea in this direction is the recent proposal of a functional approach to field theory which is probabilistically well defined also in Lorentzian spacetime. This approach, introduced in [1, 9–11], is referred to as *"symplectic quantization"*. In this framework, for a given quantum field $\phi(x)$, where $x = (ct, \mathbf{x})$ represents a point in four-dimensional space-time, the field is assumed to depend also on an additional time parameter τ, which labels the sequence of quantum fluctuations at each given point of the discretized space-time lattice. An historically well known approach which also makes use of an additional time variable is the Parisi-Wu stochastic quantization approach [12, 13], still well-defined only for Euclidean theories. The use of deterministic dynamics with additional conjugated momenta to model quantum fields fluctuations is also something not new in the literature: it is for instance one of the components of the Hybrid Monte Carlo approach to lattice gauge theory with fermions [14]. But, once more, this has been implemented only for Euclidean theories. The innovative proposal of symplectic quantization, as discussed in [1, 9, 10], is to use the deterministic dynamics to sort quantum fluctuations directly in Lorentzian space-time, by incorporating the conjugate momenta of fields with respect to τ directly into the relativistic action with a Minkowski signature. The idea that some physics can be reproduced only within a microcanonical ensemble approach has been strongly inspired by recent statistical mechanics results for the Discrete Non-Linear Schrödinger Equation, for which the localized phases lives only in the microcanonical [15, 16].

2 The formalism

The symplectic quantization approach to field theory assumes, consistently with the existence of an intrinsic time τ, the existence of conjugated momenta of the kind $\pi(x, \tau) = \dot{\phi}(x, \tau)$ and of a dynamics generated by the following generalized Lagrangian

$$\mathbb{L}(\phi, \dot{\phi}) = \int d^d x \left[\frac{1}{2}\dot{\phi}^2(x) + S(\phi, \partial_\mu \phi)\right],$$

where $S(\phi, \partial_\mu \phi)$ is the standard action for a quantum field, e.g., for a real scalar field,

$$S(\phi, \partial_\mu \phi) = \int d^d x \left(\frac{1}{2}\partial_\mu \phi(x)\partial^\mu \phi(x) - V[\phi(x)]\right) = \int d^d x \left[-\frac{1}{2}\phi\, \partial_0^2 \phi + \frac{1}{2}\sum_{i=1}^{d} \phi\, \partial_i^2 \phi - V[\phi(x)]\right]$$

where we have integrated by parts and assumed a vanishing amplitude for the field at infinity. Let us then consider the potential $V[\phi(x)] = \frac{1}{2}m^2\phi^2 + \frac{1}{4}\lambda\phi^4$. By means of a Legendre transform, we then pass to the generalized Hamiltonian.

$$\mathbb{H}[\phi, \pi] = \int d^d x \left[\frac{1}{2}\pi^2(x) - S(\phi, \partial_\mu \phi)\right] = \int d^d x \left[\frac{1}{2}\pi^2(x) + \frac{1}{2}\phi\, \partial_0^2 \phi - \sum_{i=1}^{d} \phi\, \partial_i^2 \phi + V[\phi(x)]\right].$$

We assume that, within the symplectic quantization approach, the dynamics of quantum field theoretical fluctuations is the one governed by the following Hamilton equations:

$$\dot{\phi}(x) = \frac{\delta \mathbb{H}[\phi, \pi]}{\delta \pi(x)} \qquad \dot{\pi}(x) = -\frac{\delta \mathbb{H}[\phi, \pi]}{\delta \phi(x)}, \tag{1}$$

from which we derive the equation of motion for the dynamics of quantum fluctuations:

$$\ddot{\phi}(x, \tau) = -\partial_0^2 \phi(x, \tau) + \sum_{i=1}^{d} \partial_i^2 \phi(x, \tau) - \frac{\delta V[\phi]}{\delta \phi(x, \tau)}. \tag{2}$$

The effectiveness of symplectic quantization is based on the assumption that the dynamical averages along the trajectories generated by Eq. (1) can be generically replaced by statistical averages over a pseudo-microcanonical ensemble with a partition function built on the conservation of $\mathbb{H}[\phi, \pi]$

$$\Omega(A) = \int \mathcal{D}\phi\mathcal{D}\pi\, \delta\left(A - \mathbb{H}[\phi, \pi]\right) \qquad \rho[\phi(x)] = \frac{1}{\Omega(A)}\, \delta\left(A - \mathbb{H}[\phi, \pi]\right) \tag{3}$$

where ρ is the associated probability density and $\Omega(A)$ a normalisation factor, with $\mathcal{D}\phi$ and $\mathcal{D}\pi$ the standard notation for functional integration. What we mean to say is that, being $\mathbb{H}[\phi(x)]$ a given observable of the quantum fields, symplectic quantization can be related to the standard path-integral formulation of field theory by claiming that for generic initial conditions the following equivalence between ensembles holds:

$$\lim_{T\to\infty} \frac{1}{T} \int_0^T d\tau\, O[\phi(x,\tau)] = \int \mathcal{D}\phi\mathcal{D}\pi\, \rho[\phi(x)]\, O[\phi(x)]. \tag{4}$$

It is in fact possible to see that the Feynman path integral turns out to be nothing but the Fourier anti-transform of the microcanonical partition function written in Eq.(3):

$$\mathcal{Z}(\hbar) = \int_{-\infty}^{\infty} dA\, e^{-iA/\hbar}\, \Omega(A) \propto \int \mathcal{D}\phi\, e^{iS[\phi]/\hbar}, \tag{5}$$

where in the last term of Eq. (5) we have integrated away the quadratic dependence on momenta, which only yields an infinite normalization constant.

3 Numerical simulations

In order to test the effectiveness of our approach, we have run numerical simulations of the theory on a 1+1-dimensional lattice, for which the discretized Hamiltonian reads as:

$$\mathbb{H}[\phi, \pi] = \frac{1}{2} \sum_{\mathbf{i}\in\mathbb{Z}^2} \left[\pi_{\mathbf{i}}^2 + \frac{1}{a^2}\phi_{\mathbf{i}}\Delta^{(0)}\phi_{\mathbf{i}} - \frac{1}{a^2}\phi_{\mathbf{i}}\Delta^{(1)}\phi_{\mathbf{i}} + m^2\phi_{\mathbf{i}}^2 + \frac{\lambda}{4}\phi_{\mathbf{i}}^4 \right],$$

where a denotes the lattice spacing, which has been fixed to $a = 1$, while the symbol $\Delta^{(\mu)}\phi_{\mathbf{i}}$ denotes the discrete one-dimensional Laplacian along the μ-th coordinate axis: $\Delta^{(\mu)}\phi_{\mathbf{i}} = \phi_{\mathbf{i}+\mathbf{e}^\mu} + \phi_{\mathbf{i}-\mathbf{e}^\mu} - 2\phi_{\mathbf{i}}$. On the 1+1-dimensional lattice the equations of motion Eq. (2) read as:

$$\ddot{\phi}_{n,m}(t) = -\frac{\phi_{n+1,m} + \phi_{n-1,m} - 2\phi_{n,m}}{a^2} + \frac{\phi_{n,m+1} + \phi_{n,m-1} - 2\phi_{n,m}}{a^2} - m^2\phi_{n,m} - \lambda\phi_{\mathbf{i}}^3,$$

where $\phi_{n,m}(t)$ is the field at lattice point with coordinates (n, m).

3.1 Initial conditions

We have shown in Eq. (5) that the Feynman path-integral is connected to the partition function of symplectic quantization by means of a Fourier transform. The partition function $\Omega(\mathcal{A})$ is in turn connected to the Hamiltonian dynamics through an ergodic assumption. It is then legitimate to wonder how to fix the value of $\mathcal{A}$. What we known is that there is a fundamental scale for the "action per degree of freedom", which is $\hbar$. In complete analogy with the definition of temperature in the microcanonical ensemble we can therefore fix the total action $\mathcal{A}$, as suggested in [11], according to the formula:

$$\frac{1}{\hbar} = \frac{d\Sigma_{\text{sym}}(\mathcal{A})}{d\mathcal{A}}.$$

But unless one is able to explicitly compute $\Sigma(\mathcal{A})$, the above relation between $\mathcal{A}$ and $\hbar$ is not particularly useful in practice to assign the value of $\mathcal{A}$. The standard choice to fix $\mathcal{A}$ in order to have $\hbar$ as the typical scale of action is to follow the method of molecular dynamic simulations: fix the energy scale by assigning the appropriate kinetic energy in the initial conditions. Specifically, we ensure that the momentum-momentum correlation is stationary and satisfies $\langle\pi^*(k_i)\pi(k_i)\rangle = \frac{\hbar}{2}$, where k_i denotes discretized momenta. We have therefore set the initial conditions to $|\pi^*(k_i;\tau = 0)|^2 = \hbar$ for all momenta k_i, taking for simplicity $\hbar = 1$. These initial conditions were used for all the simulations presented in this work.

3.2 The role of the non-linearity

It can be easily checked that, in the framework of symplectic quantization, the free theory of a scalar field in Minkowski spacetime is unstable. In fact, considering a quadratic potential $V[\phi] = \frac{1}{2}m^2\phi^2$, we find that in Fourier space the equations of motion reads as $\ddot{\phi}(k,\tau) + \omega_k^2\phi(k,\tau) = 0$, where $\omega_k^2 = |\mathbf{k}|^2 + m^2 - k_0^2$. When $\omega_k^2 < 0$, the solution grows exponentially, leading to an unphysical increase in the field variable. These solutions, which are formally compatible with the conservation of generalized energy, cannot be handled by computers and result, numerically, in an exponential divergence also of the energy, as shown on the right of (Fig. 1). This is due to the accumulation of numerical errors in the algorithm. We observed this behavior using periodic boundary conditions, though they are not ideal for studying causal propagation of signals in the 1+1 lattice. It was then checked numerically that a non-linear attractive potential of the kind $\lambda\phi^4$ prevents the divergence of the free theory, as can be also guessed looking at Eq. (2). The non-linear quartic interaction allows the dynamics to reach a stationary state where energy is conserved within algorithmic bounds, thus allowing the last step of our analysis: the introduction of boundary conditions suitable to study the causal structure of the lattice.

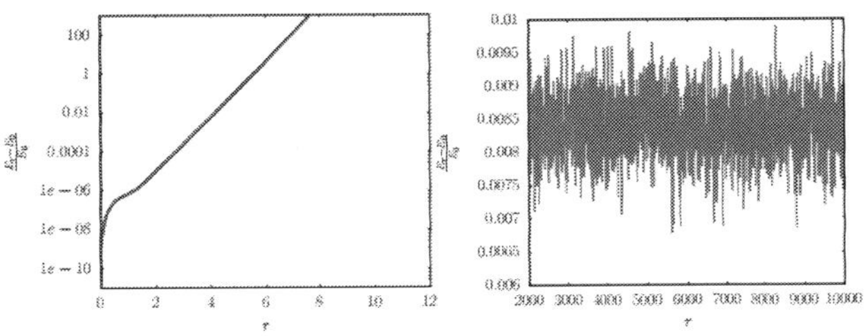

Figure 1. *Behaviour of the normalized energy $(E(\tau) - E_0)/E_0$ vs τ for a scalar theory with $m = 1.0$, $a = 1.0$, $L = 128$. On the left, the theory is free ($\lambda = 0$), and we notice an exponential growth with τ after a short transient. On the right, we added a small self-interaction term, $\lambda = 0.001$. The algorithm is stable with oscillations of order $\delta\tau$.*

3.3 Boundary conditions

An important aspect of this study is the choice of boundary conditions. We used two types of boundary conditions in the simulations. For the stability analysis (Sec. 3.2) we applied standard periodic boundary conditions. However, for the study of signal propagation (Sec. 3.4), we employed *fringe* boundary conditions [17] to simulate an infinite lattice beyond the grid. The fringe boundary conditions consist in adding to the original lattice Γ an external layer Γ_{ext} of tunable thickness, the two forming thus a larger lattice $\Gamma_f = \Gamma + \Gamma_{\text{ext}}$, which we call the *fringe* lattice, to which periodic boundary conditions are finally applied. The main idea of fringe boundary conditions is to have a free transimission of signals at the boundary of Γ and a damping across Γ_{ext}: this effect is achieved by considering different Hamiltonians for Γ and Γ_{ext}. The fringe Hamiltonian is given by $\mathbb{H}_f[\pi,\phi] = \mathbb{H}[\pi,\phi] + \mathbb{H}_{\text{ext}}[\pi,\phi]$ where $\mathbb{H}[\pi,\phi]$ is the original system Hamiltonian and $\mathbb{H}_{\text{ext}}[\pi,\phi]$ includes a damping term to control signal propagation:

$$\mathbb{H}_{\text{ext}}[\pi,\phi] = \frac{1}{2}\sum_{x\in\Gamma_{\text{ext}}}\left[\pi(x)^2 + m^2\phi(x)^2 + \frac{\lambda}{4}\phi(x)^4 + \alpha\left(\frac{1}{a^2}\phi(x)\Delta^{(0)}\phi(x) - \frac{1}{a^2}\phi(x)\Delta^{(1)}\phi(x)\right)\right],$$

with $\alpha \ll 1$. This setup allows free signal propagation across Γ and damping in Γ_{ext}, ensuring accurate results for the Feynman propagator (Sec. 3.4).

3.4 Results: the Feynman propagator

In this section, we will present numerical evidence showing that for small values of the non-linearity coefficient λ, we qualitatively recover the correct form of the free Feynman propagator. Our approach is straightforward: by setting the non-linear interaction coefficient to a small finite value, $\lambda = 0.001$, we run the symplectic dynamics with fringe boundary conditions until the system reaches a stationary state, let us say at time τ_{eq}. We then study the Fourier spectrum of the two-point correlation function $G(k) = \langle\phi^*(k)\phi(k)\rangle$ by computing the ensemble average as a sliding window average along the dynamics for times larger than the equilibration time, $\tau > \tau_{\text{eq}}$, i.e., according to the definition:

$$\langle\phi^*(k)\phi(k)\rangle = \frac{1}{\Delta\tau}\sum_{i=0}^{M}\phi^*(k,\tau_{\text{eq}}+\tau_i)\phi(k,\tau_{\text{eq}}+\tau_i),$$

where $\tau_i = i\cdot\delta\tau$ and $\Delta\tau = M\delta\tau$. In (Fig. 2), we show (right) the Fourier spectrum of $G(k)$ for $m = 3.0$ and $\lambda = 0.001$, comparing it (left) to the theoretical free Feynman propagator $G_{\text{th}}(k_0,k_1) = \left[\frac{4}{a^2}\sin^2\left(\frac{ak_0}{2}\right) - \frac{4}{a^2}\sin^2\left(\frac{ak_1}{2}\right) - m^2\right]^{-1}$. The numerical Feynman propagator qualitatively matches the theoretical prediction, exhibiting a saddle shape. The causal structure is even more evident in the real-space correlation function, where we expect undamped oscillations in time-like directions and exponential decay in space-like directions for the Feynman propagator. In fact we should obtain $\Delta_F(x-y) \sim e^{im|x-y|}$ along the time-axis and $\Delta_F(x-y) \sim e^{-m|x-y|}$ along the space-axis. The extracted mass, $m \sim 2.06 \pm 0.04$, differs from the input mass $m = 1.0$, likely due to finite-size effects from the fringe boundary conditions, which we will explore in future works.

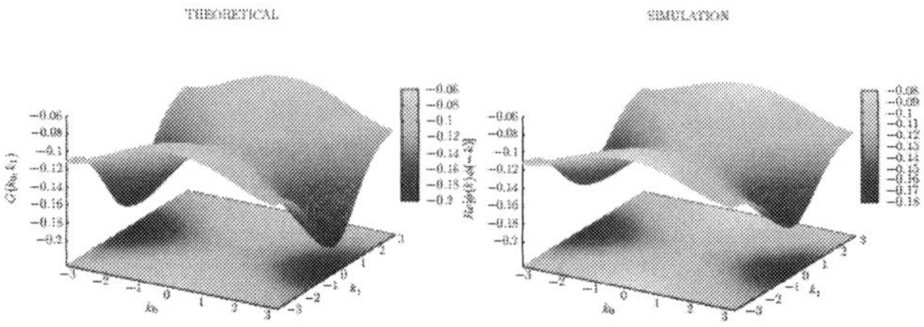

Figure 2. *Real part of the two-point correlation function in Fourier space, for a $\lambda\phi^4$ theory.* ***Left:*** *theoretical value of the free propagator with lattice spacing $a = 1.0$, lattice size $L = 128$, and mass $m = 3.0$.* ***Right:*** *numerical result from the interacting theory $\lambda = 0.001$. No unstable modes are present, meaning $\omega_k^2 > 0$ for all k.*

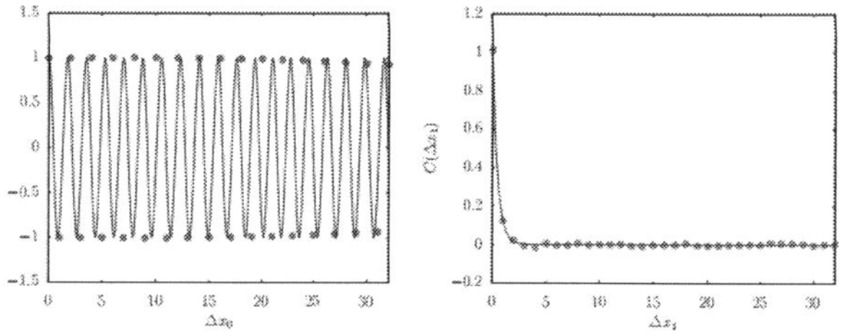

Figure 3. *Two-point correlation function in real space for a $\lambda\phi^4$ theory in 1 + 1 dimensions with fringe boundary conditions, lattice spacing $a = 1.0$, lattice size $L = 128$, mass $m = 1.0$, and nonlinearity $\lambda = 0.001$.* ***Right:*** *Exponential decay along the x_1 (spatial) axis.* ***Left:*** *Oscillations along the $x_0 = ct$ (temporal) axis.*

References

[1] M. Giachello, G. Gradenigo, Symplectic quantization and the Feynman propagator: a new real-time numerical approach to lattice field theory (2024), 2403.17149

[2] K.G. Wilson, Confinement of Quarks, Phys. Rev. D **10**, 2445 (1974). 10.1103/PhysRevD.10.2445

[3] M. Creutz, Quarks, Gluons and Lattices (Oxford University Press, 1983), ISBN 978-1-00-929039-5, 978-1-00-929038-8, 978-1-00-929037-1, 978-0-521-31535-7

[4] I. Montvay, G. Munster, Quantum fields on a lattice, Cambridge Monographs on Mathematical Physics (Cambridge University Press, 1997), ISBN 978-0-521-59917-7, 978-0-511-87919-7

[5] Z. Fodor, C. Hoelbling, Light hadron masses from lattice qcd, Reviews of Modern Physics **84**, 449–495 (2012). 10.1103/revmodphys.84.449

[6] C. Ratti, Lattice qcd and heavy ion collisions: a review of recent progress, Reports on Progress in Physics **81**, 084301 (2018). 10.1088/1361-6633/aabb97

[7] L. Maiani, A. Polosa, V. Riquer, The new pentaquarks in the diquark model, Physics Letters B **749**, 289–291 (2015). 10.1016/j.physletb.2015.08.008

[8] A. Esposito, A. Pilloni, A. Polosa, Multiquark resonances, Physics Reports **668**, 1–97 (2017). 10.1016/j.physrep.2016.11.002

[9] G. Gradenigo, R. Livi, Symplectic quantization i: Dynamics of quantum fluctuations in a relativistic field theory, Foundations of Physics **51** (2021). 10.1007/s10701-021-00470-9

[10] G. Gradenigo, Symplectic quantization ii: Dynamics of space–time quantum fluctuations and the cosmological constant, Foundations of Physics **51** (2021). 10.1007/s10701-021-00468-3

[11] G. Gradenigo, R. Livi, L. Salasnich, Symplectic quantization iii: Non-relativistic limit (2024), 2401.12355

[12] G. Parisi, Y.s. Wu, Perturbation Theory Without Gauge Fixing, Sci. Sin. **24**, 483 (1981).

[13] P.H. Damgaard, H. Huffel, Stochastic Quantization, Phys. Rept. **152**, 227 (1987). 10.1016/0370-1573(87)90144-X

[14] D.J.E. Callaway, A. Rahman, Lattice Gauge Theory in Microcanonical Ensemble, Phys. Rev. D **28**, 1506 (1983). 10.1103/PhysRevD.28.1506

[15] G. Gradenigo, S. Iubini, R. Livi, S.N. Majumdar, Localization transition in the discrete nonlinear schrödinger equation: ensembles inequivalence and negative temperatures, Journal of Statistical Mechanics: Theory and Experiment **2021**, 023201 (2021). 10.1088/1742-5468/abda26

[16] G. Gradenigo, S. Iubini, R. Livi, S.N. Majumdar, Condensation transition and ensemble inequivalence in the discrete nonlinear schrödinger equation, The European Physical Journal E **44** (2021). 10.1140/epje/s10189-021-00046-5

[17] J. Nordström, N. Nordin, D. Henningson, The fringe region technique and the fourier method used in the direct numerical simulation of spatially evolving viscous flows, SIAM Journal on Scientific Computing **20**, 1365 (1999). 10.1137/S1064827596310251

EPJ Web of Conferences **314**, 00029 (2024) https://doi.org/10.1051/epjconf/202431400029
QCD@Work 2024

Feynman integrals: Synergies between particle physics and gravitational waves

Manoj Kumar Mandal[1,*]

[1]Dipartimento di Fisica e Astronomia "G. Galilei", Università di Padova, and INFN, Sezione di Padova, Via Marzolo 8, I-35131, Padova, Italy

Abstract. Feynman integrals are essential for computing scattering amplitudes. Linear relations among these integrals, through Integral-By-Parts (IBP) identities, reduce them to a smaller set of independent integrals, known as master integrals (MIs). In twisted de-Rham cohomology, Feynman integrals form a vector space with an inner product, called the intersection number, which simplifies this reduction process. These methods have been applied in particle physics and recently extended to gravitational wave physics, notably in modeling binary black hole mergers. This proceedings highlights the synergy between these fields, showcasing how advanced techniques from Feynman integrals enable high-precision results in both areas.

1 Introduction

Feynman integrals play a fundamental role in the computation of scattering amplitudes, which encode all the physical information of a scattering process. As perturbation theory advances to higher orders, involving multiple loops, the complexity of these integrals increases significantly, necessitating efficient techniques to manage the vast number of integrals. In typical scattering amplitude evaluations, there can be on the order of $O(10^5)$ Feynman integrals, many of which are not independent. These integrals can be systematically reduced using Integration-By-Parts (IBP) identities [1], which reduce the large set of integrals to a smaller set of fundamental integrals, known as Master Integrals (MIs). The core idea behind IBP identities is that total derivatives with respect to loop momenta vanish in dimensional regularization, leading to a system of algebraic relations among integrals. This system can then be solved to express a large number of integrals in terms of a finite set of MIs.

Recently, it has been recognized that Feynman integrals exhibit an underlying vector space structure, particularly within parametric representations such as the Baikov or Feynman-Schwinger parametrizations. This structure is framed in terms of intersection theory within twisted de Rham cohomology [2–4]. In this framework, a bilinear pairing between differential forms, known as intersection numbers [5, 6], defines a scalar product on the vector space of Feynman integrals. This scalar product enables the projection of any multi-loop integral onto a basis of master integrals. In this context, the evaluation of intersection numbers is

*e-mail: manojkumar.mandal@pd.infn.it

This proceedings is based on the work done in collaboration with Giacomo Brunello, Seva Chestnov, Giulio Crisanti, Hjalte Frellesvig, Federico Gasparotto, Stefano Laporta, Pierpaolo Mastrolia, Luca Mattiazzi, Sebastian Mizera, Raj Patil, and Jan Steinhoff.

a key step, and various efficient methods have been proposed to compute these numbers [4, 7–13].

In recent years, methods used for Feynman integral computations have found surprising and impactful applications in the study of gravitational waves [14–24]. Gravitational wave phenomena, particularly from astrophysical sources such as binary black hole mergers, require precise theoretical models to compare with observational data. See [25, 26] for reviews. Remarkably, the same mathematical techniques developed for particle physics—such as IBP reduction, master integral evaluation, and dimensional regularization—have proven useful in calculating the conservative potential [27, 28], energy flux [29, 30], and waveforms [31, 32].

In this context, we have focused on the Effective Field Theory (EFT) techniques developed in [14, 27] to carry out post-Newtonian analysis in the conservative sector of the two-body problem, using an extension of the diagrammatic approach introduced in [17].

In this proceedings, we first discuss the decomposition of Feynman integrals using intersection theory, and then explore how the techniques for evaluation of Feynman integrals are applied to compute the 2-body Hamiltonian for the binary compact objects.

2 Decomposition of Feynman Integrals using Intersection theory

We consider an integral of the following form:

$$I = \int_C u(\mathbf{z})\, \varphi(\mathbf{z}) \,, \tag{1}$$

where $u(\mathbf{z})$ is a multi-valued function, and $\varphi(\mathbf{z}) = \hat{\varphi}(\mathbf{z})\, d^n\mathbf{z}$ represents a single-valued differential n-form. We impose the condition that $u(\mathbf{z})$ vanishes at the boundaries of the integration contour C, i.e., $u(\partial C) = 0$. This ensures the vanishing of surface terms. Assuming that $u(\mathbf{z})$ regulates all boundaries, we apply Stokes' theorem:

$$0 = \int_C d\,(u\,\xi) = \int_C (du \wedge \xi + u\, d\xi) = \int_C u \left(\frac{du}{u} \wedge + d \right) \xi \equiv \int_C u\, \nabla_\omega \xi, \tag{2}$$

where ξ is a differential $(n-1)$ form and $\nabla_\omega \equiv d + \omega\wedge$, with $\omega \equiv d \log u = \sum_{i=1}^{n} \hat{\omega}_i dz_i$. Following this, we obtain:

$$\int_C u\, \varphi = \int_C u\, (\varphi + \nabla_\omega \xi) \,, \tag{3}$$

which allows us to define an equivalence class for φ as: ${}_\omega\langle\varphi| \,:\; \varphi \sim \varphi + \nabla_\omega \xi$. In other words, this leads to consider the following

$$H^n_\omega = \{n\text{-forms } \varphi \,|\, \nabla_\omega \varphi = 0\} \,/\, \{\nabla_\omega \xi\}. \tag{4}$$

This is known as the *twisted cohomology group* and its elements are called *twisted cocycles*. The integration domain C, along with the integrand defined on its branch, forms what is called the *twisted cycle*. One can show that there exists an equivalence class of twisted cycles, which form the *twisted homology group*. Within this framework, the integral I can then be interpreted as the pairing between a twisted cycle and a twisted co-cycle:

$$I = \int_C u(\mathbf{z})\, \varphi(\mathbf{z}) \equiv \langle \varphi | C]. \tag{5}$$

Then, Eq. 4 can be considered as the *vector space* of Feynman integrals. Additionally, we introduce a dual space of twisted co-cycles, the so called *dual twisted cohomology group*: $(H^n_\omega)^* = H^n_{-\omega}$ where ∇_ω is replaced by $\nabla_{-\omega}$.

We denote the *dimension* of the (dual) twisted cohomology group by ν, which is the *number* of MIs; thanks to complex Morse (Picard-Lefschetz) theory, ν is determined as the *number of critical points* of the function $\log u(\mathbf{z})$ [4, 33]

$$\nu = \dim\left(H^n_{\pm\omega}\right) = \text{number of solutions of:} \quad \hat{\omega}_i = \partial_{z_i} \log u(\mathbf{z}) = 0 \qquad 1 \le i \le n. \tag{6}$$

We consider the basis of forms, denoted as $|e_i\rangle$ for $i = 1, 2, \ldots, \nu$ and basis of the dual forms $|h_i\rangle$ to decompose an arbitrary twisted co-cycle $\langle\varphi|$. Using these twisted co-cycles and their duals, we introduce a bilinear non-degenerate pairing $\langle e_i|h_j\rangle$, known as the "intersection number." This leads to the construction of the metric matrix $\mathbf{C} = \langle e_i|h_j\rangle$, whose entries are the intersection numbers between basis forms from the twisted cohomology group and it's dual. Following [2], we obtain the master decomposition formula:

$$\langle\varphi| = \sum_{i,j=1}^{\nu} \langle\varphi|h_j\rangle \left(\mathbf{C}^{-1}\right)_{ji} \langle e_i|, \tag{7}$$

which projects $\langle\varphi|$ onto a basis of twisted co-cycles $\langle e_i|$. The integral I can then be expressed as:

$$I = \int_C u\,\varphi = \sum_{i,j=1}^{\nu} \langle\varphi|h_j\rangle \left(\mathbf{C}^{-1}\right)_{ji} \int_C u\, e_i = \sum_{i,j=1}^{\nu} \langle\varphi|h_j\rangle \left(\mathbf{C}^{-1}\right)_{ji} J_i = \sum_{i}^{\nu} c_i J_i, \tag{8}$$

where J_i are the independent Feynman integrals, known as master integrals (MIs) and c_i are the rational coefficient of these MIs.

Following [2–4, 6–8, 34], the intersection numbers can be defined as

$$\langle e_i|h_j\rangle = \frac{1}{(2\pi i)^n} \int_X e_{i,c} \wedge h_j, \tag{9}$$

where $e_{i,c}$ is the twisted form with compact support. This integral localizes on the poles of the connection ω and can be expressed as sum of residues, employing Cauchy's formula. These methods have led to improvements in the reduction of Feynman integrals [11–13, 35, 36]. Recently, they have been applied to complete the reduction of two-loop five-point integrals [37], using polynomial reduction and companion tensor algebra.

3 Feynman Integrals and Post-Newtonian approximation

In the context of a bound system of two compact objects, we encounter three distinct length scales: the Schwarzschild radius R_s, associated with each compact object; the orbital radius r, describing the distance between the two objects; and the wavelength λ of the emitted gravitational waves. Assuming that the velocities of the objects are small compared to the speed of light and the objects are far apart, the system can be treated as propagating on a flat background, i.e., $g_{\mu\nu} = \eta_{\mu\nu} + h_{\mu\nu}$, where the gravitational interaction between the two objects is mediated by the graviton field $h_{\mu\nu}$. This scenario leads to the following hierarchy of length scales: $\lambda \gg r \gg R_s$. Our primary interest is in the long-distance physics at the wavelength scale λ. To simplify the analysis, we decompose the graviton field into two distinct components: short-distance potential gravitons $H_{\mu\nu}$, which scale as $(k_0, \mathbf{k}) \sim (v/r, 1/r)$, and long-distance radiation gravitons $\bar{h}_{\mu\nu}$, scaling as $(k_0, \mathbf{k}) \sim (v/r, v/r)$[14]. Using the virial theorem (or equivalently, Kepler's third law), we have $v^2 \sim 1/r$ for bound orbits. Thus, the dimensionless expansion parameter can be expressed as $v^2/c^2 \sim G_N M/rc^2$, which formally scales as $1/c^2$. Consequently, following standard post-Newtonian (PN) literature, we adopt a formal expansion in powers of $1/c$, with one PN order corresponding to $1/c^2$.

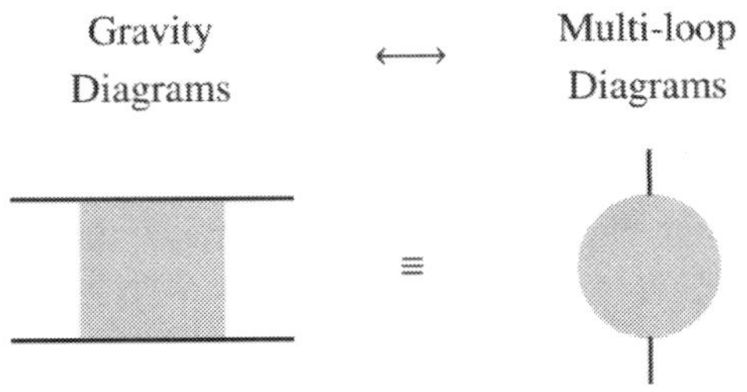

Figure 1. The diagrammatic correspondence between the four-point EFT-Gravity graphs and the two-point quantum-field-theory (QFT) graphs.

To compute the conservative binding potential of the two-body system, we neglect the radiation modes and focus on the potential modes using the Kaluza-Klein (KK) decomposition [38, 39]. In this decomposition, the components of the metric $g_{\mu\nu} = \eta_{\mu\nu} + H_{\mu\nu}$ are parameterized in terms of three fields: a scalar field $\boldsymbol{\phi}$, a 3-dimensional vector field $\boldsymbol{A}_i$, and a 3-dimensional symmetric rank-2 tensor field $\boldsymbol{\sigma}_{ij}$. The effective action for the two-body system is obtained by integrating out the gravitational degrees of freedom. This is expressed as:

$$\exp\left[\mathbf{i}\int dt\,\mathcal{L}_{\text{eff}}\right] = \int D\boldsymbol{\phi} D\boldsymbol{A}_i D\boldsymbol{\sigma}_{ij}\, e^{\mathrm{i}(S_{\text{EH}}+S_{\text{pp}})}\,, \tag{10}$$

where, S_{EH} is the Einstein-Hilbert action describing the gravitational field and S_{pp} is the worldline point particle action [40]. $\mathcal{L}_{\text{eff}}$ is the effective Lagrangian, further decomposed into kinetic and potential terms as: $\mathcal{L}_{\text{eff}} = \mathcal{K}_{\text{eff}} - \mathcal{V}_{\text{eff}}$, with $\mathcal{K}_{\text{eff}}$ being the kinetic term and $\mathcal{V}_{\text{eff}}$ representing the effective potential due to gravitational interactions. The effective potential $\mathcal{V}_{\text{eff}}$ can be computed from connected, classical, 1-particle irreducible (1PI) scattering amplitudes as follows:

$$\mathcal{V}_{\text{eff}} = \mathbf{i}\lim_{d\to 3}\int_{\mathbf{p}} e^{\mathbf{i}\mathbf{p}\cdot(\mathbf{x}_{(1)}-\mathbf{x}_{(2)})} \quad [\text{box diagram}]\,, \tag{11}$$

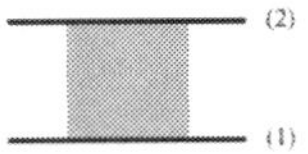

where $\mathbf{p}$ represents the momentum transfer between the two particles, and the box diagram refers to all possible Feynman diagrams with gravitons $\boldsymbol{\phi}, \boldsymbol{A}_i, \boldsymbol{\sigma}_{ij}$ mediating the gravitational interaction between the two point particles, depicted by the two solid black lines. To obtain the effective potential from the diagrammatic approach as shown in equation (11), we begin by generating all the relevant generic topologies contributing at different orders of G_N. Within the EFT framework, the sources remain static and as a result the generated Feynman diagrams are mapped to two-point multi-loop Feynman diagrams with mass-less internal lines and an external momentum (the momentum transferred between two sources) as shown in figure 1. We translate these Feynman diagrams to their corresponding Feynman amplitudes after performing the tensor algebra using by means of an in-house code that uses tools from `EFTofPNG` [41] and `xTensor` [42].

We perform the reduction of the multi-loop tensor integrals to scalar integrals by applying a set of projectors exploiting the Lorentz invariance. The generated scalar integrals are reduced by using the `LiteRED` [43] to the MIs. After the evaluation of the MIs, we perform the Fourier transform to obtain the effective potential by expanding the complete scattering amplitude as Laurent series in ϵ around $d = 3$.

The computation of the effective potential starting from the generation of the required Feynman diagrams, expressing them in multi-loop integrands, performing IBP reduction, and then applying the Fourier transformations have been automated through an in-house code, elaborating on some of the ideas implemented in `EFTofPNG` [41], and using `xTensor` [42] for tensor algebra manipulations as well as successful interface to `LiteRed` [43] for the IBP reduction. Employing this framework, we computed the complete evaluation of N^3LO Post-Newtonian (PN) correction to the the spin-orbit Hamiltonian [44], quadratic-in-spin Hamiltonian [45] for rapidly rotating compact objects, within the effective field theory diagrammatic approach of General Relativity. Additionally, we derived an effective Hamiltonian that describes the dynamical gravitoelectric tidal interaction between two nonspinning compact objects up to the 3PN order [46, 47].

4 Conclusion

Thus, the study of Feynman integrals forms a crucial bridge between particle physics and gravitational wave research. By leveraging the synergies between these fields, we not only improve our understanding of the fundamental forces of nature but also advance our ability to detect and interpret the ripples in spacetime caused by massive astrophysical events. This proceedings discusses the recently found connections, focusing on the computational techniques that make these breakthroughs possible and highlighting the ongoing interplay between Feynman integral and gravitational wave physics.

References

[1] K.G. Chetyrkin, F.V. Tkachov, Integration by Parts: The Algorithm to Calculate beta Functions in 4 Loops, Nucl. Phys. **B192**, 159 (1981). 10.1016/0550-3213(81)90199-1

[2] P. Mastrolia, S. Mizera, Feynman Integrals and Intersection Theory, JHEP **02**, 139 (2019), `1810.03818`. 10.1007/JHEP02(2019)139

[3] H. Frellesvig, F. Gasparotto, S. Laporta, M.K. Mandal, P. Mastrolia, L. Mattiazzi, S. Mizera, Decomposition of Feynman Integrals on the Maximal Cut by Intersection Numbers, JHEP **05**, 153 (2019), `1901.11510`. 10.1007/JHEP05(2019)153

[4] H. Frellesvig, F. Gasparotto, M.K. Mandal, P. Mastrolia, L. Mattiazzi, S. Mizera, Vector Space of Feynman Integrals and Multivariate Intersection Numbers, Phys. Rev. Lett. **123**, 201602 (2019), `1907.02000`. 10.1103/PhysRevLett.123.201602

[5] K. Cho, K. Matsumoto, Intersection theory for twisted cohomologies and twisted Riemann's period relations I, Nagoya Math. J. **139**, 67 (1995). 10.1017/S0027763000005304

[6] S. Mizera, Scattering Amplitudes from Intersection Theory, Phys. Rev. Lett. **120**, 141602 (2018), `1711.00469`. 10.1103/PhysRevLett.120.141602

[7] S. Mizera, Aspects of Scattering Amplitudes and Moduli Space Localization (2019), `1906.02099`

[8] H. Frellesvig, F. Gasparotto, S. Laporta, M.K. Mandal, P. Mastrolia, L. Mattiazzi, S. Mizera, Decomposition of Feynman Integrals by Multivariate Intersection Numbers, JHEP **03**, 027 (2021), `2008.04823`. 10.1007/JHEP03(2021)027

[9] V. Chestnov, F. Gasparotto, M.K. Mandal, P. Mastrolia, S.J. Matsubara-Heo, H.J. Munch, N. Takayama, Macaulay matrix for Feynman integrals: linear relations and intersection numbers, JHEP **09**, 187 (2022), `2204.12983`. 10.1007/JHEP09(2022)187

[10] V. Chestnov, H. Frellesvig, F. Gasparotto, M.K. Mandal, P. Mastrolia, Intersection numbers from higher-order partial differential equations, JHEP **06**, 131 (2023), `2209.01997`. 10.1007/JHEP06(2023)131

[11] S. Caron-Huot, A. Pokraka, Duals of Feynman Integrals. Part II. Generalized unitarity, JHEP **04**, 078 (2022), 2112.00055. 10.1007/JHEP04(2022)078

[12] S. Caron-Huot, A. Pokraka, Duals of Feynman integrals. Part I. Differential equations, JHEP **12**, 045 (2021), 2104.06898. 10.1007/JHEP12(2021)045

[13] S. Weinzierl, On the computation of intersection numbers for twisted cocycles, J. Math. Phys. **62**, 072301 (2021), 2002.01930. 10.1063/5.0054292

[14] W.D. Goldberger, I.Z. Rothstein, An Effective field theory of gravity for extended objects, Phys. Rev. D **73**, 104029 (2006), hep-th/0409156. 10.1103/PhysRevD.73.104029

[15] J.B. Gilmore, A. Ross, Effective field theory calculation of second post-Newtonian binary dynamics, Phys. Rev. D **78**, 124021 (2008), 0810.1328. 10.1103/PhysRevD.78.124021

[16] S. Foffa, R. Sturani, Conservative dynamics of binary systems to fourth Post-Newtonian order in the EFT approach I: Regularized Lagrangian, Phys. Rev. D **100**, 024047 (2019), 1903.05113. 10.1103/PhysRevD.100.024047

[17] S. Foffa, P. Mastrolia, R. Sturani, C. Sturm, Effective field theory approach to the gravitational two-body dynamics, at fourth post-Newtonian order and quintic in the Newton constant, Phys. Rev. D **95**, 104009 (2017), 1612.00482. 10.1103/PhysRevD.95.104009

[18] J. Blümlein, A. Maier, P. Marquard, G. Schäfer, Fourth post-Newtonian Hamiltonian dynamics of two-body systems from an effective field theory approach, Nucl. Phys. B **955**, 115041 (2020), 2003.01692. 10.1016/j.nuclphysb.2020.115041

[19] S. Foffa, P. Mastrolia, R. Sturani, C. Sturm, W.J. Torres Bobadilla, Static two-body potential at fifth post-Newtonian order, Phys. Rev. Lett. **122**, 241605 (2019), 1902.10571. 10.1103/PhysRevLett.122.241605

[20] J. Blümlein, A. Maier, P. Marquard, G. Schäfer, The fifth-order post-Newtonian Hamiltonian dynamics of two-body systems from an effective field theory approach (2021), 2110.13822

[21] N.E.J. Bjerrum-Bohr, P.H. Damgaard, G. Festuccia, L. Planté, P. Vanhove, General Relativity from Scattering Amplitudes, Phys. Rev. Lett. **121**, 171601 (2018), 1806.04920. 10.1103/PhysRevLett.121.171601

[22] C. Cheung, I.Z. Rothstein, M.P. Solon, From Scattering Amplitudes to Classical Potentials in the Post-Minkowskian Expansion, Phys. Rev. Lett. **121**, 251101 (2018), 1808.02489. 10.1103/PhysRevLett.121.251101

[23] Z. Bern, C. Cheung, R. Roiban, C.H. Shen, M.P. Solon, M. Zeng, Black Hole Binary Dynamics from the Double Copy and Effective Theory, JHEP **10**, 206 (2019), 1908.01493. 10.1007/JHEP10(2019)206

[24] G. Kälin, Z. Liu, R.A. Porto, Conservative Dynamics of Binary Systems to Third Post-Minkowskian Order from the Effective Field Theory Approach, Phys. Rev. Lett. **125**, 261103 (2020), 2007.04977. 10.1103/PhysRevLett.125.261103

[25] R.A. Porto, The effective field theorist's approach to gravitational dynamics, Phys. Rept. **633**, 1 (2016), 1601.04914. 10.1016/j.physrep.2016.04.003

[26] M. Levi, Effective Field Theories of Post-Newtonian Gravity: A comprehensive review, Rept. Prog. Phys. **83**, 075901 (2020), 1807.01699. 10.1088/1361-6633/ab12bc

[27] R.A. Porto, Post-Newtonian corrections to the motion of spinning bodies in NRGR, Phys. Rev. D **73**, 104031 (2006), gr-qc/0511061. 10.1103/PhysRevD.73.104031

[28] M. Levi, A.J. Mcleod, M. Von Hippel, N^3LO gravitational spin-orbit coupling at order G^4, JHEP **07**, 115 (2021), 2003.02827. 10.1007/JHEP07(2021)115

[29] G. Cho, B. Pardo, R.A. Porto, Gravitational radiation from inspiralling compact objects: Spin-spin effects completed at the next-to-leading post-Newtonian order, Phys. Rev. D **104**, 024037 (2021), 2103.14612. 10.1103/PhysRevD.104.024037

[30] G. Cho, R.A. Porto, Z. Yang, Gravitational radiation from inspiralling compact objects: Spin effects to the fourth post-Newtonian order, Phys. Rev. D **106**, L101501 (2022), 2201.05138. 10.1103/PhysRevD.106.L101501

[31] D. Bini, T. Damour, S. De Angelis, A. Geralico, A. Herderschee, R. Roiban, F. Teng, Gravitational waveforms: A tale of two formalisms, Phys. Rev. D **109**, 125008 (2024), 2402.06604. 10.1103/PhysRevD.109.125008

[32] G. Brunello, S. De Angelis, An improved framework for computing waveforms, JHEP **07**, 062 (2024), 2403.08009. 10.1007/JHEP07(2024)062

[33] R.N. Lee, A.A. Pomeransky, Critical points and number of master integrals, JHEP **11**, 165 (2013), 1308.6676. 10.1007/JHEP11(2013)165

[34] K. Matsumoto, Intersection numbers for 1-forms associated with confluent hypergeometric functions, Funkcial. Ekvac. **41**, 291 (1998)

[35] G. Fontana, T. Peraro, Reduction to master integrals via intersection numbers and polynomial expansions, JHEP **08**, 175 (2023), 2304.14336. 10.1007/JHEP08(2023)175

[36] G. Brunello, V. Chestnov, G. Crisanti, H. Frellesvig, M.K. Mandal, P. Mastrolia, Intersection numbers, polynomial division and relative cohomology, JHEP **09**, 015 (2024), 2401.01897. 10.1007/JHEP09(2024)015

[37] G. Brunello, V. Chestnov, P. Mastrolia, Intersection Numbers from Companion Tensor Algebra (2024), 2408.16668

[38] B. Kol, M. Smolkin, Non-Relativistic Gravitation: From Newton to Einstein and Back, Class. Quant. Grav. **25**, 145011 (2008), 0712.4116. 10.1088/0264-9381/25/14/145011

[39] B. Kol, M. Smolkin, Classical Effective Field Theory and Caged Black Holes, Phys. Rev. D **77**, 064033 (2008), 0712.2822. 10.1103/PhysRevD.77.064033

[40] M. Levi, J. Steinhoff, Spinning gravitating objects in the effective field theory in the post-Newtonian scheme, JHEP **09**, 219 (2015), 1501.04956. 10.1007/JHEP09(2015)219

[41] M. Levi, J. Steinhoff, EFTofPNG: A package for high precision computation with the Effective Field Theory of Post-Newtonian Gravity, Class. Quant. Grav. **34**, 244001 (2017), 1705.06309. 10.1088/1361-6382/aa941e

[42] J.M.M. García, xact: Efficient tensor computer algebra for mathematica, http://www.xact.es

[43] R.N. Lee, LiteRed 1.4: a powerful tool for reduction of multiloop integrals, J. Phys. Conf. Ser. **523**, 012059 (2014), 1310.1145. 10.1088/1742-6596/523/1/012059

[44] M.K. Mandal, P. Mastrolia, R. Patil, J. Steinhoff, Gravitational spin-orbit Hamiltonian at NNNLO in the post-Newtonian framework, JHEP **03**, 130 (2023), 2209.00611. 10.1007/JHEP03(2023)130

[45] M.K. Mandal, P. Mastrolia, R. Patil, J. Steinhoff, Gravitational quadratic-in-spin Hamiltonian at NNNLO in the post-Newtonian framework, JHEP **07**, 128 (2023), 2210.09176. 10.1007/JHEP07(2023)128

[46] M.K. Mandal, P. Mastrolia, H.O. Silva, R. Patil, J. Steinhoff, Gravitoelectric dynamical tides at second post-Newtonian order, JHEP **11**, 067 (2023), 2304.02030. 10.1007/JHEP11(2023)067

[47] M.K. Mandal, P. Mastrolia, H.O. Silva, R. Patil, J. Steinhoff, Renormalizing Love: tidal effects at the third post-Newtonian order, JHEP **02**, 188 (2024), 2308.01865. 10.1007/JHEP02(2024)188

The gravitational form factor of the pion and proton and the conformal anomaly*

Claudio Corianò[1,2,**], *Stefano* Lionetti[1,2,***], *Dario* Melle[1,2,****], and *Riccardo* Tommasi[1,2,†]

[1]Dipartimento di Matematica e Fisica, Università del Salento and INFN Sezione di Lecce, Via Arnesano 73100 Lecce, Italy

[2]National Center for HPC, Big Data and Quantum Computing, Via Magnanelli 2, 40033 Casalecchio di Reno, Italy

Abstract. We analyze the hard scattering amplitude of gravitational form factors (GFFs) of hadrons within QCD factorization at large momentum transfers, focusing on their conformal field theory (CFT) description. These form factors are key to studying quark and gluon angular momentum in hadrons, connected to Mellin moments of Deeply Virtual Compton Scattering (DVCS). The analysis uses diffeomorphism invariance and conformal symmetry in momentum space. A non-Abelian TJJ 3-point function at $O(\alpha_s^2)$ reveals a dilaton interaction in the t-channel. We present a parameterization relevant for future DVCS/GFF experiments at the Electron-Ion Collider (EIC).

1 Introduction

The experimental study of the gravitational form factors (GFFs) of the proton and pion provides valuable nonperturbative insights into their interaction with the QCD energy-momentum tensor (EMT), offering a window into the internal distributions of energy, spin, pressure, and shear forces. GFFs are formulated through the matrix elements of the EMT between hadronic states, allowing for a detailed characterization of the hadron's internal structure. Quantum corrections induce a breaking of conformal symmetry due to the trace anomaly of the stress-energy tensor

$$\hat{T}^{\mu}_{\mu} \equiv \beta(g)\, F^{a,\mu\nu} F^{a,}_{\mu\nu} + (1+\gamma_m) \sum_q m_q \bar{\psi}_q \psi_q \,, \tag{1}$$

where $\beta(g)$ is the β-function of QCD and γ_m is the anomalous dimension of the mass operator. The matrix elements of the EMT for a spin 1/2 hadron with momentum P can be expressed in terms of the GFFs as

*Presented at QCD@Work by Dario Melle and Riccardo Tommasi, 18 - 21 June 2024, Trani, Italy

**e-mail: claudio.coriano@le.infn.it
***e-mail: stefano.lionetti@unisalento.it
****e-mail: dario.melle@le.infn.it
†e-mail: riccardo.tommasi@unisalento.it

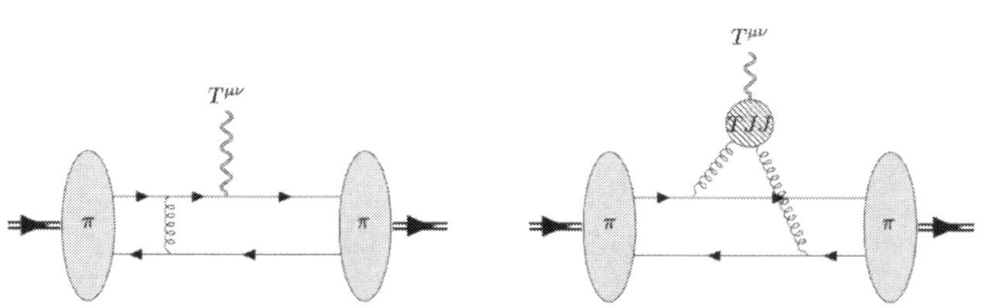

Figure 1. Factorization picture of the GFF of the pion. Leading order (left). The contribution of the conformal anomaly revealed by the TJJ vertex (right).

$$\langle p', s'|T_{\mu\nu}(0)|p, s\rangle = \bar{u}' \left[A(t)\, \frac{\gamma_{\{\mu}P_{\nu\}}}{2} + B(t)\, \frac{i\, P_{\{\mu}\sigma_{\nu\}\rho}\Delta^{\rho}}{4M} + D(t)\, \frac{\Delta_{\mu}\Delta_{\nu} - g_{\mu\nu}\Delta^2}{4M} + M \sum_{\hat{a}} \bar{c}^{\hat{a}}(t)\, g_{\mu\nu} \right] u \tag{2}$$

where $u(p)$ and $\bar{u}(p')$ are proton spinors, P is the average momentum, and Δ is the momentum transfer. At large momentum transfers, the amplitude decreases rapidly in accordance with quark and gluon counting rules. The general idea behind quark-gluon counting rules is to count the minimum number of constituents (quarks and gluons) participating in the hard scattering process, and then determine how the amplitude of the process scales with the momentum transfer, at large values of momentum transfer $t = q^2$, where q is the momentum of the graviton. Nevertheless, it remains accessible within the factorization framework using pion and proton distribution amplitudes, as this approach can be adapted from studies of electromagnetic form factors [1]. One of the major advantages of the EIC is its exceptionally high luminosity, which is expected to reach up to $10^{34}cm^2s^{-1}$. This high luminosity compensates for the rapid decrease in cross-sections typical of exclusive processes, allowing for a significant number of exclusive events to be detected even at large momentum transfers, where the cross-sections are very small. The large number of collisions ensures that even rare events are observable.

We investigate the hard scattering behavior of gravitational form factors (GFFs) at large momentum transfers and higher perturbative orders, where the factorization framework for exclusive processes applies. As shown in Fig. 1, this method uncovers a conformal anomaly at the TJJ (graviton/gluon-gluon) vertex at one loop, marked by the emergence of a dilaton interaction mediated by the anomaly. Our goal is to indicate an optimal parameterization of the hard scattering amplitude identifying the key form factors from which the dilaton interaction, though subleading in the pertubative expansion, can be singled out at future experiments. Additional details and references of our analysis can be found in [2].

At first sight, a key challenge, though, is the absence of a graviton to directly probe such matrix elements. Gravitational form factors (GFFs) are, however, closely linked to deeply virtual Compton scattering (DVCS) processes, as described in [3]. In DVCS (see Fig. 2), an electron (e) scatters off a nucleon (N) in the reaction $eN \to e'N'\gamma$, producing a photon in the final state, controlled by nonforward parton distributions [3] [4]. This process draws significant interest for both pions and protons. In DVCS, a high-energy electron interacts with a hadron, such as a proton or pion, via the exchange of a virtual photon, which then radiates a real photon. By measuring the DVCS invariant amplitudes it is possible to extract informations on the form factors of (2). We examine this perturbative correction extending the formalism of conformal field theory in momentum space (CFT_p). We propose a decomposition of the hard scattering amplitude that enables the extraction of the dilaton contribution to these matrix elements. The dilaton can be interpreted as the Nambu Goldstone mode of broken conformal invariance, whose effective interaction can be summarised at phenomenological level in the local form ($\Lambda^{-1}(\varphi F F)$), with Λ a typical hadronic scale. Our analysis is built around the exact expression of this interaction in perturbative QCD, which is nonlocal, and can be investigated using the conformal anomaly effective action. We apply methods developed for the analysis of such effective action in gravity [5], specifically adapted to QCD.

2 The CFT-induced parameterization of the pion and proton form factors

Dilaton-like t-channel exchanges in the TJJ correlator were initially studied perturbatively in QED [6, 7] and QCD [8] to explore the coupling of gravity to Standard Model fields [9]. In this work, we focus on the off-shell expansion of the non-Abelian TJJ correlator, previously examined in momentum space through perturbative QCD for on-shell gluons, as detailed in [8]. For a more recent discussion of the trace anomaly in QCD, we refer to [10]. In the CFT framework, J denotes a non-Abelian vector current in four dimensions ($d = 4$), while T represents the gauge-fixed stress-energy tensor in QCD. To integrate this interaction into the generalized form factors (GFFs) of hadrons, we extend the on-shell analysis from [8], employing the CFT_p formalism. We consider the decomposition of the operators T and J in terms of their transverse traceless part and longitudinal (local) one, separating the quark and gluon parts, that, as we are going to see, behave differently under the application of the conformal constraints. We define, for generic momenta p_i, ($i = 1, 2, 3$)

$$T^{\mu_i\nu_i}(p_i) \equiv t^{\mu_i\nu_i}(p_i) + t_{loc}^{\mu_i\nu_i}(p_i), \tag{3}$$

$$J^{a_i\mu_i}(p_i) \equiv j^{a_i\mu_i}(p_i) + j_{loc}^{a_i\mu_i}(p_i), \tag{4}$$

where

$$t^{\mu_i\nu_i}(p_i) = \Pi^{\mu_i\nu_i}_{\alpha_i\beta_i}(p_i)\, T^{\alpha_i\beta_i}(p_i), \qquad t_{loc}^{\mu_i\nu_i}(p_i) = \Sigma^{\mu_i\nu_i}_{\alpha_i\beta_i}(p)\, T^{\alpha_i\beta_i}(p_i), \tag{5}$$

$$j^{a_i\mu_i}(p_i) = \pi^{\mu_i}_{\alpha_i}(p_i)\, J^{a_i\alpha_i}(p_i), \qquad j_{loc}^{a_i\mu_i}(p_i) = \frac{p_i^{\mu_i}\, p_{i\,\alpha_i}}{p_i^2}\, J^{a_i\alpha_i}(p_i). \tag{6}$$

having introduced the transverse-traceless (Π), transverse (π), longitudinal (Σ) projectors. Their explicit expressions can be found in [2].
We adopt the following convention for the momenta: the momenta of the two gluons are denoted by p_1 and p_2, while the momentum of the graviton coupling to the energy-momentum tensor T is denoted by q. Additionally, we define the tensor $\hat{\pi}^{\mu\nu} \equiv q^2 g^{\mu\nu} - q^\mu q^\nu$, and introduce the notation $\pi^{\mu\nu} \equiv (1/q^2)\,\hat{\pi}^{\mu\nu}$. We can enumerate all possible tensor that can appear in the transverse traceless part preserving the symmetry of the correlator

$$\begin{aligned}\langle t^{\mu\nu}(p_1) j^{a\alpha}(p_2) j^{b\beta}(p_3)\rangle_q = {}& \Pi_1{}^{\mu\nu}_{\mu_1\nu_1}\pi_2{}^{\alpha}_{\alpha_1}\pi_3{}^{\beta}_{\beta_1}\Big(A_1^{(q)ab}\; p_2^{\mu_1} p_2^{\nu_1} p_3^{\alpha_1} p_1^{\beta_1} + A_2^{(q)ab}\; \delta^{\alpha_1\beta_1} p_2^{\mu_1} p_2^{\nu_1} + \\ & A_3^{(q)ab}\; \delta^{\mu_1\alpha_1} p_2^{\nu_1} p_1^{\beta_1} + A_3^{(q)ab}(p_2 \leftrightarrow p_3)\delta^{\mu_1\beta_1} p_2^{\nu_1} p_3^{\alpha_1} + A_4^{(q)ab}\; \delta^{\mu_1\beta_1}\delta^{\alpha_1\nu_1}\Big)\end{aligned} \tag{7}$$

and decompose the TJJ contribution in the form

$$\begin{aligned}\langle T^{\mu\nu}(q) J^{a\alpha}(p_1) J^{b\beta}(p_2)\rangle = {}& \langle t^{\mu\nu}(q) j^{a\alpha}(p_1) j^{b\beta}(p_2)\rangle + \langle t^{\mu\nu}(q) j_{loc}^{a\alpha}(p_1) j^{b\beta}(p_2)\rangle_g + \langle t^{\mu\nu}(q) j^{a\alpha}(p_1) j_{loc}^{b\beta}(p_2)\rangle_g \\ & + 2\mathcal{I}^{\mu\nu\rho}(q)\left[\delta^{\beta}_{[\rho} p_{2\sigma]}\langle J^{a\alpha}(p_1) J^{b\sigma}(-p_1)\rangle + \delta^{\alpha}_{[\rho} p_{1\sigma]}\langle J^{b\beta}(p_2) J^{a\sigma}(-p_2)\rangle\right] + \frac{1}{3\,q^2}\hat{\pi}^{\mu\nu}(q)\left[\mathcal{A}^{\alpha\beta ab} + \mathcal{B}_g^{\alpha\beta ab}\right],\end{aligned} \tag{8}$$

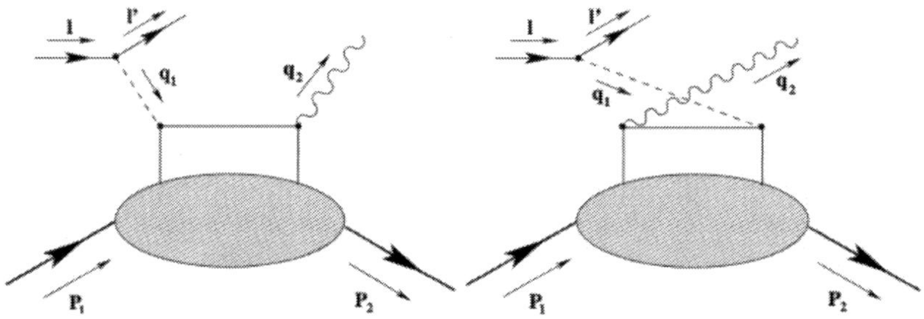

Figure 2. Factorization in the DVCS process.

where the traceless sector is

$$\langle T^{\mu\nu}(q)J^{a\alpha}(p_1)J^{b\beta}(p_2)\rangle_{tls} = \langle t^{\mu\nu}(q)j^{a\alpha}(p_1)j^{b\beta}(p_2)\rangle + \langle t^{\mu\nu}(q)j^{a\alpha}_{loc}(p_1)j^{b\beta}(p_2)\rangle_g + \langle t^{\mu\nu}(q)j^{a\alpha}(p_1)j^{b\beta}_{loc}(p_2)\rangle_g + 2\mathcal{I}^{\mu\nu\rho}(q)\left[\delta^{\beta}_{[\rho}p_{2\sigma]}\langle J^{a\alpha}(p_1)J^{b\sigma}(-p_1)\rangle + \delta^{\alpha}_{[\rho}p_{1\sigma]}\langle J^{b\beta}(p_2)J^{a\sigma}(-p_2)\rangle\right], \tag{9}$$

and the trace part is

$$\langle T^{\mu\nu}(q)J^{a\alpha}(p_1)J^{b\beta}(p_2)\rangle_{tr} = \frac{1}{3\,q^2}\hat{\pi}^{\mu\nu}(q)\left[\mathcal{A}^{\alpha\beta ab} + \mathcal{B}^{\alpha\beta ab}_g\right]. \tag{10}$$

The trace part contains the anomaly contribution $\mathcal{A}^{\alpha\beta ab} = \mathcal{A}^{\alpha\beta}\delta^{ab}$ (a and b are color indices) and a second term proportional to the gluon equations of motion $\mathcal{B}^{\alpha\beta ab}_g$. This is not to be considered part of the trace anomaly although it is part of the trace of the correlator since the trace operation with a Minkowski metric $\eta_{\mu\nu}$ gives

$$\eta_{\mu\nu}\langle T^{\mu\nu}(q)J^{a\alpha}(p_1)J^{b\beta}(p_2)\rangle_{tr} = \left[\mathcal{A}^{\alpha\beta ab} + \mathcal{B}^{\alpha\beta ab}_g\right]. \tag{11}$$

The first tensor term $\mathcal{A}$ defines the anomaly contribution, as given by the first term on the right-hand-side of (1)

$$\mathcal{A}^{\alpha\beta ab}(p_1, p_2) = A_n\delta^{ab}u^{\alpha\beta}(p_1, p_2), \tag{12}$$

and explicitly as

$$\mathcal{A}^{\alpha\beta ab} = \frac{1}{3}\;\frac{g_s^2}{16\pi^2}\,(11C_A - 2n_f)\delta^{ab}u^{\alpha\beta}(p_1, p_2), \tag{13}$$

which is proportional to the QCD β function for n_f massless flavours. We have defined the tensor structure

$$u^{\alpha\beta}(p_1, p_2) \equiv (p_1\cdot p_2)g^{\alpha\beta} - p_2^{\alpha}p_1^{\beta}, \tag{14}$$

that summarizes the conformal anomaly contribution, being the Fourier transform of the anomaly term at $O(g^2)$

$$u^{\alpha\beta}(p_1, p_2) = -\frac{1}{4}\int d^4x\int d^4y\; e^{ip_1\cdot x + ip_2\cdot y}\;\frac{\delta^2\{F^a_{\mu\nu}F^{a\mu\nu}(0)\}}{\delta A_\alpha(x)A_\beta(y)}\bigg|_{A=0}. \tag{15}$$

The anomaly form factor characterising the exchange of a dilaton pole in the TJJ is then distilled from (10) in the form

$$\frac{\beta}{q^2}\delta^{ab} \subset \langle T^{\mu\nu}(q)J^{a\alpha}(p_1)J^{b\beta}(p_2)\rangle, \tag{16}$$

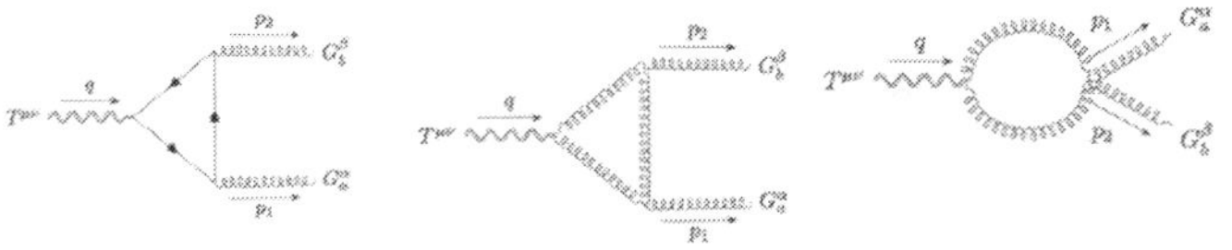

Figure 3. Typical contributon to the TJJ vertex at one loop, present in the quark and gluon sectors.

where the $1/q^2$ anomaly pole, interpreted as the signature of a dilaton exhange in perturbative QCD, has been extracted from the longitudinal projector $\pi^{\mu\nu}$ of the last term on the right-hand side of (8). The analysis of the perturbative amplitude is constrained by ordinary Ward identities, Slavnov-Taylor identities (STIs) and anomalous conformal Ward identities (CWIs) that we discuss in our work. Eq. (8), which is one of the results of our analysis, shows that the structure of the effective vertex corresponding to the TJJ correlator is modified compared to the ordinary CFT case, with modifications affecting both the longitudinal and the trace sectors. In the non-Abelian case, the gauge-fixing and ghost sectors of QCD break conformal symmetry. Consequently, the reconstruction of the correlator using CFT$_p$ permits the inclusion of non-conformal terms. These terms do not appear in the decomposition of non-Abelian correlators, as Slavnov-Taylor identities (STIs) are not required in the conformal framework. This distinction arises because the general CFT analysis is not tied to a specific Lagrangian, unlike in perturbative QCD.
The pole for off-shell gluons can be extracted from the trace sector by following the steps we have outlined. For the quark contributions, we utilize CFT$_p$, which is constrained solely by the standard Ward identities (WIs) and conformal Ward identities (CWIs). These serve as a guiding framework for the full expansion of the correlator. Additionally, we account for the extra contributions arising from the exchange of virtual gluons and ghosts within the loops of the TJJ vertex. One may define the quantum averaged stress energy tensor, with just the pole-term included due to the $1/q^2$ exchange

$$T^{\mu\nu}_{anom}(z) = \beta(g)\,(g^{\mu\nu}\Box - \partial^\mu\partial^\nu)_z \int d^4x'\, \Box^{-1}_{z,x'} \left[F^a_{\alpha\beta}F^{a\alpha\beta}\right]_{x'}, \tag{17}$$

from which one may extract the anomaly-induced vertex at trilinear level in the specific form

$$\begin{aligned}\Gamma^{\mu\nu\alpha\beta\,ab}_{\rm anom}(p_1,p_2) &= \int d^4x \int d^4y\, e^{ip_1\cdot x + ip_2\cdot y}\, \frac{\delta^2 T^{\mu\nu}_{anom}(0)}{\delta A^a_\alpha(x) A^b_\beta(y)} \\ &= \beta(g)\frac{1}{3\,q^2}\left(g^{\mu\nu}q^2 - q^\mu q^\nu\right)u^{\alpha\beta}(p_1,p_2)\delta^{ab}\end{aligned} \tag{18}$$

with a trace anomaly

$$g_{\mu\nu}\Gamma^{\mu\nu\alpha\beta ab}(p,q)\big|_{A=0} = \beta(g)\,u^{\alpha\beta}(p_1,p_2)\delta^{ab}. \tag{19}$$

By differentiating $T^{\mu\nu}_{\rm anom}(0)$ to higher orders with respect to the external classical gluon field, we derive analogous relations for the 4-point function ($TJJJ$) appearing in the gauge covariant expansion of the anomaly contribution [2].

3 Conclusions

The conformal anomaly interaction is described through a longitudinal, transverse, and trace sector decomposition of the relevant vertices, with a pole emerging in the trace channel.

This parameterization of hard scattering processes, as presented, can be seamlessly extended to the hadronic level, specifically for protons and pions, offering a potential phenomenological framework for analysing the role of this anomaly in the GFFs of pion and proton. We have shown that conformal anomalies are intrinsically tied to the presence of effective dilaton degrees of freedom in hard scattering processes. In the perturbative framework, these anomalies manifest as the emergence of a dilaton pole during hard scattering. This phenomenon is accompanied by a sum rule, satisfied by the anomaly form factor, which we have checked at the one-loop level in perturbation theory, in analogy to other anomaly form factors for chiral interactions. Details of this specific analysis can be found in [2].

This work is partially funded by the European Union, Next Generation EU, PNRR project "National Centre for HPC, Big Data and Quantum Computing", project code CN00000013; by INFN, inziativa specifica *QG-sky* and by the grant PRIN 2022BP52A MUR "The Holographic Universe for all Lambdas" Lecce-Naples.

References

[1] H.n. Li, G.F. Sterman, The Perturbative pion form-factor with Sudakov suppression, Nucl. Phys. B **381**, 129 (1992). 10.1016/0550-3213(92)90643-P

[2] C. Corianò, S. Lionetti, D. Melle, R. Tommasi, The Gravitational Form Factors of Hadrons from CFT in Momentum Space and the Dilaton in Perturbative QCD (2024), 2409.05609

[3] X. Ji, Deeply Virtual Compton Scattering, Phys. Rev. D **55**, 7114 (1997). 10.1103/PhysRevD.55.7114

[4] A.V. Radyushkin, Nonforward parton distributions, Phys. Rev. D **56**, 5524 (1997). 10.1103/PhysRevD.56.5524

[5] C. Corianò, M.M. Maglio, Conformal field theory in momentum space and anomaly actions in gravity: The analysis of three- and four-point function, Phys. Rept. **952**, 2198 (2022), 2005.06873. 10.1016/j.physrep.2021.11.005

[6] M. Giannotti, E. Mottola, The Trace Anomaly and Massless Scalar Degrees of Freedom in Gravity, Phys. Rev. **D79**, 045014 (2009), 0812.0351. 10.1103/PhysRevD.79.045014

[7] R. Armillis, C. Corianò, L. Delle Rose, Conformal Anomalies and the Gravitational Effective Action: The TJJ Correlator for a Dirac Fermion, Phys. Rev. **D81**, 085001 (2010), 0910.3381. 10.1103/PhysRevD.81.085001

[8] R. Armillis, C. Corianò, L. Delle Rose, Trace Anomaly, Massless Scalars and the Gravitational Coupling of QCD, Phys. Rev. **D82**, 064023 (2010), 1005.4173. 10.1103/PhysRevD.82.064023

[9] C. Corianò, L. Delle Rose, M. Serino, Gravity and the Neutral Currents: Effective Interactions from the Trace Anomaly, Phys.Rev. **D83**, 125028 (2011), 1102.4558. 10.1103/PhysRevD.83.125028

[10] Y. Hatta, A. Rajan, K. Tanaka, Quark and gluon contributions to the QCD trace anomaly, JHEP **12**, 008 (2018), 1810.05116. 10.1007/JHEP12(2018)008

New calculation of collision integrals for cosmological phase transitions

Carlo Branchina[1,2,*], *Angela* Conaci[1,2,**], *Stefania* De Curtis[3,***], *Luigi* Delle Rose[1,2,****], *Andrea* Guiggiani[3,†], *Ángel* Gil Muyor[4,‡], and *Giuliano* Panico[3,§]

[1]Dipartimento di Fisica, Università della Calabria, I-87036 Arcavacata di Rende, Cosenza, Italy
[2]INFN, Gruppo Collegato di Cosenza, Arcavacata di Rende, I-87036, Cosenza, Italy
[3]INFN Sezione di Firenze and Dipartimento di Fisica e Astronomia, Università di Firenze, Via G. Sansone 1, I-50019 Sesto Fiorentino, Italy
[4]IFAE and BIST, Universitat Autònoma de Barcelona, 08193 Bellaterra, Barcelona, Spain

Abstract. First order phase transitions in the early universe may have left a variety of experimentally accessible imprints. The dynamics of such transitions is governed by the density perturbations caused by the propagation of the bubble wall in the false vacuum plasma, conveniently described by a Boltzmann equation. The determination of the bubble wall expansion velocity is crucial to determine the experimental signatures of the transition. We report on the first full (numerical) solution to the Boltzmann equation. Differently from traditional ones, our approach does not rely on any ansatz. The results significantly differ from the ones obtained within the fluid approximation and large differences for the friction acting on the bubble wall are found. The wall velocity is calculated in a singlet extension of the Standard Model, including out-of-equilibrium contributions from both the top quark and the electroweak gauge bosons.

1 Introduction

The recent observation of gravitational waves has fueled a lively interest in the study of cosmological phase transitions (PhTs) and their dynamics. If a first order phase transition towards the electroweak (EW) vacuum did occur in the early universe, the latter is expected to have produced a source of gravitational waves that could be detected by future space-based interferometers [1–8]. This would make theories of the EW phase transition (EWPT) falsifiable and, combined with collider experiments, could play a key role in our quest for beyond the Standard Model (BSM) physics. As it is well known, in fact, the Standard Model cannot account for a first order EWPT, that can only be realized in BSM scenarios.

Besides the gravitational wave background, a first order PhT would have left many other imprints, and could for instance be responsible for the abundance of dark matter, the matter-antimatter asymmetry, and many other phenomena. In this respect, given a BSM model and

*e-mail: carlo.branchina@unical.it
**e-mail: angela.conaci@unical.it
***e-mail: stefania.decurtis@fi.infn.it
****e-mail: luigi.dellerose@unical.it
†e-mail: andrea.guiggiani@unifi.it
‡e-mail: agil@ifae.es
§e-mail: giuliano.panico@unifi.it

the transition pattern from the high temperature vacuum to the EW one, an accurate modelling of the PhT dynamics is crucial to quantitatively assess its impact. The transition is induced by a non-trivial scalar profile that describes a spherically symmetric solution [9–11] where the scalar field is close to its true vacuum value in a region of space and tends to its false vacuum one at larger distances. When the energy difference between the two vacua is sufficiently small, the scalar field varies rapidly in a small region, and the solution can be thought of as describing a bubble of true vacuum separated by a wall from a false vacuum background. The transition then proceeds by nucleation and expansion of these bubbles.

One of the fundamental quantities that characterizes the gravitational wave spectrum, as well as other quantities of interest, is the expansion speed of the PhT front, typically referred to as the bubble wall (terminal) velocity v_w. In the steady state regime, the latter arises from the competition of two different contributions: the external friction exerted by particles in the false vacuum sea (modelled as a plasma) hitting the wall, and the internal pressure caused by the potential difference between the false and true vacuum phases. The expanding bubble drives the plasma out of equilibrium, inducing a backreaction that slows down its propagation. This is not the only source of friction. In fact, the non-trivial spatial profile of the plasma temperature and velocity hinders the acceleration of the bubble, irrespective of the species in the plasma being out of equilibrium or not.

The wall velocity at equilibrium was first studied in [12] and has recently re-attracted a large interest. Comparing the results obtained assuming local thermal equilibrium to those found including out-of-equilibrium effects from the top quark for some benchmark points in a singlet extension of the Standard Model (SSM), Ref. [13] advances that calculations of the wall velocity at equilibrium may provide a good approximation to the full result. Although it was later shown by some of the present authors that this expectation is not really fulfilled [14, 15], the equilibrium determination of v_w certainly remains of interest, especially in connection to the iterative procedure that some of us presented in [16], through which out-of-equilibrium contributions can be systematically quantified. The wall velocity at equilibrium represents the zeroth order approximation of this calculation, and can be used as a first estimate to scan the parameter space of BSM theories [17].

As first shown in [18, 19], perturbations around equilibrium can be described by an effective Boltzmann equation. The authors developed an approach, dubbed here "old formalism" (OF), where the perfect fluid ansatz is taken for distribution functions and the Boltzmann equation is solved by taking suitable (weighted) moments that recast it into a system of ordinary differential equations. The fluid approximation has two main drawbacks, namely the appearance of a singularity in the friction for sonic walls and an unphysical dependence on the choice of weights: different weights lead to different results for the friction, and hence for the out-of-equilibrium effects. It has been recently argued that the singularity at the speed of sound is an artefact of the approximations used to solve the Boltzmann equation [20, 21], leading the authors to conclude that no critical behaviour is related to a sonic bubble wall.

Although the "new formalism" (NF) developed in [20, 21] and an extended fluid approximation [22, 23] successfully removed the unphysical singularity, both these strategies rely on an ansatz to solve the Boltzmann equation. The dependence of the results on the various ansatz urges to find a complete solution that does not rely on any assumption. In [16] some of us presented for the first time a numerical solution to the full Boltzmann equation. To deal with the collision integral that appears in the equation, technically its most challenging part, an iterative procedure was introduced. The efficiency of the numerical algorithm was significantly improved in two subsequent works [14, 15], where it was observed that the symmetries of the collision integral allow it to be interpreted as an Hermitian operator. In turn, this makes it possible to study it through its spectral decomposition, where a hierarchy between eigenvalues was observed and exploited to simplify the calculation.

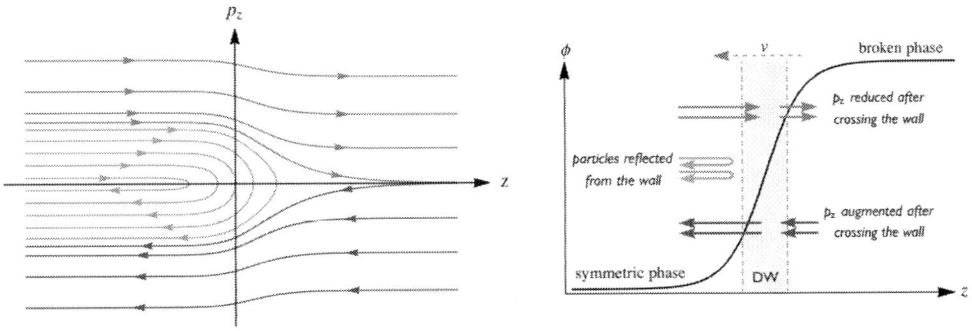

Figure 1. Left panel: Paths in the collisionless limit with constant energy $E = \sqrt{p_\perp^2 + p_z^2 + m^2(z)}$ and transverse momentum in the $z - p_z$ phase space for $m(z) \propto 1 + \tanh(z/L)$. The red, green and blue curves denote three different classes of paths, as explained in the main text. The arrows show the direction of propagation the phase space. Right panel: Schematic representation of the particle's behaviour across the domain wall (DW).

2 The Boltzmann equation

Large bubbles can be conveniently described taking a planar approximation for the wall. Throughout this manuscript, we assume the dynamics has reached a steady state regime where the wall moves at a constant velocity v_w. Orienting the z-axis in the direction opposite to the propagation of the PhT front, in the wall reference frame the Boltzmann equation for the distribution function f of a particle of mass $m(z)$ reads

$$\mathcal{L}[f] \equiv \left(\frac{p_z}{E}\partial_z - \frac{(m^2(z))'}{2E}\partial_{p_z}\right) f = -C[f], \tag{1}$$

where $\mathcal{L}$ is the Liouville operator and C is the collision integral that accounts for microscopic interactions between particles in the plasma.

The collision integral ensures that at sufficiently large distances from the wall (local) thermal equilibrium is restored, so that for $z \to \pm\infty$ the distribution function tends to the Fermi or Bose–Einstein one,

$$f_v = \frac{1}{e^{\beta\gamma(E - v p_z)} \pm 1}, \tag{2}$$

with v the fluid velocity, $\beta = 1/T$ and $\gamma = (1 - v^2)^{-1/2}$. Deviations from equilibrium are expected to be significant only close to the wall. When they are small enough, writing $f = f_v + \delta f$, the Boltzmann equation can be linearized in δf to give

$$\left(\frac{p_z}{E}\partial_z - \frac{(m^2(z))'}{2E}\partial_{p_z}\right)\delta f + \overline{C}[\delta f] = \frac{(m^2(z))'}{2E}\partial_{p_z} f_v = \beta\gamma v \frac{(m^2(z))'}{2E} f_v', \tag{3}$$

where we defined

$$f_v' \equiv -\frac{e^{\beta\gamma(E - v p_z)}}{(e^{\beta\gamma(E - v p_z)} \pm 1)^2}, \tag{4}$$

and $\overline{C}[\delta f]$ denotes the linearized collision integral. The right hand side of (3), the source term, gives sizable contributions close to the wall and rapidly decays away from it.

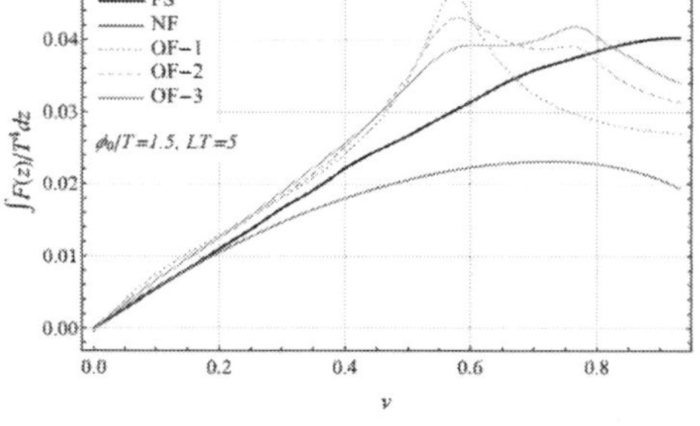

Figure 2. Friction acting on the bubble wall as a function of the velocity v. The black curve shows the solution to the full Boltzmann equation (FS), the green curves the results obtained with the old formalism at order 1, 2 and 3, respectively, and the red curve the result obtained with the new formalism [21].

2.1 Flow paths and the collision integral

Along paths where both the transverse momentum $p_\perp$ and the combination $p_z^2 + m^2(z)$ are constant, the Liouville operator reduces to a total derivative,

$$\mathcal{L} = \left(\frac{p_z}{E}\partial_z - \frac{(m^2(z))'}{2E}\partial_{p_z}\right) \quad \rightarrow \quad \frac{p_z}{E}\frac{d}{dz}. \tag{5}$$

These paths are nothing but the trajectories in the $(p_\perp, p_z, z)$ phase space in the collisionless limit $C \rightarrow 0$, along which both the energy and the momentum parallel to the wall are conserved. Three classes of solutions are found for $C = 0$. They are schematically shown in fig. 1, where, to simulate the case of a particle whose mass is generated by the field's value inside the bubble, $m(z)$ was modeled as $m(z) = m_0(1 + \tanh(z/L))$, with L the wall thickness. Red curves describe particles that travel towards the bubble and have enough momentum to cross the wall, $p_z(-\infty) > m_0$. After crossing it, they lose part of their momentum in favour of the mass $m(z)$. Green curves describe particles that travel towards the wall with momentum $p_z(-\infty) < m_0$ and get reflected from it. Finally, blue curves describe particles that travel in the opposite direction and exit from the bubble, gaining momentum as the mass decreases to 0 ($p_z(-\infty)$ will thus be $p_z(-\infty) < -m_0$).

The collision integral for $2 \rightarrow 2$ processes is

$$C[f] = \sum_i \frac{1}{4N_p E_p} \int \frac{d^3\mathbf{k}\, d^3\mathbf{p}'\, d^3\mathbf{k}'}{(2\pi)^5 2E_k 2E_{p'} 2E_{k'}} |\mathcal{M}_i|^2 \delta^4(p + k - p' - k')\mathcal{P}[f], \tag{6}$$

where $\mathcal{P}[f]$ is the population factor (the ± signs are for bosons and fermions, respectively)

$$\mathcal{P}[f] = f(p)f(k)(1 \pm f(p'))(1 \pm f(k')) - f(p')f(k')(1 \pm f(p))(1 \pm f(k)), \tag{7}$$

N_p are the degrees of freedom of the incoming particle with momentum p whose distribution function we aim to calculate, k is the momentum of the other incoming particle, and p', k' are the momenta of the outgoing particles. The sum runs over all $2 \rightarrow 2$ scattering processes with squared amplitude $|\mathcal{M}_i|^2$. As mentioned in the introduction, the expression (6) for the collision integral can be greatly simplified by exploiting its symmetries to first write it as an Hermitian operator, and to express its eigenfunctions in spherical harmonics. This allows to swap the nine-dimensional integral for a matrix multiplication and to use a multipole expansion to identify the eigenfunctions that truly contribute to the collision integral. For details,

Table 1. *First row:* m_s, λ_{hs} and λ_s individuate the benchmark points BP1 and BP2. T_n and T_c are the nucleation and critical temperatures, T_+ and T_- the temperatures in front and behind the DW. *Second row*: terminal values of the parameters v_w, δ_s, L_h, L_s for the two benchmark points. In each column, the first number is the result found including both the top quark and EW gauge bosons out-of-equilibrium perturbations. Numbers in square brackets are obtained neglecting the gauge bosons contribution, while the numbers in parentheses are obtained neglecting out-of-equilibrium perturbations altogether.

	m_s (GeV)	λ_{hs}	λ_s	T_n (GeV)	T_c (GeV)	T_+ (GeV)	T_- (GeV)
BP1	103.8	0.72	1	129.9	132.5	130.1	129.9
BP2	80.0	0.76	1	95.5	102.8	96.7	95.5

	v_w	δ_s	L_hT_n	L_sT_n
BP1	0.28 [0.39] (0.57)	0.78 [0.79] (0.75)	9.2 [9.7] (8.1)	7.4 [7.7] (6.7)
BP2	0.41 [0.47] (0.61)	0.81 [0.81] (0.81)	5.1 [5.2] (4.7)	4.2 [4.3] (4.1)

we refer the reader to [14, 15]. In passing, let us also note that in [15] a strategy to go beyond the linear approximation (3) and deal with non-linear terms was presented and implemented. We will refrain from discussing non-linear terms in this manuscript.

3 Numerical analysis

Writing as before $f = f_v + \delta f$, the collision integral can be split in two parts: one proportional to $\delta f(p)$, and another one where δf depends on the integrated momenta k, p', k'. The linearized Boltzmann equation then takes the form

$$\frac{d}{dz}\delta f - \frac{Q(p)}{p_z}\frac{f_v}{f'v}\delta f = S + \frac{f_v}{pz}\langle\delta f\rangle \tag{8}$$

where S is the δf-independent source term (see (3)), and $\langle\delta f\rangle$ collectively denotes all the contributions where δf is integrated.

To solve the Boltzmann equation, a numerical iterative method was developed in [16], where $\langle\delta f\rangle$ terms are first ignored and, starting from $n = 1$, the solution at iteration $n + 1$ is obtained inserting the solution for δf found at iteration n in $\langle\delta f\rangle$. The strategy was first tested considering a single species in the plasma, the top quark, the state with the largest coupling to the Higgs, and taking the tanh ansatz for the Higgs field

$$h(z) = \frac{h_0}{2}\left(1 + \tanh\frac{z}{L_h}\right), \tag{9}$$

where L_h is the thickness of the bubble wall and h_0 is the Higgs VEV in the broken phase. The numerical analysis in [16] was carried out fixing the transition temperature $T = 100$ GeV, $L_h = 5/T$ and $h_0 = 150$ GeV.

The out-of-equilibrium friction acting on the bubble wall (N are the degrees of freedom),

$$F(z) = \frac{dm^2}{dz}N\int\frac{d^3\mathbf{p}}{(2\pi)^3 2E}\delta f(p), \tag{10}$$

was numerically evaluated. The result is reported in Fig. 2, where the integrated friction thus found is shown as a function of the wall velocity (black curve) together with previous results

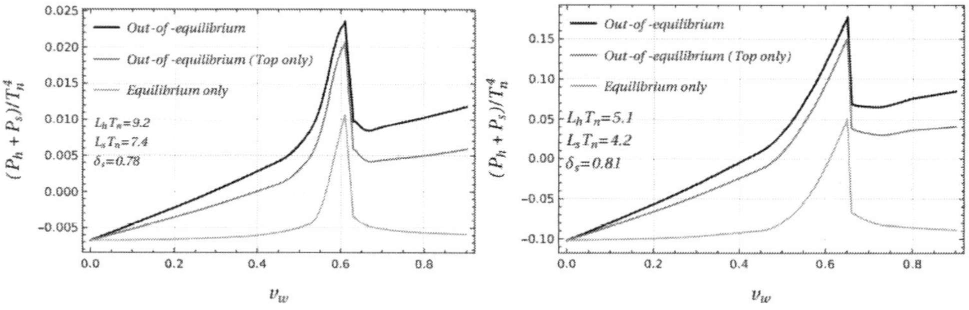

Figure 3. Total pressure $P_h + P_s$ as a function of the wall velocity v_w for the two benchmark points BP1 (*left panel*) and BP2 (*right panel*) in Table 1. The green curves are obtained assuming all species are at equilibrium, the red ones including out-of-equilibrium effects from the top quark, and the black ones including out-of-equilibrium effects from the top quark and EW gauge bosons.

(green and red curves). In particular, green curves show the integrated friction obtained in the old formalism [19] when the fluid approximation is improved with higher-order corrections (1, 2 and 3) [22]. The red curve shows the result obtained in the new formalism [21]. A fair numerical agreement is found between all the solutions for $v_w \lesssim 0.3$. At higher velocities, the OF's solution develops peaks that correspond to zero eigenvalues of the Liouville operator. The result for the full solution confirms that the latter are an artifact of the weight strategy. The NF's solution does not show these peaks but turns out to be in agreement with the full solution only for small velocities $v_w \lesssim 0.3$.

The bubble wall velocity was later calculated in the SSM (for two benchmark points) including the top quark perturbations only [14], and adding the EW gauge boson contributions [15]. Two-step phase transitions where the EW vacuum is reached from an intermediate phase with non-vanishing singlet were analysed, and the tanh ansatz was taken for both the Higgs and the singlet ($s(z) = s_0(1 + \tanh(z/L_s + \delta_s))$ supplements Eq. (9)). Hydrodynamic equations are used together with the scalar equations of motion (EOM) to determine v_w, the tilt δ_s in the scalar profiles and the wall thicknesses L_h and L_s as the values that make the pressure exerted on the wall and its gradient vanish, allowing for a steady state solution. The pressure on the wall in the Higgs and in the singlet direction are denoted with P_h and P_s, respectively. At zeroth order in the proposed iterative procedure, the calculation of the wall velocity through the combined solution of the hydrodynamic equations and of the scalar EOM coincides with the equilibrium determination of v_w. The numerical algorithm we developed to solve these equations already allows, at this stage, to calculate the wall velocity at equilibrium in the full parameter space of BSM theories in a fair amount of time [17].

The results for v_w are summarized in Fig. 3, where we plot the total pressure in terms of v_w, and in Table 1, where the chosen benchmark points are individuated through the values of the singlet mass m_s in the EW vacuum, of the portal coupling λ_{hs}, and of the singlet's self-coupling λ_s. The impact of out-of-equilibrium contributions from both the top and the EW gauge bosons can be appreciated.

4 Conclusions and Outlook

In [16] it was presented for the first time a numerical solution to the full Boltzmann equation describing the distribution function of particles during a first order phase transition. Differently from other approaches in the literature, we did not rely on any ansatz for the out-of-equilibrium distribution functions. The results were critically compared with those found using the traditional strategies put forward in the literature, namely the fluid approximation [19], its extended version [22, 23], and the new formalism [21]. The efficiency of the algorithm developed in [16] was later improved in [14] and [15], where we also calculated the wall velocity for benchmark points in the singlet extension of the Standard Model.

These results represent a great advancement toward a reliable characterization of the bubble wall dynamics. The improved computational performances obtained with the refinements in [14] and [15] of the algorithm proposed in [16] open the door to the possibility of scanning the parameter space of BSM models and studying experimentally falsifiable signals (gravitational waves, baryon asymmetry and so on) therein. This ambitious program will be inaugurated by the equilibrium calculation of v_w in several BSM models of phenomenological interest [17].

References

[1] C. Caprini et al., JCAP **04**, 001 (2016), 1512.06239
[2] C. Caprini et al., JCAP **03**, 024 (2020), 1910.13125
[3] S. Kawamura et al., Class. Quant. Grav. **23**, S125 (2006)
[4] S. Kawamura et al., Class. Quant. Grav. **28**, 094011 (2011)
[5] W.R. Hu, Y.L. Wu, Natl. Sci. Rev. **4**, 685 (2017)
[6] W.H. Ruan, Z.K. Guo, R.G. Cai, Y.Z. Zhang, Int. J. Mod. Phys. A **35**, 2050075 (2020), 1807.09495
[7] J. Luo et al. (TianQin), Class. Quant. Grav. **33**, 035010 (2016), 1512.02076
[8] P. Athron, C. Balázs, A. Fowlie, L. Morris and L. Wu, Prog. Part. Nucl. Phys. **135** (2024), 104094 2305.02357
[9] S. R. Coleman, Phys. Rev. D **15**, 2929-2936 (1977) [erratum: Phys. Rev. D **16**, 1248 (1977)]
[10] C. G. Callan, Jr. and S. R. Coleman, Phys. Rev. D **16**, 1762-1768 (1977)
[11] A. D. Linde, Phys. Lett. B **70**, 306-308 (1977)
[12] J. Ignatius, K. Kajantie, H. Kurki-Suonio and M. Laine, Phys. Rev. D **49**, 3854-3868 (1994) astro-ph/9309059
[13] B. Laurent, J. M. Cline, Phys. Rev. D **106**, 023501 (2022), 2204.13120
[14] S. De Curtis, L. Delle Rose, A. Guiggiani, Á. Gil Muyor, G. Panico, JHEP **05**, 194 (2023) 2303.05846
[15] S. De Curtis, L. Delle Rose, A. Guiggiani, Á. Gil Muyor, G. Panico, JHEP **05**, 009 (2024), 2401.13522
[16] S. De Curtis, L.D. Rose, A. Guiggiani, A.G. Muyor, G. Panico, JHEP **03**, 163 (2022), 2201.08220
[17] C. Branchina, et al., *in preparation*
[18] G.D. Moore, T. Prokopec, Phys. Rev. Lett. **75**, 777 (1995), hep-ph/9503296
[19] G.D. Moore, T. Prokopec, Phys. Rev. D **52**, 7182 (1995), hep-ph/9506475
[20] J.M. Cline, K. Kainulainen, Phys. Rev. D **101**, 063525 (2020), 2001.00568
[21] B. Laurent, J.M. Cline, Phys. Rev. D **102**, 063516 (2020), 2007.10935
[22] G.C. Dorsch, S.J. Huber, T. Konstandin, JCAP **08**, 020 (2021), 2106.06547
[23] G.C. Dorsch, S.J. Huber, T. Konstandin (2021), 2112.12548

EPJ Web of Conferences **314**, 00032 (2024) https://doi.org/10.1051/epjconf/202431400032
QCD@Work 2024

BPS multi-vortices in chiral perturbation theory

Aldo Vera[1,*]

[1]Núcleo Matemáticas Física y Estadística, Universidad Mayor, Avenida Manuel Montt 367, Santiago, Chile

Abstract. We show that a BPS bound can be derived from chiral perturbation theory minimally coupled to electrodynamics at a finite isospin chemical potential. At a critical value of the isospin chemical potential, and for a fixed pressure value, a system of coupled first-order differential field equations can be derived for the gauge potential and the hadronic profile. This BPS system is solved by magnetic multi-vortices with quantized flux supported by a superconducting current. The topological charge density is related to the magnetic flux, but it is dressed by the hadronic profile. We discuss some of the main physical properties of this novel BPS bound. In particular, we show that a critical value for the magnetic field can be deduced.

1 Introduction

A very important region of the phase diagram of quantum chromodynamics (QCD) corresponds to low temperatures, vanishing baryonic chemical potential ($\mu_B = 0$) and finite isospin chemical potential ($\mu_I \neq 0$) [1]-[4]. In this sector there is no sign problem, so that detailed results on the phase diagram are available. Indeed, in the seminal paper by Son and Stephanov [1], it has been shown that a transition occurs from the hadronic phase to a superfluid phase and, eventually, to a superconducting phase. It is expected that at low isospin density, the ground state is a pion condensate, while a Fermi liquid with Cooper pairing should appear at higher density.

On the other hand, from the Ginzburg-Landau theory (GL theory) of superconductivity, we know that the transition to type-I to type-II superconductors depends on the value of the phase of the Higgs coupling. Even more, exactly at the transition point, one can saturate a Bogomol'nyi–Prasad–Sommerfield bound (BPS bound) for the free energy (per unit of length), where the magnetic flux plays the role of the topological charge, and multi-vortices solutions (where the vortices can be placed at arbitrary locations) appear [5], [6].

Now, since μ_I is responsible for the Cooper pairing in the sector of QCD mentioned above playing a similar role to the Higgs coupling in the GL theory, one would expect that at a special value of μ_I it should be possible to saturate a suitable BPS bound in such a way that multi-vortices solutions should appear. Here we show that this is indeed the case. For a critical value of μ_I, a novel BPS bound can be derived from the Gibbs free energy of the system and whose solutions are multi-vortices.

We will work with chiral perturbation theory (χPT), which describes the low energy limit of QCD [7], using the analytical tools introduced in Refs. [8]-[12].

*e-mail: aldo.vera@umayor.cl

2 The gauged chiral perturbation theory

The low-energy limit of QCD can be described by χPT. Up to order $O(p^2)$ in the momentum expansion, and considering the coupling with Maxwell's electrodynamics, the action is

$$I[U,A] = \frac{1}{4e^2}\int_M d^4x\,\sqrt{-g}\left(K\mathrm{Tr}\left[L^\mu L_\mu\right] - F_{\mu\nu}F^{\mu\nu}\right)\,, \tag{1}$$
$$L_\mu = U^{-1}D_\mu U = L^j_\mu t_j\,,\quad F_{\mu\nu} = \partial_\mu A_\nu - \partial_\nu A_\mu\,,\quad t_j = i\sigma_j\,.$$

In Eq. (1), $U(x)\in SU(2)$ is the pionic field, A_μ is the Maxwell potential and σ_i are the Pauli matrices. The coupling K is a positive constant given by $K = (\frac{ef_\pi}{2})^2$, being f_π the pions decay constant and e the electric charge. Here we will use $f_\pi = 93$ MeV and $e = \sqrt{4\pi\alpha} = 0.303$. In our convention $c = 1$ and $\hbar = 1$. The covariant derivative, D_μ, is defined as

$$D_\mu U = \nabla_\mu U + A_\mu U\hat{O}\,,\quad \hat{O} = U^{-1}\left[t_3, U\right]\,, \tag{2}$$

being ∇_μ the partial derivative. The isospin chemical potential can be introduced through the covariant derivative in the following form

$$D_\mu U \to \bar{D}_\mu U = D_\mu U + \mu_I[t_3, U]g_{\mu t}\,, \tag{3}$$

where μ_I is the value of the isospin chemical potential. The electromagnetic current is

$$J_\mu = -\frac{K}{2}\mathrm{Tr}(\hat{O}L_\mu)\,. \tag{4}$$

3 The matter fields and the Gibbs free energy

The pionic field can be written in the exponential representation as

$$U = \cos(\alpha)\mathbf{1} + \sin(\alpha)n_i t^i\,, \tag{5}$$
$$n_i = \{\sin\Theta\cos\Phi, \sin\Theta\sin\Phi, \cos\Theta\}\,,$$

where $\alpha = \alpha(x^\mu)$, $\Theta = \Theta(x^\mu)$, $\Phi = \Phi(x^\mu)$ are the three degrees of freedom of the U field in $SU(2)$, and $\mathbf{1}$ denotes the 2×2 identity matrix. In terms of this parametrization, the covariant derivative $D_\mu U$ in Eq. (2) reads

$$D_\mu\alpha = \partial_\mu\alpha\,,\quad D_\mu\Theta = \partial_\mu\Theta\,,\quad D_\mu\Phi = \partial_\mu\Phi - 2A_\mu\,. \tag{6}$$

One can see that the degree of freedom $\Phi(x^\mu)$ plays the role of the phase of the complex Higgs field in the GL theory (while α and Θ are not affected by the covariant derivative).

Since we are interested in multi-vortices with quantized magnetic field along the third spatial direction, the natural Ansatz is

$$\alpha = \alpha(x_1, x_2)\,,\quad \Phi = \Phi(x_1, x_2)\,,\quad \Theta = \frac{\pi}{2}\,, \tag{7}$$
$$A_\mu dx^\mu = A_1\,dx_1 + A_2\,dx_2\,,\qquad A_i = A_i(x_1, x_2)\,, \tag{8}$$

which leads to a magnetic field in the x_3 direction.

The Gibbs free energy density is defined as

$$\boldsymbol{G} = \boldsymbol{F} + PV\,,\qquad \boldsymbol{F} = \int d^3x F\,,\quad \boldsymbol{G} = \int d^3x G\,. \tag{9}$$

Here F is the free energy density (defined as $F = -\mathcal{L}$, being $\mathcal{L}$ the lagrangian density), V is the volume and P the pressure.

Replacing the Ansatz in Eqs. (5), (7) and (8) (taking into account the non-trivial isospin chemical potential in Eq. (3)), into Eq. (1), we obtain the Gibbs free energy density, G, of the system

$$e^2 G = \frac{K}{2}\left\{\left(\vec{\nabla}\alpha\right)^2 + \sin^2(\alpha)\left(\vec{D}\Phi\right)^2\right\} + \frac{1}{2}(\vec{B}^2) + 2K\mu_I^2\cos^2(\alpha)\,, \tag{10}$$

where we have denoted

$$\left(\vec{\nabla}\alpha\right)^2 = \sum_{i=1}^{2}(\partial_i\alpha)^2\,, \quad \left(\vec{D}\Phi\right)^2 = \sum_{i=1}^{2}(D_i\Phi)^2\,.$$

Note that the Gibbs free energy density is positive definite, as expected. To achieve this, it was necessary to fix the pressure in terms of the chemical potential to the value $P = 2K\mu_I^2$. Thus, once the isospin chemical potential is fixed, the value of the pressure can be obtained. This is different from what happens in Ref. [13], where magnetic vortices were constructed at zero external pressure.

From Eq. (10), we will derive a BPS bound of the form

$$\begin{aligned} G &= (\text{BPS equations})^2 + d\omega \\ &\Rightarrow \boldsymbol{G} \geq \omega\,. \end{aligned} \tag{11}$$

4 A single vortex

Here we will consider the simple case of a single vortex. Let us start with the flat metric in cylindrical coordinates

$$ds^2 = -dt^2 + dr^2 + r^2 d\theta^2 + dz^2\,.$$

The requirement to have a finite Gibbs free energy per unit of length (in this case, in the z direction)

$$\frac{\boldsymbol{G}}{L} = \int dr\, d\theta\, G\,, \tag{12}$$

gives rise to the following boundary conditions for the fields at infinity;

$$\alpha \underset{r\to\infty}{\longrightarrow} \frac{\pi}{2} + c_1\pi\,,\ c_1 \in \mathbb{N}\,, \qquad \vec{D}\Phi \underset{r\to\infty}{\longrightarrow} 0 \Leftrightarrow \vec{A} \underset{r\to\infty}{\longrightarrow} \frac{1}{2}\vec{\partial}\Phi\,. \tag{13}$$

On the other hand, close to the origin, regularity requires

$$\alpha \underset{r\to 0}{\longrightarrow} c_2\pi\,,\ c_2 \in \mathbb{N}\,, \qquad \vec{A} \underset{r\to 0}{\longrightarrow} 0\,. \tag{14}$$

Note that, the condition in Eq. (14) means that

$$U \underset{r\to 0}{\longrightarrow} \pm\mathbf{1}\,. \tag{15}$$

Once the boundary conditions have been fixed, one can derive the quantization of the magnetic flux. The U field in Eq. (5) must be single valued, that is,

$$U(r,\theta) = U(r,\theta+2\pi)\,. \tag{16}$$

Now, taking into account Eq. (7), it follows that

$$\Phi(r,\theta) = \Phi(r,\theta+2\pi) + 2n\pi\,,\quad n \in \mathbb{N}\,, \tag{17}$$

which implies that the following integral is an integer

$$\frac{1}{2\pi}\oint_{\Gamma_\infty}\vec{\partial}\Phi\cdot d\vec{l}=n\,, \tag{18}$$

where Γ_∞ is the circle at spatial infinity. Now, as the finiteness of the Gibbs free energy requires Eq. (13), due to the Stokes theorem (as in the case of the GL theory) one obtains that n is exactly the magnetic flux, and it is quantized.

For one vortex, the pionic field and the Maxwell potential must take the form

$$\alpha=\alpha(r)\,,\quad A_r=A_z=0\,,\quad A_\theta=A(r)\,,\quad \Phi=n\theta\,, \tag{19}$$

where the spatial coordinates have the ranges: $0<r<\infty$, $0\leq\theta<2\pi$, $-\infty<z<\infty$.

The Gibbs free energy for one vortex is the integral of the expression in Eq. (10) considering the Ansatz in Eq. (19). Supplementing and subtracting the same appropriate terms to obtain an expression in the form of Eq. (11), we lead to

$$\begin{aligned}\frac{\boldsymbol{G}}{L}=&2\pi\int\frac{1}{2e^2r}\{K(r\alpha'+\sin(\alpha)(n-2A))^2+(A'+2\mu_I\sqrt{K}r\cos(\alpha))^2\}dr\\&-\frac{2\pi}{e^2}\int\left\{K\sin(\alpha)(n-2A)\alpha'+2\mu_I\sqrt{K}\cos(\alpha)A'\right\}dr\,.\end{aligned} \tag{20}$$

We can see that the second term in Eq. (20) is a total derivative for a particular value of the isospin chemical potential, that is $\mu_I^c=\sqrt{K}$. In fact, for this value

$$-K\sin(\alpha)(n-2A)\alpha'-2\mu_I\sqrt{K}\cos(\alpha)A'=\{K\cos(\alpha)(n-2A)\}'\,.$$

Therefore, it follows from Eq. (20) that the Gibbs free energy is minimized when the following BPS equations are satisfied

$$r\alpha'+\sin(\alpha)(n-2A)=0\,, \tag{21}$$

$$A'+2Kr\cos(\alpha)=0\,. \tag{22}$$

From the above, the topological charge for this field configuration reads

$$\omega=\frac{2\pi}{e^2}\int\{K\cos(\alpha)(n-2A)\}'dr=\frac{2\pi K}{e^2}n\,. \tag{23}$$

Here we can see that the topological charge density is different from the magnetic flux density. However, once the boundary conditions are implemented, the topological charge is in fact proportional to the magnetic flux (times a dimensional constant that depends on the couplings of the theory), as can be seen from Eqs. (18) and (23).

Figue 1 shows the Gibbs energy density of the vortex for different values of the topological charge. To integrate Eqs. (21) and (22) we have used the boundary conditions in Eqs. (13) and (14), with $c_1=0$ and $c_2=1$.

5 Multi-vortices

We can generalize the above results to describe multi-vortices configurations. First, it is important to note that the boundary conditions in Eqs. (13) and (14) must be satisfied at the locations of all the vortices, as one can always choose a local coordinates system where the position of the j–th vortex is the origin, that is

$$U\underset{r\to r_j}{\to}\pm\mathbf{1}\,,\quad\forall\, j=1,..,N\,, \tag{24}$$

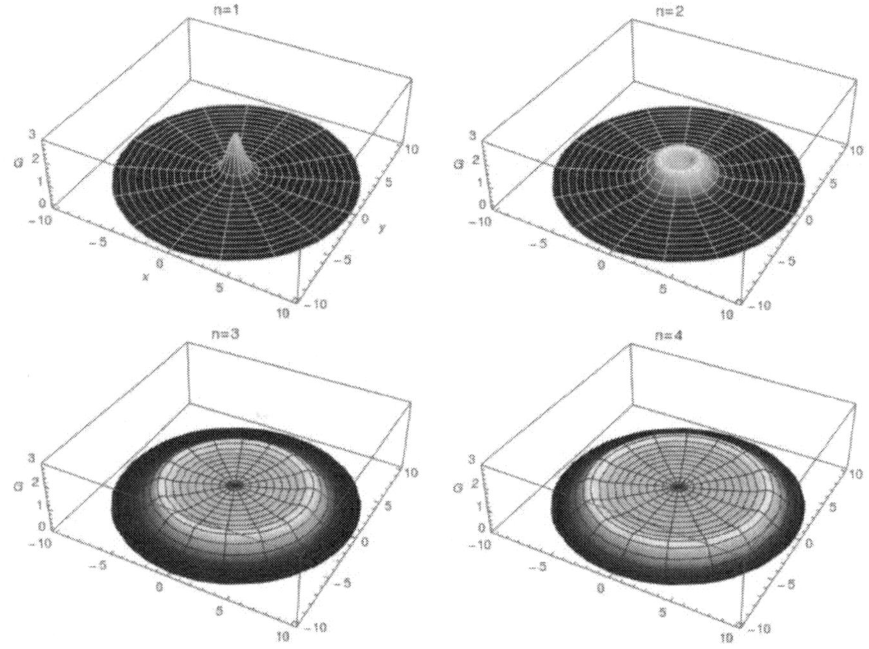

Figure 1. Gibbs free energy density of one vortex for different values of the topological charge.

where N is the total number of vortices and r_j is the position of the j−th vortex. In the same way, Eq. (18) would be an integer for any closed curve Γ enclosing all the points $\left\{r_j\right\}_{j=1,\dots,N}$, where $\sin(\alpha)$ vanishes.

With this in mind, let us return to the general expression for the Gibbs free energy density presented in Eq. (10). Here we can follow the same steps shown for the single vortex, but for multi-soliton configurations, it is convenient to introduce an improper gauge transformation. Indeed, locally, we can gauge away the function Φ as follows

$$A_\mu \to \widehat{A_\mu} = A_\mu + \frac{1}{2}\partial_\mu \Phi \,, \tag{25}$$

$$\Phi \to \widehat{\Phi} = \Phi - \Phi = 0 \,. \tag{26}$$

Although this is very convenient from the computational point of view, it is very important to emphasize that Φ is a multi-valued function precisely due to the non-vanishing magnetic flux (see Eqs. (17) and (18)). In fact, if one decides to "gauge away" the Φ function in a configuration with N vortices, it becomes necessary to include explicitly in the magnetic field the "Aharonov-Bohm contributions" from all the vortices. Thus, Eqs. (25) and (26) do not define a proper gauge transformation, but are very useful to derive the BPS bound. In fact, considering Eqs. (25) and (26), one can rewrite the Gibbs free energy density as

$$\begin{aligned} e^2 G =& \frac{K}{2}\Big((\partial_1\alpha + \sin(\alpha)\widehat{A}_2)^2 + (\partial_2\alpha - \sin(\alpha)\widehat{A}_1)^2\Big) + \frac{1}{2}\Big(\widehat{B}_z - 2\mu_I\sqrt{K}\cos(\alpha)\Big)^2 \\ &- K\sin\alpha(\partial_1\alpha\widehat{A}_2 - \partial_2\alpha\widehat{A}_1) + 2\mu_I\sqrt{K}\cos(\alpha)\widehat{B}_z \,. \end{aligned} \tag{27}$$

Again, the last two terms form a total derivative for the critical value $\mu_I^c = \sqrt{K}$. In fact, this critical isospin chemical potential allows that

$$-K \int dx^1 dx^2 \Big[\sin(\alpha)(\partial_1\alpha\,\widehat{A}_2 - \partial_2\alpha\,\widehat{A}_1) - \cos(\alpha)\widehat{B}_z\Big] = \int d\omega\,,$$
$$\omega = \cos(\alpha)(\widehat{A}_1 dx^1 + \widehat{A}_2 dx^2)\,, \tag{28}$$

where we have used that $\widehat{B}_z = F_{12} = \partial_1\widehat{A}_2 - \partial_2\widehat{A}_1$. Thus, the Gibbs free energy density is minimized when the following first order equations are satisfied

$$\partial_1\alpha + \sin(\alpha)\,\widehat{A}_2 = 0\,, \tag{29}$$
$$\partial_2\alpha - \sin(\alpha)\,\widehat{A}_1 = 0\,, \tag{30}$$
$$\widehat{B}_z - 2K\cos(\alpha) = 0\,. \tag{31}$$

6 Some physical implications from the BPS bound

Naturally, the system of equations presented above can be integrated numerically to obtain the profiles. However, many of the interesting properties of these configurations can be read directly from the BPS equations and the topological charge without solving the system. Here we list some of its main physical implications.

1. The topological charge density is not simply the magnetic flux density, it is a "dressed magnetic flux" modulated by the hadronic profile (see Eqs. (23) and (28)).

2. The multi-vortices are minima of the Gibbs free energy at critical values of the isospin chemical potential and pressure. These values are

$$\mu_I^c = \sqrt{K} = \frac{ef_\pi}{2} = 14,1\,\text{MeV}\,, \tag{32}$$
$$P = 2K(\mu_I^c)^2 = 858.474\,(\text{MeV})^4\,. \tag{33}$$

3. The BPS bound allows to find very easily the maximum value for the magnetic field beyond which the condensate ceases to exist. Indeed, looking at Eq. (31), one gets

$$B_{\text{max}} = 2K = 2\left(\frac{ef_\pi}{2}\right)^2 = 397,03\,(\text{MeV})^2 = 2,04\times 10^{14}\,\text{G}\,. \tag{34}$$

4. The magnetic field is generated by a self-sustained current, given by

$$J_\mu = 2K\sin^2(\alpha)D_\mu\Phi\,. \tag{35}$$

This current is not-null even when the electromagnetic field is suppressed. In fact, there is a persistent current generated by the coupling with pions, given by $J_\mu^{(0)} = 2K\sin^2(\alpha)\partial_\mu\Phi$.

5. As expected, the first order BPS equations imply the following second order system

$$\Delta\alpha - \sin(\alpha)\cos(\alpha)\Big((\vec{D}\Phi)^2 - 4\mu_I^2\Big) = 0\,,$$
$$\Delta\Phi - 2\,\vec{\nabla}\cdot\vec{A} + 2\cot(\alpha)\vec{\nabla}\alpha\cdot\vec{D}\Phi = 0\,,$$
$$\partial_j F^{ij} - 2K\sin^2(\alpha)D^i\Phi = 0\,,$$

obtained through the variation of the free energy with respect to the field α, Φ and A_μ.

6. The present formalism can be applied even when the gauged χPT includes a pions mass term [14]. In particular, considering

$$\mathcal{L}_{\text{mass}} = -\frac{2Km_\pi^2}{e^2}\left(1 - \frac{1}{4}\text{Tr}[U]^2\right) = \frac{2Km_\pi^2}{e^2}\sin^2(\alpha) ,$$

the free energy density of the system becomes

$$e^2F = \frac{K}{2}\left\{\left(\vec{\nabla}\alpha\right)^2 + \sin^2(\alpha)\left(\vec{D}\Phi\right)^2\right\} + \frac{1}{2}(\vec{B}^2) + 2K\left(\mu_I^2 - m_\pi^2\right)\cos^2(\alpha) - 2K\left(\mu_I^2 - m_\pi^2\right) .$$

Consequently, the inclusion of the mass term for the pions manifests itself in a shift: $\mu_I^2 \to \mu_I^2 - m_\pi^2$. Hence, all the previous results still hold. The critical value for the isospin chemical potential when the pions mass is taken into account reads

$$\mu_I^c = \sqrt{K + m_\pi^2} < 1.1 m_\pi .$$

7 Conclusions

We have shown that the gauged χPT at critical values of the isospin chemical potential and pressure admit superconducting multi-vortices solutions. This has been achieved through the derivation of a novel BPS bound whose topological charge density is related to the magnetic flux density, but it is dressed by the hadronic profile. These multi-vortices are supported by a persistent current that does not vanish when the Maxwell field is suppressed, showing that the condensate has superconducting characteristics. Also, using the BPS equations, we have predicted an upper bound on the intensity of the magnetic field. More details about this construction can be found in Ref. [15].

References

[1] D.T. Son, M. A. Stephanov, Phys. Rev. Lett. 86 (2001) 592-595
[2] T. Kojo, D. Suenaga, Phys. Rev. D 105 (2022) 7, 076001
[3] R. D. Pisarski and F. Wilczek, Phys. Rev. D **29**, 338-341 (1984)
[4] B. B. Brandt, G. Endrodi and S. Schmalzbauer, EPJ Web Conf. **175**, 07020 (2018)
[5] A. A. Abrikosov, Sov. Phys. JETP **5**, 1174-1182 (1957)
[6] E. J. Weinberg, Phys. Rev. D 19 (1979) 3008
[7] S. Scherer, Adv. Nucl. Phys., vol. 27, p. 277, 2003
[8] F. Canfora, Eur. Phys. J. C **78**, no.11, 929 (2018)
[9] F. Canfora, S. H. Oh and A. Vera, Eur. Phys. J. C **79**, no.6, 485 (2019)
[10] F. Canfora, M. Lagos and A. Vera, Eur. Phys. J. C **80**, no.8, 697 (2020)
[11] F. Canfora, S. Carignano, M. Lagos, M. Mannarelli and A. Vera, Phys. Rev. D **103**, no.7, 076003 (2021)
[12] F. Canfora, JHEP **11**, 007 (2023)
[13] P. Adhikari, Phys. Lett. B **790**, 211-217 (2019)
[14] S. B. Gudnason and M. Nitta, Phys. Rev. D **94**, no.6, 065018 (2016)
[15] F. Canfora, M. Lagos and A. Vera, [arXiv:2405.08082 [hep-th]]

Silica aerogel characterization for the ePIC dRICH detector

Anna Rita Altamura[1,*], on behalf of the ePIC dRICH Collaboration

[1]Istituto Nazionale di Fisica Nucleare, sezione di Bari, Italy

Abstract. The ePIC detector is specifically designed to address the entire physics program at the Electron-Ion Collider (EIC) and features several sub-detectors, including the dual-radiator Ring Imaging Cherenkov (dRICH) detector for high-momentum particle identification. Silica aerogel is chosen as the solid radiator due to its crucial optical properties. Ongoing research aims to optimize these characteristics, focusing on the transmittance measurements of aerogel tiles with varying refractive indices and the associated measurement setup.

1 Introduction

Ring Imaging Cherenkov (RICH) detectors identify charged particles by utilizing Cherenkov radiation, which occurs when charged particles move faster than the phase speed of light in a medium[1]. Traditional gas and liquid radiators have refractive indices either below 1.0018 (C_5F_{12}) or above 1.27 (liquid C_6F_{14})[1]. Silica aerogel emerges as the only practical solution to bridge this refractive index gap, avoiding the challenges of high-pressure gases or liquefied forms. Consisting of 99.8% air and 0.2% silicon dioxide, silica aerogel offers a compact structure with good optical transmittance and a low refractive index, making it the preferred choice for many RICH detectors.

2 Aerogel optical properties

The transmittance of silica aerogel quantifies how much light can pass through the material without absorption or scattering. Its unique structure features a three-dimensional lattice of interconnected nanopores, with diameters ranging from 2 to 50 nm[2]. This architecture allows visible light to travel with minimal scattering and absorption, enabling propagation with low attenuation. Light in silica aerogel mainly experiences Rayleigh scattering, where light is elastically scattered by particles much smaller than its wavelength.

Transmittance typically peaks in the near-infrared spectrum, where the silica network shows minimal radiation absorption. The relationship between aerogel transmittance T and the wavelength λ of the impinging radiation traversing a sample with thickness t is usually described using the Hunt formula[3]:

$$T(\lambda) = e^{-\frac{t}{\Lambda_T}} = Ae^{-\frac{C \times t}{\lambda^4}} \quad (1)$$

*e-mail: anna.rita.altamura@cern.ch

EPJ Web of Conferences **314**, 00033 (2024) https://doi.org/10.1051/epjconf/202431400033
QCD@Work 2024

Table 1. List of the tiles under tests with their refractive index and nominal thickness *t*.

Tile	*n*	t [cm]	Tile	*n*	t [cm]
1		20.7	23		2.05
2		2.08	24	1.02(*2021*)	2.08
3		2.01	25		2.08
4	1.03	2.05	26		2.08
5		2.04	27		2.05
6		0.98	28		2.06
7		0.97	29		2.04
8		2.03	30		1.95
9		2.05	31		1.99
10	1.04	2.03	32		2.17
11		2.04	33		2.14
12		2.05	34		2.14
13		2.05	35	1.02(*2022*)	2.13
14		2.07	36		2.12
15	1.05	2.06	37		1.91
16		2.06	38		1.94
17		2.08	39		2.03
18		2.00	40		2.03
19	1.005	2.06	41		2.04
20		2.06	42		1.97
21	1.03	2.02			
22		2.03			

Here, C accounts for aerogel clarity, A for light absorption, and Λ_T is the "transmission length," indicating the distance light can travel with minimal absorption or scattering. High optical quality samples have A close to 1 and C near 0. Silica aerogel exhibits low absorption due to the minimal impurities in its silica network.

Additionally, the tiles studied are hydrophobic, repelling water and thus maintaining transparency in humid conditions. This property prevents water infiltration, which can scatter light and reduce transmittance.

The scattering and absorption capabilities can be assessed through the transmission length Λ_T:

$$\frac{1}{\Lambda_T} = \frac{1}{\Lambda_{scat}} + \frac{1}{\Lambda_{abs}}, \tag{2}$$

where Λ_{scat} is the scattering length and Λ_{abs} is the absorption length of the sample.

The Hunt formula (Eq. 1) assumes a Rayleigh scattering cross-section dependence on λ^4 and an absorption cross-section independent of the radiation wavelength. Consequently, A remains unaltered with variations in the wavelength of photons.

3 Aerogel characterization

We have characterized several 10×10 cm^2 and 15×15 cm^2 aerogel tiles with various refractive indeces at CERN in July 2022. Table 1 reports a list of the tested samples, where for each tile its refractive index and expected thickness in the center point of the tile are reported. Tiles from 1 to 17 were produced by Aerogel Factory Co. Ltd (Chiba, Japan) and delivered in March 2021[4][5], except tiles 6 and 7 which belongs to a production manufactured by Matsushita Electric Works (Japan) in 2000.

For practical reasons, this section will primarily focus on the results obtained for tile 1 and tile 2, which are composed of aerogel with a refractive index of 1.03. Any difference between tiles 1 and 2 and the other tiles will be highlighted.

3.1 Transmittance

Transmittance measurements were performed using a Perkin Elmer Lambda 650 spectrophotometer with an integrating sphere. A deuterium light source covered the spectrum from 250

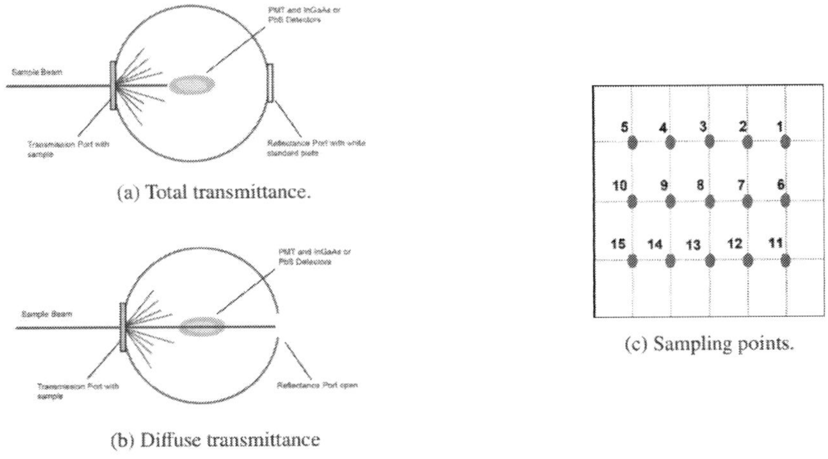

(a) Total transmittance.

(b) Diffuse transmittance

(c) Sampling points.

Figure 1. Schematic of total and diffuse measurement setup (left) and sampling points identification on the tile (right)

to 300 nm, while a halogen source was used for 300 to 800 nm, with measurements taken at 5 nm intervals. The aerogel sample was positioned in the transmission port at the sphere's entrance. As light passes through, it undergoes partial forward scattering, backward scattering, or absorption, with detection performed by a photomultiplier tube (PMT) at the sphere's bottom.

Total transmittance was measured with the reflectance port closed to prevent beam escape, as shown in Fig. 1a. The diffuse contribution was then measured by removing the reflectance port, allowing for the detection of only the diffuse light component, as depicted in Fig. 1b. Finally, the linear transmittance was calculated as the difference between total and diffuse transmittance.

Transmittance was measured at 15 different sampling points on each tile (identified in Fig. 1c) to assess its dependence on thickness and light wavelength (ranging from 250 to 800 nm). The overall transmittance for each tile at a given wavelength was calculated by averaging the values from the sampling points. This average was then fitted using the Hunt formula outlined in Eq. 1. However, small discrepancies were observed in the fit, as shown in Fig. 2. These discrepancies may stem from the assumption of a constant absorption term across the wavelength spectrum. To address this, a new term was introduced to account for absorption, considering the absorption length to vary with λ^8 as proposed in [3], leading to the *extended Hunt formula*:

$$T(\lambda) = e^{-\frac{t}{\Lambda_{abs}}} e^{-\frac{t}{\Lambda_{scat}}} = A e^{-\frac{B \times t}{\lambda^8}} e^{-\frac{C \times t}{\lambda^4}} \tag{3}$$

A comparison between the fit using the standard Hunt formula and its extended version is presented in Fig. 2 for tile 1 with a refractive index of $n = 1.03$. It is evident that the extended Hunt formula provides a significantly better fit to the data compared to Eq. 1.

3.2 Thickness

In Fig. 3, a schematic illustrates the metrologic measurements taken at various points on tile 1, highlighting its concave shape. Indeed, both sides (A and B) of the tile have a curvature

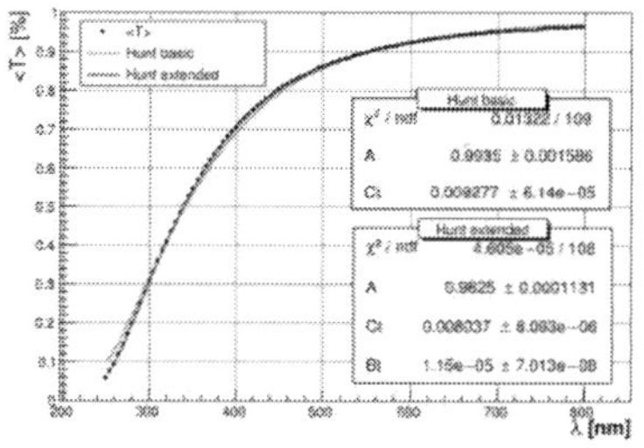

Figure 2. Transmission data of tile 1 (n = 1.03) fitted by both basic Hunt formula (green line) and its extended version (red line) including an additional absorption term.

Figure 3. Concave geometry of tile 1 obtained with metrology measurement (left) and vectors normal to the aerogel tile 1 on the exit side of the tile (right).

and this might have an impact on the resolution of the Cherenkov angle measurement in RICH applications, which can be slightly affected by the geometry and shape of the aerogel tile. The exit side's curvature, visible in Fig. 3, is crucial for estimating the normal vectors to the surface, which can enhance the accuracy of photon direction simulations.

The influence of the sample's shape on transparency is illustrated in Fig. 4, which shows tile 1 alongside the transmittance dispersion at the sampling points. The tile exhibits a thickness variation of approximately 0.5 mm, with greater thickness near the edges. As seen in Fig. 4, thicker areas correspond to lower transmittance, while thinner sections yield higher values. However, the overall variations in transmittance across the tile are minimal, with a spread of less than 1%, indicating a fairly uniform transmittance despite the tile's concave shape.

Considering a concave-shaped configuration, it is expected that point 8 (located at the center of the tile, as shown in Fig. 1c) would exhibit the smallest thickness. Consequently, when fitting the measured transmittance at point 8 using the Hunt extended formula, the thickness was constrained to its minimum value. Subsequently, the fit parameters (A, B, and C) obtained from this process were utilized to determine the thickness at all other sampling points across the entire tile.

3.3 Transmission length

Another way to address the impact of scattering and absorption on transmittance is through the transmission length Λ_{T}. This parameter is determined by both the absorption length Λ_{abs} and the scattering length Λ_{scat}, as illustrated in Eq. 2. In our model, we made the assumption that the scattering length varies with λ^4, and the absorption length is dependent on λ^8.

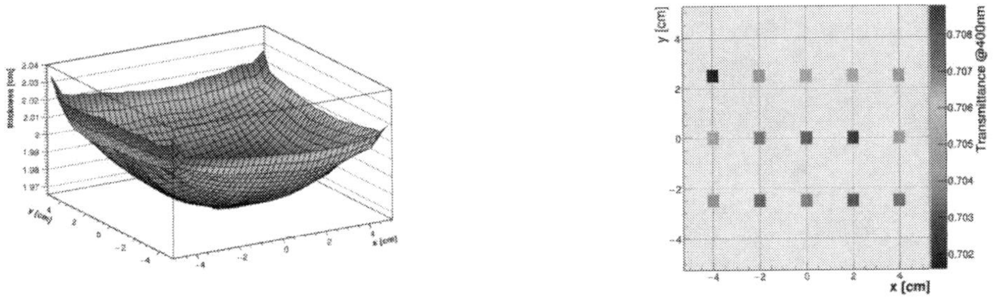

Figure 4. Comparison between tile thickness (left) and transparency dispersion (right). The measurements refer to tile 1 with n=1.03.

The lengths associated with transmission, scattering, and absorption were determined from the transmittance measurements. These parameters were obtained by fitting the transmittance curve as a function of wavelength, as described in Eq. 3:

$$\Lambda_{\mathrm{T}} = -\frac{t}{ln(T(\lambda))}, \qquad \Lambda_{\mathrm{scat}} = \frac{\lambda^4}{C}, \qquad \Lambda_{\mathrm{abs}} = \frac{\lambda^8 t}{Bt - \lambda^8 lnA}$$

In this case, we used the average thickness value obtained from the metrology measurements to estimate the transmission, absorption, and scattering lengths.

3.4 Comparison

While we have focused on presenting the results for tiles 1 and 2, a comprehensive comparison of the results obtained from all the tested tiles can yield valuable insights into how the optical properties of the aerogel vary as a function of the refractive index.

In Fig. 5, we present the outcomes of the transmittance measurements for all the tested tiles, plotted as a function of the wavelength. These results show an overall increase in transmittance, even though distinctions can be found among tiles with different refractive indices.

The transmittance measurements reveal that tiles with a refractive index of $n = 1.03$ (tiles 1-5) generally exhibit higher transmittance values at 400 nm compared to all the other tiles. However, two notable exceptions are observed: tiles 6 and 7, featuring larger transmittance than the other tiles with $n = 1.03$ due to their thickness, and the tiles with a very low refractive index of $n = 1.005$, which exhibit significantly lower transmittance values due to the old manufacturing technology used.

A clear distinction among the tiles is further emphasized in Fig. 5, where the transmission length at 400 nm is plotted as a function of the refractive index. Tiles with a refractive index of $n = 1.03$ exhibit notably larger transmission lengths compared to the other tiles. As already mentioned, tiles with $n = 1.03$ and a thickness of 1 cm (tiles 6-7) show significantly shorter transmission lengths if compared to the ones with same refractive index and 2 cm.

We can also compare the transmission length values obtained through our evaluation with those provided by the manufacturers (nominal values). In particular, our extracted values tend to be slightly lower than the nominal values for nearly all the tiles. This variance could potentially arise from the thickness value employed in our calculations or from uncertainties related to the assumption that the transmittance value is an average taken from the 15 sampling points.

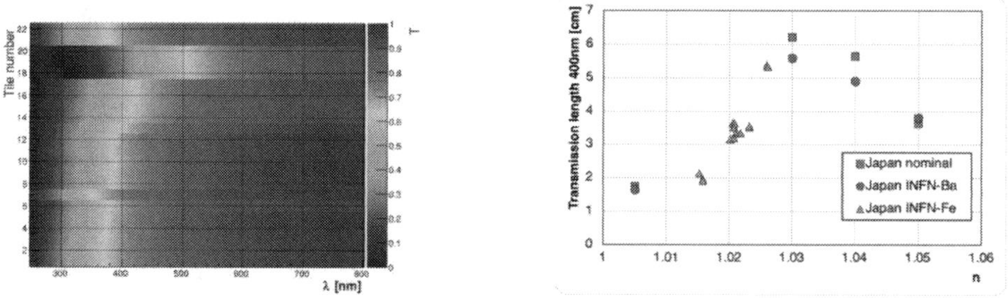

Figure 5. Transmittance as a function of the wavelength for all the tiles (left) and Λ_T at 400 nm as a function of the refractive index.

4 Conclusion

This study investigates the optical properties of several silica aerogel tiles with varying refractive indices. Initially, the transmittance curve was fitted as a function of wavelength, necessitating a new approach for improved accuracy. An additional term was incorporated into the basic Hunt formula to account for photon absorption, significantly enhancing the fit quality. The extended Hunt formula was then used to calculate the transmission, absorption, and scattering lengths for each tile, while the average thickness was determined from multiple measurements across each tile.

The results indicate that aerogel samples with refractive index n=1.03 demonstrates higher transmittance and longer transmission lengths. Indeed, at this refractive index, transmission lengths reach their maximum, while scattering and absorption probabilities are minimized. Notably, transmission lengths exceeded tile thickness significantly, demonstrating aerogel's potential as a radiator in Cherenkov detectors. Additionally, the absorption length was much greater than the scattering length, suggesting that absorption's impact is negligible.

Overall, advancements in high-quality hydrophobic silica aerogel make it an excellent candidate for Cherenkov applications, balancing a low momentum threshold for emission and a high photon yield for improved angular resolution in particle identification.

References

[1] E. Nappi and J. Seguinot. "Ring imaging Cherenkov detectors: The state of the art and perspectives". In: *Riv. Nuovo Cim.* 28.8-9 (2005), pp. 1–130. DOI: 10.1393/ncr/i2006-10004-6

[2] Jyoti Gurav et al. "Silica Aerogel: Synthesis and Applications". In: *Journal of Nanomaterials* 2010 (Aug. 2010). DOI: 10.1155/2010/409310

[3] E. Nappi. "Aerogel and its applications to RICH detectors". In: *Nucl. Phys. B Proc. Suppl.* 61 (1998). Ed. by E. Borchi et al., pp. 270–276. DOI: 10.1016/S0920-5632(97) 00573-2

[4] Ichiro Adachi. "Status of high-quality silica aerogel radiators". In: *NIM-A* 952 (2020), p. 161919. ISSN: 0168-9002. DOI: https://doi.org/10.1016/j.nima.2019.02.046

[5] Makoto Tabata et al. "Development of transparent silica aerogel over a wide range of densities". In: *NIM-A* 623.1 (2010), pp. 339–341. ISSN: 0168-9002. DOI: https://doi.org/10.1016/j.nima.2010.02.241

Characterization of Monolithic Active Pixel Sensors for future collider experiments

Angelo Colelli[1,2,*], *Francesco* Barile[1,2], *Giuseppe Eugenio* Bruno[1,2], *Domenico* Di Bari[1,2], *Antonio* Franco[2], *Shyam* Kumar[2], *Cosimo* Pastore[2], *Rajendra* Nath Patra[2,3], and *Triloki* Triloki[1,2]

[1]Università degli Studi di Bari Aldo Moro, Dipartimento Interateneo di Fisica *"M. Merlin"*, Bari, Italy
[2]INFN, sezione di Bari, Italy
[3]University of Jammu, India

Abstract. In high-energy physics experiments, Monolithic Active Pixel Sensors (MAPS) have become crucial components of vertex and tracking detectors over the past decade due to the integration of readout circuitry with the detection volume in a single chip. The requirement to achieve precise tracking and vertexing capabilities for upgrade of HEP experiments, such as ALICE at LHC and ePIC at EIC, has implied a strong R&D towards an ultra-thin (a few tens of μm), bent, wafer-scale silicon sensors produced with stitching technology. Recent ongoing activities on CMOS silicon sensor testing performed at the INFN Laboratory in Bari will be described. The characterization of analogue silicon pixel sensors of 65 nm CMOS technology using electrical test pulsing and ^{55}Fe as a soft X-ray source will be discussed. Furthermore, a study on timing performance will be presented.

1 Introduction

Monolithic Active Pixel Sensors (MAPS) represent nowadays a promising technology for vertex detectors in high-energy physics, demonstrating capabilities to revolutionize tracking in future experiments [1, 2]. The integration of the detection volume, together with the readout circuitry in a single chip, and the availability of commercial fabricated chips make possible the realization of ultra-thin, large-scale vertex detectors foreseen in the upgrade of the A Large Ion Collider Experiment (ALICE) at the Large Hadron Collider (LHC) at CERN or in the future electron-Proton/Ion Collider (ePIC) detector at the Electron-Ion Collider (EIC) at BNL.

The requirement of a good separation of primary and secondary interaction vertices at high interaction rate [3], demands pixel detectors having enhanced spatial resolution and being placed closer to the interaction point as well as superior timing capabilities. In this context, the 65 nm CMOS Image Sensor Processor (ISP) has been chosen as candidate technology to achieve this goal [4]. In fact, this technology enables large wafer size along with high circuit densities and consequently a more advanced readout circuitry that can be mounted in the same pixel pitch, allowing for precise timing measurements.

*e-mail: angelo.colelli@ba.infn.it

An Analog Pixel Test Structure (APTS) based on Tower Partners Semiconductor Co. (TPSCo) has been developed to assess this technology [4]. In this work, the test results of the APTS equipped with fast individual Operational Amplifier (OA)-based buffering will be presented. In the following section a description of the chip characteristics together with the experimental setup is provided. Afterwards laboratory measurements and ongoing activities performed at the INFN High Technology Laboratory in Bari are presented in a dedicated section.

2 The 65 nm CMOS analogue pixel test structure

The APTS has a die size of 1.5×1.5 mm^2 and consists of a 4×4 squared active pixel matrix. In order to minimise the electric field distortion the matrix is surrounded by a ring of pixels not involved in the readout, effectively occupying an area corresponding to 6×6 pixels.
It has been produced in four different pixel pitches as described in Table 1. Each pixel has been equipped with an OA to maximize the speed capability and thus characterize charge collection time in the sub-nanosecond range. For this aim, an analogue readout of the 4×4 central submatrix has been implemented allowing the acquisition of the full signal waveform.

Table 1. APTS main characteristics.

APTS Chip	
matrix size	4×4 pixels
pixel pitch	10, 15, 20, 25 μm
Output Drivers	Operational Amplifier
Readout	direct analogue
Coupling	DC/AC
Biasing	up to -4.8V (+50V) for DC (AC).
Design Process	Standard
	Modified
	Modified with Gap

2.1 Sensor variants

Three different pixel designs have been produced to optimize the charge collection and reduce the charge sharing. The **standard** design (Fig. 1, left) consists of a n-well collection electrode (1 μm diameter) on a high-resistivity p-type epitaxial layer having a low-resistivity p-type layer as substrate. Applying a reverse bias voltage, it is possible to deplete a volume extending from the small n-well collection electrode to the bottom of the epitaxial layer. Since the depleted volume is balloon-shaped, outside it the charge collection is mostly due to diffusion resulting in long and non uniform collection time together with charge sharing phenomena. The presence of a deep p-well (bright green area in the Fig. 1), placed at about 2 μm from the collection electrode, allows the use of both the PMOSs and NMOSs in the pixel matrix preventing n-wells from collecting charge because they are well separated from the epitaxial layer.
The **modified** design (Fig. 1, middle) is similar to the standard one except for the addition of a low-dose n-type implant in the epitaxial layer. A planar junction deep in the sensor is obtained thus achieving full depletion of the epitaxial layer with some reverse bias, and signal charge collection by drift.
In **modified with gap** design (Fig. 1, right) a 2.5 μm gap at the pixel boundaries has been

introduced. According to the simulations [5] it enhances the lateral electric field that accelerate the charge collection in the depletion volume with a reduction of both collection time and charge sharing.

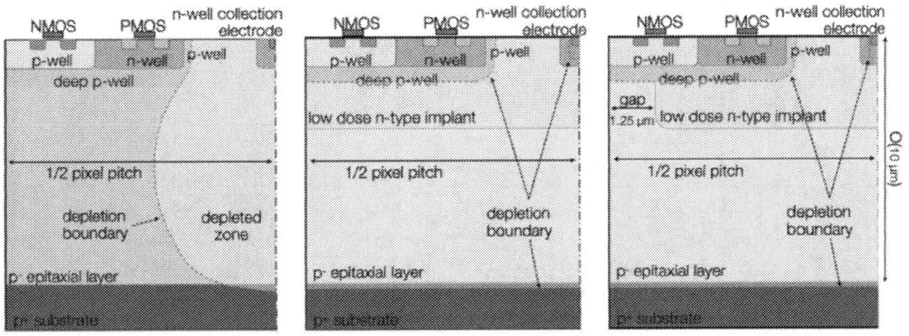

Figure 1. APTS pixel design variants. From left to right the standard, the modified (low dose n-type implant) and the modified with gap (same as the modified with the addition of a gap at the pixel's edges) [6].

2.2 Analog circuit

Two different methods for reverse biasing the sensor have been developed [7]. In the first case (Fig. 2, left), an AC-coupling capacitor is placed between n-well collection electrode and the gate of the input transistor. In this way, the sensing node can be reverse biased by a voltage greater than 5 V (typical diode breakdown voltage in Silicon) and it is reset via a diode while the substrate is directly connected to the ground. In the second case (Fig. 2, middle), a DC-coupling has been developed using a direct connection between the sensing node and the input transistor gate. A PMOS transistor resets the collection electrode to a reference voltage value and a reverse bias voltage is applied both to the substrate and to the bulk of the NMOS transistors inside the pixel matrix. At the periphery of the pixel matrix high speed Operational Amplifiers (OA) buffer the pixel analog outputs. At the end, an output stage (50 Ω resistor) makes the signal readable on board or on oscilloscope. The aim is to exploit the speed capabilities offered by the OA to study the charge collection time in the sub-nanosecond range.

3 The experimental setup

The experimental setup for operating with the APTS-OA (Fig.3) consists of a data acquisition board (DAQ), a proximity board and a carrier board. The DAQ is powered at 5V PC-controlled via USB interface. It hosts the firmware control FPGA together with the connections for trigger, busy and reverse bias. In the proximity board the analog to digital converters (ADCs) provide the external pixels readout as well sending power and biasing the chip. On the carrier board the chip is glued and bonded. Four high bandwidth SMA connectors are directly soldered with the four innermost pixels output lines allowing for a direct read out with the oscilloscope (Rohde & Schwarz RTO 1044 - Sampling Rate 20 GSa/s and 4 GHz). The 12 external pixels are read out by the ADCs (Sampling Rate 4 MHz). Figure 4 represents a typical waveform acquired by the oscilloscope at 20 GSa/s and 50 ns timeframe. The collected data are processed using an offline algorithm that allows for the calculation

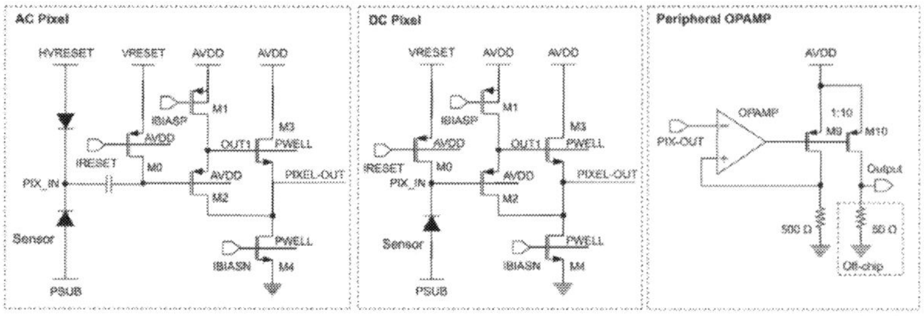

Figure 2. APTS readout chain. Left and middle: front-end circuit with the circuitry inside the AC and DC pixel. The AC coupling (left) has been realized by putting a capacitor between n-well collection electrode (PIX_IN) and the gate of the input transistor (MO); while the DC coupling (middle) by a direct connection between them. Right: Operational Amplifier circuitry located at the periphery of the matrix. The off chip line (dotted rectangle) to the measurement system is represented.

of amplitude and fall-time. The baseline has been calculated as the average of the first 100 points - corresponding to 5 ns - while the underline as the average of 100 points 5 ns after the signal falling down. The amplitude is calculated from the difference between the baseline and the amplitude. The fall-time of the signal is defined as difference between the times at which the signal reduce by 10% and 50% of the amplitude. These values have been extracted by linear interpolation of the signal fall edge.

4 Chip characterization

Laboratory activities have been performed to assess the performances of the APTS-OA AC coupled chip in the modified with gap version (Fig. 1, right) with 10 μm as pixel pitch and it can be divided into two steps. In the first one, the operating point has been optimized for the fastest front-end response. In the second one, the chip matrix has been tested for charge collection studies and timing using a radioactive ^{55}Fe source emitting 5.9 keV (Mn-K$_\alpha$) and 6.5 keV (Mn-K$_\beta$) photons.

In the electronic calibration, the output voltage (baseline) has been measured at different values of the pixel voltage reset (V_{reset}) that sets the collection electrode voltage to a known reference level. In Figure 5 (left) the baseline as a function of V_{reset} for the AC coupled pixel matrix at HV = 3.6 V and $V_{sub} = V_{pwell} = 0$ V bias is shown. The baseline has been calculated as the average of 100 trigger events acquired by the DAQ and the oscilloscope for the outer and the inner pixels, respectively. The V_{reset} has been incremented at step of 10 mV starting from 20 mV up to 890 mV. In the range of 200–600 mV, an increasing trend of the baseline as a function of the V_{reset} is notable. In order to define a V_{reset} range in which the baseline is stable the ratio between the variation of the baseline and the V_{reset} (Δbaseline/ΔV_{reset}) has been calculated. Figure 5 (right) shows that in the range of 200–400 mV, the Δbaseline/ΔV_{reset} reaches almost stable values so representing the best circuitry response. Similar plots have been obtained for increasing values of HV while maintaining the V_{sub} and V_{pwell} fixed at 0 V. Moreover, no improvements or differences have been observed with values of V_{sub} and V_{pwell} different from 0 V.

As the second step, the ^{55}Fe energy spectrum has been acquired for a direct characterization of the chip with a particular investigation of the signal fall time. For this reason, only the four innermost pixels have been read out and used as trigger by the oscilloscope. The charge sharing effect has been investigated by reconstructing the cluster in each event. Figure 6 shows

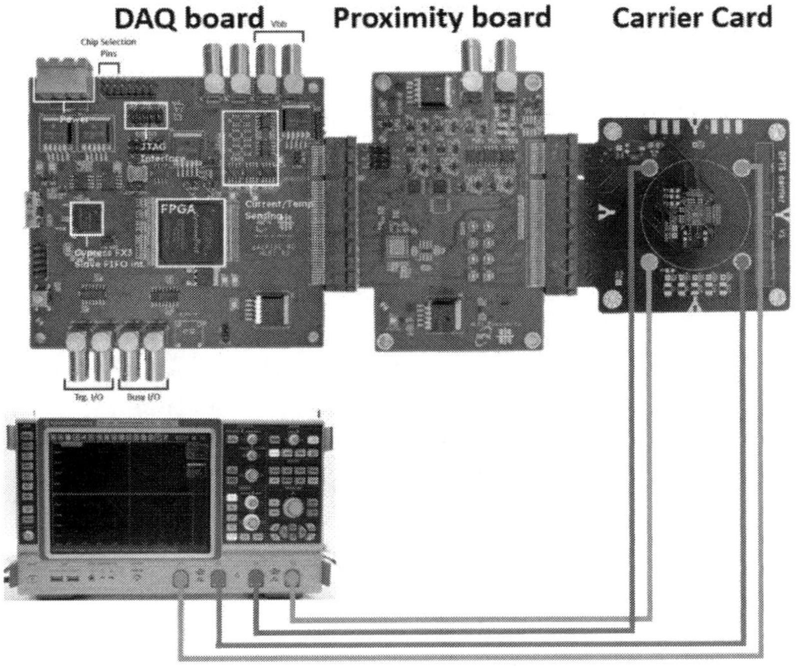

Figure 3. Experimental setup composed by FPGA-based DAQ board (left), to control the test sequence and reading out the pixel chip, the proximity board (middle), to provide power supply and analogue biases to the chip and carrier card (right) that host the chip. In the bottom a sketch of the connection of the four SMA cables to the oscilloscope for a direct measurements of the four innermost pixels of the chip matrix.

the energy spectra at different bias voltages for cluster size 1 events (i.e. events with one fired pixel). Here it is clear that if HV increases the Mn-K_α peaks start to be almost overlapped. This means that the effect of the electric field in the epitaxial layer is so intense that any further increase does not improve the charge collection performance.
Similar analysis has been performed as a function of different cluster size (i.e. 2, 3 and 4). As shown in Figure 7, more than 80% of the events belong to cluster size 1 while cluster size 2 events drop to less than 20%. Cluster size 3 and 4 events are almost negligible. This results could be a possible indication of low charge sharing for the modified with gap version of the chip matrix [3].
The fall time versus the amplitude distribution has been derived, as shown in Figure 8. Two spots corresponding at Mn-K_α and Mn-K_β emission peaks having a fall time value around 110 ps are clearly visible. Moreover, a fraction of events showing larger fall time and low amplitude are also present. A possible explanation could be that the phenomena is due to photons impinging the pixel far from the collection electrode but further investigations and studies are ongoing.

5 Conclusions

The Analogue Pixel Test Structures in the modified with gap version AC coupled and equipped with fast Operational Amplifier have been preliminary tested in the INFN High

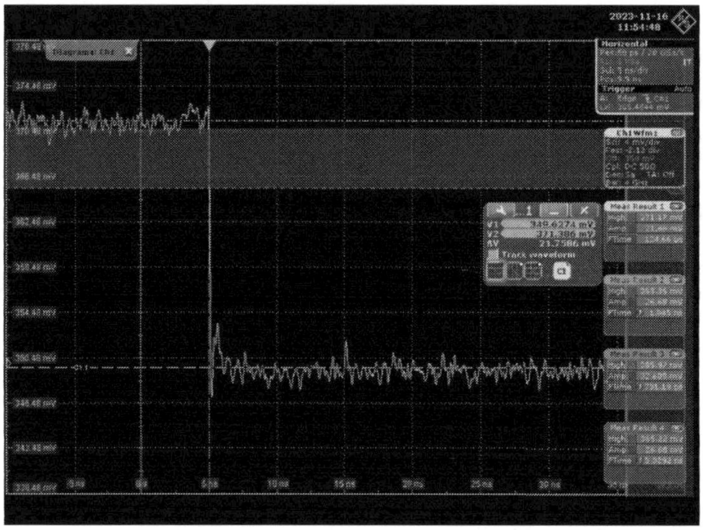

Figure 4. Typical signal acquired by the oscilloscope, Rohde & Schwarz RTO 1044 with a sampling rate of 20 GSa/s, bandwidth of 4 GHz, and timeframe of 50 ns.

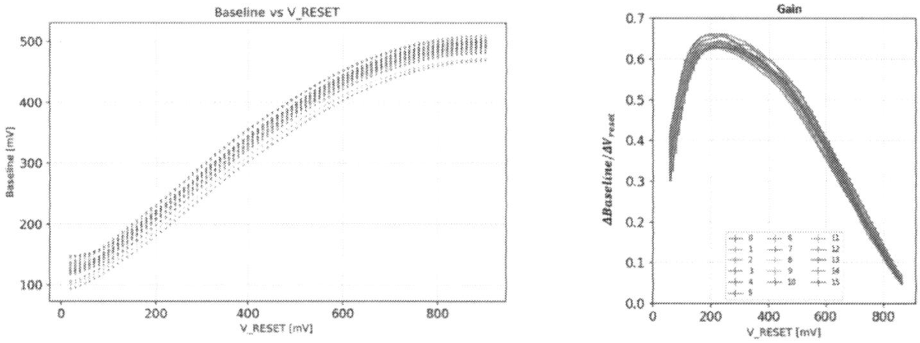

Figure 5. Baseline (left) and ΔBaseline/ΔV_{reset} (right) as a function of V_{reset} for AC coupled pixel matrix of 16 pixels at HV = 3.6 V and V_{sub}=V_{pwell} = 0 V as bias.

Technology Laboratory in Bari to validate the 65 nm CMOS technology as candidate for future collider experiments that require precise tracking and vertexing capabilities. The electronic characterization of the pixel matrix has been made revealing an almost stable working condition in a wide range of the V_{reset} voltage. Moreover, a fall time of about 110 ps has been obtained from a direct pixels characterization using ^{55}Fe as source. These results show good performance in terms of charge collection and time response.

To further investigate about the APTS AC coupled sensor the data of the test beam campaign at CERN SPS will be analyzed. In fact, the aim is to compare these results with to the ones already obtained with the APTS-OA DC coupled for which a time resolution of 63 ps, a detection efficiency exceeding 99% and a spatial resolution of 2 μm have been demonstrated in Ref.[6].

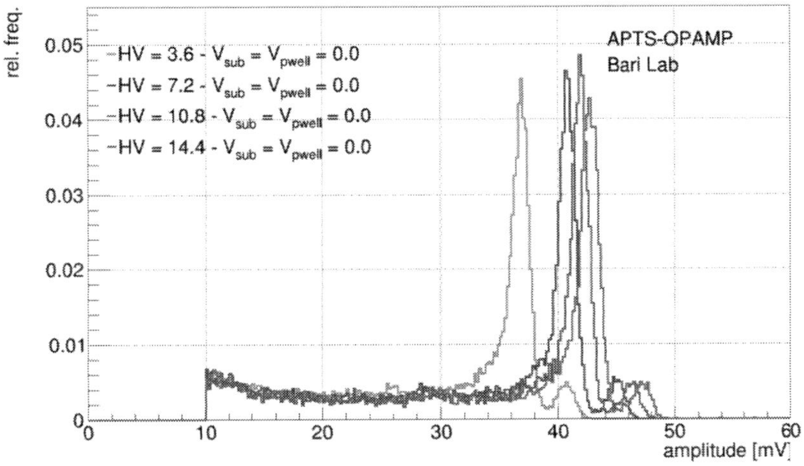

Figure 6. [55]Fe energy spectrum cluster size 1 measured with a modified with gap AC coupled chip at different bias configurations.

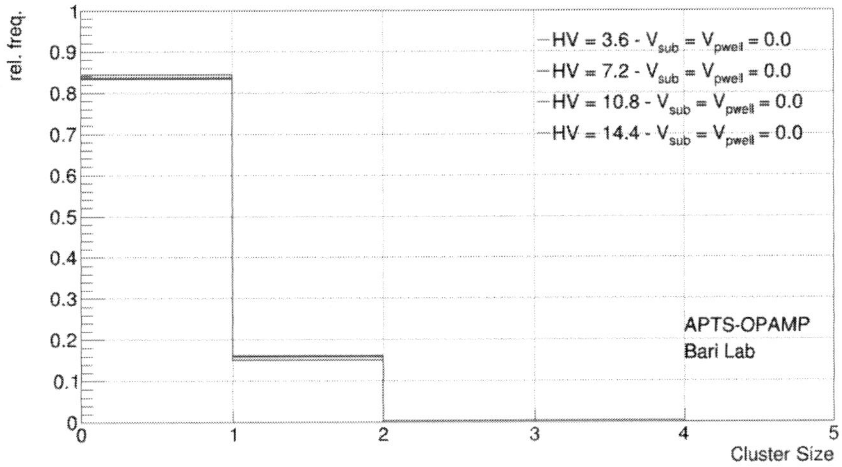

Figure 7. Cluster size distribution for modified with gap AC coupled chip at different bias configurations.

References

[1] F. Reidt, Nuclear Instruments and Methods in Physics Research Section A: Accelerators, Spectrometers, Detectors and Associated Equipment **1032**, 166632 (2022)

[2] M. Mager, Nuclear Instruments and Methods in Physics Research Section A: Accelerators, Spectrometers, Detectors and Associated Equipment **824**, 434 (2016), frontier Detectors for Frontier Physics: Proceedings of the 13th Pisa Meeting on Advanced Detectors

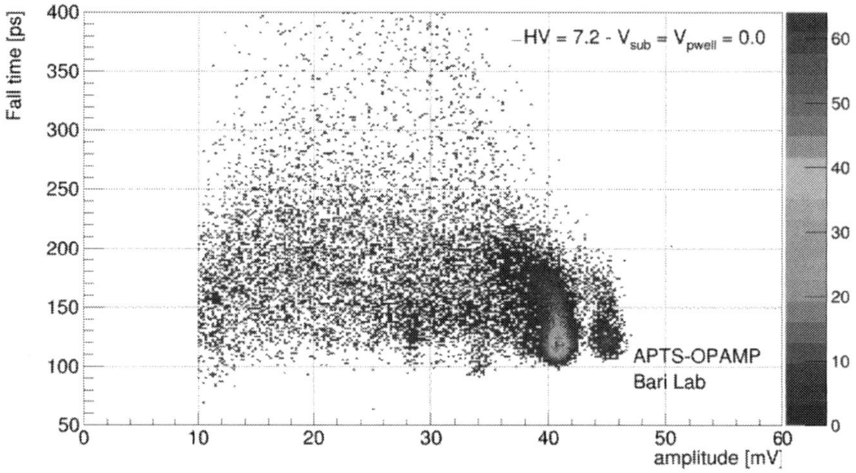

Figure 8. Fall time versus Amplitude distribution (Cluster size 1) for modified with gap AC coupled chip at HV= 7.2 V and V_{sub}=V_{pwell} = 0 V as bias.

[3] O. Aberle, I. Béjar Alonso, O. Brüning, P. Fessia, L. Rossi, L. Tavian, M. Zerlauth, C. Adorisio, A. Adraktas, M. Ady et al., *High-Luminosity Large Hadron Collider (HL-LHC): Technical design report*, CERN Yellow Reports: Monographs (CERN, Geneva, 2020), https://cds.cern.ch/record/2749422

[4] L. Musa, Tech. rep., CERN, Geneva (2019), https://cds.cern.ch/record/2703140

[5] M. Munker, M. Benoit, D. Dannheim, A. Fenigstein, T. Kugathasan, T. Leitner, H. Pernegger, P. Riedler, W. Snoeys, Journal of Instrumentation **14**, C05013–C05013 (2019)

[6] G.A. Rinella, L. Aglietta, M. Antonelli, F. Barile, F. Benotto, S.M. Beolè, E. Botta, G.E. Bruno, F. Carnesecchi, D. Colella et al., *Time performance of analog pixel test structures with in-chip operational amplifier implemented in 65 nm cmos imaging process* (2024), 2407.18528, https://arxiv.org/abs/2407.18528

[7] W. Deng, G.A. Rinella, M. Aresti, J. Baudot, F. Benotto, S. Beole, W. Bialas, G. Borghello, S. Bugiel, M. Campbell et al., Journal of Instrumentation **18**, C01065 (2023)

EPJ Web of Conferences **314**, 00035 (2024) https://doi.org/10.1051/epjconf/202431400035
QCD@Work 2024

Heavy-flavour production as a function of event activity in pp collisions with the ALICE experiment

László Gyulai[1,2,*] for the ALICE Collaboration

[1]Budapest University of Technology and Economics, 3 Műegyetem rkp., H-1111 Budapest, Hungary
[2]HUN-REN Wigner Research Centre for Physics, 29–33 Konkoly–Thege Miklós út, H-1121 Budapest, Hungary

Abstract. Due to their large masses, the production of heavy-flavour quarks can be computed perturbatively, thus providing a powerful tool to test the corresponding QCD calculations. Additionally, measurements of heavy-flavour hadrons are useful to reveal the details of heavy-quark fragmentation in pp collisions at LHC energies. Event-activity-dependent measurements of heavy-flavour production may shed light on the mechanisms of interplay between soft and hard processes. In this contribution, we present recent measurements of the ALICE experiment on charm-hadron production as a function of charged-particle multiplicity in pp collisions at various energies, including the measurements of charm baryon-to-meson production yield ratios. New results of D-meson production as a function of the transverse spherocity of the event, as well as of the transverse event-activity classifier R_T, are also presented.

1 Introduction

Heavy-flavour (c and b) quarks are produced in the initial stages of high-energy hadronic collisions in hard partonic scatterings. Their production and fragmentation significantly differ from light-flavour partons because of colour-charge and mass-related effects [1, 2]. The most common theoretical approach to describe heavy-flavour production is based on the perturbative quantum chromodynamics (QCD) and the factorisation theorem [3]. Measurements at the ALICE and CMS experiments [4, 5] found, however, that charm baryon-to-meson ratios are significantly enhanced compared to measurements in e^+e^- and ep collisions, and underpredicted by the models based on fragmentation functions estimated in those collision systems, hence questioning the universality of fragmentation.

Recently, it was observed that pp collisions with large final-state multiplicity exhibit similar features to proton–nucleus (pA) and even nucleus–nucleus (AA) collisions, such as collective dynamics and strangeness enhancement [6, 7]. In the case of AA collisions, the observed phenomena are usually related to the presence of the quark–gluon plasma (QGP) – state of matter in which quarks and gluons can be found in a deconfined state. However, the QGP is not expected to be created in pp collisions. Hence, theoretical works try to explain the observed similarities between small and large collision systems with vacuum quantum chromodynamics (QCD) processes at the soft–hard boundary, such as mini-jet production [8] or multiple-parton interactions (MPI) [9, 10]. The study of heavy-flavour hadron production as

*e-mail: gyulai.laszlo@wigner.hun-ren.hu

a function of event-activity and event-shape observables allows us to investigate the interplay between hard and soft particle production and the role of MPI in heavy-quark production processes.

2 Results

2.1 Charm-hadron yield ratios as a function of multiplicity

The ALICE Collaboration recently reported a measurement of the $p_{\rm T}$-differential cross sections of prompt Λ_c^+ baryons to D^0 mesons at midrapidity ($|y| < 0.5$) in pp collisions at $\sqrt{s} = 13$ TeV. Data was reported down to $p_{\rm T} = 0$ GeV/c and the corresponding ratio of both spectra was evaluated [11]. The results were compared with various model calculations: different tunes of PYTHIA 8 [12, 13], the Catania [14] and QCM [15] models implementing quark recombination, the SHM+RQM model [16] based on statistical hadronisation accounting for additional excited charm-baryon states yet unobserved, and the POWLANG model [17] which assumes the formation of a small, deconfined, and expanding fireball even in pp collisions.

As shown in Fig. 1, the data suggests a significant increase of Λ_c^+/D^0 ratio at low and intermediate transverse momenta. This behaviour is different from the corresponding ratios measured in electron–positron collisions at the LEP, where the average ratio over the whole $p_{\rm T}$ range is $0.113 \pm 0.013 \pm 0.006$ [18]. PYTHIA 8 with the Monash tune fails to describe

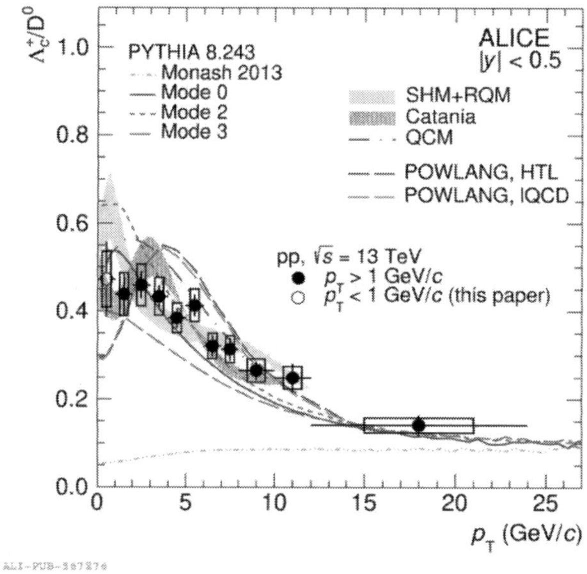

Figure 1. Ratio between the $p_{\rm T}$-differential cross sections at midrapidity ($|y| < 0.5$) of prompt Λ_c^+ baryons and D^0 mesons in pp collisions at $\sqrt{s} = 13$ TeV [11] compared with the predictions from PYTHIA Monash tune [12], PYTHIA CR-BLC Mode 0, 2 and 3 [13], Catania [14], QCM [15], SHM+RQM [16], and POWLANG [17] models.

the observed trends, as this tune is based on the measurements performed in e^+e^- and e^-p collisions. By adding colour junctions to the colour-reconnection implementation in PYTHIA as shown by the CR-BLC Mode 0, 2 and 3, the trend of ALICE data is reproduced. The other considered models also describe the measurement within uncertainties in the full p_T range, but the current precision of the measurement does not allow for a precise discrimination among the models.

The ALICE experiment also measured the Λ_c^+/D^0 ratio as a function of the event multiplicity [19]. The bottom panels of Fig. 2 show the results for the lowest and the highest multiplicity classes compared to several models. The data shows a significant (5.3σ) enhancement

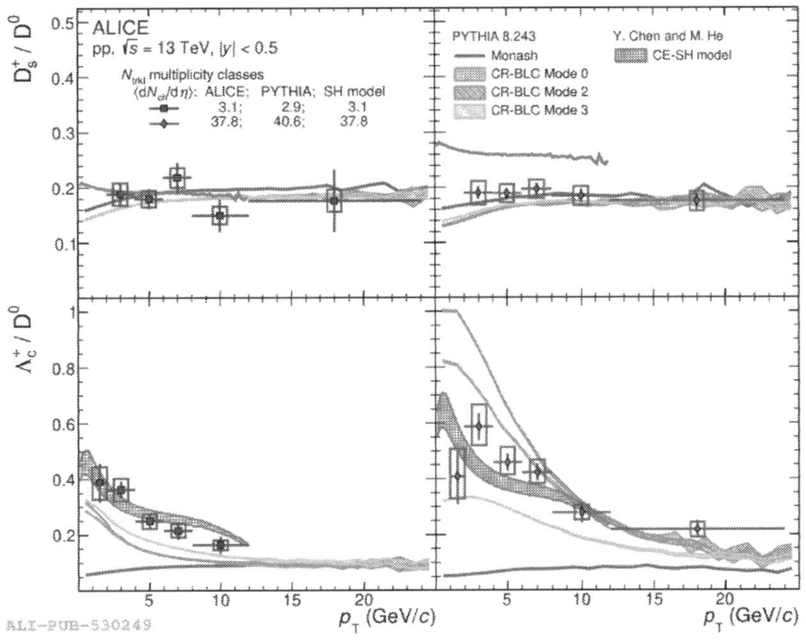

Figure 2. The D_s^+/D^0 (top) and Λ_c^+/D^0 (bottom) ratios measured in pp collisions with $\sqrt{s}$ = 13 TeV centre-of-mass energy at midrapidity for the lowest (left) and highest (right) event multiplicity classes [19]. The measurements are compared to PYTHIA predictions with the Monash tune [12] and the CR-BLC tunes [13], and the CE-SH model [20], evaluated in the similar multiplicity classes.

from low- to high-multiplicity events in the $1 < p_T < 12$ GeV/c interval. The PYTHIA Monash tune fails to reproduce the p_T dependence of Λ_c^+/D^0 ratio and does not show any dependence on the multiplicity. On the other hand, the CR-BLC tunes describe the Λ_c^+/D^0 decreasing trend with p_T, are overall closer in magnitude, as well as show a clear dependence on multiplicity, qualitatively reproducing the trend observed in data. The measurements are also compared with the predictions of a canonical-ensemble statistical hadronisation (CE-SH) model [20], which is a generalisation of the grand-canonical statistical hadronisation model. The CE-SH model calculations describe well the Λ_c^+/D^0 ratios, reproducing also the multiplicity dependence which is connected to the strict baryon-number conservation in this model.

The ratio between strange and non-strange mesons (D_s^+ and D^0 respectively) as a function of multiplicity was also studied by the ALICE experiment. There is no enhancement observed in the data for neither multiplicity interval, and the D_s^+/D^0 ratio is independent of p_T and multiplicity. This behaviour is well described by all the considered PYTHIA tunes. The CE-SH model, while reproducing the p_T-trend of the results, predicts an unobserved increase in the D_s^+/D^0 ratio towards high multiplicities due to the strangeness-number conservation.

2.2 D-meson self normalised yields as a function of spherocity

Event-shape observables are connected to how the particle transverse momentum distributes in a collision, and help to study the perturbative and non-perturbative aspects of QCD. Event shape observables help in isolating the more jetty-like (containing high-p_T jets) or isotropic (with parton scattering with low Q^2) events. A recent measurement of the ALICE Collaboration utilised the transverse spherocity S_0 [21] to study the production of D-meson yields. Low values of S_0 represent hard, jetty events, while $S_0 \approx 1$ corresponds to isotropic events dominated by soft QCD processes.

Figure 3 presents the average D-meson (including D^0, D^+, and D^{*+} mesons) self normalised yields in different p_T intervals, at low and high multiplicity as a function of transverse spherocity. Focusing on the high-pT part, the data suggest an increase in the self normalised

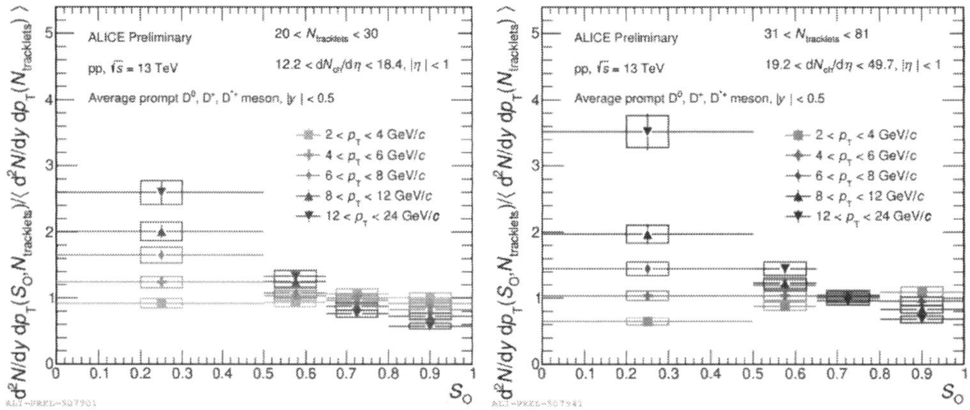

Figure 3. Average D-meson self normalised yields at midrapidity as a function of spherocity S_0 in pp collisions at $\sqrt{s} = 13$ TeV in different p_T intervals. The yields are presented for tracklet-multiplicity range 20–30 (left) and 31–81 (right).

yields in the jetty (low-S_0) events towards high multiplicity. For the isotropic events, no significant dependence on multiplicity is observed. The results demonstrate that at higher transverse momentum the D-meson production is shifted towards jetty events, while at low-p_T the production of D mesons is more isotropic.

2.3 D^0 yields as a function of transverse event activity

Another event-activity observable recently studied at the ALICE experiment is the transverse-event activity classifier R_T [22]. This observable considers the multiplicity only in regions which are distant in azimuthal angle from the direction of hard-scattering products. This allows for the investigation of soft processes, which are mostly independent of hard scattering and are related to the production of particles in the underlying event.

The ALICE Collaboration studied the production of D^0 mesons as a function of transverse event activity in pp collisions at $\sqrt{s} = 13$ TeV [23]. The results reported in Fig. 4 are presented as a ratio between the yields in low-activity ($R_T < 1$) and high-activity ($R_T > 1$) events with respect to all the events ($R_T > 0$). The left panel shows the D-meson yields in an azimuthal region close to the leading track trigger with $p_T > 5$ GeV/c ($|\Delta\varphi| < \pi/3$). In this region, a dependence on R_T is suggested by the data in the lowest p_T interval. Towards higher p_T values, the results are consistent with unity, as these D^0 mesons are produced in

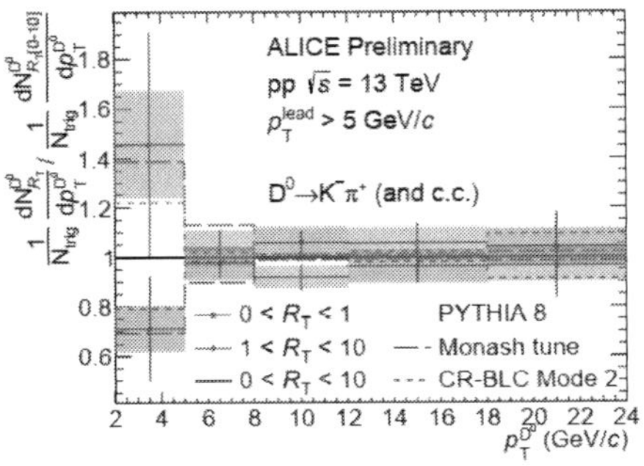

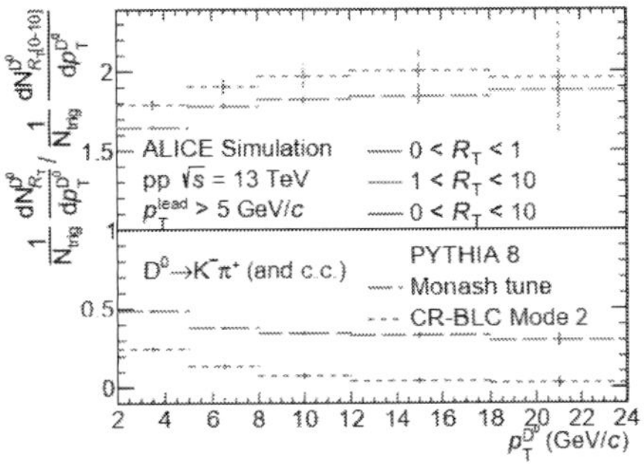

Figure 4. Left: Ratio of the R_T-dependent per-event D^0 meson yields to the R_T-integrated yield in the azimuthal region centered around a leading $p_T > 5$ GeV/c track trigger, as a function of p_T, in pp collisions at $\sqrt{s} = 13$ TeV. The results are compared to model calculations from the PYTHIA Monash [12] and CR-BLC Mode 2 [13] tunes. Right: Ratio of the R_T-dependent per-event D^0 meson yields to the R_T-integrated yield in the azimuthal region distant from the leading-track trigger, as a function of p_T in pp collisions at $\sqrt{s} = 13$ TeV simulated by PYTHIA 8 with the Monash and CR-BLC Mode 2 tunes.

hard scatterings. The results were also compared to PYTHIA with the Monash tune [12] and CR-BLC Mode 2 [13]. The simulations reproduce the measurement within uncertainties.

Due to the limited size of the dataset collected during LHC Run 2, the measurement of D^0 yields in the region azimuthally distant from the leading trigger track ($\pi/3 < |\Delta\varphi| < 2\pi/3$) was not possible. PYTHIA, however, predicts an R_T dependence of D^0 production over the whole p_T range (see right panel of Fig. 4). This behavior signals that the processes in this region could be indeed connected to the underlying event. The data collected during the LHC Run 3 data taking period might make the measurement in this region possible.

3 Summary

Measurements of charm-hadron production as a function of event activity and event-shape observables can help distinguish between particles produced in hard scatterings and in softer processes. This allows us to investigate the interplay between these processes and particle production in high-energy collisions. Additionally, by measuring the strange-to-nonstrange meson production yield ratios (like D_s^+/D^0) and baryon-to-meson production yield ratios (Λ_c^+/D^0), one can study the hadronisation mechanisms in the charm sector and the effects of strangeness enhancement. The size of the datasets collected during the LHC Run 2 allows for testing different QCD models, however, it usually does not allow us to favour one model over the others due to the statistical limitations of the current LHC Run 2 data. The ongoing LHC Run 3 data taking period (from which large datasets have already been collected and are available) is expected to provide much more precise measurements, which will further constrain the QCD models and will allow for more differential evaluation of the event-activity dependence of heavy-flavour production and fragmentation.

This work has been supported by the NKFIH grant OTKA FK131979 and K135515, as well as by the 2021-4.1.2-NEMZ_KI-2024-00034 projects.

References

[1] A. Andronic et al., Heavy-flavour and quarkonium production in the LHC era: from proton–proton to heavy-ion collisions, Eur. Phys. J. C **76**, 107 (2016), 1506.03981. 10.1140/epjc/s10052-015-3819-5

[2] S. Acharya et al. (ALICE), Direct observation of the dead-cone effect in quantum chromodynamics, Nature **605**, 440 (2022), [Erratum: Nature 607, E22 (2022)], 2106.05713. 10.1038/s41586-022-04572-w

[3] J.C. Collins, D.E. Soper, G.F. Sterman, Factorization of Hard Processes in QCD, Adv. Ser. Direct. High Energy Phys. **5**, 1 (1989), hep-ph/0409313. 10.1142/9789814503266_0001

[4] S. Acharya et al. (ALICE), Measurement of Prompt D^0, Λ_c^+, and $\Sigma_c^{0,++}(2455)$ Production in Proton–Proton Collisions at $\sqrt{s}$ = 13 TeV, Phys. Rev. Lett. **128**, 012001 (2022), 2106.08278. 10.1103/PhysRevLett.128.012001

[5] A.M. Sirunyan et al. (CMS), Production of Λ_c^+ baryons in proton–proton and lead–lead collisions at $\sqrt{s_{NN}}$ = 5.02 TeV, Phys. Lett. B **803**, 135328 (2020), 1906.03322. 10.1016/j.physletb.2020.135328

[6] J. Adam et al. (ALICE), Enhanced production of multi-strange hadrons in high-multiplicity proton–proton collisions, Nature Phys. **13**, 535 (2017), 1606.07424. 10.1038/nphys4111

[7] S. Acharya et al. (ALICE), Investigations of Anisotropic Flow Using Multiparticle Azimuthal Correlations in pp, p–Pb, Xe–Xe, and Pb–Pb Collisions at the LHC, Phys. Rev. Lett. **123**, 142301 (2019), 1903.01790. 10.1103/PhysRevLett.123.142301

[8] A. Leonidov, D. Ostrovsky, Angular pattern of minijet transverse energy flow in hadron and nuclear collisions, Eur. Phys. J. C **16**, 683 (2000), hep-ph/9812416. 10.1134/1.1481483

[9] C. Bierlich, Microscopic collectivity: The ridge and strangeness enhancement from string–string interactions, Nucl. Phys. A **982**, 499 (2019), 1807.05271. 10.1016/j.nuclphysa.2018.07.015

[10] A. Ortiz, G. Bencédi, H. Bello, Revealing the source of the radial flow patterns in proton–proton collisions using hard probes, J. Phys. G **44**, 065001 (2017), 1608.04784. 10.1088/1361-6471/aa6594

[11] S. Acharya et al. (ALICE), Charm production and fragmentation fractions at midrapidity in pp collisions at $\sqrt{s}$ = 13 TeV, JHEP **12**, 086 (2023), 2308.04877. 10.1007/JHEP12(2023)086

[12] P. Skands, S. Carrazza, J. Rojo, Tuning PYTHIA 8.1: the Monash 2013 Tune, Eur. Phys. J. C **74**, 3024 (2014), 1404.5630. 10.1140/epjc/s10052-014-3024-y

[13] J.R. Christiansen, P.Z. Skands, String Formation Beyond Leading Colour, JHEP **08**, 003 (2015), 1505.01681. 10.1007/JHEP08(2015)003

[14] V. Minissale, S. Plumari, V. Greco, Charm hadrons in pp collisions at LHC energy within a coalescence plus fragmentation approach, Phys. Lett. B **821**, 136622 (2021), 2012.12001. 10.1016/j.physletb.2021.136622

[15] J. Song, H.h. Li, F.l. Shao, New feature of low p_T charm quark hadronization in pp collisions at $\sqrt{s}$ = 7 TeV, Eur. Phys. J. C **78**, 344 (2018), 1801.09402. 10.1140/epjc/s10052-018-5817-x

[16] M. He, R. Rapp, Charm-Baryon Production in Proton–Proton Collisions, Phys. Lett. B **795**, 117 (2019), 1902.08889. 10.1016/j.physletb.2019.06.004

[17] A. Beraudo, A. De Pace, D. Pablos, F. Prino, M. Monteno, M. Nardi, Heavy-flavor transport and hadronization in pp collisions, Phys. Rev. D **109**, L011501 (2024),

2306.02152. 10.1103/PhysRevD.109.L011501

[18] L. Gladilin, Fragmentation fractions of c and b quarks into charmed hadrons at LEP, Eur. Phys. J. C **75**, 19 (2015), 1404.3888. 10.1140/epjc/s10052-014-3250-3

[19] S. Acharya et al. (ALICE), Observation of a multiplicity dependence in the p_{T}-differential charm baryon-to-meson ratios in proton–proton collisions at $\sqrt{s}$ = 13 TeV, Phys. Lett. B **829**, 137065 (2022), 2111.11948. 10.1016/j.physletb.2022.137065

[20] Y. Chen, M. He, Charged-particle multiplicity dependence of charm-baryon-to-meson ratio in high-energy proton–proton collisions, Phys. Lett. B **815**, 136144 (2021), 2011.14328. 10.1016/j.physletb.2021.136144

[21] A. Ortiz, G. Paić, E. Cuautle, Mid-rapidity charged hadron transverse spherocity in pp collisions simulated with Pythia, Nucl. Phys. A **941**, 78 (2015), 1503.03129. 10.1016/j.nuclphysa.2015.05.010

[22] T. Martin, P. Skands, S. Farrington, Probing Collective Effects in Hadronisation with the Extremes of the Underlying Event, Eur. Phys. J. C **76**, 299 (2016), 1603.05298. 10.1140/epjc/s10052-016-4135-4

[23] Physics Preliminary Summary: D^0 meson yields as a function of the underlying event activity in pp collisions at $\sqrt{s}$ = 13 TeV, ALICE-PUBLIC-2024-001 (2024)

EPJ Web of Conferences **314**, 00036 (2024) https://doi.org/10.1051/epjconf/202431400036
QCD@Work 2024

Physics programme of ALICE 3: A next-generation heavy-ion detector for LHC Run 5 and beyond

Nicola Nicassio[1,*], on behalf of the ALICE Collaboration

[1]Dipartimento Interateneo di Fisica 'M. Merlin' and Sezione INFN, Via Orabona 4, Bari, Italy

Abstract. The ALICE Collaboration is designing a completely new apparatus, ALICE 3, for the LHC Runs 5 and 6. The detector comprises a large pixel-based tracking system covering eight units of pseudorapidity, complemented by various particle identification systems including silicon time-of-flight layers, a ring-imaging Cherenkov detector, a muon identification system, an electromagnetic calorimeter and a forward conversion tracker. ALICE 3 will, on the one hand, enable novel studies of the quark-gluon plasma (QGP) and, on the other hand, open up important physics opportunities in other areas of QCD and beyond. The main new studies in the QGP sector focus on low-p_{T} heavy-flavour production, including beauty hadrons, multi-charm baryons and charm-charm correlations, as well as on precise multi-differential measurements of dielectrons to probe the mechanism of chiral-symmetry restoration and the time-evolution of the QGP temperature. Besides QGP studies, ALICE 3 can uniquely contribute to hadronic physics, with femtoscopic studies of the interaction potentials between charm mesons and searches for nuclei with charm, and to fundamental physics, with tests of the Low theorem for ultra-soft photon emission. This contribution will cover the latest detector concept and the state-of-the-art projections for the resulting physics performance in dielectron and heavy-flavour measurements.

1 Introduction

ALICE (A Large Ion Collider Experiment) is a general-purpose experiment at the CERN LHC designed to study the microscopic dynamics of the strongly interacting matter produced in heavy-ion collisions and, in particular, the quark-gluon plasma (QGP), a state of matter where quarks and gluons are deconfined. The LHC Runs 1 and 2 led to significant advances in our understanding of the QCD phase diagram and of the QGP [1]. With the current major upgrade to ALICE 2 and the future upgrade to ALICE 2.1, which will see an upgrade of the inner silicon tracker layers (ITS3) and a new system of forward electromagnetic and hadronic calorimeters (FoCal), LHC Runs 3 and 4 will enable further systematic measurements of fundamental properties for our understanding of the QGP [2]. However, despite the rich planned physics programme, fundamental questions will remain open, such as the QGP properties driving its constituents to equilibration, the hadronization mechanisms of the medium, the partonic equation of state and its temperature dependence, and the underlying dynamics of chiral symmetry restoration. Addressing such questions requires substantial improvements in terms of detector performance and rate capabilities, calling for a next-generation experiment, ALICE 3, planned for LHC Run 5 and beyond [3].

*e-mail: Nicola.Nicassio@ba.infn.it

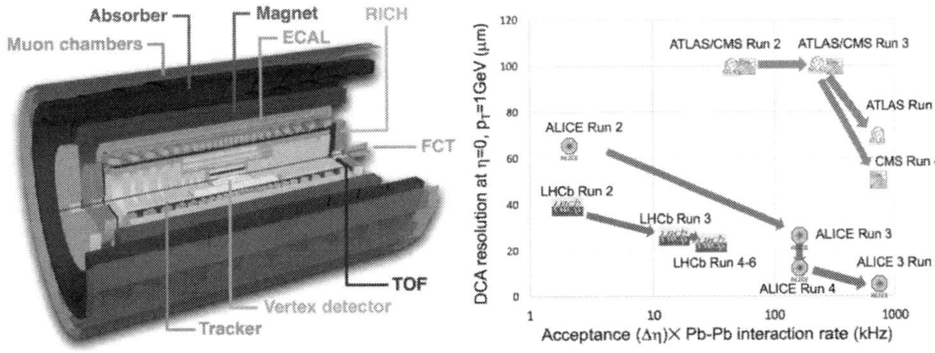

Figure 1. Left: ALICE 3 state-of-the-art detector concept. Right: Improvement of ALICE pointing resolution and expected effective statistics in Pb-Pb collisions from Run 2 to Run 6 compared with LHCb, CMS and ATLAS current projections.

2 ALICE 3 physics goals

The ALICE 3 primary goal is the understanding of the mechanisms ruling the different stages of heavy-ion collisions, from the medium formation to the heavy-flavour diffusion, thermalization and hadronization in the QGP. The ALICE 3 design is optimized for measurements of low transverse momentum (p_T) heavy-flavour production, including charm-charm correlations, multi-charm baryon and beauty hadron production, as well as measurements of net-quantum number fluctuations to constrain the susceptibilities of the QGP and dileptons to probe the time-evolution of the QGP and the mechanism of chiral symmetry restoration in the medium. Besides QGP studies, ALICE 3 can uniquely contribute to hadronic physics, with femtoscopic studies of the interaction potentials between charm mesons and searches for nuclei with charm. ALICE 3 can also contribute to fundamental physics, with precision experimental tests of Low's theorem for ultra-soft photon emission and searches for axion-like particles in ultra-peripheral heavy-ion collisions via light-by-light scattering measurements.

3 ALICE 3 detector concept

All the measurements outlined in Sect. 2 set requirements on vertexing, tracking, particle identification and rate capabilities. The combination of these requirements led to the state-of-the-art detector concept sketched in the left panel of Fig. 1. The key requirements are a retractable vertex detector featuring an unprecedented pointing resolution, a compact and lightweight all-silicon tracker embedded in a superconducting solenoidal magnet, efficient identification of γ, $e^{\pm}$, $\mu^{\pm}$, $\pi^{\pm}$, $K^{\pm}$, p and $\bar{p}$, acceptance over eight units of pseudorapidity ($|\eta| < 4$) and forward subsystems for event characterisation as well as continuous read-out combined with online data processing. The uniqueness of ALICE 3 in terms of pointing resolution and effective statistics is demonstrated in the right panel of Fig. 1, which shows the corresponding expected improvement by order of magnitudes from ALICE 2 (Run 2) to ALICE 3 (Runs 5 and 6) compared with LHCb, CMS and ATLAS current projections.

The heart of ALICE 3 is a retractable vertex detector made of three layers in the barrel and three disks at each end-cap based on curved wafer-scaled ultra-thin silicon Monolithic Active Pixel Sensors (MAPS) sensors installed in secondary vacuum inside the beampipe with the inner layer placed at 5 mm distance from the beam line after beam stabilisation. The resulting pointing resolution for pions at $\eta = 0$ is shown as a function of p_T in the left panel of Fig. 2. A pointing resolution better than 10 μm is ensured for p_T larger than 200 MeV.

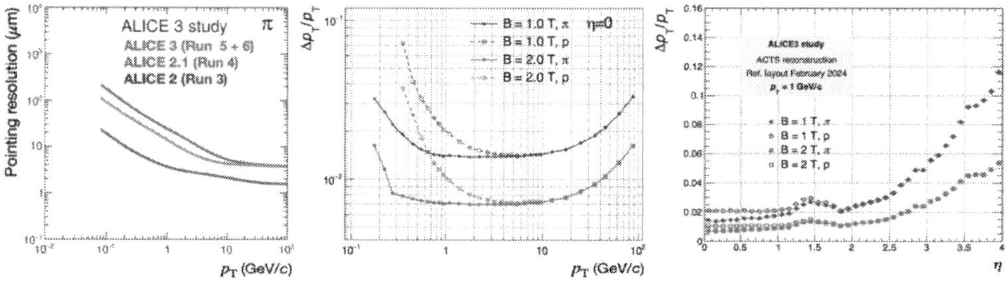

Figure 2. Left: Pointing resolution for pions as a function of p_T. Middle and right: Relative p_T resolution for pions and protons as a function of both p_T and η assuming $B = 1$ T and $B = 2$ T.

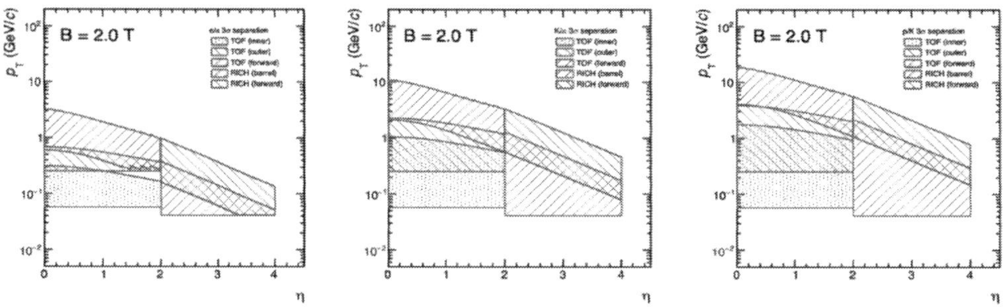

Figure 3. Expected η-p_T regions with larger than 3σ separation using ALICE 3 TOF and RICH systems for $B = 2$ T. Electron/pion, pion/kaon and kaon proton sepatation plots are shown from left to right.

The ALICE 3 tracking system consists of eight cylindrical layers and nine forward disks at each end-cap equipped with MAPS installed in a volume of 7 m length and 80 cm radius. The middle and right panels of Fig. 2 show the resulting relative p_T resolution for pions and protons as a function of p_T at $\eta = 0$ and as a function of η at $p_T = 1$ GeV/c. The curves obtained assuming a magnetic field of $B = 1$ T and $B = 2$ T are compared, with $B = 2$ T being the baseline. A better than 2% relative p_T resolution is ensured over a wide η range.

The particle identification system consists of Time-of-Flight (TOF) and Ring-Imaging Cherenkov (RICH) subsystems, an electromagnetic calorimeter (ECAL), a Muon Identification system (MID) and a Forward Conversion Tracker (FCT). With a time resolution of 20 ps, an inner TOF layer at 19 cm radius (iTOF), an outer layer at 85 cm radius (oTOF) and forward disks at ±3.70 m distance from the interaction point (fTOF), better than 3σ e/π, π/K and K/p separation is achieved up to 0.7, 2 and 4 GeV/c momentum, respectively. Low Gain Avalanche Diodes (LGADs), both hybrid and monolithic in CMOS technology, and Silicon Photomultipliers (SiPMs) are under study as viable technology options for the TOF sensors. The oTOF and fTOF are followed by a projective barrel RICH layer (bRICH) and forward RICH disks (fRICH). With aerogel as Cherenkov radiator and a SiPM-based photodetector in a proximity-focusing configuration, the RICH extends the 3σ e/π, π/K and K/p separation up to 3, 10 and 17 GeV/c momentum, respectively. The resulting TOF and RICH particle identification coverage in the η-p_T plane is shown in Fig. 3. The ECAL, MID and the FCT are designed to provide identification of high-energy photons and electrons up to hundreds of GeV for $-4 < \eta < 1.5$, muons down to $p_T \approx 1.5$ GeV/c for $|\eta| < 1.3$ and soft photons with p_T down to 1 MeV/c for $4 < \eta < 5$, respectively. The ECAL is based on Pb-scintillator sampling technology and includes a high-energy resolution segment based on $PbWO_4$ crystals at $\eta = 0$ with SiPM readout. The MID consists of a steel hadron absorber followed by dedicated muon chambers made of scintillator bars readout using SiPMs. Alternative options based on Multi-Wire Proportional Chambers (MWPCs) and Resistive Plate Chambers (RPCs) are also under study. Finally, the FCT consists of forward disks equipped with MAPS.

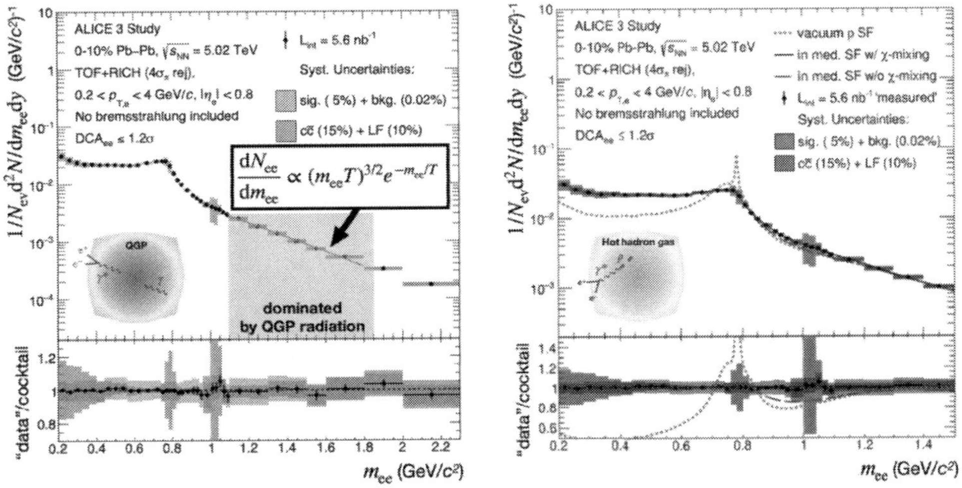

Figure 4. Simulated background-subtracted dielectron invariant mass spectrum in central (0-10%) Pb-Pb collisions at $\sqrt{s_{NN}}$ = 5.02 TeV. Left: Fit of the spectrum with an exponential model in the m_{ee} range 1.1-1.8 GeV/c^2 to extract the early-stage QGP temperature. Right: Comparison of the spectrum obtained using the ρ spectral function in medium under the hypothesis of ρ-a_1 mixing with expectations for the ρ spectral function in vacuum and in medium without chiral mixing.

4 ALICE 3 physics performance

In this section, a selection of simulations on the expected ALICE 3 physics performance is reported highlighting the impact of each subsystem specifications on the physics programme. The projections take into account that, throughout LHC Runs 5 and 6, ALICE 3 will collect integrated luminosities of 18 fb^{-1} and 35 nb^{-1} for pp and Pb-Pb collisions, respectively.

4.1 Dielectrons

Dileptons are produced in all the different stages of a heavy-ion collision. Since they are not interacting via the strong force, they leave the collision unscattered after production, making them ideal messengers to study the whole evolution of the collision [4]. Fig. 4 shows the the ALICE 3 projection for the expected background-subtracted thermal dielectron invariant mass (m_{ee}) spectrum for central Pb-Pb collisions at center-of-mass energy per nucleon pair $\sqrt{s_{NN}}$ = 5.02 TeV. In the region above 1 GeV/c^2, the slope of the spectrum provides direct information on the QGP temperature, as shown in the left panel of Fig. 4. With the very large expected statistics, ALICE 3 will also open up the possibility of probing the time dependence of QGP temperature using double differential dilepton spectra of m_{ee} and dielectron transverse momentum $p_{T,ee}$. For m_{ee} < 1.2 GeV/c^2, the shape of the spectrum is dominated by thermal production of ρ mesons. Their spectral function (SF) is modified in the hot medium where chiral symmetry restoration is predicted. The expected high-precision measurement of the m_{ee} spectrum in central Pb–Pb collision with ALICE 3 would enable the study of chiral symmetry restoration mechanisms like ρ-a_1 mixing [5], as shown in the right panel of Fig. 4. The key ALICE 3 ingredients to achieve the desired statistical and systematic precision are the excellent $e^{\pm}$ identification down to low p_T achieved with TOF and RICH, the small detector material budget to reduce the combinatorial background from γ conversion or π^0 Dalitz decays and the excellent pointing resolution to efficiently suppress the dominant background from heavy-flavour hadron decays.

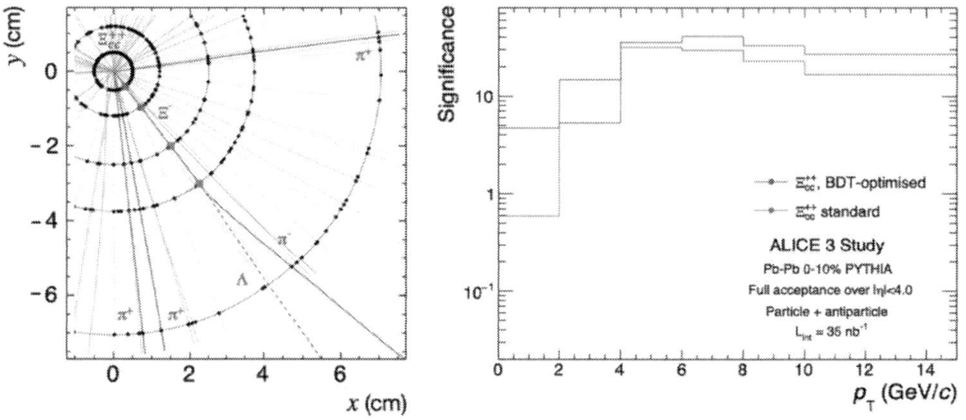

Figure 5. Left: ALICE 3 strangeness tracking in $\Xi_{cc}^{++} \rightarrow \Xi_c^+ + \pi^+$ with $\Xi_c^+ \rightarrow \Xi^- + 2\pi^+$. Right: Simulated Ξ_{cc}^{++} significance in central (0-10%) Pb-Pb collisions at $\sqrt{s_{NN}}$ = 5.52 TeV as a function of p_T using standard selections (lower curve) and machine learning (upper curve).

4.2 Multi-charm baryons

Multi-charm baryon yields are powerful probes of the hadron formation mechanisms since they are expected to be produced through combination of uncorrelated charm produced in heavy-ion collisions [6]. ALICE 3 will enable systematic measurements of multi-charm states currently inaccessible, such as Ξ_{cc}^{++} through decay to $\Xi_c^+\pi^+$ with the subsequent decay $\Xi_c^+ \rightarrow \Xi^- \pi^+ \pi^+$. With the first tracking layer at 5 mm from the beam axis, strange baryons like Ξ^- can be directly tracked before they decay, as shown in the left panel of Fig. 5, thus improving the capability to identify weak decays from primary and secondary sources. The resulting expected significance in central (0-10%) Pb-Pb at $\sqrt{s_{NN}}$ = 5.52 TeV as a function of p_T using standard selections and machine learning is shown in the right panel of Fig. 5.

4.3 Heavy-quark correlations

The angular correlation of heavy-flavour hadrons pairs is a powerful probe of QGP scattering, since it is sensitive to the energy loss mechanisms and the degree of thermalization of heavy-quarks in the QGP [7]. The left panel of Fig. 6 shows the ALICE 3 projection for $D^0\bar{D}^0$ correlation for minimum-bias Pb-Pb collisions at $\sqrt{s_{NN}}$ = 5.52 TeV. The momentum correlation of heavy-flavour hadron pairs is a powerful probe of the nature of exotic bound states. The nature of recently discovered exotic bound states, like T_{cc}^+ [8], is still unclear. The hypotheses of compact objects like tetraquarks or molecular $c\bar{c}$ states like D^0D^{*+} for T_{cc}^+ are both not excluded. Measurements of the source size dependence of the two-particle momentum correlation function in different colliding systems, from pp to Pb-Pb, provide direct insight into the properties of the DD interaction potential [9]. The projections for the D^0D^{*+} correlation function $C_{D^0D^{*+}}$ as a function of the invariant relative momentum k^* in the hypothesis of the presence of a bound state are shown in the right panel of Fig. 6. These measurements will be out of reach in LHC Runs 3 and 4, since they require excellent pointing resolution, very high identification purity and reconstruction efficiency, as well as large η acceptance, and they will be only possible with ALICE 3.

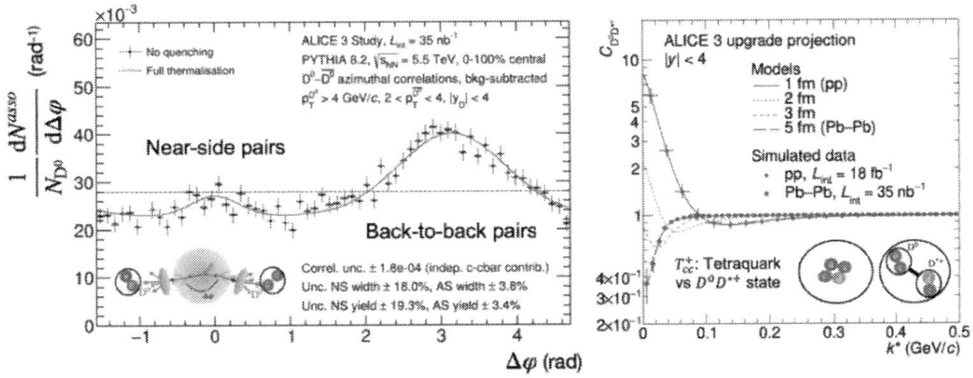

Figure 6. Left: Simulated $D^0\bar{D}^0$ azimuthal correlation distribution in minimum-bias Pb-Pb collisions with ALICE 3 at $\sqrt{s_{\rm NN}} = 5.52$ TeV. Right: Simulated (dots) and predicted (lines) D^0D^{*+} correlation $C_{D^0D^{*+}}$ as a function of the invariant relative momentum k^* evaluated for different source sizes from pp to Pb-Pb collisions in the hypothesis of the presence of a bound state.

5 Conclusions

ALICE 3 is a next-generation multipurpose detector at the LHC, proposed as a follow-up to the present ALICE experiment. Its main goal is to unravel the microscopic dynamics of the QGP produced in heavy-ion collisions and to address fundamental open questions in QCD physics and beyond. An innovative detector concept is required to allow high precision measurements of electromagnetic probes and heavy-flavour hadrons that will remain inaccessible in Runs 3 and 4. Dedicated R&D efforts are ongoing, with significant implications also for future high-energy and nuclear physics experiments.

References

[1] ALICE Collaboration, The ALICE experiment: a journey through QCD, Eur. Phys. J. C **84**, 813 (2024). https://doi.org/10.1140/epjc/s10052-024-12935-y

[2] Z. Citron et al., Future physics opportunities for high-density QCD at the LHC with heavy-ion and proton beams, CERN-LPCC-2018-07 (2018). https://cds.cern.ch/record/2650176

[3] ALICE Collaboration, Letter of intent for ALICE 3: A next-generation heavy-ion experiment at the LHC, CERN-LHCC-2022-009 (2022). https://cds.cern.ch/record/2803563

[4] R. Rapp, Dilepton Spectroscopy of QCD Matter at Collider Energies, Advances in High Energy Physics **2013**, 148253 (2013). https://doi.org/10.1155/2013/148253

[5] P. M. Hohler et al., Is ρ-meson melting compatible with chiral restoration?, Physics Letters B **731**, 103-109 (2014). https://doi.org/10.1016/j.physletb.2014.02.021

[6] A. Andronic et al., The multiple-charm hierarchy in the statistical hadronization model, J. High Energ. Phys. **2021**, 35 (2021). https://doi.org/10.1007/JHEP07(2021)035

[7] M. Nahrgang et al., Azimuthal correlations of heavy quarks in Pb + Pb collisions at $\sqrt{s}$ = 2.76 TeV at the CERN Large Hadron Collider, Phys. Rev. C **90**, 024907 (2014) https://doi.org/10.1103/PhysRevC.90.024907

[8] LHCb Collaboration. Observation of an exotic narrow doubly charmed tetraquark. Nat. Phys. **18**, 751–754 (2022). https://doi.org/10.1038/s41567-022-01614-y

[9] Y. Kamiya et al., A. Femtoscopic study on DD^* and $D\bar{D}^*$ interactions for T_{cc} and $X(3872)$. Eur. Phys. J. A **58**, 131 (2022). https://doi.org/10.1140/epja/s10050-022-00782-y

EPJ Web of Conferences **314**, 00037 (2024) https://doi.org/10.1051/epjconf/202431400037
QCD@Work 2024

How the hadron gas affects $X(3872)$ and $\psi(2S)$ production in heavy ion collisions

Luciano M. Abreu[1,*], *Fernando* S. Navarra[1,**], and *Hildeson* P.L. Vieira[1,***]

[1]Instituto de Física, Universidade de São Paulo, Rua do Matão, 1371, CEP 05508-090, São Paulo, SP, Brazil

Abstract. The $X(3872)$ to $\psi(2S)$ yield ratio ($N_X/N_{\psi(2S)}$) has been recently measured in *pp*, *pPb* and *PbPb* collisions by the LHCb and CMS Collaborations at the LHC. It was found that this ratio grows with the system size and this growth was attributed to the formation of quark gluon plasma and/or of a hot hadron gas. Here we focus on the effects of the hadron gas on the abundance of $X(3872)$ and $\psi(2S)$. The interaction of charmonium states with the light mesons in the gas can be studied with an effective Lagrangian formalism. One can calculate the relevant cross sections and use them as input in rate equations. With this formalism one can follow the time evolution of the $X(3872)$ and $\psi(2S)$ abundances until the final freeze-out of the gas. The initial conditions for this evolution are given by the coalescence model, which depends on the spatial configuration of the states and yields quite different numbers for compact tetraquarks and extended meson molecules. It has been found that the interactions in the hadron gas yield a suppression in most of the states and, most importantly, preserve the difference between the number of tetraquarks and the number of meson molecules, formed at the beginning of the hadronization. We discuss the predictions made for the ratio $N_X/N_{\psi(2S)}$ in *PbPb* collisions at $\sqrt{s_{NN}} = 5.02$ TeV to be measured by the ALICE Collaboration in the Run 3. The existing calculations suggest that the molecular configuration generates a ratio compatible with the CMS data, whereas the ratio obtained with the tetraquark configuration is 50 times smaller.

1 Introduction

In the last two decades, several newly discovered hadrons have exhibited properties that conflict with quark model predictions, leading them to be classified as nonconventional states. For a recent review, see [1]. The characteristics of these states are still the focus of heated debate, making the study of exotic quarkonium spectroscopy a prominent research area. The central question is: what is their internal structure? These states could be weakly bound hadronic molecules, compact multiquark systems, kinematic cusps, excited conventional hadrons, glueballs, hybrids, or even a mix of these possibilities [1]. To date, there is no definitive answer. A quintessential example is the $X(3872)$, the first and most well-known exotic state. Its true nature remains controversial. The two most widely explored configurations in the literature are an open-charm meson weakly bound state and a compact tetraquark.

* e-mail: luciano.abreu@ufba.br
** e-mail: navarra@if.usp.br
*** e-mail: hildeson.paulo@ufba.br

A new phase in the exploration of exotic charmonium states began with the first observation of $X(3872)$ in relativistic heavy-ion collisions, as recently reported by the CMS Collaboration [2]. The data were given in terms of the ratio $\mathcal{R} = N_{X(3872)}/N_{\psi(2S)}$ These two particles have similar masses and the same decay channel. In their ratio some experimental uncertainties are cancelled. The comparison of the ratios measured in pp, pPb and Pb-Pb collisions [3] shows a growth with the system size, which may be seen as an evidence of a medium effect.

In nucleus-nucleus collisions we expect to observe the effects of three media: the color glass condensate (CGC), a dense system of partons formed at the very initial state of the collisions, the quark gluon plasma (QGP) formed later, after the thermalization, and the hadron gas (HG), formed at the late stage of the collisions, after the expansion and cooling of the QGP and the phase conversion from deconfined quarks to hadrons [4]. The interpretation of the LHC data must be done cautiously because of the different p_T ranges measured by the different experiments. The bulk of particles with low and moderate transverse momenta were formed at the end of the QGP and lived in the hadron gas until freeze-out. They carry information about both media. On the other hand, particles with high p_T, and in particular those measured by the CMS Collaboration, were produced in the very beginning of the collision, before thermalization, and hence they carry information about the CGC. Information from the QGP and HG will come from the ALICE data, which will include particles with low p_T.

Looking at Fig. 1a we observe that $\mathcal{R}$ grows with the system size. This behavior can be attributed either to a suppression of $\psi(2S)$ or to an enhancement of $X(3872)$ or to both effects. Assuming that the QGP and HG play a role, this suppression/enhancement can come from the hadronization process (at the end of the QGP), from the interactions between $X(3872)$ and $\psi(2S)$ with the light mesons of the hadron gas and from the hadron gas time evolution. This latter aspect was shown to only amplify the effects of the former two.

The hadron gas lives for about 10 fm and freezes out generating the observed particles. Conventional and exotic hadrons formed at the end of the mixed phase can interact with the (mostly light) particles in the hadron gas and their multiplicities may experience modifications due to production and absorption processes. In the case of exotic hadrons the final multiplicity will depend on the interaction cross sections, which, in turn, depend on the spatial configuration of the states. Meson molecules are larger, and therefore have greater cross sections and stronger interaction with the hadronic medium than compact tetraquarks. We will discuss this issue in the next section.

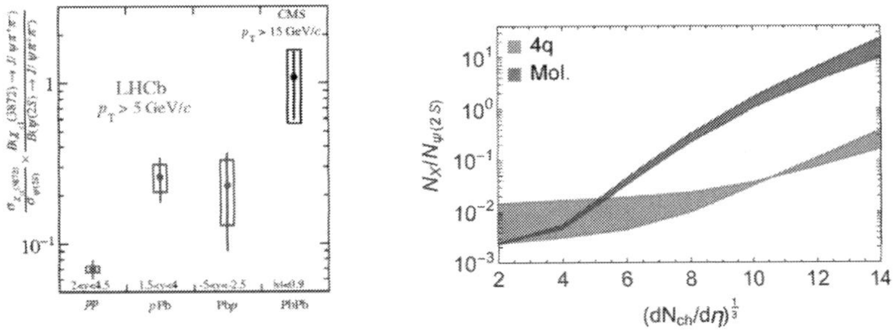

(a) (b)

Figure 1. a) Ratio $\mathcal{R}$ observed at the LHCb and CMS. Figure taken from Ref. [3]. b) Ratio $\mathcal{R}$ as a function of the system size (represented by $(dN/d\eta)^{1/3}$) calculated in [8].

2 Particle interactions in the hadronic medium

In [5] we improved the effective Lagrangian formalism introduced in [6] to study the interactions of the $X(3872)$ with the π's and ρ's of the hadron gas. In the case of the $J/\psi - \pi$ interactions it was possible to check that our cross sections were in agreement with those obtained previously with QCD sum rules [7]. More recently, in [8], we introduced the analogous formalism for the interactions of the $\psi(2S)$. In both cases, we employed form factors to take into account the finite size of hadrons. In the case of the $\psi(2S)$, due to the similarities with the J/ψ, we assumed that the form factors are given by the same parametrizations used for the J/ψ, which had already been computed via QCD sum rules [9]. The only difference is in the coupling constant. It is well known that the $\psi(2S)$ is larger than the fundamental charmonium state by a factor of about 2 and hence, based on geometrical arguments, we expect the $\psi(2S)$ to have cross sections 4 times bigger than the J/ψ. Following this reasoning, in [8] the form factor parametrizations obtained for the J/ψ were used for the $\psi(2S)$ vertices with the coupling constants varying in the range $[g, 2g]$.

With the effective Lagrangians one can compute the vacuum cross sections and then take the average over the momenta of the incoming particles, which follow the Bose-Einstein distribution. In [8] it was found that the resulting "thermally averaged cross sections" show a remarkably weak dependence on the temperature. Moreover, it was found that, for the studied particles, the absorption cross sections were larger than the production ones. However, as it will be seen in the next sections this does not automatically imply a suppression in the final yields.

3 Time evolution

With the thermally-averaged cross sections discussed in the previous section one can solve the rate equation and compute the time evolution of the hadron abundances. The rate equation is written as:

$$\frac{dN_h(\tau)}{d\tau} = \sum_{a,b,c} \left[\langle \sigma_{ab\to ch} v_{ab} \rangle n_a(\tau) N_b(\tau) - \langle \sigma_{ch\to ab} v_{ch} \rangle n_c(\tau) N_h(\tau) \right], \tag{1}$$

where $N_h(\tau)$ represents the multiplicity of the state of type h ($h = \psi(2S), X(3872)$); $n_i(\tau)$ and $N_i(\tau)$ denote the density and the number of the meson of type i at a given time τ, respectively. The particles of the hadron gas are open charm mesons (a, b) and pions and ρ's (c). They are assumed to be in equilibrium, implying that in the Maxwell-Boltzmann approximation $n_i(\tau)$ becomes

$$n_i(\tau) = \frac{1}{2\pi^2} \gamma_i g_i m_i^2 T(\tau) K_2\left(\frac{m_i}{T(\tau)}\right), \tag{2}$$

where γ_i, g_i and m_i is the fugacity, degeneracy factor and mass of the particle of type i, respectively. The multiplicity $N_i(\tau)$ is then obtained by multiplying $n_i(\tau)$ by the volume $V(\tau)$. The hadron gas expansion is often modeled [10] by the boost invariant Bjorken flow with an accelerated transverse expansion. The volume and temperature profiles as a function of the proper time τ are as follows [10]:

$$\begin{aligned} V(\tau) &= \pi\left[R_C + v_C(\tau - \tau_C) + \frac{a_C}{2}(\tau - \tau_C)^2\right]^2 \tau c, \\ T(\tau) &= T_C - (T_H - T_F)\left(\frac{\tau - \tau_H}{\tau_F - \tau_H}\right)^{\frac{4}{5}}, \end{aligned} \tag{3}$$

where R_C, v_C, a_C and T_C are the transverse size, transverse velocity, transverse acceleration and temperature at the time τ_C, respectively; $T_H(T_F)$ is the temperature at the hadronization (kinetic freeze-out) time $\tau_H(\tau_F)$. The parameters in Eq. (3) are fixed for a hadronic medium formed in central Pb Pb collisions at $\sqrt{s_{NN}} = 5.02$ TeV [10].

Having fixed all these parameters, one can solve Eq. (1), once the initial conditions are chosen, as will be discussed in the next section.

4 Initial conditions

The conversion of quarks from the QGP into hadrons is a non-perturbative process and it is not well understood. There are two models which are widely used to describe this process: the coalescence model (CM) [10] and the statistical hadronization model (SHM) [11]. According to the coalescence model, the probability of producing multiquark hadrons from quarks in the QGP is given by the overlap of the Wigner function of the produced hadron with the phase-space distribution of the constituents in the medium. The Wigner function is obtained from the wave function of the state. Hence the CM is sensitive to the spatial structure of the state and yields different results for compact (tetraquark) and extended (molecules) objects. From the coalescence model one obtains the following multiplicities for $\psi(2S)$ and $X(3872)$ at the end of the mixed phase [8, 10]:

$$N_{\psi(2S)}(\tau_H) \approx 1.8 \times 10^{-4}, \qquad N_X^{(4q)}(\tau_H) \approx 1.8 \times 10^{-4}, \qquad N_X^{(Mol)}(\tau_H) \approx 1.1 \times 10^{-2}, \quad (4)$$

As it can be seen, the coalescence mechanism generates initial conditions in which molecules are more abundant than compact tetraquarks by a factor of about 50, reflecting the fact that forming a loosely bound state is easier than a compact tetraquark. With these initial condi-

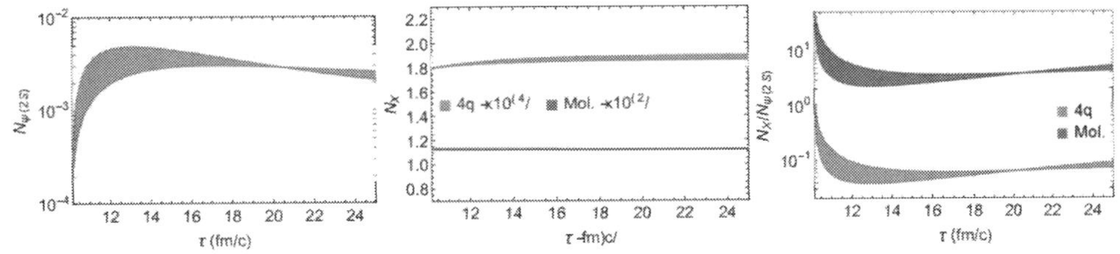

(a) (b) (c)

Figure 2. Multiplicity of $\psi(2S)$ (a), of $X(3872)$ (b) and their ratio (c) as a function of the proper time in central Pb Pb collisions at $\sqrt{s_{NN}} = 5.02$ TeV.

tions, Eq. (1) can be solved, as will be discussed in the next section.

5 Multiplicities and system size dependence

5.1 The time evolution of the ratio $\mathcal{R}$

In Fig. 2a one can see the time evolution of the $\psi(2S)$ multiplicity, which, from the beginning to the end of the hadron gas phase, increases by about one order of magnitude. This means that the gain terms in the evolution equation (1) play a dominant role at early times and higher temperatures. Indeed, at the beginning of the hadron gas phase the number of $\psi(2S)$'s is very small and so is the second term in the right hand side of Eq. (1). As a result, the derivative $dN_h/d\tau$ becomes large and positive leading to a strong growth of $N_{\psi(2S)}$. As time evolves, the gain and loss terms become of the same order and the $\psi(2S)$ multiplicity suffers just a slight reduction. Fig. 2b shows a very important result, namely that the abundance of both multiquarks states, tetraquarks and molecules, is nearly conserved during the hadron gas phase. In Fig. 2c one can see the time evolution of $\mathcal{R}$. As expected from the discussion above, one observes a strong variation of this ratio in the early times and after that a relative stabilization. Interestingly, the molecular configuration generates a ratio greater than 1, which seems compatible with the value obtained by the CMS Collaboration for Pb Pb collisions at $\sqrt{s_{NN}} = 5.02$ TeV within the rapidity and transverse momentum ranges $|y| < 1.6$ and $15 < p_T < 50$ GeV. Since the data have been collected at high-transverse momenta and in a specific rapidity range whereas the results from [8] have been computed considering the full range in p_T and y, a direct comparison between theoretical results and data is not possible yet.

5.2 The system size dependence

The size of the source can be related to a measurable quantity, the charged-particle pseudorapidity density at mid-rapidity, $[dN_{ch}/d\eta\,(|\eta| < 0.5)$. This quantity can be related to the freeze-out temperature by means of the empirical formula [12, 13]:

$$T_F = T_{F0}\, e^{-b\mathcal{N}}, \tag{5}$$

where $T_{F0} = 132.5\,MeV$, $b = 0.02$, and $\mathcal{N} \equiv [dN_{ch}/d\eta\,(|\eta| < 0.5)]^{1/3}$. Assuming that the hadron gas undergoes a Bjorken cooling, i.e. $T = T_h\,(\tau_h/\tau)^{1/3}$, then the freeze-out time τ_F can be written in terms of $\mathcal{N}$ as:

$$\tau_F = \tau_H \left(\frac{T_H}{T_{F0}}\right)^3 e^{3b\mathcal{N}}. \tag{6}$$

A larger source generates a bigger $\mathcal{N}$, which, from the equation above, implies a bigger τ_F, i.e. a longer hadron gas phase. As a consequence, the use of Eq. (6) in (1) will give rise to a dependence of the multiplicity N_h on the size of the source. As shown in Ref. [14], similar empirical formulas relating $\mathcal{N}$ with the volume of the system (V), charm quark number (N_c) and light quark number (N_q) can also be obtained. They are [14]:

$$V = 2.82\,\mathcal{N}^3 \qquad N_c = 7.9\times 10^{-5}\,\mathcal{N}^{4.8} \qquad N_u = N_d = 0.37\,\mathcal{N}^3. \tag{7}$$

In these estimates it is assumed that the charm quark number and the number of D mesons (N_D) are proportional. Hence, the use of Eq. (7) in the formula of the coalescence model to estimate the dependence of the initial conditions with $\mathcal{N}$ (as done in [14]) together with (6) in (1), generates a dependence of N_h on $\mathcal{N}$. Using the above expressions to compute $N_{X(3872)}$ and $N_{\psi(2S)}$, it is possible to obtain their ratio, $\mathcal{R}$, which is shown in Fig. 1b. In the figure one can see that the ratio $\mathcal{R}$ grows with $\mathcal{N}$, in qualitative agreement with the data shown in Fig. 1a. From the results shown in Fig. 2 one can conclude that $\mathcal{R}$ grows mostly because $\psi(2S)$ is suppressed. Unfortunately, as mentioned before, a direct comparison is not yet possible. Another limitation of this approach is that for very small systems it looses validity, as it becomes unlikely to produce quark gluon plasma and even a hadron gas, which would live long enough to produce the effects discussed here. Therefore, the curves shown on the left part of the plot shown in Fig. 1b, i.e. in the region $2 < \mathcal{N} < 4$, are at the limit of their validity.

6 Final remarks

Taken together, the results in Figs. 1, 2, suggest that in central Pb Pb collisions at $\sqrt{sNN} =$ 5.02 TeV the ratio $\mathcal{R}$ for minimum bias events is:

$$\frac{N_X}{N_{\psi(2S)}} \simeq 5 \quad \text{for molecules} \qquad \frac{N_X}{N_{\psi(2S)}} \simeq 0.1 \quad \text{for tetraquarks} \tag{8}$$

This result obtained in [8] is a prediction to be tested at the run 3 of the ALICE Collaboration, where it will be possible to measure $X(3872)$ and $\psi(2S)$ at low transverse momentum, i.e. $2 < p_T < 8$ GeV. In this kinematical region the hydrodynamical approach discussed here should be applicable. With these forthcoming data, it will be possible, for the first time, to investigate the medium effects on $X(3872)$, on $\psi(2S)$ and on their ratio and check the predictions made in [8]. Moreover, in view of the numbers above one may have a good chance to discriminate between molecules and tetraquarks. From the figures one can conclude that is easier to identify the different configurations in larger systems and in more central collisions.

References

[1] H. X. Chen, W. Chen, X. Liu, Y. R. Liu and S. L. Zhu, Rept. Prog. Phys. **86**, 026201 (2023)

[2] A. M. Sirunyan *et al.* [CMS], Phys. Rev. Lett. **128**, 032001 (2022)

[3] R. Aaij *et al.* [LHCb]; Phys. Rev. Lett. **132**, 242301 (2024)

[4] K. Adcox *et al.* [PHENIX], Nucl. Phys. A **757**, 184 (2005)

[5] A. M. Torres, K. P. Khemchandani, F. S. Navarra, M. Nielsen, L. M. Abreu, Phys. Rev. D **90**, 114023 (2014); L. M. Abreu, K. P. Khemchandani, A. Martinez Torres, F. S. Navarra, and M. Nielsen, Phys. Lett. B **761**, 303 (2016)

[6] S. Cho and S. H. Lee, Phys. Rev. C **88**, 054901 (2013)

[7] F. S. Navarra, M. Nielsen, R. S. Marques de Carvalho and G. Krein, Phys. Lett. B **529**, 87 (2002); F. O. Duraes, S. H. Lee, F. S. Navarra and M. Nielsen, Phys. Lett. B **564**, 97 (2003)

[8] L. M. Abreu, F. S. Navarra and H. P. L. Vieira, Phys. Rev. D **110**, 014011 (2024)

[9] M. E. Bracco, M. Chiapparini, F. S. Navarra and M. Nielsen, Prog. Part. Nucl. Phys **67**, 1019 (2012)

[10] S. Cho *et al.* [EXHIC], Prog, Part. Nucl. Phys. **95**, 279 (2017)

[11] A. Andronic, P. Braun-Munzinger, M. K. Köhler, K. Redlich and J. Stachel, Phys. Lett. B **797**, 134836 (2019)

[12] B. Abelev et al. [ALICE Collaboration], Phys.Rev. C **88**, 044910 (2013)

[13] C. Le Roux, F. S. Navarra and L. M. Abreu, Phys. Lett.B **817**, 136284 (2021)

[14] L. M. Abreu, H. P. L. Vieira and F. S. Navarra, Phys. Rev. D **105**, 116029 (2022)

EPJ Web of Conferences **314**, 00038 (2024) https://doi.org/10.1051/epjconf/202431400038
QCD@Work 2024

Thermal field fluctuations in a chiral lagrangian with broken scale invariance

L. Passarella[1,2,*], *A.* Lavagno[1,2], and *A.* Drago[3,4]

[1]Department of Applied Science and Technology, Politecnico di Torino, 10129, Torino, Italy
[2]INFN, Sezione di Torino, 10126, Torino, Italy
[3]Department of Physics and Earth Science, University of Ferrara, 44122 Ferrara, Italy
[4]INFN, Sezione di Ferrara, 44122 Ferrara, Italy

Abstract. We study the finite temperature equation of state by using an effective lagrangian in which a dilaton field reproduces the breaking of scale symmetry in QCD. We start by extending a previous investigation in the pure gauge sector, where the dynamics of the gluon condensate, expressed in terms of a dilaton lagrangian, is dominated below the critical phase transition temperature, while at greater temperatures the condensate evaporates in the form of quasi-free gluons. In this context, we study the role of the inclusion of thermal fluctuations of the dilaton field and compare our results with the lattice QCD data. Moreover, we take into account of the meson sector at zero chemical potential by means of an effective lagrangian which incorporates broken scale in addition to spontaneously broken chiral symmetry. Beyond the mean-field approximation, the relevance of the thermal fluctuations of the scalar glueball, other than the contribution of the σ and π meson fields is considering by following a general technique. In this framework, we investigate the thermodynamic nature of the phase transition.

1 Pure gauge sector

Severals effective Lagrangians in nuclear physics are constructed to replicate the behavior of QCD in the strong interaction regime, characterized by the breaking of chiral and scale symmetries [1–3]. Chiral symmetry breaking is linked to the generation of fermionic masses, whereas the breaking of scale invariance arises from quantum corrections, known as the scale or trace anomaly, and is likely related to the development of a gluon condensate. More precisely, the breaking of scale invariance in QCD is manifested in the non-conservation of the dilation current. In order to reproduce the QCD scale anomaly a dilaton field has been introduced, whose Lagrangian reads [4]

$$\mathcal{L} = \frac{1}{2}\partial_\mu\phi\partial^\mu\phi - \mathcal{V}\,, \tag{1}$$

where

$$\mathcal{V} = \frac{B}{4}\left(\phi_0^4 - \phi^4 + 4\phi^4 \ln\frac{\phi}{\phi_0}\right) = \frac{B_0}{4}\left(1 - \chi^4 + 4\chi^4 \ln\chi\right), \tag{2}$$

[*]e-mail: luca_passarella@polito.it

with $\chi = \phi/\phi_0$ and $B_0 = B\phi_0^4$. In the description of the pure gauge sector, we consider three distinct contributions to the thermodynamic potential [4]. The first one, $\langle \mathcal{V}(\chi)\rangle - P_{\rm dil}(\chi, T)$, arises from the gluon condensate, whose dynamics are expressed in terms of a dilaton Lagrangian (the angle brackets denote a thermal average). The second one, $P_{\rm q\text{-}free}(\chi, T)$, is due to the quasi-free gluons: when the value of the gluon condensate is large, at $T < T_C$, most of the gluons are frozen, while, for $T > T_C$, the condensate evaporates and gluons become quasi-free particles. The last one, $P_{\rm int}(\chi, T)$, is due to the perturbative gluon-gluon interaction. Taking into account the presence of thermal fluctuations, it is possible to divide the glueball into two distinct components, the mean field $\bar{\chi}$ and the fluctuating part Δ_χ, i.e., $\chi = \bar{\chi} + \Delta_\chi$, with $\langle \Delta_\chi \rangle = 0$. The thermal fluctuation for a generic meson field at zero chemical potential is given by the following equation:

$$\langle \Delta_i^2 \rangle = \frac{n_i}{2\pi^2} \int_0^\infty \frac{k^2}{\epsilon_i} \frac{1}{e^{\beta\epsilon_i} - 1}\, dk\,, \tag{3}$$

where n_i the degeneracy factor, $\epsilon_i^2 = k^2 + m_i^{*2}$ and m_i^* is the effective mass. The thermal average of a complicated functions of χ, $f(\chi)$, can be evaluated following the technique proposed in Ref.s [5, 6] by performing integrations with the following Gaussian weighting function

$$P_\chi(z) = \sqrt{\frac{1}{2\pi\langle \Delta_\chi^2\rangle}} \exp\left(-\frac{z^2}{2\langle \Delta_\chi^2 \rangle}\right), \tag{4}$$

finding that

$$\langle f(\chi)\rangle = \int_{-\infty}^{\infty} P_\chi(z) f(\bar{\chi} + z)\, dz\,. \tag{5}$$

The thermodynamical potential in the mean field approximation is:

$$\frac{\Omega(\chi, T)}{V} = \langle \mathcal{V}(\chi)\rangle - P_{\rm dil}(\chi, T) - P_{\rm q\text{-}free}(\chi, T) - P_{\rm int}(\chi, T) + \frac{B_0}{4} - \frac{1}{2} m_\chi^{*2} \langle \Delta_\chi^2 \rangle\,, \tag{6}$$

where

$$P_{\rm dil}(\chi, T) = \frac{1}{6\pi} \int_0^\infty dk \frac{k^4}{\sqrt{k^2 + m_\chi^{*2}}} \frac{1}{e^{\beta\sqrt{k^2 + m_\chi^{*2}}} - 1}\,, \tag{7}$$

$$P_{\rm q\text{-}free}(\chi, T) = -2(N_c^2 - 1) T \int \frac{d^3k}{(2\pi)^3} \ln\left(1 - e^{-k/T}\right) \Theta\left(k - K(\chi)\right)\,, \tag{8}$$

$$\begin{aligned} P_{\rm int}(\chi, T) = {}& g^2 N_c (N_c^2 - 1) \Bigg\{(-3)\left[\int \frac{d^3k}{(2\pi)^3} \frac{1}{k} N_B\left(\frac{k}{T}\right) \Theta\left(k - K(\chi)\right)\right]^2 \\ & + \int \frac{d^3k_1}{(2\pi)^3} \frac{d^3k_2}{(2\pi)^3} \frac{1}{k_1 k_2} N_B\left(\frac{k_1}{T}\right) N_B\left(\frac{k_2}{T}\right) \Theta\left(k_1 - K(\chi)\right) \\ & \times \Theta\left(k_2 - K(\chi)\right) \left[\frac{9}{4}\Theta\left(|\boldsymbol{k}_1 + \boldsymbol{k}_2| - K(\chi)\right) - \frac{1}{4}\Theta\left(|\boldsymbol{k}_1 - \boldsymbol{k}_2| - K(\chi)\right)\right]\Bigg\}\,. \end{aligned} \tag{9}$$

At high temperatures, the primary contribution to thermodynamic observables should be attributed to quasi-free gluons, rather than to glueballs, which are dilaton field excitations. Conversely, quasi-free gluons are expected to be suppressed below the critical temperature T_c, the temperature at which the dilaton field exhibits discontinuity, thus below T_c, we assume they are frozen within the gluon condensate. Whereas, above T_c, the condensate evaporates, allowing gluons to become quasi-free particles. In order to take this behaviour into account, we introduce in the above last equation an infrared cut-off K in the gluon distribution function. This ensures that only gluons with momentum exceeding K contribute to the thermodynamics of the system. Furthermore, at extremely high temperatures, the influence of the dilaton field on the thermodynamic quantities should be negligible. To this end, an ultraviolet cutoff, equal to $3K$, will be incorporated into the expression of $P_{\rm dil}$. Following Ref. [4], we assume that K is a function of the gluon condensate, which is defined as the expectation value of the dilaton field:

$$K(\chi) = \frac{A}{(1-\chi)^{\alpha}} . \tag{10}$$

Consequently, if $\chi \to 1$, then $K \to \infty$, while if $\chi \to 0$, then $K \to A$. We use $A = 1.015$ GeV and $\alpha = 0.5$. In Eq.(9), $N_B(x) = (e^x - 1)^{-1}$ is the Bose–Einstein distribution and g^2 is the temperature dependent running coupling-constant

$$g^2(T) = \frac{48\pi^2}{11N_c \ln\left(\frac{T^2+S^2}{\Lambda^2}\right)} , \tag{11}$$

where S is a regulator, related to the existence of a minimal momentum K for the propagating gluons. In particular, we assume $S = 7.15$ GeV and $\Lambda = T_c$, where T_c is the value of the critical temperature computed in the model.

In order to realize consistent thermodynamics, it is necessary to ensure that Ω is a minimum with respect to variations in the mean field $\bar{\chi}$. The pressure is $P = -\Omega/V$. In Fig. 1, we present the behavior of P and I as a function of temperature, where $I = (\epsilon - 3P)/T^4$ is the trace anomaly and $\epsilon = T\frac{dP}{dT} - P$. The green curves shown in Fig.1 is obtained by solving the

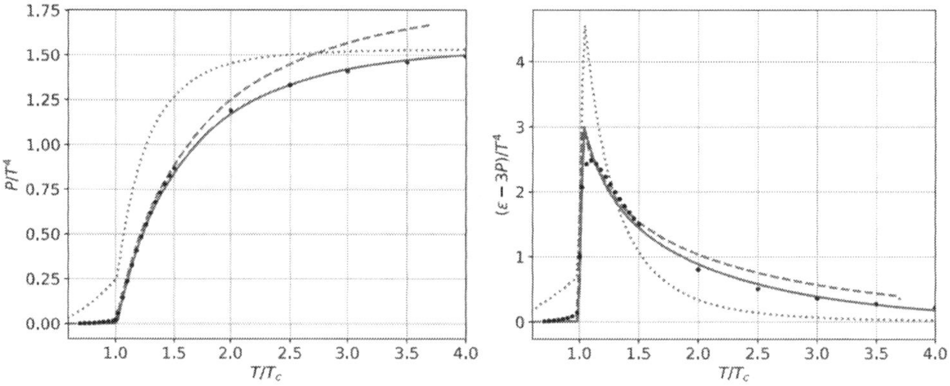

Figure 1. The left panel shows P/T^4, compared with the lattice data [7]. The red dashed curve represents a scenario where, as in Ref. [4], the thermal fluctuations related to the χ field are not considered while in the solid blue the thermal fluctuations are taken into account. The green dotted curves represents a result obtained by assuming that the constituent gluons are described by an Abelian vector field, as in Ref. [5]. The right panel shows the trace anomaly I.

field equation, derived from the following Lagrangian $\mathcal{L}_{\mathcal{G}}$, defined as in Ref. [5]

$$\mathcal{L}_{\mathcal{G}} = \mathcal{L} + \mathcal{L}_{\mathcal{A}} , \tag{12}$$

where $\mathcal{L}$ is the same of (1) and $\mathcal{L}_{\mathcal{A}}$ is

$$\mathcal{L}_{\mathcal{A}} = -\frac{1}{4}\boldsymbol{A}_{\mu\nu} \cdot \boldsymbol{A}^{\mu\nu} + \frac{1}{2}G^2\phi^2\boldsymbol{A}_{\mu} \cdot \boldsymbol{A}^{\mu} . \tag{13}$$

Following a standard procedure, the field equation of the dilaton field, as well as the masses of the glueball and gluon field, m^*_χ and m^*_A, respectively, can be solved in a self-consistent way.

2 Meson sector at zero chemical potential

The present investigation can be extended by considering the meson sector at zero chemical potential through the introduction of an effective Lagrangian that incorporates broken scale in addition to spontaneously broken chiral symmetry [8]

$$\tilde{\mathcal{L}} = \frac{1}{2}\partial_\mu\sigma\partial^\mu\sigma + \frac{1}{2}\partial_\mu\boldsymbol{\pi}\cdot\partial^\mu\boldsymbol{\pi} + \frac{1}{2}\partial_\mu\phi\partial^\mu\phi - \tilde{\mathcal{V}}\,, \tag{14}$$

where

$$\begin{aligned}\tilde{\mathcal{V}} =& \mathcal{V}(\chi) - \frac{1}{2}B_0\delta\chi^4\ln\frac{\sigma^2+\pi^2}{\sigma_0^2} + \frac{1}{2}B_0\delta\chi^2\left[\frac{\sigma^2+\pi^2}{\sigma_0} - \frac{\chi^2}{2}\right]\\ &-\frac{1}{4}\epsilon_1'\chi^2\left[\frac{4\sigma}{\sigma_0} - 2\left(\frac{\sigma^2+\pi^2}{\sigma_0^2}\right) - \chi\right] - \frac{3}{4}\epsilon_1'\,;\end{aligned} \tag{15}$$

here $\delta = 4/33$, $\epsilon_1' = \sigma_0^2 m_\pi^2$ is the explicit symmetry breaking term and $\mathcal{V}(\chi)$ is defined as in Eq.(2). The approach adopted in the previous section can be extended to the case of σ and $\boldsymbol{\pi}$ fields, by decomposing them into mean field and fluctuating parts, where the mean value of the $\boldsymbol{\pi}$ field is zero. As in the previous case, we can define the following Gaussian weight functions [6]

$$P_\sigma(z) = \sqrt{\frac{1}{2\pi\langle\Delta_\sigma^2\rangle}}\exp\left(-\frac{z^2}{2\langle\Delta_\sigma^2\rangle}\right), \tag{16}$$

and

$$P_{\boldsymbol{\pi}}(y) = \sqrt{\frac{2}{\pi}}\left(\frac{3}{\langle\Delta_{\boldsymbol{\pi}}^2\rangle}\right)^{(3/2)}\exp\left(-\frac{3y^2}{2\langle\Delta_{\boldsymbol{\pi}}^2\rangle}\right), \tag{17}$$

in order to evaluate the average of a complicated function $O(\bar{\sigma}+\Delta_\sigma, \boldsymbol{\pi}^2)$, over the fluctuating fields σ and $\boldsymbol{\pi}$, in the following manner

$$\langle O(\bar{\sigma}+\Delta_\sigma,\boldsymbol{\pi}^2)\rangle = \int_{-\infty}^{\infty} dz P_\sigma(z)\int_0^\infty dy y^2 P_{\boldsymbol{\pi}}(y) O(\bar{\sigma}+z,y^2)\,. \tag{18}$$

The thermodynamical potential in the mean field approximation is

$$\begin{aligned}\frac{\tilde{\Omega}(\chi,\sigma,\boldsymbol{\pi},T)}{V} =& \langle\tilde{\mathcal{V}}\rangle + \frac{T}{2\pi^2}\int_0^\infty dk k^2\left[\ln(1-e^{-\beta\omega_\sigma}) + 3\ln(1-e^{-\beta\omega_\pi})\right]\\ &-\frac{1}{2}m_\chi^{*2}\langle\Delta_\chi^2\rangle - \frac{1}{2}m_\sigma^{*2}\langle\Delta_\sigma^2\rangle - \frac{1}{2}m_{\boldsymbol{\pi}}^{*2}\langle\Delta_{\boldsymbol{\pi}}^2\rangle\\ &-P_{\rm dil}(\chi,T) - P_{\rm q\text{-}free}(\chi,T) - P_{\rm int}(\chi,T) + \frac{1}{4}[B_0(1-\delta)+\epsilon_1']\,.\end{aligned} \tag{19}$$

Here a constant term has been added so that $\langle\tilde{\mathcal{V}}\rangle$ is zero in the vacuum and the terms in the second line are subtracted, thus avoiding double counting.

In Fig.2, we present the behavior of the mean fields $\bar{\chi}$ and $\bar{\sigma}/\sigma_0$ as a function of temperature.

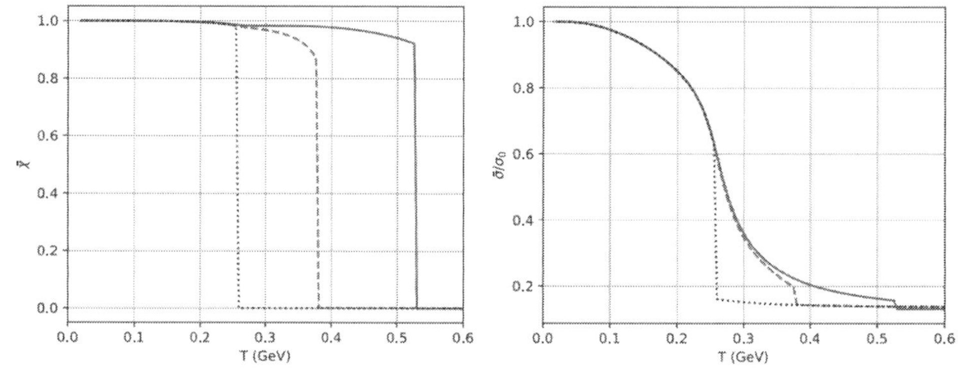

Figure 2. In the left panel, the mean glueball field $\bar{\chi}$ is presented as a function of temperature. The blue solid curve is derived by considering only the thermal fluctuations on the σ and π fields. The red dashed curve is obtained by incorporating fluctuations in the χ field, while the black dotted curve is obtained by including gluons in the same way of Ref. [4]. In the right panel, we report the mean sigma field $\bar{\sigma}/\sigma_0$ as a function of temperature.

3 Conclusions and Outlook

In this study, we extend a previous work in which the equation of state at finite temperature was investigated by using an effective Lagrangian in which a dilaton field reproduces the breaking of the scale symmetry in QCD. This is based on an evaporation model in which the gluons are frozen at low temperatures in the non-perturbative condensate, while at high temperature they escape from the condensate and behave as almost free particles. We have shown that it is possible to reproduce the main results of the EOS lattice QCD for thermodynamic quantities. By taking into account the thermal fluctuations in the dilaton field χ, we are able to reproduce more accurately the pure gauge lattice data, particularly at high temperatures ($T > 2.0\ T_c$), as shown in Fig. 1. Furthermore, the incorporation of the thermal fluctuations and the use of the rigorous technique introduced in Ref. [6], allows us to avoid the main problems encountered in Ref. [4], i.e., the appearance of an imaginary part in the thermodynamical potential, previously assumed as the presence of a system instability. In this framework, the model shows a first order transition, obtained by minimizing the thermodynamical potential for $T_c = 0.270$ GeV.

Afterthat, we also extend this technique to the meson sector at zero chemical potential. In order to obtain the vacuum pion mass, we have chosen the explicit symmetry breaking parameter $\epsilon_1' = \sigma_0^2 m_\pi^2$. We took $\sigma_0 = 0.110$ GeV and $m_\pi = 0.138$ GeV. The thermal fluctuations, as defined in Ref. [6], were employed for the σ, π fields and the dilaton field χ. Regarding Fig. 2, the introduction of fluctuations in the dilaton field χ results in a reduction of the critical temperature, from $T_c = 0.530$ GeV to $T_c = 380$ GeV. The fluctuations in the χ field also have a significant impact on the masses: in the absence of fluctuations, the masses of the σ and π fields vanish after T_c, however, in the presence of fluctuations, they continue to increase linearly with increasing temperature. Fig. 2 shows that at high temperatures the solution for $\chi \sim 1$ becomes unstable and χ become equal to zero. The fourth panel of Fig. 3 shows how the minimum of the potential shifts from a value around $\chi \sim 1$ at low temperatures to $\chi = 0$ at higher temperatures. The two minima of the potential have the same value at $T = 0.530$ GeV. This feature can be interpreted as a first-order phase transition at a critical temperature T_c, which restores scale symmetry.

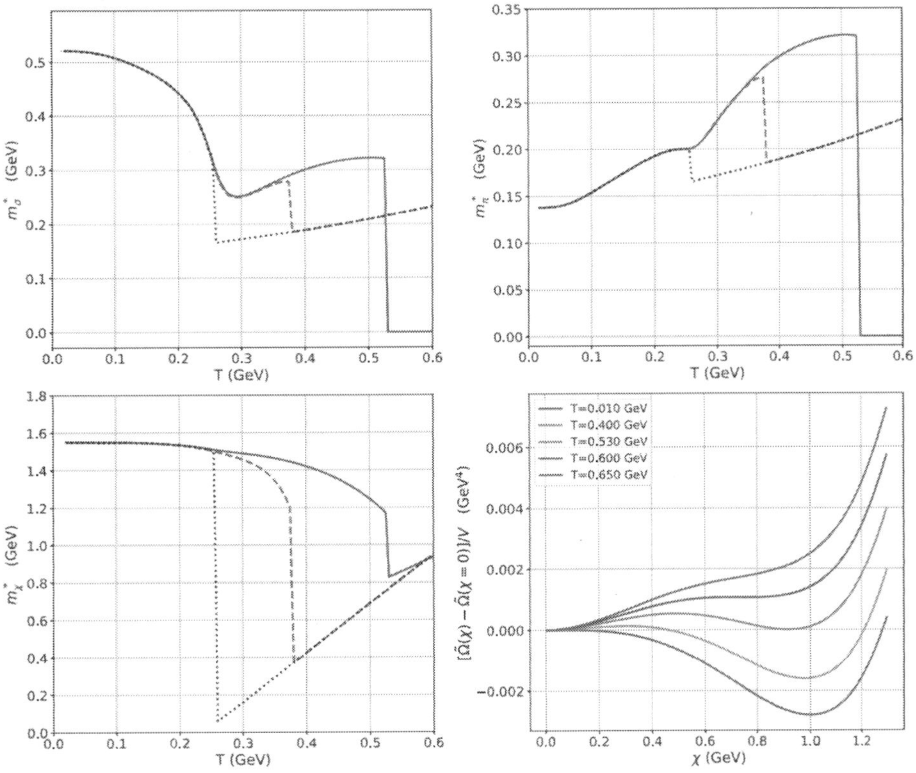

Figure 3. The effective masses m^*_σ, m^*_π and m^*_χ are shown as a function of the temperature in the top-left, top-right and bottom-left panels respectively. The blue solid curve is derived by considering only the thermal fluctuations on the σ and $\boldsymbol{\pi}$ fields. The red dashed curve is obtained by incorporating also the fluctuations in the χ field, while the black dotted curve corresponds to the gluons considered in the same way of Ref. [4]. The lower right-hand panel displays the behavior of the thermodynamic potential $\tilde{\Omega}$ as a function of χ for different values of temperature. In this last panel, only fluctuations in the σ and $\boldsymbol{\pi}$ fields are considered (the potential is normalised to zero at the origin).

A further extension of this work will be the study of a chiral Lagrangian which takes into account of the nucleons and quark degrees of freedom, by considering investigations on high energy astrophysical phenomena and relatistivistic heavy ion collisions [9, 10].

References

[1] S. Benic, I. Mishustin, C. Sasaki, Phys. Rev. D **91**, 125034 (2015)
[2] L. Bonanno, A. Drago, A. Lavagno, Phys. Rev. Lett. **99**, 242301 (2007)
[3] L. Bonanno, A. Drago, Phys. Rev. C **79**, 045801 (2009)
[4] A. Drago et al., Nuclear Physics A **742**, 165 (2004)
[5] G.W. Carter et al., Phys. Rev. C **61**, 045206 (2000)
[6] Á. Mócsy et al., Phys. Rev. C **70**, 015204 (2004)
[7] Sz. Borsányi et al., Journal of High Energy Physics **07**, 256 (2012)
[8] G.W. Carter et al., Nuclear Physics A **618**, 317 (1997)
[9] F. Di Clemente, A. Drago, G. Pagliara, ApJ **967**, 159 (2024)
[10] A. Lavagno, D. Pigato, Eur. Phys. J. A **58**, 237 (2022)

EPJ Web of Conferences **314**, 00039 (2024) https://doi.org/10.1051/epjconf/202431400039
QCD@Work 2024

Exploring QGP signature in small system: Insights from ALICE in p–Pb and pp collisions

Rajendra Nath Patra[1,2,*] for the ALICE Collaboration

[1]INFN Bari, Bari, Italy
[2]University of Jammu, Jammu, India

Abstract. The formation of quark-gluon plasma (QGP) has been confirmed in ultra-relativistic heavy-ion (i.e. Pb-Pb) collisions. Results from small collision systems, such as proton-ion (p–Pb) and proton-proton (pp) collisions at LHC energies, have revealed several notable features: a hardening of the p_{T}-spectra and evidence of radial flow, an enhancement in (multi-)strange hadron production with increasing multiplicity, and non-zero anisotropic flow coefficients. These observations suggest the presence of collective effects, similar to those found in the QGP in heavy-ion collisions. In the searching of QGP formation in small collision systems, in particular, high-multiplicity (HM) p–Pb and pp collisions are essential as their charged-particle multiplicity is comparable to that of peripheral heavy-ion collisions. This proceeding presents the LHC Run 2 results of light-flavour hadron (π, K, p) production in high-multiplicity pp collisions at $\sqrt{s}$ = 13 TeV with a focus on transverse momenta (p_{T}) spectra, mean transverse momenta ($\langle p_{\mathrm{T}} \rangle$), ratio of p_{T}-spectra and the ratio of the integrated yields of kaon-to-pion and proton-to-pion. The results are compared with that of heavy-ion collisions. The anisotropic flow (v_2) measurement in high-multiplicity p–Pb collisions at $\sqrt{s_{\mathrm{NN}}}$ = 5.02 TeV and pp collisions at $\sqrt{s}$ = 13 TeV are also shown and compared with the results from Pb–Pb collisions.

1 Introduction

The heavy-ion collisions (A–A) at the Relativistic Heavy Ion Collider (RHIC) and the Large Hadron Collider (LHC) have confirmed the formation of the quark-gluon plasma (QGP), a short-lived new state of nuclear matter where quarks and gluons are deconfined [1]. The production of particles with high transverse momentum (p_{T}) or containing heavy-quarks can be described by perturbative quantum chromodynamics (pQCD). The high-p_{T} partons and heavy quarks lose energy through interacting in the QGP , therefore used as probes of the QGP. On the other-side the production dominated by light-flavour hadrons (π, K, p) at low transverse momentum (p_{T}), created from the soft interactions among the partons, carry useful information of collective behaviour and freeze-out property of the QGP, spatial anisotropy of the collision geometry and anisotropic expansion of medium.

The ALICE experiment characterizes the properties of the QGP medium using heavy-ion collisions (i.e., Pb–Pb, Xe–Xe), while pp collisions are typically used as a reference [2–6]. The collective phenomena such as radial flow, long-range angular correlation and anisotropic

*e-mail: rajendra.nath.patra@cern.ch,rajendrapatra07@gmail.com

flow [4–9], are typically observed in A–A collisions, where the QGP medium is formed. The recent measurements of light-flavour hadrons production shows mass dependent hardening of the p_T-spectra, an increase in mean transverse momenta ($\langle p_T \rangle$) and an enhancement of baryon-to-meson ratio at intermediate-p_T in small collision systems at large multiplicity depicted similarity with the results in heavy-ion collisions where QGP is formed [4, 10, 11]. The enhancement of strange hadron production in heavy-ion collisions is one of the key signatures of QGP formation. The ratios of (multi-)strange hadrons to non-strange hadrons exhibit a clear trend of smooth increase with multiplicity until reaching saturation. The observation of the smooth evolution of strange to non-strange yield ratios across different collision systems and centere-of-mass energies suggests physics of hadronization follows a mechanism that depends only on the multiplicity. [11, 12]. The origin of strangeness production in small collision systems has been studied through a topological analysis of production towards- and transverse-to-leading particle (a proxy for the jet axis) [13]. Strange hadron production from underlying event (soft scatterings) dominates over hard scattering, and the relative production of strangeness is favoured in transverse-to-leading (i.e. soft scatterings) processes. The measurements of long range angular correlations and non-zero anisotropic flow in heavy-ion collisions are well explained by collective effects of the QGP medium [14]. A similar "ridge"-like structure in angular correlations and non-zero anisotropic flow coefficients has also been observed in p–Pb and pp collisions [15, 16].

The above mentioned observations in small systems (p–Pb, pp) require a comprehensive explanation of the mechanisms behind these phenomena. This proceedings presents the latest measurements from LHC Run 2 data on light-flavour production in high-multiplicity pp collisions at $\sqrt{s}$ = 13 TeV [17]. The elliptic flow coefficient (v_2) of identified particles in high-multiplicity p–Pb and pp collisions at $\sqrt{s_{NN}}$ = 5.02 TeV and $\sqrt{s}$ = 13 TeV, respectively, are presented to address the question of a unified description of collective phenomena observed from small to large collision systems.

2 Results

The results presented here are based on data samples; of pp collisions at $\sqrt{s}$ = 13 TeV and p–Pb collisions at $\sqrt{s_{NN}}$ = 5.02 TeV collected by the ALICE experiment during the LHC Run 2 data-taking campaign 2015-2018. The minimum bias (MB) trigger requiring hits on both V0A and V0C detectors. The MB events with at least one charged-particle produced in the pseudorapidity region $|\eta| < 1$ are defined as pp inelastic INEL>0 event class. The different V0M multiplicity classes are defined from the sum of all signals in V0M detector (combined signals of V0A and V0C detectors). Different multiplicity classes in pp collision are from the V0M detector signals. For example, the V0M-I and V0M-X multiplicity classes are corresponding to events containing 0–0.92% highest multiplicity and 64.5–100% lowest multiplicity, respectively [11]. For pp collisions at $\sqrt{s}$ = 13 TeV, an additional high-multiplicity (HM) trigger, with a threshold of relative V0M amplitude with respect of average value V0M/$\langle$V0M$\rangle$ > 4.9, is applied, selecting the top 0.1% of the MB events [18].

The p_T-differential spectra of the π, K and p for three HM classes, corresponding to the 0-0.01%, 0.01-0.05% and 0.05-0.1% of the MB events of pp collisions at $\sqrt{s}$ = 13 TeV, are shown in Fig. 1 along with the reference MB spectra for the INEL>0 [19]. The bottom panels of the figure show the ratios of the HM spectra to the INEL>0 class. For all three identified hadrons, the ratios exhibit a hardening trend with increasing multiplicity, with the the effect being more pronounced for protons. The observation of hardening of the p_T-spectra with increasing multiplicity, along with mass ordering, is inline with the previously reported results in pp, p–Pb and Pb–Pb collisions where they have been attributed to the hydrodynamical flow [4, 10, 11].

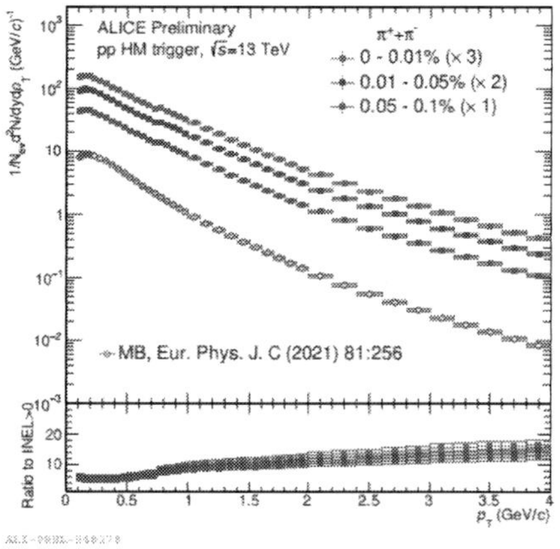

(a) $\pi^+ + \pi^-$

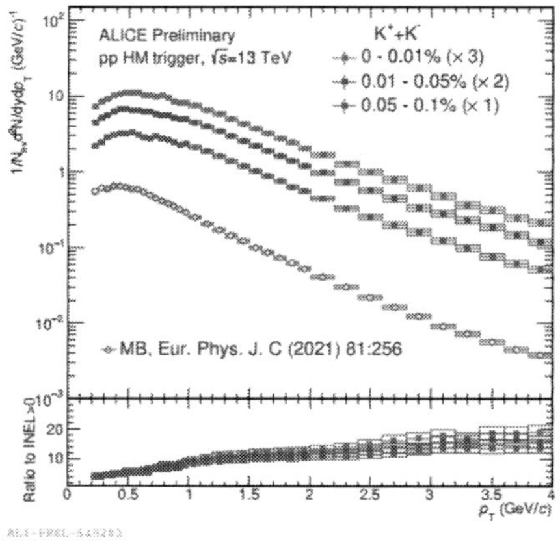

(b) $\mathrm{K}^+ + \mathrm{K}^-$

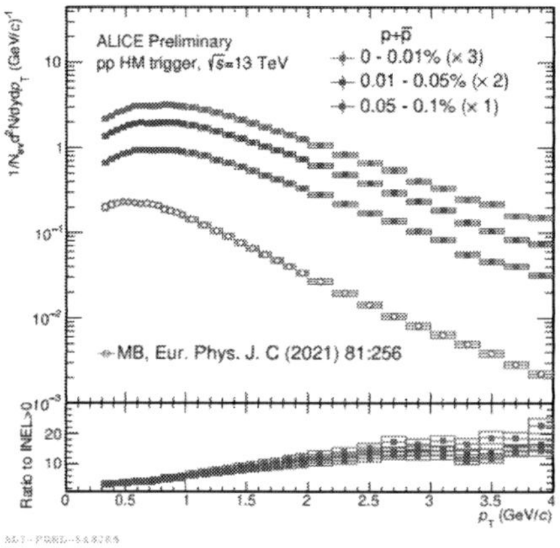

(c) $\mathrm{p} + \bar{\mathrm{p}}$

Figure 1. p_{T} differential spectra of π, K and p for three different high-multiplicity classes, compared with the reference minimum bias result [19]. The spectra are scaled for better visibility. The ratios to INEL>0 are shown in the bottom panels.

The measurement of mean transverse momentum $\langle p_{\mathrm{T}} \rangle$ as a function of charged-particle multiplicity density $\langle \mathrm{d}N_{\mathrm{ch}}/\mathrm{d}\eta \rangle$ for three particles (π, K and p) is shown in Fig. 2. The results are compared with previously reported $\langle p_{\mathrm{T}} \rangle$ measurements in pp collisions at $\sqrt{s} = 7$ and 13 TeV [11, 20]. A continuously increasing trend of $\langle p_{\mathrm{T}} \rangle$ with increasing of $\langle \mathrm{d}N_{\mathrm{ch}}/\mathrm{d}\eta \rangle$ can be observed, with a steeper rise for heavier hadrons. Similar observations have been reported in pp [21] and p–Pb [10] collisions at lower collision energies. The qualitative trend of increasing $\langle p_{\mathrm{T}} \rangle$ with charged-particle multiplicity, along with the mass ordering, is consistent with the results observed in Pb–Pb collisions, though in a different multiplicity region [4]. This observation can be explained by the presence of a stronger radial flow effect at higher multiplicities, which is more pronounced for particles with larger mass, across all three collision systems.

The relative production of different charged particles with transverse momenta can be studied by examining particle ratios across various multiplicity classes. Fig. 3 shows K/π and

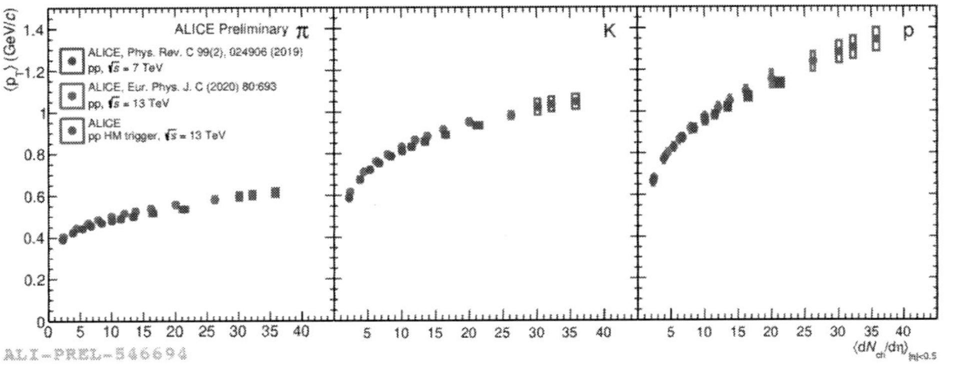

Figure 2. Mean transverse momentum $\langle p_T \rangle$ as a function of charged-particle multiplicity density ($\langle dN_{ch}/d\eta \rangle$) of π, K and p in high-multiplicity pp collisions at $\sqrt{s}$ = 13 TeV and compared with results from pp collisions at $\sqrt{s}$ = 7 and 13 TeV [11, 20].

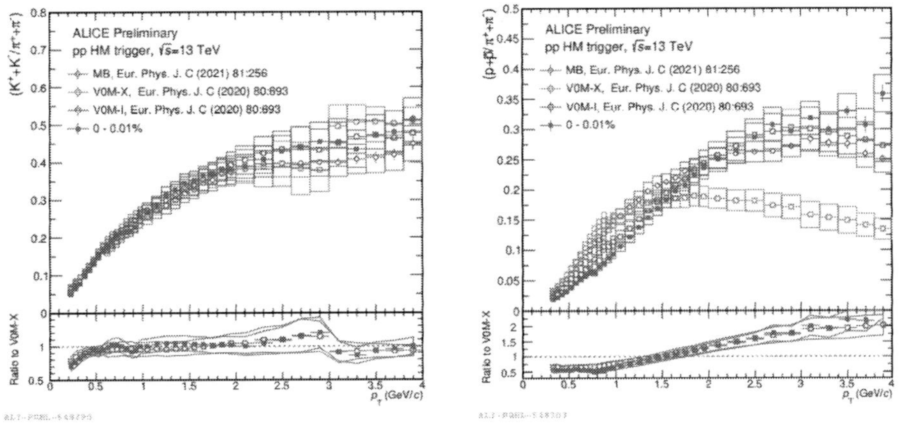

Figure 3. Ratios K/π (left) and p/π (right) of the p_T-spectra. The reference results of the V0M and minimum bias classes as shown for comparison [11, 19]. The ratio to V0M-X are shown in the bottom panels.

p/π ratios as a function of p_T for the high-multiplicity class 0-0.01%, along with published V0M multiplicity classes and minimum bias results for the same collision system (pp) at $\sqrt{s}$ = 13 TeV [11, 19]. The lower panels of the figure show a comparison of the ratios to the V0M-X class. The K/π ratio shows no significant dependence on multiplicity within the explored p_T range. However, the p/π ratio exhibits depletion at low p_T ($\sim$ 0.8 GeV/c) and an enhancement with increasing p_T, similar to the trend obtained in the V0M-I class. This enhancement of baryon-to-meson ratio has been reported in all three collision systems–pp, p–Pb and Pb–Pb, attributed to radial flow boosting heavier particle at mid p_T [20]. A similar pattern of K/π and p/π ratios is also observed in the other high-multiplicity classes, 0.01-0.05% and 0.05-0.1%.

The p_T-integrated particle ratios, kaon- and proton-to-pion, as a function of multiplicity $\langle dN_{ch}/d\eta \rangle$ is shown in Fig. 4 alongside published results from pp, p–Pb and Pb–Pb collisions [4, 6, 10, 11, 20]. The high-multiplicity data points are in good agreement with ratios from the p–Pb and Pb–Pb collisions at similar multiplicities. The ratios exhibit a smooth evolution with increasing charged-particle multiplicity, indicating that the multiplicity is the key driving force of the hadrochemistry, irrespective of the collision energy and colliding system. The increasing trend of K/π ratio with increasing multiplicity could be due to the fact that the strange to non-strange hadron production increases with increasing multiplicity. The small decrease of the p/π ratio with increasing multiplicity supports the hypothesis of antibaryon-baryon annihilation in the hadronic phase, a process which is less significant in low multiplicity or peripheral collisions [22].

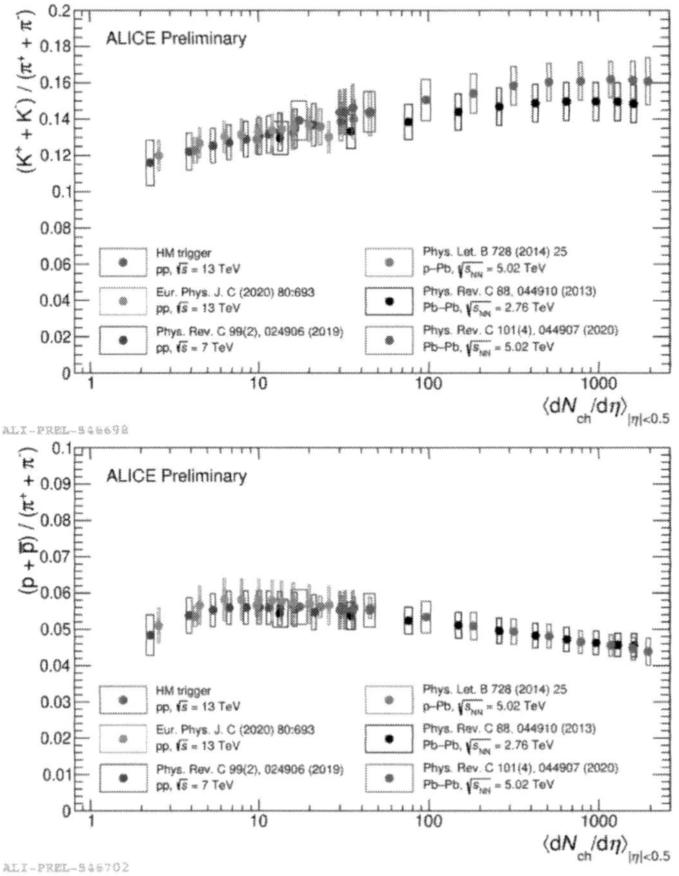

Figure 4. Integrated yield ratios K/π (top) and p/π (bottom)as a function of charged-particle multiplicity density ($\langle dN_{ch}/d\eta \rangle$) for high-multiplicity pp collisions at $\sqrt{s}$ = 13 TeV, compared with results from pp, p–Pb and Pb–Pb collisions [4, 6, 10, 11, 20].

The anisotropic dynamic in small collision systems is studied by measuring the elliptic flow coefficient v_2 as a function of p_T in high-multiplicity p–Pb and pp collisions at $\sqrt{s_{NN}}$ = 5.02 TeV and $\sqrt{s}$ = 13 TeV, respectively, as shown in Fig 5. The figure presents measurement of v_2 for identified mesons ($\pi^{\pm}$, $K^{\pm}$, K^0_S) and baryons ($p(\bar{p})$, $\Lambda(\bar{\Lambda})$) up to p_T < 7.5 GeV/c. The behaviour of v_2 with p_T is found similar among all three collision systems: Pb–Pb [23], p–Pb and pp. The coefficient v_2 is found to decrease gradually from Pb–Pb to pp collision systems, which can be attributed to higher anisotropy in larger systems. For p_T < 2–3 GeV/c, v_2 of lighter hadrons is larger than heavier ones at the same p_T, which is known as mass ordering of particle species. This behaviour provides significant evidence of collective radial flow, which gives isotropic boost to all particles equally, in addition to the anisotropic expansion of the system. Beyond the low-p_T region, the particles are grouped according to the number of their constituent quarks, known as baryon-meson v_2 grouping. This observation of baryon-meson v_2 grouping at intermediate-p_T in heavy-ion collisions is explained by the hypothesis of particle production via the quark coalescence mechanism [24, 25].

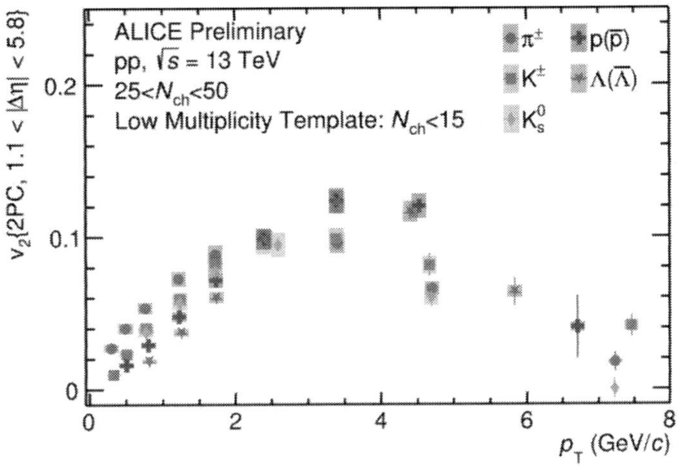

ALI-PREL-573050

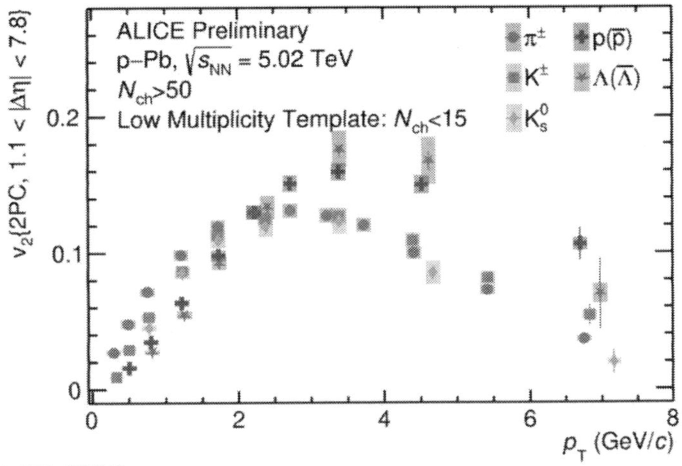

ALI-PREL-573065

Figure 5. Elliptic flow v_2 as a function of p_T for mesons ($\pi^{\pm}, K^{\pm}, K^0_s$) and baryons ($p(\bar{p}), \Lambda(\bar{\Lambda})$) in high-multiplicity p–Pb collisions at $\sqrt{s_{NN}}$ = 5.02 TeV (top) and pp collisions at $\sqrt{s}$ = 13 TeV (bottom).

3 Conclusions

High-multiplicity p–Pb and pp collisions exhibit shown properties that resemble collective behaviour attributed to the formation of QGP in Pb–Pb collisions. In particular, the transverse momentum spectra of the π, K and p in high-multiplicity pp collisions at $\sqrt{s}$ = 13 TeV exhibit multiplicity and mass-dependent hardening similar to what is observed in heavy-ion collisions [4, 6]. The mean transverse momentum $\langle p_{\mathrm{T}} \rangle$ increases with multiplicity and its mass-dependent steepness can be attributed to a stronger radial flow effect with increasing multiplicity and particle masses. The ratios K/π and p/π of the p_{T}-spectra confirm a similar trend as observed in the V0M-I multiplicity class of pp collisions at $\sqrt{s}$ = 13 TeV [11]. The ratios of p_{T}-integrated yields of K/π and p/π indicate that hadronchemistry depends on multiplicity only, regardless of collision systems and colliding energies. The elliptic flow coefficient v_2 measured in p–Pb and pp collisions at $\sqrt{s_{\mathrm{NN}}}$ = 5.02 TeV and $\sqrt{s}$ = 13 TeV, respectively, exhibit similar features for Pb–Pb but with smaller v_2 values. The mass ordering at low-p_{T} and baryon-meson grouping at intermediate-p_{T} observed in high-multiplicity pp and p–Pb systems are consistent with the elliptic flow measurement in the Pb–Pb collisions. A comprehensive comparison of various observables across different collision systems suggest that collective effects can be commonly present in both small and large systems.

References

[1] W. Busza, K. Rajagopal, W. van der Schee, Heavy Ion Collisions: The Big Picture, and the Big Questions, Ann. Rev. Nucl. Part. Sci. **68**, 339 (2018), [arXiv]1802.04801. 10.1146/annurev-nucl-101917-020852

[2] K. Aamodt et al. (ALICE), The ALICE experiment at the CERN LHC, JINST **3**, S08002 (2008). 10.1088/1748-0221/3/08/S08002

[3] B.B. Abelev et al. (ALICE), Performance of the ALICE Experiment at the CERN LHC, Int. J. Mod. Phys. A **29**, 1430044 (2014), [arXiv]1402.4476. 10.1142/S0217751X14300440

[4] S. Acharya et al. (ALICE), Production of charged pions, kaons, and (anti-)protons in Pb-Pb and inelastic pp collisions at $\sqrt{s_{NN}}$ = 5.02 TeV, Phys. Rev. C **101**, 044907 (2020), [arXiv]1910.07678. 10.1103/PhysRevC.101.044907

[5] S. Acharya et al. (ALICE), Production of pions, kaons, (anti-)protons and ϕ mesons in Xe–Xe collisions at $\sqrt{s_{\mathrm{NN}}}$ = 5.44 TeV, Eur. Phys. J. C **81**, 584 (2021), [arXiv]2101.03100. 10.1140/epjc/s10052-021-09304-4

[6] B. Abelev et al. (ALICE), Centrality dependence of π, K, p production in Pb-Pb collisions at $\sqrt{s_{NN}}$ = 2.76 TeV, Phys. Rev. C **88**, 044910 (2013), [arXiv]1303.0737. 10.1103/PhysRevC.88.044910

[7] B.B. Abelev et al. (ALICE), Multi-strange baryon production at mid-rapidity in Pb-Pb collisions at $\sqrt{s_{NN}}$ = 2.76 TeV, Phys. Lett. B **728**, 216 (2014), [Erratum: Phys.Lett.B 734, 409–410 (2014)], [arXiv]1307.5543. 10.1016/j.physletb.2014.05.052

[8] G. Aad et al. (ATLAS), Observation of Long-Range Elliptic Azimuthal Anisotropies in $\sqrt{s}$ =13 and 2.76 TeV pp Collisions with the ATLAS Detector, Phys. Rev. Lett. **116**, 172301 (2016), [arXiv]1509.04776. 10.1103/PhysRevLett.116.172301

[9] S. Acharya et al. (ALICE), Investigations of Anisotropic Flow Using Multiparticle Azimuthal Correlations in pp, p-Pb, Xe-Xe, and Pb-Pb Collisions at the LHC, Phys. Rev. Lett. **123**, 142301 (2019), [arXiv]1903.01790. 10.1103/PhysRevLett.123.142301

[10] B.B. Abelev et al. (ALICE), Multiplicity Dependence of Pion, Kaon, Proton and Lambda Production in p-Pb Collisions at $\sqrt{s_{NN}}$ = 5.02 TeV, Phys. Lett. B **728**, 25 (2014), [arXiv]1307.6796. 10.1016/j.physletb.2013.11.020

[11] S. Acharya et al. (ALICE), Multiplicity dependence of π, K, and p production in pp collisions at $\sqrt{s}$ = 13 TeV, Eur. Phys. J. C **80**, 693 (2020), [arXiv]2003.02394. 10.1140/epjc/s10052-020-8125-1

[12] J. Adam et al. (ALICE), Enhanced production of multi-strange hadrons in high-multiplicity proton-proton collisions, Nature Phys. **13**, 535 (2017), [arXiv]1606.07424. 10.1038/nphys4111

[13] S. Acharya et al. (ALICE), Investigating strangeness enhancement with multiplicity in pp collisions using angular correlations (2024), [arXiv]2405.14511

[14] J. Adam et al. (ALICE), Anisotropic flow of charged particles in Pb-Pb collisions at $\sqrt{s_{\mathrm{NN}}}$ = 5.02 TeV, Phys. Rev. Lett. **116**, 132302 (2016), [arXiv]1602.01119. 10.1103/PhysRevLett.116.132302

[15] V. Khachatryan et al. (CMS), Evidence for collectivity in pp collisions at the LHC, Phys. Lett. B **765**, 193 (2017), [arXiv]1606.06198. 10.1016/j.physletb.2016.12.009

[16] S. Acharya et al. (ALICE), Multiplicity and event-scale dependent flow and jet fragmentation in pp collisions at $\sqrt{s}$ = 13 TeV and in p–Pb collisions at $\sqrt{s_{\mathrm{NN}}}$ = 5.02 TeV, JHEP **03**, 092 (2024), [arXiv]2308.16591. 10.1007/JHEP03(2024)092

[17] R.N. Patra, Collective phenomena study in small systems using the bulk of particle production in high-multiplicity pp collisions at $\sqrt{s}$ =13 TeV with ALICE, in *Proceedings of The European Physical Society Conference on High Energy Physics — PoS(EPS-HEP2023)* (2024), Vol. 449, p. 214, [arXiv]2310.20613

[18] S. Acharya et al. (ALICE), Pseudorapidity distributions of charged particles as a function of mid- and forward rapidity multiplicities in pp collisions at $\sqrt{s}$ = 5.02, 7 and 13 TeV, Eur. Phys. J. C **81**, 630 (2021), [arXiv]2009.09434. 10.1140/epjc/s10052-021-09349-5

[19] S. Acharya et al. (ALICE), Production of light-flavor hadrons in pp collisions at $\sqrt{s}$ = 7 and $\sqrt{s}$ = 13 TeV, Eur. Phys. J. C **81**, 256 (2021), [arXiv]2005.11120. 10.1140/epjc/s10052-020-08690-5

[20] S. Acharya et al. (ALICE), Multiplicity dependence of light-flavor hadron production in pp collisions at $\sqrt{s}$ = 7 TeV, Phys. Rev. C **99**, 024906 (2019), [arXiv]1807.11321. 10.1103/PhysRevC.99.024906

[21] S. Chatrchyan et al. (CMS), Study of the Inclusive Production of Charged Pions, Kaons, and Protons in *pp* Collisions at $\sqrt{s}$ = 0.9, 2.76, and 7 TeV, Eur. Phys. J. C **72**, 2164 (2012), [arXiv]1207.4724. 10.1140/epjc/s10052-012-2164-1

[22] J. Steinheimer, J. Aichelin, M. Bleicher, Nonthermal p/π Ratio at LHC as a Consequence of Hadronic Final State Interactions, Phys. Rev. Lett. **110**, 042501 (2013), [arXiv]1203.5302. 10.1103/PhysRevLett.110.042501

[23] S. Acharya et al. (ALICE), Anisotropic flow of identified particles in Pb-Pb collisions at $\sqrt{s}_{\mathrm{NN}}$ = 5.02 TeV, JHEP **09**, 006 (2018), [arXiv]1805.04390. 10.1007/JHEP09(2018)006

[24] W. Zhao, C.M. Ko, Y.X. Liu, G.Y. Qin, H. Song, Probing the Partonic Degrees of Freedom in High-Multiplicity $p-Pb$ collisions at $\sqrt{s_{NN}}$ = 5.02 TeV, Phys. Rev. Lett. **125**, 072301 (2020), [arXiv]1911.00826. 10.1103/PhysRevLett.125.072301

[25] Y. Wang, W. Zhao, H. Song, Exploring the partonic collectivity in small systems at the LHC (2023), [arXiv]2401.00913

EPJ Web of Conferences **314**, 00040 (2024) https://doi.org/10.1051/epjconf/202431400040
QCD@Work 2024

Holographic phases of QCD

Akash Singh[1,*] and *K.P.* Yogendran[1,**]

[1]Department of Physical Sciences, IISER Mohali, Sector 81, Knowledge City, Punjab, 140306, India

Abstract. In this presentation, we will discuss the QCD phase diagram at finite temperature and finite baryon chemical potential using a hardwall holographic AdS/QCD model. We study bulk gravity solutions with appropriate boundary conditions that represent strongly interacting nuclear matter, deconfined quarks, and various possible condensates over a range of baryon chemical potential and temperature.

1 Introduction

The QCD phase diagram involves a variety of phases in the spectrum of temperature and baryon chemical potential. These phases can be observed in the form of hadrons (baryons and mesons) in the gaseous and liquid phases under low to intermediate conditions. In extreme conditions, they manifest as the quark-gluon plasma in heavy-ion collisions at very high temperatures, or as baryonic pairs (superconductors/superfluids), meson condensates, or quark pairs inside neutron stars at supranuclear densities [1]. However, it is challenging to compute the properties of the matter in these phases due to the strong nature of interactions among the constituents at low energies and/or low temperatures.

Lattice QCD has demonstrated its importance as a tool for making predictions at finite temperatures and zero chemical potential. However, it encounters limitations at finite chemical potentials due to the infamous sign problem. Alternatively, effective field theories such as chiral perturbation theory, the Nambu-Jona-Lasinio (NJL) model, the Hadron Resonance Gas (HRG) model, and others rely heavily on approximations, resulting in constrained regions of validity.

A modern method for understanding the QCD phase diagram is based on the AdS/CFT correspondence. There are two main approaches: top-down and bottom-up. The top-down approach is rooted in string theory, whereas the bottom-up approach is more phenomenological. Some examples of the top-down approach are the D3/D7 setup and the Witten-Sakai-Sugimoto models. The bottom-up approach includes Hardwall models, Softwall models, VQCD models, and more. The hardwall AdS/QCD holography has mostly been limited to the finite density *deconfined* phases, lacking a description of a finite density confined phase. In our work, we focus on a family of background geometries in a 5-dimensional hardwall framework, which we call the charged AdS geometry (CAdS). These have nonzero electric fields but do not have a black hole.

In section 2, we describe our 5-dimensional hardwall model. The possible solutions are discussed in section 3. We detail the phase diagram in the $\mu - T$ plane that we obtain with

*e-mail: akashsingh@iisermohali.ac.in
**e-mail: yogendran@iisermohali.ac.in

different choices of boundary conditions in section 4. Section 5 study the probe baryonic condensates. We conclude with a discussion and future directions in section 6.

2 Holographic hardwall model

The hardwall model is based on imposing an IR cutoff in the bulk radial coordinate. This breaks the conformal symmetry and enables the possibility of a confinement-deconfinement transition [2]. This work focuses on transitions from confining geometries to deconfining geometries at finite chemical potential. In this section, we discuss a 5-D holographic hardwall model based on AdS/QCD models. The model incorporates a bulk metric field $g_{\mu\nu}$, which is dual to the conserved boundary stress-energy tensor T_{ij} (here $\mu, \nu = (x_0, x_1, x_2, x_3, z)$ and i, j restricted to boundary directions only). The model also includes a $U(1)$ gauge field A_μ corresponding to the conservation of the baryon number in boundary theory (or the global $U(1)_B$ symmetry). Therefore, the minimally coupled Einstein-Maxwell action with Gibbons-Hawking and the boundary counter term in 5-dimensions is

$$S = \frac{N_c^2}{8\pi^2 L^3} \int d^5x \sqrt{g}\,(R - 2\Lambda) - \frac{\theta N_f N_c}{24\pi^2 L} \int d^5x \sqrt{g}\frac{F^2}{4} - \frac{N_c^2}{4\pi^2 L^3} \int d^4x \sqrt{\gamma}\left(\Theta - \frac{3}{L}\right) \quad (1)$$

with N_f, N_c, and L are the number of flavors, colors and AdS radius, respectively. The cosmological constant in the Einstein-Hilbert part of the action is set to be $2\Lambda = -\frac{12}{L^2}$ so that the asymptotic solution is AdS_5. Here γ is the induced metric on the UV surface $z = \epsilon$, and Θ is the trace of the extrinsic curvature associated with this surface.

The parameter θ is a new idea that needs to be explained. It allows for different choices of the internal compactification manifold of the Dp-branes as well as dilaton couplings to the Dp-brane. The normalization we consider in the F^2 term is such that when $\theta = 6$, we get the Maxwell action by expansion of the DBI action for the D7-brane.

3 Solutions/phases

In order to simplify the equations of motion and derive solutions with the appropriate boundary conditions, the first crucial step is to carefully select an ansatz for the fields. We consider a diagonal metric ansatz,

$$ds^2 = -g(z)\frac{L^2}{z^4}dt^2 + \frac{L^2}{z^2}d\vec{x}^2 + \frac{L^2}{g(z)}dz^2 \quad (2)$$

which implies that the (conformal) boundary is at $z = 0$. This ansatz respects the translational and rotational symmetry of the phases in which we are interested. The gauge field ansatz is $A_0 = \phi(z)$, $A_i = 0$ where $i = 1, 2, 3$ with gauge condition $A_z = 0$. The equations of motion with these ansatz are:

$$g'(z) - \frac{6g(z)}{z} + 4z - \zeta z^5\left(\frac{\phi'(z)^2}{2}\right) = 0\,; \qquad \phi''(z) - \frac{\phi'(z)}{z} = 0 \quad (3)$$

where the parameter $\zeta = \frac{\theta N_f}{9N_c}L^2$ controls the effect of the gauge field on the background (large N_c being the probe approximation).

There are three boundary conditions: one for g and two for ϕ. The functions $g(z)$ and $\phi(z)$ with suitable boundary conditions describe geometric solutions. There are three possible geometries:

1. **Thermal AdS:** zero density, confined phase.

$$g(z) = z^2 \; ; \qquad \phi(z) = \mu \tag{4}$$

2. **Charged Black Hole:** finite density, deconfined phase.

$$g(z) = z^2\left(1 - cz^4 + \frac{\zeta\mu^2}{z_H^4 L^2} z^6\right) ; \qquad \phi(z) = \mu\left(1 - \frac{z^2}{z_H^2}\right) \tag{5}$$

   where z_H is the black hole horizon. The parameter c is fixed by the condition that z_H is the smallest root of $g(z_H)$ for each μ (quark chemical potential).

3. **Charged AdS:** finite density, confined phase.

   The charged AdS is an overcharged horizonless geometry in the bulk that represents a finite-density confined phase of the boundary theory [3].

$$g(z) = z^2\left(1 - cz^4 + \frac{\zeta Q}{L^2} z^6\right) ; \quad \phi(z) = \pm\mu + Qz^2; \tag{6}$$

   with $\frac{\zeta^2 Q^4}{L^4 c^3} > \frac{4}{27}$ implying that the function $g(z)$ never has a root. Here, Q and c (or equivalently, $g(z_0)$) are free parameters that have to be fixed to completely determine the solutions. In Section 4, we will discuss the possible choices of boundary conditions to fix these parameters.

The thermodynamic quantities for these geometric phases (as shown in Table 1) can be obtained as follows: Finite quark chemical potential $\mu_Q = \mu = \frac{\mu_B}{N_c}$ as the non-normalizable mode of the gauge field $\phi(z)$. Following [2] or [4], we can set the temperature of the thAdS and CAdS to any finite value, the temperature in CBH is fixed by the regularity of the metric near the horizon, the pressure is computed using the Euclidean on-shell action,

Table 1. Thermodynamic Quantities.

	ThAdS	**CBH**	**CAdS**
Chemical potential	$\phi(\epsilon)$	$\phi(\epsilon)$	$\phi(\epsilon)$
Temperature	T	$\frac{1}{\pi z_H}\left\lvert 1 - \frac{\zeta\mu^2 z_H^2}{2L^2}\right\rvert$	T
Pressure	$\frac{N_c^2}{4\pi^2 z_0^4}$	$\frac{N_c^2}{8\pi^2}\left(\frac{1}{z_H^4} + \frac{\zeta\mu^2}{z_H^2 L^2}\right)$	$\frac{N_c^2}{8\pi^2}\left(\frac{z_0^2+g(z_0)}{z_0^6}\right)$

4 Boundary conditions and phase diagram

In CAdS geometries, we need to determine the boundary conditions at the IR cutoff that will result in equations of state. These different boundary conditions significantly change the boundary theory phases and, thus, the phase diagram.

4.1 A simple boundary condition

A simple choice of boundary conditions on the gauge field is to set $\phi(z_0) = 0$ which leads to an equation of state $Q(\mu) = \frac{\mu}{z_0^2}$. The other boundary condition $g(z_0)$ (or $\bar{g}_0$) is fixed via the

ρ-meson mass as a normalizable (isospin) gauge field plane wave fluctuation. For a detailed discussion, see [3].

This allows us to extend the confinement/deconfinement phase transition of [2] in the finite chemical potential region. At $T = 0$, the deconfined phase is preferred for a large chemical potential, as described by the following inequality,

$$\left(\frac{\mu\alpha_g}{M_g}\right)^2 > \frac{6N_c}{\theta N_f}\left(1 + \sqrt{4 + 3\alpha_\rho\left(\frac{M_\rho}{\alpha_g M_g} - 1.766\right)^4}\right) \tag{7}$$

where α_g and α_ρ are numerical constants.

4.2 Physically motivated boundary condition

Next, we will explore the IR boundary conditions such that the hardwall model represents the equation of state of a baryonic system at low to intermediate densities as determined by the effective field theory (EFT) methods.

That is to say, we choose $g(z_0)$ (and $\phi(z_0)$), so that the pressure (and the number density) is the pressure (and the number density) of baryonic degrees of freedom computed by EFT techniques.

$$\rho = \frac{\theta N_f N_c}{12\pi^2} Q\ ; \qquad g(z_0) = \bar{g}_0 + \frac{8\pi^2}{N_c^2}\left(p - \frac{2\pi^2 z_0^2}{\theta N_f N_c}\rho^2\right) \tag{8}$$

At low densities, a gas of interacting baryons might be described by in-medium chiral perturbation theory with the Van der Waals type phase [5]. At higher densities and/or temperatures, nuclear physics suggests the appearance of a quarkyonic phase, which is described by the two-flavor NJL model [6]. Therefore, we chose these descriptions and compared them with the deconfined (CBH) phase to obtain the phase diagram.

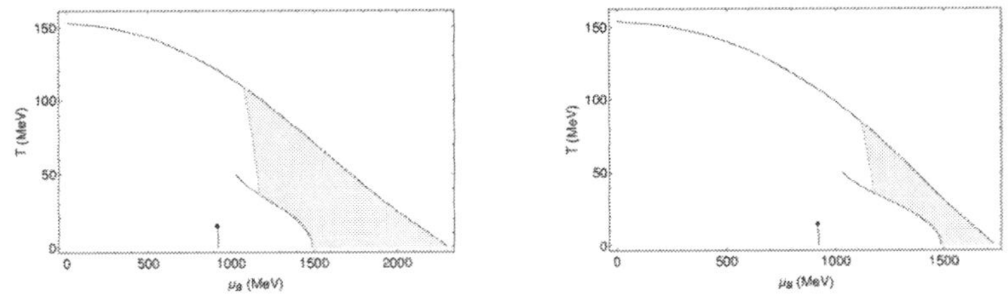

(a) $\zeta = 0.77$, $g_0 = 6$ (b) $\zeta = 1$, $g_0 = 6$

Figure 1. Phase diagram in $\mu - T$ plane for two different values of ζ.

The phase diagram for physical IR boundary conditions with van der Waals and quarkyonic phases is shown in figure 1. The van der Waals type phase is favored at low density. A liquid-gas phase transition represented by a solid green line exists, which ends at a critical point. As the chemical potential increases, the van der Waals phase makes a transition to the quarkyonic phase at intermediate density (blue line). Since the vdW description is valid only up to $T \sim 50$ MeV, we do not extend this coexistence line further. The quarkyonic matter melts into a deconfinement phase (charged black hole) at high densities (shown by the red line). The yellow region marks a phase where the chiral condensate σ vanishes indicated by $\frac{\sigma(T,\mu)}{\sigma(0,0)} < 0.1$.

5 Baryonic condensate solutions

The AdS/CFT dictionary allows us to map the nonzero vev of a condensate operator with the scaling dimensions Δ to a non-trivial massive complex scalar field profile without a node and vanishing non-normalizable mode. Therefore, further developing our idea, we include a complex scalar field ψ charged under $U(1)_B$. This scalar field is dual to the baryon condensate given the charge q of the field in the bulk and the scaling dimension of the condensate operator Δ in the boundary. The action for the Einstein-Maxwell-Scalar system with appropriate counter terms is,

$$S = S_0 - \lambda_s \int d^5x \sqrt{g}\left(|D\psi|^2 + m^2|\psi|^2\right) - \lambda_s \int d^4x \sqrt{\gamma}\frac{d-\Delta}{2L}|\psi|^2 \tag{9}$$

where S_0 is given in (1). The equation of motion for the scalar field in the *confined* backgrounds discussed in section 3 is,

$$\psi''(z) + \psi'(z)\left(-\frac{5}{z} + \frac{g'(z)}{g(z)}\right) - \left(\frac{m^2}{g(z)} - \frac{q^2 z^4 \phi(z)^2}{g(z)^2}\right)L^2\psi(z) = 0 \tag{10}$$

At the on-set of condensation, the scalar field will have small amplitude. Therefore, the on-set of condensation can be reliably estimated by studying the scalar field in the probe approximation. An interesting outcome of the probe analysis in CAdS geometries (at $T = 0$) is the vanishing of the condensate at high density signaled by the node appearing for the scalar field profile as shown in figure 2

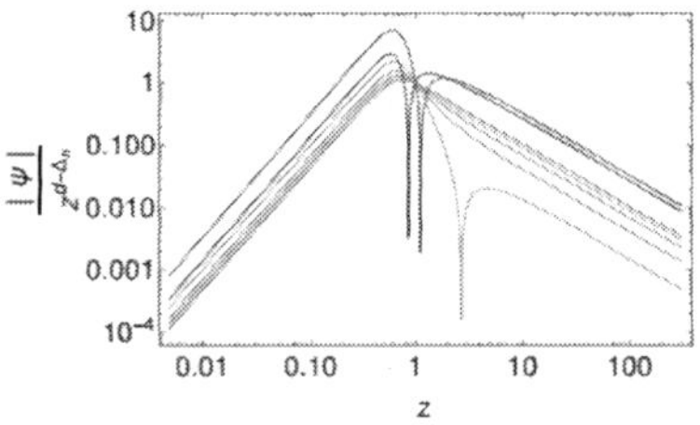

Figure 2. Extended absolute scalar field solution in a LOGLOG plot in the NJL phase. Different colors represent varying values of μ, with orange indicating the lowest and violet the highest.

It should be noted that in these cases, the value of the scalar field at the IR cutoff is being held fixed. The table 5 shows the on-set and off-set of the condensates in the Van der Waals and NJL phase in the probe approximation.

Table 2. The on-set and off-set of condensate.

Phase	Δ	q	sign	On-set μ (MeV)	Off-set μ (MeV)
VdW	4	6	-ve	1096.2	1139.7
VdW	4.25	6	-ve	1218.0	1261.5
VdW	4.5	6	-ve	1331.1	1374.6
NJL	3	2	+ve	1722.6	1914.0

The choice of q and Δ in the NJL phase is justified because the boundary condensate operator is expected to be composed of two quark fields in the quarkyonic phase. In contrast, in the VdW phase, the charge is $2N_c$, but the scaling dimension does not need to be $3N_c$ due to anomalous dimension corrections. Therefore, we show condensation in the VdW phase for a few values of Δ. It is noteworthy that the on- and off-set is shifting; however, the difference is always $\simeq 40$ MeV.

In an upcoming study, a comprehensive analysis of condensates with different values of Δ in the VdW phase and backreacted solutions will be presented.

6 Conclusion and Discussion

A simple 5-D holographic hardwall model provided low density confined phase, resulting in a complete phase diagram. By implementing the physical IR boundary conditions, we were able to match equations of state with the low density effective field theory results. The probe analysis of baryon condensates in the confined phase at zero temperature identifies a critical chemical potential beyond which the condensate disappears. The study for backreacted condensate solutions with physical boundary conditions is ongoing. The backreaction of the scalar field leads to a new set of boundary conditions at the IR cutoff, which results in interesting outcomes. The enhancement of the model by incorporating isospin and chiral condensate will be interesting for the QCD phenomenology. Another interesting question is to construct a top-down D3/D7 hardwall framework in 10-D with backreaction that could provide a finite density confined phase with baryons as solitons localized behind the cutoff.

References

[1] C. J. Pethick, T. Schaefer and A. Schwenk, "Bose-Einstein condensates in neutron stars," [arXiv:1507.05839 [nucl-th]]

[2] C. P. Herzog, "A Holographic Prediction of the Deconfinement Temperature," Phys. Rev. Lett. **98** (2007), 091601 doi:10.1103/PhysRevLett.98.091601 [arXiv:hep-th/0608151 [hep-th]]

[3] A. Singh and K. P. Yogendran, "Confined phases at finite density in the Hardwall model," [arXiv:2408.10986 [hep-th]]

[4] A. Singh and K. P. Yogendran, "Phases of a 10-D holographic hard wall model," JHEP **02** (2023), 168 doi:10.1007/JHEP02(2023)168 [arXiv:2208.09387 [hep-th]]

[5] S. Fiorilla, N. Kaiser and W. Weise, "Chiral thermodynamics of nuclear matter," Nucl. Phys. A **880** (2012), 65-87 doi:10.1016/j.nuclphysa.2012.01.003 [arXiv:1111.2791 [nucl-th]]

[6] S. P. Klevansky, "The Nambu-Jona-Lasinio model of quantum chromodynamics," Rev. Mod. Phys. **64** (1992), 649-708 doi:10.1103/RevModPhys.64.649

EPJ Web of Conferences **314**, 00041 (2024) https://doi.org/10.1051/epjconf/202431400041
QCD@Work 2024

Hagedorn temperature in confining gauge theories from holography

Francesco Bigazzi[1,*], *Tommaso* Canneti[1,2,**], *Federico* Castellani[1,2,***], and *Aldo L.* Cotrone[1,2,****]

[1]Istituto Nazionale di Fisica Nucleare, Sezione di Firenze, Via G. Sansone 1; 50019 Sesto Fiorentino (Firenze), Italy
[2]Dipartimento di Fisica e Astronomia, Università di Firenze, Via G. Sansone 1; 50019 Sesto Fiorentino (Firenze), Italy

Abstract. We provide a short overview on recent results on the Hagedorn temperature of confining gauge theories having a dual holographic string description.

1 Introduction

A quantum system displays a Hagedorn behavior if at high energies E the density of states grows exponentially like the energy, $\rho(E) \sim e^{E/T_H}$, where T_H is called the Hagedorn temperature. The thermal partition function of such a system diverges when $T \to T_H^-$. Examples include both quantum field theories (QFT) like Yang-Mills or quenched QCD and string theory. The occurrence of such kind of behavior in the realm of strong interactions was actually first pointed out by Rolf Hagedorn [1], who argued for an asymptotically exponential hadron mass spectrum with a limiting temperature which hence took his name. Today we would say that this is expected to hold for the confining phase of $SU(N)$ Yang-Mills or quenched (large N) QCD: these theories (for $N > 2$) display first order deconfinement transitions at some critical temperature $T_c < T_H$ so that the Hagedorn regime is actually never reached on a stable branch.

Computing the Hagedorn temperature of a quantum field theory is not an easy issue, in general. For instance we do not have, yet, first principle computations of T_H in Yang Mills theories from lattice studies, although the latter provide clear indications and indirect hints on the Hagedorn temperature both at small [2] and large N [3].

The Hagedorn temperature for bosonic and fermionic strings in flat spacetime is exactly known. What about string theory on curved backgrounds? This is an interesting question in view of the holographic string/gauge theory correspondence, which, for instance, maps large N confining gauge theories at strong coupling, to weakly coupled string theories on curved gravity backgrounds with fluxes. As a corollary of such a correspondence, the gauge theory Hagedorn temperature corresponds to that of the dual string theory. Can we thus compute the Hagedorn temperature of strongly coupled large N confining gauge theories from the

*e-mail: bigazzi@fi.infn.it
**e-mail: canneti@fi.infn.it
***e-mail: castellani@fi.infn.it
****e-mail: cotrone@fi.infn.it

holographic dual description? At first sight, this appears as a daunting task since, apart from very special examples,[1] we do not know how to compute the partition function of string theory on curved backgrounds with fluxes. A possible way out is suggested by a flat space analysis that we are going to review in the following section. This will provide the basis for our study of the Hagedorn temperature in holographic confining gauge theories. Remarkably, in cases where complementary QFT results at strong coupling are available, we will be able to present non trivial cross-checks of our results.

2 Strings in flat spacetime

Consider type II superstrings in flat ten-dimensional spacetime. It is a textbook result that the number of particle species with mass less than M asymptotes to

$$d(M) \sim M^{-10} e^{2\pi\sqrt{2\alpha'}M}, \tag{1}$$

from which we can read the Hagedorn temperature

$$T_H = \sqrt{\frac{T_s}{4\pi}} = \frac{1}{2\pi\sqrt{2\alpha'}}, \tag{2}$$

where $T_s = (2\pi\alpha')^{-1}$ is the string tension. In fact, when $T \to T_H^-$, the torus (one-loop) partition function of string theory diverges, signaling the occurrence of a tachyonic mode in the spectrum. The latter can be directly detected by considering the mass shell condition for string theory on a background obtained from the ten-dimensional flat one by compactifying the (Euclidean) time direction on a circle of length $\beta = 1/T$. As it was shown in e.g. [5], strings winding the thermal circle have a tachyonic mode when the size of the circle is small enough. In fact, the ground state with winding number ± 1 has mass

$$m_W^2 = \frac{2}{\alpha'}\left(\frac{1}{8\pi^2\alpha' T^2} - 1\right), \tag{3}$$

which actually becomes tachyonic when $T > T_H$. For T close to T_H this state corresponds to an almost massless complex scalar field χ (ofter referred to as the thermal scalar field) in target space, which can be accounted for in the low energy effective action as [5, 6]

$$S_\chi = \frac{1}{T}\int d^9x \left(\partial_i\chi^*\partial^i\chi + m_W^2\chi^*\chi\right). \tag{4}$$

What we want to review here is how the extension of the above results to string theory on certain classes of curved backgrounds allows to obtain information about the Hagedorn temperature of the dual confining gauge theories in the planar limit at strong coupling [7–10]. To this aim, the interplay between the worldsheet perspective and the thermal scalar effective description turns out to play a crucial role.

3 Strings on confining backgrounds

In certain classes of theories, the $T > 0$ confining phase of $(q+1)$-dimensional QFTs at low energy, is holographically mapped to closed string theory on backgrounds with metric asymptotically going as

$$ds^2 \approx 2\pi\alpha' T_s\left(1 + \frac{r^2}{l^2}\right)(dt^2 + \eta_{ij}dx^i dx^j) + dr^2 + r^2 d\Omega_{d-1}^2 + ds_M^2, \quad (r \ll l) \tag{5}$$

[1] See [4] for recent results on the asymptotic density of states of string theory on exact pp-wave backgrounds.

where $i, j = 1, \cdots q$, $t \sim t + \beta$, r is the radial direction dual to the field theory renormalization group energy scale, l is the curvature radius (with the $r \sim 0$ limit mapped to the infra-red regime of the dual QFT), T_s is the confining string tension of the gauge theory and $\mathcal{M}$ is a transverse, typically compact, $(9 - d - q)$-dimensional space. Notice the presence of a flat Euclidean d-dimensional subspace in the background.

The backgrounds typically include non trivial Ramond-Ramond field strengths and may also feature a running dilaton field and a non-zero Kalb-Ramond field. Let us assume here that the latter has no legs along the time direction (the general case has been treated as well in [10]), and consider a closed string at $r = 0$ and at a point in $\mathcal{M}$, winding once along the time circle so that its embedding includes

$$t(\tau, \sigma) = \frac{\beta}{2\pi}\sigma + \xi^0(\tau, \sigma), \tag{6}$$

where $\sigma \in [0, 2\pi]$ and τ are the string worldsheet coordinates.

Now, according to the results presented in [7–10], it is possible to obtain, at the quantum level, the spectrum of quadratic fluctuations of the worldsheet fields around that classical embedding, finding that the ground state of the winding string has a mass which becomes tachyonic when $T > T_H$, with T_H being the solution of

$$\frac{T_s}{2}\beta_H^2 = 2\pi[\Delta(\mu) + \delta\epsilon], \tag{7}$$

where

$$\mu = \frac{\beta_H}{2\pi}\frac{\sqrt{2\pi\alpha' T_s}}{l}, \tag{8}$$

is the mass of the d scalar fields y^I, with $y^I y^I = r^2$, at $T = T_H$, and

$$\Delta(\mu) = 1 - \frac{d}{2}\mu + d\mu^2 \log 2 + O(\mu^4), \tag{9}$$

is the zero-point energy of the worldsheet sigma model to second order in the mass parameter μ. The finiteness of $\Delta(\mu)$ is guaranteed by a mass-matching condition $\sum_b m_b^2 = \sum_f m_f^2$ involving the bosonic ($m_b = \mu$, with $b = 1, \cdots, d$)) and the fermionic (m_f, with $f = 1, \cdots, 8$) masses.

The $O(\mu)$ term in the expansion of $\Delta(\mu)$ arises from the contribution of the massive bosonic zero modes. As such it can also be captured by the thermal scalar effective approach, after suitably extending eq. (4) to the class of curved backgrounds considered here [11]. In general, within that framework, the Hagedorn temperature is obtained requiring that, neglecting the backreaction on the background, the linearized equation of motion for the thermal scalar admits a normalizable solution.

The $O(\mu^2)$ term in (9), instead, is due to the non-zero modes. The two-derivative effective approach misses such terms.[2] However, it provides another contribution, at the same order, which can be related to quartic terms in the bosonic zero modes in the worldsheet Hamiltonian. By construction, this contribution is not captured by the sigma model approach if one limits the analysis to quadratic order. The $\delta\epsilon$ piece in (7) precisely accounts for such a missing term (and higher order ones) which, being due to contributions of the bosonic zero modes, can be captured by the effective action for the thermal scalar field.

Solving eq. (7) we can determine T_H in a perturbative expansion in

$$\frac{\sqrt{\alpha'}}{l} \sim \lambda^{-k} \sim \frac{M_{gl}}{\sqrt{T_s}}, \quad k > 0, \quad \lambda \gg 1, \tag{10}$$

[2]Curvature corrections to the effective action can capture such kind of contributions [12]; however they come with coefficients which are not a-priori determined.

where λ is a 't Hooft-like coupling and M_{gl} is the glueball mass scale. To leading order we get [7]

$$T_H = \sqrt{\frac{T_s}{4\pi}}, \tag{11}$$

which is a simple generalization of the flat space relation (2), where T_s, the tension of the confining gauge theory, replaces the string tension in flat space.

To next-to-leading order (NLO) we get [8]

$$T_H = \sqrt{\frac{T_s}{4\pi}}\left[1 + \frac{d}{2\sqrt{2}}\frac{\sqrt{\alpha'}}{l}\right]. \tag{12}$$

This result coincides with the one obtained from the effective action for the winding string scalar mode [11].

As we have anticipated, we can go beyond NLO using an interplay of worldsheet and effective methods. The latter rely on a perturbative analysis of the equation of motion for the thermal scalar χ, whose low energy dynamics is captured by the generalization to curved space of the action in (4). In particular, the higher derivative corrections to the mass m_W^2 (see [12]) match with what we can single out from (7).

As an example, let us consider backgrounds involving a global AdS_{d+1} factor in the metric. Near the center of AdS, the latter can be written in the form (5), with $q = 0$, $l = R_{AdS}$ and $2\pi\alpha' T_s = 1$. The dual QFTs are conformal field theories (CFT) compactified on $S^{(d-1)}$ spheres. They are "confining" in the sense explained in [13], since they experience a first order transition between a low temperature phase with $O(1)$ degrees of freedom and a high temperature "deconfined" phase with $O(N^2)$ degrees of freedom. The critical temperature T_c for such a transition is set by the inverse radius of the $S^{(d-1)}$ sphere, hence, in turn, by R^{-1}_{AdS}. Setting $R_{AdS} = 1$, the Hagedorn temperature for such theories turns out to be given by [9, 14, 15]

$$T_H = \sqrt{\frac{g}{2\pi}} + \frac{d}{8} + \frac{d^2 + d - 8d\log 2}{32\sqrt{2}\pi^{3/2}\sqrt{g}} + \frac{4d^3 + 7d^2 - 2d}{1024\pi^2 g} + O(g^{-3/2}), \tag{13}$$

where, in the above mentioned units, $g = 1/4\pi\alpha'$ is proportional to some positive power of the 't Hooft coupling λ. Hence, T_H is parametrically larger than T_c. Remarkably, the NNLO coefficient of (13), can be entirely computed from a pure sigma model perspective including interaction terms in the worldsheet Hamiltonian which are quartic in the bosonic zero modes [10]. In principle, we could envisage the full result to be fixed without any ambiguity by a generalization of this method. Unfortunately, the latter is very demanding and the interplay with the effective approach turns out to be very convenient.

Notably, in two relevant cases, QFT results for T_H at strong coupling are available, via integrability and quantum spectral curve methods, which allow us to test the validity of the result in (13).

When $d = 4$ and the dual field theory is $\mathcal{N} = 4$ super-Yang-Mills on S^3, we have that $g = \lambda/4\pi$. Numerical quantum spectral curve results for T_H at strong coupling, as collected in [16] and, more recently in [14], show an impressive agreement with our analytic result (13) at NNNLO.

An analogous impressive agreement to NNNLO is found when $d = 3$ and the dual gauge theory is ABJM [17] on S^2. Here the numerical quantum spectral curve results at strong coupling have been found in [15].

4 Further extensions

The above holographic results on T_H have been extended in [18] in the following directions. From one side the effect of N_f quenched flavors on the Hagedorn temperature has been considered, finding that, in the scheme where the 't Hooft coupling is fixed, $T_H/\sqrt{T_s}$ decreases when N_f increases. Analogously, in the same scheme, $T_H/\sqrt{T_s}$ has been found to increase for increasing value of the Yang-Mills θ-angle.

Finally, embedding the string theory setup in M-theory, where the winding string is replaced by a double-winding $M2$-brane, it has been possible to compute, in certain cases, both non-perturbative instanton-like corrections to the value of T_H discussed above, and the value of T_H in a regime where the effective string coupling gets large, along the lines of what has been done in flat space in [19].

References

[1] R. Hagedorn, "Statistical thermodynamics of strong interactions at high energies. 3. Heavy-pair (quark) production rates," Nuovo Cim. Suppl. **6** (1968), 311-354 CERN-TH-751

[2] M. Caselle, A. Nada and M. Panero, "Hagedorn spectrum and thermodynamics of SU(2) and SU(3) Yang-Mills theories," JHEP **07**, 143 (2015) [erratum: JHEP **11**, 016 (2017)] https://doi.org/10.1007/JHEP07(2015)143

[3] B. Bringoltz and M. Teper, "In search of a Hagedorn transition in SU(N) lattice gauge theories at large-N," Phys. Rev. D **73**, 014517 (2006) https://doi.org/10.1103/PhysRevD.73.014517

[4] T. Canneti, "On the asymptotic density of states in solvable models of strings," [arXiv:2406.08405 [hep-th]]

[5] J. J. Atick and E. Witten, "The Hagedorn Transition and the Number of Degrees of Freedom of String Theory," Nucl. Phys. B **310** (1988), 291-334. https://doi.org/10.1016/0550-3213(88)90151-4

[6] G. T. Horowitz and J. Polchinski, "Selfgravitating fundamental strings," Phys. Rev. D **57** (1998), 2557-2563 https://doi.org/10.1103/PhysRevD.57.2557

[7] F. Bigazzi, T. Canneti and A. L. Cotrone, "On the Hagedorn temperature in holographic confining gauge theories," JHEP **01** (2023), 034 https://doi.org/10.1007/JHEP01(2023)034

[8] F. Bigazzi, T. Canneti and W. Mück, "Semiclassical quantization of the superstring and Hagedorn temperature," JHEP **08** (2023), 185 https://doi.org/10.1007/JHEP08(2023)185

[9] F. Bigazzi, T. Canneti and A. L. Cotrone, "Higher Order Corrections to the Hagedorn Temperature at Strong Coupling," JHEP **10** (2023), 056 https://doi.org/10.1007/JHEP10(2023)056

[10] F. Bigazzi, T. Canneti, F. Castellani, A. L. Cotrone and W. Mück, "Hagedorn temperature in holography: world-sheet and effective approaches," JHEP **09** (2024), 193 https://doi.org/10.1007/JHEP09(2024)193

[11] E. Y. Urbach, "The black hole/string transition in AdS_3 and confining backgrounds," JHEP **09** (2023), 156 https://doi.org/10.1007/JHEP09(2023)156

[12] T. Harmark, "Hagedorn temperature from the thermal scalar in AdS and pp-wave backgrounds," JHEP **06**, 140 (2024) https://doi.org/10.1007/JHEP06(2024)140

[13] E. Witten, "Anti-de Sitter space, thermal phase transition, and confinement in gauge theories," Adv. Theor. Math. Phys. **2** (1998), 505-532 https://doi.org/10.4310/ATMP.1998.v2.n3.a3

[14] S. Ekhammar, J. A. Minahan and C. Thull, "The asymptotic form of the Hagedorn temperature in planar super Yang-Mills," J. Phys. A **56** (2023) no.43, 435401 https://doi.org/10.1088/1751-8121/acf9d0

[15] S. Ekhammar, J. A. Minahan and C. Thull, "The ABJM Hagedorn Temperature from Integrability," JHEP **10** (2023), 066 https://doi.org/10.1007/JHEP10(2023)066

[16] T. Harmark and M. Wilhelm, "Solving the Hagedorn temperature of AdS_5/CFT_4 via the Quantum Spectral Curve: chemical potentials and deformations," JHEP **07** (2022), 136 https://doi.org/10.1007/JHEP07(2022)136

[17] O. Aharony, O. Bergman, D. L. Jafferis and J. Maldacena, "N=6 superconformal Chern-Simons-matter theories, M2-branes and their gravity duals," JHEP **10** (2008), 091 https://doi.org/10.1088/1126-6708/2008/10/091

[18] F. Bigazzi, T. Canneti, A. L. Cotrone and J. M. Penín, "Holographic Hagedorn Temperature: Strong Coupling, Flavor and θ-angle Effects," [arXiv:2407.07943 [hep-th]]

[19] J. G. Russo, "Free energy and critical temperature in eleven-dimensions," Nucl. Phys. B **602** (2001), 109-131 https://doi.org/10.1016/S0550-3213(01)00128-6

EPJ Web of Conferences **314**, 00042 (2024) https://doi.org/10.1051/epjconf/202431400042
QCD@Work 2024

QCD critical point: Recent developments

Mikhail Stephanov[1,2,*]

[1]Department of Physics, University of Illinois, Chicago, Illinois 60607, USA
[2]Kadanoff Center for Theoretical Physics, University of Chicago, Chicago, Illinois 60637, USA

Abstract. Recent developments aimed at mapping QCD phase diagram and the search for the QCD critical point in heavy-ion collisions are briefly reviewed.

1 Introduction

Mapping the phase diagram of QCD is a fascinating theoretical challenge with experimental implications ranging from the early Universe to neutron stars and to heavy-ion collision experiments. This report focuses on the recent developments in the search for the QCD critical point – the point where the crossover at low baryon chemical potential becomes the first-order phase transition separating the hadron gas and the quark-gluon plasma phases of QCD. The search for the critical point is a major component of the heavy-ion collision experiments at RHIC [1, 2] (see Fig. 1). A comprehensive theoretical understanding of critical point physics and its manifestations in heavy-ion collisions is crucial for the success of the search. Such a theory effort is underway, as reviewed, e.g., in [3, 4].

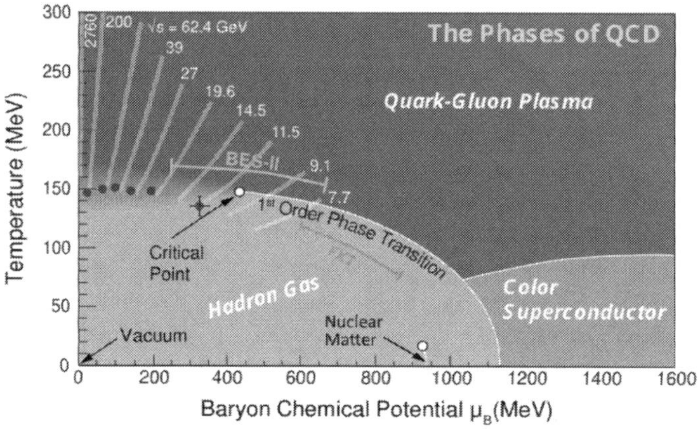

Figure 1. QCD phase diagram and beam energy scan experiments, Ref.[4].

*e-mail: misha@uic.edu

Determination of the location of the critical point by a first-principles theoretical calculation is a longstanding challenge due to the well-known sign problem in lattice Monte Carlo simulations at *finite* μ_B. The most recent results on this front have been obtained by using a broad range of theoretical techniques: Padé resummations of the Taylor expansion in μ_B, whose coefficients are calculated on the lattice [5], also using conformal maps [6], the hybrid lattice plus gauge-gravity correspondence approach (black-hole engineering) [7] and the functional renormalization group approach [8]. These are different methods with different potential sources of systematic errors. It is, therefore, notable, that they all point to the existence of the critical point in the same region of the QCD phase diagram: $T_c \sim 100-110$ MeV and $\mu_c \sim 420-650$ MeV.

2 Expected signatures and BES-II results

At a recent CPOD 2024 meeting in Berkeley, the STAR collaboration at RHIC presented the results of the latest beam energy scan (BES-II) experiment [9]. The experiment explored the region of the phase diagram marked BES-II in Figure 1 (where the collision energies $\sqrt{s}$ and corresponding fireball expansion histories are also indicated). To appreciate the significance of these results it is helpful to recall the expected signatures of the critical point, see e.g., review [3]. Figure 2 illustrates the expected contribution of the critical point to the baryon density susceptibility (left column), reflecting the QCD equation of state, and to the factorial cumulants of the proton multiplicity fluctuations (middle column), measured experimentally. While the approximate proportionality relationship between the order $n = 2$ (Gaussian) cumulants of these quantities has been known previously, a similar relationship for higher order cumulants ($n = 3$ and 4) has been established recently via the maximum entropy method in Ref.[10], summarized in Section 6.

The purpose of Figure 2 is to highlight the expected *qualitative* features of the *non-monotonic* dependence of these quantities on the collision energy $\sqrt{s}$. In summary: one expects a peak (or bump) in the 2nd and 3rd cumulants and a dip followed by a peak in the 4th cumulant as μ_B is increased, i.e., $\sqrt{s}$ is *decreased* [3].

The quantitative characteristics of these non-monotonic features — such as the position, height, and width of the peaks — are sensitively dependent on the quantitative properties of the QCD equation of state near the critical point — such as the location, strength and shape of the critical point singularity. While qualitative features of this singularity are universal, the quantitative properties can be described in terms of the non-universal parameters introduced in Ref.[12], see Section 3. The sensitivity to these unknown parameters makes it difficult to predict the critical point signatures quantitatively [13]. Conversely, observation of such signatures would allow us to determine or tightly constrain these important parameters of the QCD equation of state, which is the goal of the experiments.

While quantitative comparison to experimental data has to wait for more quantitative theoretical calculations taking into account not only the equilibrium expectations, but also important non-equilibrium effects, it might still be helpful to analyze the qualitative features of the recent experimental results shown in the *right* column of Figure 2 from the prospective of the equilibrium theory expectations (the *middle* column):

(1) The apparent dominant feature of the measured $n = 2$ cumulant is its monotonic decrease towards lower $\sqrt{s}$ (increasing μ_B). This effect has been already observed in BES-I data and attributed to baryon number conservation (see, e.g., Ref.[11]). What is notable is that this monotonic decrease is interrupted at lower collision energies $\sqrt{s} \lesssim 11$ GeV. Remarkably such an *excess* over the baseline would be consistent with the presence of the critical point as shown in the sketch for ω_2 in the middle column.

https://doi.org/10.1051/epjconf/202431400042

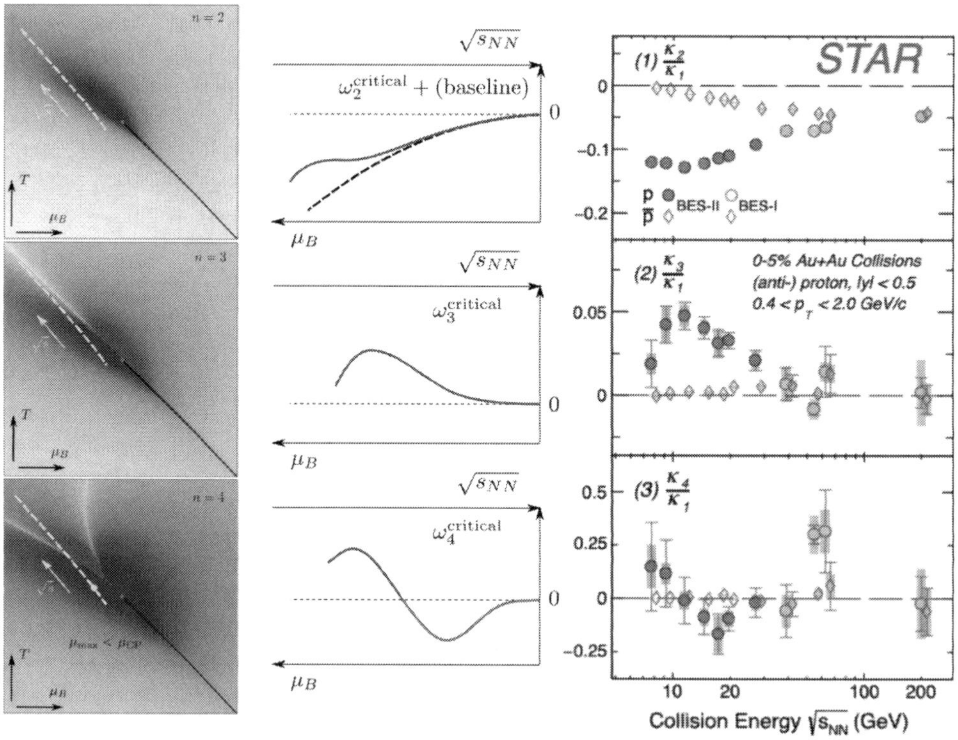

Figure 2. The signatures of the critical point expected in experimentally measured factorial cumulants of proton multiplicity. Each row in the *left column* shows a sketch of the QCD phase diagram region near the critical point (red circle) with the density plot of the baryon density cumulant/susceptibility χ_n for $n = 2, 3, 4$ [3]. The cumulants are positive/negative in the blue/red regions. The yellow dashed line is the freezeout trajectory – a set of points where the heavy-ion collision fireball freezes out as a function of the collision energy $\sqrt{s}$ (as $\sqrt{s}$ increases in the direction of the arrow). The yellow circle marks the point where the cumulant reaches maximum value. The *middle column* shows qualitatively the corresponding expectation for the collision energy dependence of the corresponding (normalized) factorial cumulant of the proton multiplicity $\omega_n \equiv \kappa_n/\kappa_1$ due to the critical fluctuations. For $n = 2$ also the noncritical baseline (evident in the experimental data from BES-I and understood as the consequence of baryon number conservation [11]) is added. The *right column* shows experimental results from BES-II at RHIC reported by STAR [9].

(2) The dominant feature of the measured $n = 3$ factorial cumulant is the *peak* at around $\sqrt{s} = 11$ GeV. This feature is also in apparent qualitative agreement with the equilibrium expectations for ω_3 shown in the middle column.

(3) The dominant feature of the $n = 4$ factorial cumulant data is the *dip* at $\sqrt{s} \approx 19$ GeV. This feature also appears to be in qualitative agreement with the equilibrium critical point predictions (middle column) [3, 14, 15]. However, to establish the *peak* feature expected from the critical point at lower $\sqrt{s}$, the data at $\sqrt{s} < 7.7$ GeV would be necessary. Such data could be provided by the fixed target (FXT) component of BES-II (see Fig. 1) and/or future dedicated experiments.

One must keep in mind that the *non-critical* baseline should be established in order to reach definitive conclusions about the origin of the apparent features of the data, such as, e.g., the peak in ω_3 or the dip in ω_4.[1] In this regard, the non-monotonicity is a key property of critical fluctuations, which distinguishes them from the monotonic baseline contributions.

It is important to highlight that the critical point on the phase diagram is located at a higher chemical potential μ_B compared to the position of the maximum for each of the cumulants: $\mu_{\rm CP} > \mu_{\rm max}$, as indicated in Fig. 2, left column. Therefore, *if* the features of the experimental data discussed above are due to the critical point, this critical point has to be located at higher μ_B than 420 MeV — the freezeout point at the lower end of the collision energy range reported in Ref.[9], where the interesting non-monotonic features of the data discussed above appear. Notably, such a critical point location would be also consistent with the recent theoretical estimates [5–8] briefly discussed in Section 1.

It is important to remember that the expectations discussed above are based on the assumption of (local) thermal equilibrium of critical fluctuations. Non-equilibrium effects are important in heavy-ion collisions, of course, and particularly so for determining the critical fluctuation signatures, see e.g., Ref.[16, 17]. The recent developments in describing the non-equilibrium evolution of fluctuations are reviewed in Ref.[4, 18] and, briefly, in Section 5.

The connection between the QCD phase diagram and experimental observables runs through hydrodynamic evolution with a given equation of state, which includes also the evolution of fluctuations, followed by freezeout to convert these hydrodynamic fluctuations into the observable fluctuations of particle multiplicities. The following sections highlight recent advances along this outline.

3 Equation(s) of state of QCD for hydrodynamic calculations

Since the equation of state of QCD in the finite μ_B region where one searches for the critical point is not known, it is necessary to have an efficient parametrization of possible such equations of state, which could then be discriminated by experimental data. Some information about the equation of state is available from lattice calculations, but this information is limited to the region of small μ_B, typically $\mu_B \lesssim 2 - 3T$ by the estimated convergence radius of Taylor expansion. On the other hand, universality of critical singularity constrains the equation of state near the critical point itself up to a few unknown non-universal parameters. Such parameters were introduced and standardized in Ref.[12], where a parametric family of equations of state, which match the lattice data at $\mu_B = 0$ and include a universal critical point singularity, was constructed.

In a more recent work, Ref.[19], a significantly improved parametrization of the QCD equation of state was developed. Its main additional ingredients include the use of the T' expansion scheme, introduced in Ref.[20]. This scheme amounts, essentially, to reorganization of the Taylor expansion in powers of μ_B from being an expansion at fixed T into an expansion at fixed $T'(\mu_B, T)$, where $T'(\mu_B, T) = T_c$ is the equation for the crossover line. In addition, the mapping of T and μ_B on the QCD phase diagram to r and h parameters on the Ising model phase diagram takes into account the C-symmetry $\mu_B \to -\mu_B$:

$$\frac{T'(\mu_B, T) - T_c}{T_T' T_c} = -hw\frac{\sin(\alpha_1 - \alpha_2)}{\cos\alpha_1}\,; \quad \frac{\mu_B^2 - \mu_{B,c}^2}{2\mu_{B,c} T_c} = -w(r\rho\cos\alpha_1 + h\cos\alpha_2)\,. \qquad (1)$$

where the parameters, introduced in Ref.[12], control the location $(T_c, \mu_{B,c})$, the strength (w) and the shape $(\rho, \alpha_{1,2})$ of the otherwise universal critical point singularity.

[1]The various non-critical baselines presented in Ref.[9] fail to adequately explain the data across all collision energies. Generally, the non-critical baselines are monotonic, which prevents them from fully capturing the non-monotonic features of the data.

As a result, the parametric equation of state proposed in Ref.[19] can be used for hydrodynamic calculations in a wider region of the QCD phase diagram, importantly, including μ_B up to 700 MeV, and for a wider range of the critical point parameters.

4 Hydrodynamic trajectories near QCD critical point

Hydrodynamics is a very universal theory which describes a wide range of physical systems from astrophysical, such as neutron stars, to subatomic, such as heavy-ion collisions. Given the equation of state and initial conditions it governs the evolution of hydrodynamic variables – conserved densities characterizing the local (near) equilibrium state of the fluid. Since mapping the QCD phase diagram and determining the equation of state is the goal of heavy-ion collisions, hydrodynamics is the primary tool used to connect the experimental measurements to the underlying QCD equation of state – see Ref.[4] for a recent review.

In order to predict the signatures of the critical point taking into account non-equilibrium effects it is crucial to understand the trajectories on the phase diagram traced by expanding and cooling fireball during its evolution towards freezeout. Under often used, and reasonable, zeroth order approximation of ideal hydrodynamics these trajectories follow the lines of constant specific entropy $\hat{s} \equiv s/n$, where s is the entropy density and n is the baryon density. The singularity of the equation of state near the critical point is responsible for an intricate behavior of these trajectories, which has been understood and classified recently in Ref.[21].

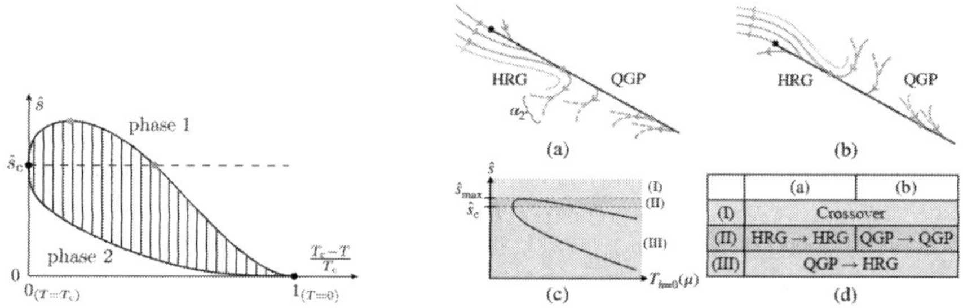

Figure 3. Left: A sketch of the specific entropy as a function of temperature along the phase coexistence line. Right: The classification of isentropic trajectories according to the way they behave in the vicintiy of the critical point. See Ref.[21].

The key to this behavior is the non-monotonicity of $\hat{s}$ on one of the branches of the coexistence line (first-order phase transition). The non-monotonicity is a robust consequence of the universality of the singularity at the critical point, which requires $\hat{s}$ to increase away from the critical point, combined with the third law of thermodynamic, which requires $\hat{s}$ to vanish at zero temperature, as illustrated in Fig.3 (left). This gives rise to an interesting pattern of the trajectories near the QCD critical point shown in Fig. 3 (right). The unusual (class II) trajectories enter and reemerge on the *same* side of the coexistence line [21]. Possible experimental consequences of such a behavior are under investigation.

5 Hydrodynamic evolution of fluctuations

In equilibrium, fluctuations of (local) thermodynamic variables are completely determined by equation of state. In particular, the universal singular behavior of fluctuations near a critical point allows us to search for the QCD critical point in experiments. However, hydrodynamic evolution is characterized by a sustained deviation of the expanding fireball from equilibrium. These deviations are manifested in the departure of the fluctuation measures from their equilibrium values. The evolution of these measures has been a subject of recent research whose progress is reviewed, e.g., in Refs.[4, 18].

A major recent development is the derivation of the deterministic equations for the evolution of the *non-gaussian* fluctuation correlators, which are crucial for the QCD critical point search [22, 23].

Fluctuations are characterized locally by the n-point correlators $H_n \equiv \langle \delta\psi(x_1)\ldots\delta\psi(x_n)\rangle$ of hydrodynamic variables, ψ. The evolution equations for these correlators can be derived from stochastic hydrodynamics. The equations can be expanded in and organized by the powers of the small parameter of hydrodynamics – the ratio of the microscopic and hydrodynamic scales – which controls the smallness of the fluctuations. In the diagrammatic representation introduced in Ref.[22] this is a loop expansion. The lowest order, tree level equations were derived in Ref.[22]. Loop corrections lead to the renormalization of hydrodynamic parameters and the so-called "long-time tails".

The quantities particularly suited for writing and solving fluctuation equations are the Wigner transforms of the correlators H_n, which require a generalization of the well-known two-point Wigner transform to $n > 2$-point correlators. Such a generalization, $W_n(x, \boldsymbol{q}_1 \ldots \boldsymbol{q}_n)$, was introduced in Ref.[22]. Physically, W_n describes the magnitude of the correlations of fluctuations with wave-vectors $\boldsymbol{q}_1, \ldots \boldsymbol{q}_n$ (in the local rest frame of the fluid at point x).

6 Freezeout and maximum entropy

The correlators of hydrodynamic fluctuations $H_n = \langle \delta\psi(x_1)\ldots\delta\psi(x_n)\rangle$ discussed in the previous section cannot be measured directly. Instead, experiments measure the correlators of particle multiplicity fluctuations characterized by the event averages $G_n \equiv \langle \delta f(x_1, \boldsymbol{p}_1)\ldots\delta f(x_n, \boldsymbol{p}_n)\rangle$ (with space-time variables x integrated over the freezeout surface), where $f(x, \boldsymbol{p})$ are the particle phase space distribution functions.

The measured particle correlators G_n are related to the hydrodynamic correlators H_n. In particular, certain integrals of the particle correlators over momentum space variables $\boldsymbol{p}_1 \ldots \boldsymbol{p}_n$ are constrained by local conservation laws (of energy, baryon number, etc.) to be equal to the corresponding hydrodynamic correlators. However, of course, there is not enough information in the conservation law constraints alone to determine the particle correlators G_n from given hydrodynamic correlators H_n.

A solution to this problem has been found recently in Ref.[10] by applying the maximum entropy method (MEM). By requiring the entropy of the resonance gas with given correlators G_n to be maximized under the conservation law constraints one obtains a unique solution.

Such a solution passes several nontrivial checks. In particular, applied to the case $n = 1$ it reproduces the well-known and time-tested Cooper-Frye method of freezeout. For critical fluctuations, the MEM matches the most singular contribution to G_n, induced by fluctuations of the effective critical field σ in an earlier approach [3, 14, 17, 24]. MEM goes further, however, and can describe also subleading and even non-critical contributions to multiplicity fluctuations within the same formalism.

In a nutshell, the key observation from the MEM is that the deviations of the hydrodynamic fluctuations from an uncorrelated ideal resonance gas correspond directly to the *factorial* cumulants of the particle multiplicities.[2] This observation highlights the importance

[2]More precisely [10], the MEM matches irreducible relative correlators (IRCs) for particle multiplicities, $\hat{\Delta}G_n$, where lower order correlations are subtracted, to similarly constructed hydrodynamic IRCs, $\hat{\Delta}H_n$. The factorial cumulants, κ_n, up to usually negligible quantum statistics effects, are equal to the phase-space integrals of $\hat{\Delta}G_n$.

of *factorial* cumulants as experimental fluctuation measures since these cumulants are more directly related to the QCD equation of state than the normal cumulants.[3]

7 Conclusions and Outlook

The release of the BES-II data by STAR represents a major step towards uncovering the structure of the QCD phase diagram. It is remarkable that the non-monotonic features of the data are in qualitative agreement with the expectations from equilibrium thermodynamics near the QCD critical point, if one assumes such a point is located at $\mu_B \gtrsim 420$ MeV. Such a location of the critical point would be consistent with recent estimates from various theoretical approaches. A comprehensive quantitative comparison of theory with experimental data necessitates dynamic simulations of heavy-ion collisions, incorporating critical fluctuations, which are currently under development.

This work is supported by the U.S. Department of Energy, Office of Science, Office of Nuclear Physics Award No. DE-FG0201ER41195.

References

[1] M.M. Aggarwal et al. (STAR), An Experimental Exploration of the QCD Phase Diagram: The Search for the Critical Point and the Onset of De-confinement (2010), 1007.2613

[2] G. Odyniec, RHIC Beam Energy Scan Program: Phase I and II, PoS **CPOD 2013**, 043 (2013). 10.22323/1.185.0043

[3] A. Bzdak, S. Esumi, V. Koch, J. Liao, M. Stephanov, N. Xu, Mapping the Phases of Quantum Chromodynamics with Beam Energy Scan, Phys. Rept. **853**, 1 (2020), 1906.00936. 10.1016/j.physrep.2020.01.005

[4] L. Du, A. Sorensen, M. Stephanov, The QCD phase diagram and Beam Energy Scan physics: a theory overview (2024), 2402.10183. 10.1142/S021830132430008X

[5] D.A. Clarke, P. Dimopoulos, F. Di Renzo, J. Goswami, C. Schmidt, S. Singh, K. Zambello, Searching for the QCD critical endpoint using multi-point Padé approximations (2024), 2405.10196

[6] G. Basar, QCD critical point, Lee-Yang edge singularities, and Padé resummations, Phys. Rev. C **110**, 015203 (2024), 2312.06952. 10.1103/PhysRevC.110.015203

[7] M. Hippert, J. Grefa, T.A. Manning, J. Noronha, J. Noronha-Hostler, I. Portillo Vazquez, C. Ratti, R. Rougemont, M. Trujillo, Bayesian location of the QCD critical point from a holographic perspective (2023), 2309.00579

[8] Y. Lu, F. Gao, Y.X. Liu, J.M. Pawlowski, QCD equation of state and thermodynamic observables from computationally minimal Dyson-Schwinger equations, Phys. Rev. D **110**, 014036 (2024), 2310.18383. 10.1103/PhysRevD.110.014036

[9] A. Pandav (STAR collaboration), plenary talk at CPOD 2024, `https://conferences.lbl.gov/event/1376/contributions/8772/`

[10] M.S. Pradeep, M. Stephanov, Maximum Entropy Freeze-Out of Hydrodynamic Fluctuations, Phys. Rev. Lett. **130**, 162301 (2023), 2211.09142. 10.1103/PhysRevLett.130.162301

[3] The importance of factorial cumulants (FC), as compared to normal cumulants (NC), has been understood before [25, 26]. In particular, FCs measure deviations of the fluctuations from a Poisson distribution – the natural baseline for *discrete* variables such as multiplicities. These deviations are due to "genuine" (irreducible) correlations between the particles. Finally, the acceptance dependence of FCs (power scaling with Δy) is much simpler than that of NCs.

[11] V. Vovchenko, V. Koch, C. Shen, Proton number cumulants and correlation functions in Au-Au collisions at sNN=7.7–200 GeV from hydrodynamics, Phys. Rev. C **105**, 014904 (2022), 2107.00163. 10.1103/PhysRevC.105.014904

[12] P. Parotto, M. Bluhm, D. Mroczek, M. Nahrgang, J. Noronha-Hostler, K. Rajagopal, C. Ratti, T. Schäfer, M. Stephanov, QCD equation of state matched to lattice data and exhibiting a critical point singularity, Phys. Rev. C **101**, 034901 (2020), 1805.05249. 10.1103/PhysRevC.101.034901

[13] J.M. Karthein, M.S. Pradeep, K. Rajagopal, M. Stephanov, Y. Yin, Equilibrium expectations for non-Gaussian fluctuations near a QCD critical point, in *21st International Conference on Strangeness in Quark Matter 2024* (2024), 2409.16249

[14] M.A. Stephanov, On the sign of kurtosis near the QCD critical point, Phys. Rev. Lett. **107**, 052301 (2011), 1104.1627. 10.1103/PhysRevLett.107.052301

[15] E. Shuryak, Four-nucleon clustering near the QCD critical point: theory versus experiment (2024), 2405.16617

[16] S. Mukherjee, R. Venugopalan, Y. Yin, Real time evolution of non-Gaussian cumulants in the QCD critical regime, Phys. Rev. C **92**, 034912 (2015), 1506.00645. 10.1103/PhysRevC.92.034912

[17] M. Pradeep, K. Rajagopal, M. Stephanov, Y. Yin, Freezing out fluctuations in Hydro+ near the QCD critical point, Phys. Rev. D **106**, 036017 (2022), 2204.00639. 10.1103/PhysRevD.106.036017

[18] M. Stephanov, QCD Critical Point and Hydrodynamic Fluctuations in Relativistic Fluids, Acta Phys. Polon. B **55**, 5 (2024), 2403.03255. 10.5506/APhysPolB.55.5-A4

[19] M. Kahangirwe, S.A. Bass, E. Bratkovskaya, J. Jahan, P. Moreau, P. Parotto, D. Price, C. Ratti, O. Soloveva, M. Stephanov, Finite density QCD equation of state: Critical point and lattice-based T' expansion, Phys. Rev. D **109**, 094046 (2024), 2402.08636. 10.1103/PhysRevD.109.094046

[20] S. Borsányi, Z. Fodor, J.N. Guenther, R. Kara, S.D. Katz, P. Parotto, A. Pásztor, C. Ratti, K.K. Szabó, Lattice QCD equation of state at finite chemical potential from an alternative expansion scheme, Phys. Rev. Lett. **126**, 232001 (2021), 2102.06660. 10.1103/PhysRevLett.126.232001

[21] M.S. Pradeep, N. Sogabe, M. Stephanov, H.U. Yee, Nonmonotonic specific entropy on the transition line near the QCD critical point, Phys. Rev. C **109**, 064905 (2024), 2402.09519. 10.1103/PhysRevC.109.064905

[22] X. An, G. Başar, M. Stephanov, H.U. Yee, Evolution of Non-Gaussian Hydrodynamic Fluctuations, Phys. Rev. Lett. **127**, 072301 (2021), 2009.10742. 10.1103/PhysRevLett.127.072301

[23] X. An, G. Basar, M. Stephanov, H.U. Yee, Non-Gaussian fluctuation dynamics in relativistic fluids, Phys. Rev. C **108**, 034910 (2023), 2212.14029. 10.1103/PhysRevC.108.034910

[24] M.A. Stephanov, Non-Gaussian fluctuations near the QCD critical point, Phys. Rev. Lett. **102**, 032301 (2009), 0809.3450. 10.1103/PhysRevLett.102.032301

[25] B. Ling, M.A. Stephanov, Acceptance dependence of fluctuation measures near the QCD critical point, Phys. Rev. C **93**, 034915 (2016), 1512.09125. 10.1103/PhysRevC.93.034915

[26] A. Bzdak, V. Koch, N. Strodthoff, Cumulants and correlation functions versus the QCD phase diagram, Phys. Rev. C **95**, 054906 (2017), 1607.07375. 10.1103/PhysRevC.95.054906

EPJ Web of Conferences **314**, 00043 (2024) https://doi.org/10.1051/epjconf/202431400043
QCD@Work 2024

Physics at future accelerators

Barbara Mele[1,*]

[1]INFN, Sezione di Roma, c/o Dip. di Fisica, Sapienza Università di Roma, 00185 Rome, Italy

Abstract. New high-energy colliders are currently under discussion, and will potentially be operational after the completion of the LHC/HL-LHC program. They could open up scenarios for the exploration of the Standard Model of fundamental interactions and its possible extensions with unprecedented levels of precision and/or effective collision energy. In this talk, the strengths of the physics potential of the various projects will be summarized and highlighted.

1 Introduction

Advancing in the fascinating exploration of the fundamental interactions in particle physics requires large infrastructures, where high-energy collisions involving fundamental matter constituents can be realized and studied. There are currently quite a few large facilities under consideration to move forward in high-energy collisions in the post CERN Large-Hadron-Collider (LHC) era [1],[2].

We will summarise the physics potential of such facilities. In particular, we will first focus on the physics case of the projects that might be operational by 2035-50 (that is FCC-ee, CepC, ILC, CLIC [1],[2]), which are all electron-positron colliders, typically dedicated to precision physics.

Then, we will discuss (less technologically mature) projects at larger c.m. collision energies $\sqrt{S}$, that might be operational by 2055 or later. They address energy-frontier physics (also named *discovery* physics). We will mainly focus on FCC-hh (a proton-proton collider at $\sqrt{S}\sim 100$ TeV) and a multi-TeV muon collider with $\sqrt{S}\sim 10+$ TeV.

In figure 1, we show the expected timelines for the facilities considered.

Before addressing its *physics case* (no doubt, of fundamental relevance in estimating a particular collider scientific value), we want to stress that assessing a future accelerator project is by now a highly *multi-dimensional* task, involving a large number of different and complicated issues, that we list in the following :

- project feasibility, maturity of technology, and technical risk
- degree of innovation
- construction/operation costs (versus constraints from funding agencies)
- power consumption and carbon footprint
- start-up time
- total operation time (connected also to staging and expandability in c.m. energy)

*e-mail: barbara.mele@roma1.infn.it

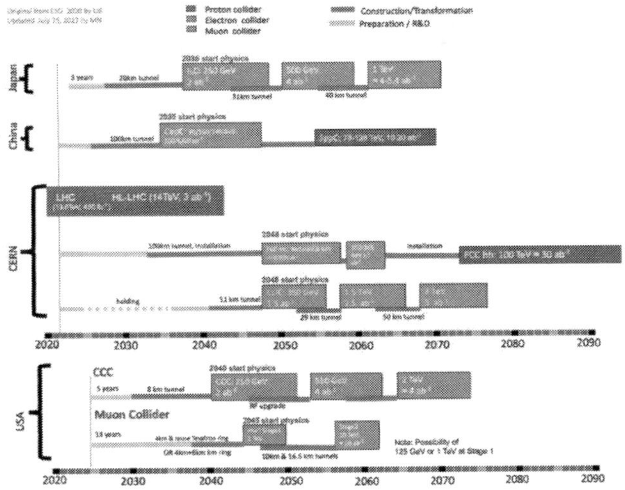

Figure 1. Projected timelines for preparation, construction, and physics operations for some of the leading proposed future collider options, as from [3].

- location versus infrastructures versus politics (discussed in a global context)
- HEP community support (both at regional and international level)
- fraction of present HEP community involved.

There is also a major caveat when discussing the physics potential for a particular project. Presently, we can just give a tentative picture of what could be the actual potential of a project that will be operational in about 15-20 years, or more. For instance, by comparing the expectations in Technical Design Reports of the initial LHC experiments with what has actually been accomplished at the LHC today, it is clear that what was initially considered *too challenging to reach* became instead *possible* in many many cases. Even more is expected during the forthcoming LHC High-Luminosity (HL-LHC) phase.

In Sect. 2 we will review the present theory status, and outline the main strategies allowing to improve our knowledge at higher energy colliders. In Sect. 3, we will go through and highlight the strengths of the physics potential for various colliders. Finally, in Sect. 4, we will comment on the forthcoming process of the update of European Strategy for Particle Physics (ESPP).

2 Our present Physics Vision

Our present PhysicsVision in particle physics is largely based on the fact that the huge amount of data collected at the LHC experiments fits the SM predictions at an amazing level of accuracy. Indeed, the Standard Model (SM) can today be considered THE THEORY correctly describing all phenomena at the LHC energy scale (~few TeV's). At the moment, we do not have any real hint of Beyond the Standard Model (BSM) physics, although a lot of work in the LHC experiments has been dedicated to search for any possible BSM signal. On the other hand, bounds on the masses of new heavy states predicted by many BSM models have been widely extended, and the simplest versions of different BSM models look quite Fine-Tuned on the basis of the LHC analyses.

Nevertheless, we have great (although quite *foggy* at the moment) expectations for new BSM phenomena at high energy colliders ! Indeed, there are presently two kinds of issues in the SM. The first is the existence of phenomena not described by the theory: notably a quantum theory of gravity, plus a series of empirical evidences (such as Dark Matter, Barion Asymmetry, neutrino masses). The second issue is the internal *poor consistency* of the model, which is connected to the ElectroWeak Symmetry Breaking (EWSB) (or Higgs) sector.
In fact, the Higgs sector looks like an effective description of something more fundamental. This is because a number of issues characterising the Higgs Lagrangian:

- the Higgs field is the only *fundamental* scalar particle in the SM; there could well be a microscopic interpretation for it that is actually still missing
- the Higgs scalar field is not protected by symmetries (in a sense, the Higgs Lagrangian is the less constrained sector of the SM Lagrangian), which gives rise to the *naturalness* problem, for which the Higgs mass is naturally driven to the energy cut off of the theory
- there are many different unexplained couplings in the Higgs Lagrangian, all fixed by masses (proliferation of parameters historically leads to breakdown in theory models...)
- there is a clear (and unexplained) fermion masses (Yukawa couplings) *hierarchy* (and neutrinos might have a special role in this respect)
- the trilinear Higgs self-coupling λ determines the shape and evolution of the Higgs potential, hence having crucial cosmological consequences.

One can wonder what is the most effective way to proceed to clear up the above issues in the post-LHC era.

It is worth stressing that presently high-energy colliders are still *by far* the most powerful instrument we know to probe physics at smaller length scales.

At colliders one can presently identify four main strategies to advance in high-energy physics (HEP):

- exploring the characteristics of the Higgs sector, and either confirming or spoiling the SM picture (which is of primary relevance, since the Higgs sector is so critical)
- searching for new heavy states coupled to the SM, acting as a cut-off for the SM, and possibly explaining the *naturalness* issue and/or non-SM phenomena (dark matter, ...); these have extensively been searched for but not yet found at LHC in minimal version of different theory models
- looking for new *dark* states (i.e., states uncoupled to SM at tree level) either in production or/and heavy-state (Higgs, top,...) decays, which can in general be characterized by elusive signatures, and may give rise to long-lived particles
- exploring $\Lambda \gg o(1)$TeV indirect effects (where Λ is the energy scale typical of the BSM framework) through high-accuracy studies of SM cross sections or kinematical distributions, and searches for rare processes, which is done via an Effective Field Theory (EFT) parametrization.

At this stage, every single method out of the above list is of fundamental importance to make progress in our understanding of fundamental interactions in collider physics.

In particular, e^+e^- colliders can have great opportunities in all sectors, also thanks to the event cleanness (implying model independence), and the high level of accuracy in the measurements. Indeed, going from hadron to lepton colliders, life gets much easier for the following reasons :

- the typical outcome of colliding pointless leptons is similar to what one would *ideally* obtain by setting the colliding parton c.m. energy at a hadron collider

- the complexity of collisions collapses
- kinematics is well defined
- there is a dramatic drop in backgrounds
- events are clean and simple
- pile up typically vanishes.

This is why there is by now a quite general consensus on an e^+e^- Higgs factory as the next collider to build, benefitting also from a more mature accelerator technology.

From what discussed, the precision needed in the Higgs coupling measurements is a critical parameter to evaluate a particular facility potential. In general, the BSM relative impact on Higgs couplings (assuming natural models not showing up by new heavy states production at LHC) can be up to a few percent [4]. Furthermore, different patterns of deviations from the SM couplings for different New Physics (NP) models are expected. The trilinear Higgs self-coupling has a special role, and in a sense is the most exposed coupling to BSM effects. It has an impact on the Higgs vacuum stability, and for particular anomalous deviations could explain Baryogenesis from a cosmological ElectroWeak (EW) phase transition. The maximal trilinear coupling deviation which would be compatible with no other BSM observation at colliders goes from few % to about 20% [5]. This is therefore the target for both theory and experimental accuracies.

3 Physics programme at future projects

The FCC research infrastructure (based on a new 90.7 km tunnel at CERN to host multiple colliders) foresees different stages [6]-[9] :

- FCC-ee: electron-positron collider @ 91, 160, 240, 365 GeV
- FCC-hh: proton-proton collider @ 100 TeV, and heavy-ions (Pb) @39 TeV
- FCC-eh: electron-proton collider @ 3.5 TeV.

There is currently a Feasibility Study ongoing for the FCC at CERN, evaluating the large amount of different aspects that should be considered in order to assess the actual feasibility of the project, according to the lines discussed in Sect. 1. The final FCC Feasibility Study Report will be ready by March 2025, to be evaluated by the next ESPP. The project approval by the CERN Council, if any, is expected around 2027-28 [6], and the total infrastructure schedule is currently represented in figure 1.

In what follows, we focus first on the physics case for the FCC-ee and FCC-hh. We than consider a multi-TeV muon collider. Note that the physics programme for FCC-ee is partly overlapped (in particular in the Higgs Factory stage) with the programme of linear colliders at $\sqrt{S} < 500$GeV.

3.1 FCC-ee (and CepC)

In figure 2, we report the total integrated luminosities corresponding at different c.m. energy stages for FCC-ee, with the corresponding total running time. The corresponding values for CepC (a similar project to be based in China [1],[2]) are also shown, along with the ratio of integrated luminosites for FCC-ee and CepC. The exact sequence and duration of stages might actually be further elaborated later on, during the colliders constructions.

The lower energy stages at 91 GeV and 160 GeV correspond to regimes previously explored at LEP, but with hugely larger statistics at FCC-ee (about 10^5 and 10^3 larger $e^+e^- \rightarrow Z$ and $e^+e^- \rightarrow WW$ samples, respectively). These will make the so-called *intensity frontier* part of the circular collider programme, by realizing

Energy ($\sqrt{s}$)	$\mathcal{L}_{\text{int}}$ (Run time)		$\mathcal{L}_{\text{FCC-ee}}/\mathcal{L}_{\text{CEPC}}$
	FCC-ee (4 IPs)	CEPC (2 IPs)	
91 GeV (Z-pole)	300 ab^{-1} (4 years)	100 ab^{-1} (2 years)	3
161 GeV ($2m_W$)	20 ab^{-1} (2 years)	6 ab^{-1} (1 year)	3.3
240 GeV	10 ab^{-1} (3 years)	20 ab^{-1} (10 years)	0.5
350 GeV	0.4 ab^{-1} (1 year)	0.2 ab^{-1}	2
365 GeV ($2m_t$)	3 ab^{-1} (4 years)	1 ab^{-1} (5 years)	3

Figure 2. Integrated luminosities currently foreseen at different c.m. energy stages for FCC-ee and CepC.

- a new generation of EW and QCD precision measurements (see figure 3 for projected accuracies on different observables)
- a Flavor Factory, based on the the huge sample of b, c, τ pairs arising from $\sim 6 \cdot 10^{12} Z$ expected decays, equivalent to an order of magnitute more than what will be available from the Belle II experiment at Super KEK
- direct searches for *light* weakly interacting new physics (axion-like particles, dark photons, heavy neutral leptons, possibly characterized by long lifetimes).

According to figure 3, statistical precisions up to 1000 times better than at LEP are obtained, with an improvement in experimental systematic precisions by a factor from 10 to 50, depending on the observable. Projected parametric uncertainties in the SM predictions are currently limited by theory systematics [11].

EW precision measurements remarkably provide an indirect probe on NP. An increase by X in accuracy gives in general an increase by $\sqrt{X}$ in indirect mass reach.
At FCC-ee one then expects an improvement by a factor 4-5 in the EFT energy reach, corresponding to explorable energy scales of up to 70 TeV via the description of BSM effects with dim 6 operators.

The FCC-ee energy stages at 240 GeV and 350 (365) GeV provide instead two brand new collision setups, named

- a Higgs Factory, where o(10^6) Higgs bosons are expected via the process $e^+e^- \to HZ$, (allowing a clean and model independent determination of the Higgs mass, width, and couplings)
- a top Factory, where o(10^6) top quark pairs are produced via the process $e^+e^- \to t\bar{t}$ (allowing an extremely accurate determination of the top mass, width, and electroweak couplings, as reported in figure 3).

The latter two stages are in principle of interest for both circular and linear e^+e^- colliders.
In figure 4, the expected accuracies on κ_i multipliers for the Higgs couplings are reported for different e^+e^- colliders, and compared to the HL-LHC potential.

Let's now assume we find a deviation δ_i in the H coupling g_{Hii} at an e^+e^- collider. In general, the NP energy scale $\mathcal{M}$ will be connected to the Higgs coupling deviations according to $\delta_i \sim v^2/\mathcal{M}^2$, where v is the Higgs vev. Then, $\delta_i \sim 6\%$ would correspond to NP at about 1 TeV, while $\delta_i \sim 0.06\%$ would imply NPs at about 10 TeV.

Hence, in order to figure out what causes a certain Higgs coupling deviation we will need an *energy-frontier* facility to explore the corresponding $\mathcal{M}$ scale in a direct way.

Technology for future *energy-frontier* colliders is presently not completely mature, and possibly new technologies are needed. The FCC-hh, a 100 TeV pp collider (the natural

Observable	Present value ± error	FCC-ee stat.	FCC-ee syst.	Comment and leading exp. error
m_Z (keV)	91186700 ± 2200	**4**	100	From Z line shape scan Beam energy calibration
Γ_Z (keV)	2495200 ± 2300	**4**	25	From Z line shape scan Beam energy calibration
$\sin^2\theta_W^{eff}$ ($\times 10^6$)	231480 ± 160	**2**	2.4	from $A_{FB}^{\mu\mu}$ at Z peak Beam energy calibration
$1/\alpha_{QED}(m_Z^2)(\times 10^3)$	128952 ± 14	**3**	Small	From $A_{FB}^{\mu\mu}$ off peak QED&EW errors dominate
R_ℓ^Z ($\times 10^3$)	20767 ± 25	**0.06**	0.2–1	Ratio of hadrons to leptons Acceptance for leptons
$\alpha_s(m_Z^2)$ ($\times 10^4$)	1196 ± 30	**0.1**	0.4–1.6	From R_ℓ^Z above
σ_{had}^0 ($\times 10^3$) (nb)	41541 ± 37	**0.1**	4	Peak hadronic cross section Luminosity measurement
$N_\nu(\times 10^3)$	2996 ± 7	**0.005**	1	Z peak cross sections Luminosity measurement
R_b ($\times 10^6$)	216290 ± 660	**0.3**	< 60	Ratio of $b\bar{b}$ to hadrons Stat. extrapol. from SLD
A_{FB}^b, 0 ($\times 10^4$)	992 ± 16	**0.02**	1–3	b-quark asymmetry at Z pole From jet charge
$A_{FB}^{pol,\tau}$ ($\times 10^4$)	1498 ± 49	**0.15**	<2	τ polarization asymmetry τ decay physics
τ lifetime (fs)	290.3 ± 0.5	**0.001**	0.04	Radial alignment
τ mass (MeV)	1776.86 ±0.12	**0.004**	0.04	Momentum scale
τ leptonic ($\mu\nu_\mu\nu_\tau$) B.R. (%)	17.38 ±0.04	**0.0001**	0.003	e/μ/hadron separation
m_W (MeV)	80350 ± 15	**0.25**	0.3	From WW threshold scan Beam energy calibration
Γ_W (MeV)	2085 ± 42	**1.2**	0.3	From WW threshold scan Beam energy calibration
$\alpha_s(m_W^2)(\times 10^4)$	1170 ± 420	**3**	Small	from R_ℓ^W
$N_\nu(\times 10^3)$	2920 ± 50	**0.8**	Small	Ratio of invis. to leptonic in radiative Z returns
m_{top} (MeV/c^2)	172740 ± 500	**17**	Small	From $t\bar{t}$ threshold scan QCD errors dominate
Γ_{top} (MeV/c^2)	1410 ± 190	**45**	Small	From $t\bar{t}$ threshold scan QCD errors dominate
$\lambda_{top}/\lambda_{top}^{SM}$	1.2 ± 0.3	**0.10**	Small	From $t\bar{t}$ threshold scan QCD errors dominate
ttZ couplings	± 30%	**0.5–1.5%**	Small	From $\sqrt{s}$ = 365 GeV run

Figure 3. Measurement of selected precision measurements at FCC-ee, compared with present precision, as from [10]. Statistical errors are indicated in bolded. The systematic uncertainties are initial estimates, ideally to be improved down to statistical errors.

kappa-0	HL-LHC	LHeC	HE-LHC	ILC_{250}	ILC_{500}	$CLIC_{380}$	$CLIC_{1500}$	$CLIC_{3000}$	CEPC	FCC-ee_{240}	FCC-ee_{365}	FCC-ee/eh/hh
κ_W (%)	1.2	0.75	0.66	1.8	0.29	0.86	0.17	0.11	1.3	1.3	0.43	0.15
κ_Z (%)	1.0	1.2	0.6	0.29	0.23	0.5	0.26	0.23	0.13	0.2	0.17	0.12
κ_g (%)	2.2	3.6	1.4	2.3	0.97	2.5	1.3	0.9	1.5	1.7	1.0	0.52
κ_γ (%)	1.7	7.5	0.98	6.7	3.4	98⋆	5.0	2.2	3.7	4.7	3.9	0.35
$\kappa_{Z\gamma}$ (%)	10	—	4.0	99⋆	86⋆	120⋆	15	6.9	8.2	81⋆	75⋆	0.7
κ_c (%)	—	4.0	—	2.5	1.3	4.3	1.8	1.4	2.2	1.8	1.3	0.95
κ_t (%)	2.8	—	2.0	—	6.9	—	—	2.6	—	—	—	1.0
κ_b (%)	2.7	2.1	1.7	1.8	0.58	1.9	0.48	0.38	1.2	1.3	0.67	0.45
κ_μ (%)	4.4	—	1.8	15	9.4	320⋆	13	5.8	8.9	10	8.9	0.42
κ_τ (%)	1.6	3.3	1.1	1.9	0.7	3.0	1.3	0.89	1.3	1.4	0.73	0.49

Figure 4. Expected relative precision on Higgs couplings (%) in the κ-0 scenario, as from [12]. Different future accelerators are compared.

follow-on to FCC-ee in the same tunnel) needs high-field (~16 Tesla) dipole magnets that are not yet available. On the other hand, a multi-TeV muon collider is based on a brand new fascinating concept which needs R&D in different sectors. In the next two subsections, we will summarise the amazing physics potential of these two possibilities.

3.2 FCC-hh

The physics potential of proton collisions at 100 TeV c.m. energy can provide a huge advancement with respect to what will be obtained at the HL-LHC [9],[13]. Assuming an integrated luminosity of about 30 ab^{-1}, one can highlight the reach of such machine by the following points

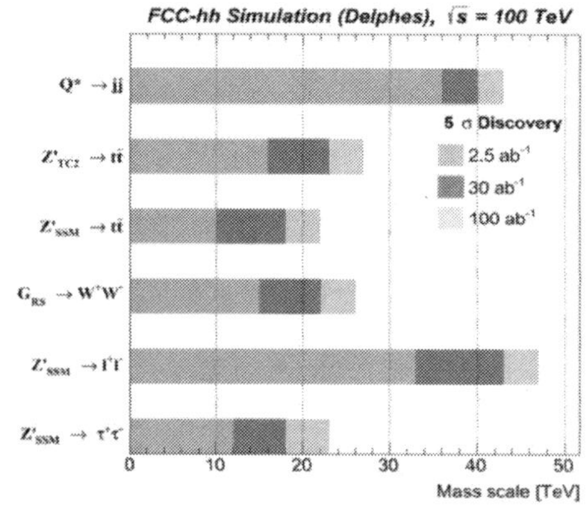

Figure 5. Resonance mass reach at FCC-hh at different integrated luminosities and for different BSM s-channel resonances [7].

- mass reach in BSM searches (4-6) times larger than $\mathcal{M}_{HL-LHC}$ (see figure 5 for resonance masses reach)
- for multiple heavy-particle final states, the statistic $n_(H, W, Z, t)$ is increased by at least two orders of magnitudes with respect to HL-LHC (e.g., about 500 larger for ttH, and 400 for HH final states)
- huge Higgs production rates (i.e., more than $10^{10}H$ and $10^7 HH$) allowing to test even most challenging Higgs couplings (such as the Higgs self-couplings)
- unique sensitivity to rear decays
- can explore extreme (hence *clean*) phase-space regions with high statistics
- much higher gain with respect to HL-LHC at high-p_T and large invariant masses.

3.3 Multi-TeV muon colliders

We reviewed in Sect. 2, the advantages of colliding leptons versus hadrons, that is pointlike versus composite objects. Such advantages hugely expand at muon colliders, when considering pointlike objects colliding at multi-TeV c.m. energies, giving rise to an incredibly rich phenomenology [14],[15].

Currently, reference designs foresee for a muon collision energy $\sqrt{S} \sim 10$+TeV an integrated luminosity $\mathcal{L} \sim (\sqrt{S}/10 \text{ TeV})^2 \times 10$ ab^{-1} [16],[17]. The cross section for pair production of heavy electroweakly interacting new particles of mass $\lesssim \sqrt{S}/2$ is of the order of the *point* cross section $\sigma(\mu^+\mu^- \to \gamma^* \to e^+e^-) \sim 4\pi\alpha^2/(3S) \sim 1\text{fb}\,(10\,\text{TeV}/\sqrt{S})^2$, which, multiplied by the integrated luminosity above will give o(10^4) events. Remarkably, such large statistics (allowing non only to discover heavy new states but also to make measurements with o(1%) statistical errors) holds for heavy particles with masses almost up to

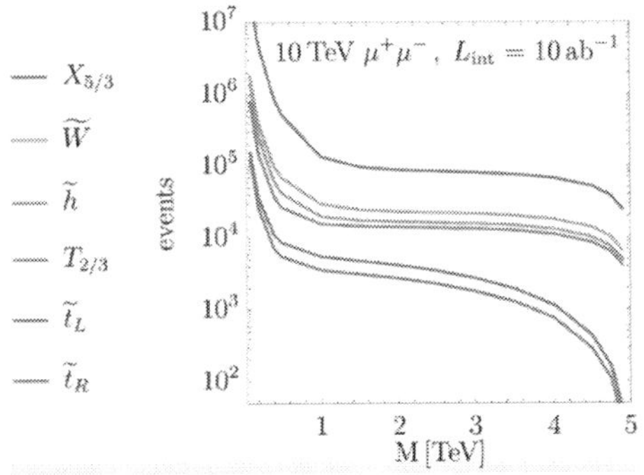

Figure 6. Number of expected events at a 10 TeV muon collider with 10 ab^{-1} of integrated luminosity, for pair production of different BSM heavy particles [15].

the production threshold, as can be seen in figure 6, where different BSM scenarios are considered. Indeed, just one hundredth of the nominal luminosity might be sufficient to produce o(100) events, and discover new pair-produced EW multi-TeV particles!

Apart from direct pair production of new heavy BSM particles, there is a vast physics case for multi-TeV muon colliders, including [14],[15]:

- single heavy-resonance production [both neutral resonances, in $\mu^+\mu^-$ annihilation with masses up to $\sim\sqrt{S}$, and charged ones, anticipated by $W^\pm$ radiation from the initial muon beams, with masses close to $\sqrt{S}$]
- at multi-TeV energies, a muon collider transforms into a "WW collider", since cross sections for vector-boson fusion processes, $\mu^+\mu^- \to W^*W^*\nu\nu \to X\nu\nu$, increase logarithmically with $\sqrt{S}$
- precision improvement for many observables, and exploration of the Higgs sector with high accuracy, including both trilinear and quartic Higgs self-couplings
- complete exploration of the WIMP hypothesis for dark matter.

Since at lepton colliders the initial c.m. energy is most effectively transferred to the point-like collision, muon colliders with $\sqrt{S}\sim$10+TeV have in general a physics potential which is competitive with a pp collider at o(100) TeV, and particularly strong in the EW sectors.

4 The European Strategy for Particle Physics Update (ESPPU)

In March 2024, CERN Council launched the next ESPPU, to be concluded in 2026 [18]. Through this process, future of CERN after the HL-LHC (planned to stop in 2041) should be defined, supporting the design and delivery of a future particle accelerator in a timely, affordable and sustainable way, so as to maintain the world-leading role of CERN in accelerator-based particle physics. To start the operation of the next large facility at CERN around 2045, construction must begin around 2032. Then, a timely definition of a scientifically sound and ambitious plan for the future of the field is crucial.

By the end of March 2025, submission of main input from the community will be completed, and at the end of June 2025 an Open Symposium will be held based on the submitted documents. After that, the collected information will be managed by the different groups organising and running the ESPP, and presented to the CERN Council by the end of 2025. A final Council decision on Update of the Strategy is expected by June 2026 [18].

References

[1] C. Adolphsen, D. Angal-Kalinin, T. Arndt, M. Arnold, R. Assmann, B. Auchmann, K. Aulenbacher, A. Ballarino, B. Baudouy and P. Baudrenghien, *et al.* CERN Yellow Rep. Monogr. **1** (2022), 1-270 doi:10.23731/CYRM-2022-001 [arXiv:2201.07895 [physics.acc-ph]], and references therein

[2] T. Roser, R. Brinkmann, S. Cousineau, D. Denisov, S. Gessner, S. Gourlay, P. Lebrun, M. Narain, K. Oide and T. Raubenheimer, *et al.* JINST **18** (2023) no.05, P05018 doi:10.1088/1748-0221/18/05/P05018 [arXiv:2208.06030 [physics.acc-ph]], and references therein

[3] M. Narain, L. Reina, A. Tricoli, M. Begel, A. Belloni, T. Bose, A. Boveia, S. Dawson, C. Doglioni and A. Freitas, *et al.* [arXiv:2211.11084 [hep-ex]]

[4] K. Fujii, C. Grojean, M. E. Peskin, T. Barklow, Y. Gao, S. Kanemura, H. Kim, J. List, M. Nojiri and M. Perelstein, *et al.* [arXiv:1710.07621 [hep-ex]]

[5] R. S. Gupta, H. Rzehak and J. D. Wells, Phys. Rev. D **88** (2013), 055024 doi:10.1103/PhysRevD.88.055024 [arXiv:1305.6397 [hep-ph]]

[6] https://home.cern/science/accelerators/future-circular-collider

[7] A. Abada *et al.* [FCC], Eur. Phys. J. C **79** (2019) no.6, 474 doi:10.1140/epjc/s10052-019-6904-3

[8] A. Abada *et al.* [FCC], Eur. Phys. J. ST **228** (2019) no.2, 261-623 doi:10.1140/epjst/e2019-900045-4

[9] A. Abada *et al.* [FCC], Eur. Phys. J. ST **228** (2019) no.4, 755-1107 doi:10.1140/epjst/e2019-900087-0

[10] A. Blondel and P. Janot, Eur. Phys. J. Plus **137** (2022) no.1, 92 doi:10.1140/epjp/s13360-021-02154-9 [arXiv:2106.13885 [hep-ex]]

[11] A. Freitas, S. Heinemeyer, M. Beneke, A. Blondel, S. Dittmaier, J. Gluza, A. Hoang, S. Jadach, P. Janot and J. Reuter, *et al.* [arXiv:1906.05379 [hep-ph]]

[12] J. de Blas, M. Cepeda, J. D'Hondt, R. K. Ellis, C. Grojean, B. Heinemann, F. Maltoni, A. Nisati, E. Petit and R. Rattazzi, *et al.* JHEP **01** (2020), 139 doi:10.1007/JHEP01(2020)139 [arXiv:1905.03764 [hep-ph]]

[13] M. L. Mangano, W. Riegler, M. Aleksa, P. P. Allport, S. Asai, A. Ball, M. I. Besana, E. R. Bielert, S. Bologna and E. Boos, *et al.* "Conceptual design of an experiment at the FCC-hh, a future 100 TeV hadron collider," doi:10.23731/CYRM-2022-002

[14] H. Al Ali, N. Arkani-Hamed, I. Banta, S. Benevedes, D. Buttazzo, T. Cai, J. Cheng, T. Cohen, N. Craig and M. Ekhterachian, *et al.* Rept. Prog. Phys. **85** (2022) no.8, 084201 doi:10.1088/1361-6633/ac6678 [arXiv:2103.14043 [hep-ph]]

[15] C. Aime, A. Apyan, M. A. Mahmoud Mohammed, N. Bartosik, F. Batsch, A. Bertolin, M. Bonesini, S. Bottaro, D. Buttazzo and R. M. Capdevilla Roldan, *et al.* "Muon Collider Physics Summary," [arXiv:2203.07256 [hep-ph]]

[16] C. Accettura, D. Adams, R. Agarwal, C. Ahdida, C. Aimè, N. Amapane, D. Amorim, P. Andreetto, F. Anulli and R. Appleby, *et al.* Eur. Phys. J. C **83** (2023) no.9, 864 [erratum: Eur. Phys. J. C **84** (2024) no.1, 36] doi:10.1140/epjc/s10052-023-11889-x [arXiv:2303.08533 [physics.acc-ph]]

[17] C. Accettura *et al.* [International Muon Collider], "Interim report for the International Muon Collider Collaboration (IMCC)," CERN, 2024, ISBN 978-92-9083-667-4, 978-92-9083-668-1 doi:10.23731/CYRM-2024-002 [arXiv:2407.12450 [physics.acc-ph]]

[18] https://europeanstrategyupdate.web.cern.ch/

QCD@Work 2024: Highlights

Giuseppe Eugenio Bruno, *Gabriele* Chiodini, *Pietro* Colangelo, *Donato Maria* Creanza, *Fulvia* De Fazio, *Floriana* Giannuzzi, *Salvatore* My, and *Stefano* Nicotri

Abstract. The highlights of QCD@Work 2024 are summarized.

The XI edition of QCD@Work - International Workshop on Quantum Chromodynamics - Theory and Experiment, was held in the year of the 50th anniversary of the discovery of the J/ψ which proved the existence of the charm quark. This was a turning point for the Standard Model of the electroweak interactions and for the theory of strong interactions QCD. The path leading to the formulation of QCD is rather recent, and one can identify two milestones reached in the seventies of the last century. The first one is understanding that nonabelian gauge theories enjoy the property of asymptotic freedom, which opened the way to build QCD as a gauge theory described by the SU(3) group. Elementary constituents of hadrons, the quarks, had already been introduced as tools to classify the wealth of observed hadronic states. However, as always in science since Galilei, the verdict came from experiments. Thanks to the observation of the Bjorken scaling in deep inelastic experiments, the presence of pointlike partons inside nucleons was established. The second milestone reached in November 1974 was the observation of the J/ψ, hence of the charm quark, which proved that quarks are real entities. This was 50 years before QCD@Work 2024.

Fascinating features of QCD, asymptotic freedom and confinement, are still under scrutiny by theory and experiment. They are both represented in this edition of the workshop with developments at the frontier of particle physics.

Asymptotic freedom allows us to perform perturbative calculations at high orders in the strong coupling constant α_s; the session "New challenges in perturbative QCD" has collected important contributions. Two eminent Italian Physicists who greatly contributed to this sector sadly passed away last year: Stefano Catani and Marcello Ciafaloni. The workshop was opened by Giovanni Ridolfi with a personal selection of Stefano's and Marcello's far-reaching scientific achievements [1].

Presented advances in theory concern the forward Higgs production, with the related difficulties in perturbative QCD [2]. New computations of the event shape distributions have been discussed [3], together with algorithms to find Cwebs in multiparton scattering amplitudes [4]. A method to measure the top quark mass using leptonic final states, with leptons coming from W and B decays, has been presented together with the results obtained by the ATLAS Collaboration using this method [5].

For proton-proton collisions, the general mass variable flavour number schemes ACOT and the simplified S-ACOT have been discussed. In particular, the calculation of Z hadronic production with at least one b quark beyond the leading order in QCD has been presented [6]. New studies of the Transverse Momentum Distributions allow to gain insights into the hadron properties. The conformal features of TMD operators have been discussed, and the result of the conformal rapidity evolution of such operators in the Sudakov region has been given [7]. The problem of the transverse Λ polarization in unpolarized production processes

(e^+e^- and pp collisions, semi-inclusive DIS) has been analyzed. The TMD approach has been illustrated, and universality tests for the TMD polarizing fragmentation function have been discussed [8].

New experimental results in pp collisions at the LHC energy were presented by the ATLAS and CMS Collaborations. In particular, in the ATLAS contribution the impact of LHC data in the PDF determination was shown to be more and more relevant [9]. The most precise determination of α_s in a single measurement from the Z p_T spectrum was presented [9], together with the newest results on a benchmark QCD process, the Drell-Yan W/Z di-lepton production [10]. Among the CMS results, the most stringent limit on rare decays of Higgs boson into a photon and a meson, the observation of quantum entanglement in top quark pair production and the first evidence of time-dependent CP violation in $B_s^0 \to J/\psi\,\phi(1020)$ decays are worth to be highlighted [11].

The nonperturbative regime of QCD, related to the quark and gluon confinement inside hadrons, was discussed in the session "Low-Energy and Nonperturbative QCD". A spectacular feature of low-energy QCD is the hadron spectrum. New results have been presented by the BESIII [12, 13], LHCb [14] and CMS Collaborations [15]. In particular, in the BESIII contribution the determination of the quantum numbers of $X(2370)$, an important glueball candidate, is discussed. A new state $X(1880)$ is observed in $J/\psi \to (3\pi^+\pi^-)\gamma$ and an exotic isoscalar state is found in $J/\psi \to \eta\eta'\gamma$ [12]. BESIII also reported results on hyperons, including the measurement of the polarizations of Ξ^- and Ξ^0, which are useful to test the CP asymmetry in hyperons with more than one strange quark [13]. Three-body decays of B mesons to light states, analyzed by LHCb, reveal a wealth of information on the light quark spectroscopy [14]. Among the experimental contributions, CMS has reported the observation of new structures in the di-charmonium mass spectrum, hinting at exotic states comprising two charm quarks [15]. The issue of conventional and nonconventional hadrons has been lively discussed, considering examples of flavour-cryptoesotic tetraquarks [16] and tetraquarks in two-dimensional QCD [17]. The use of a quantity called "configuration entropy" in holographic models to characterise the different kinds of hadrons has been developed [18].

The treatment of single flavour baryons at large N in low-energy QCD has been discussed. Since it cannot be done in terms of Skyrmions, a description has been proposed in terms of a pancake-shaped sheet. How this possibility can be realised in a holographic model of QCD has been discussed at the workshop [19].

QCD is almost everywhere in particle physics. Even while probing other kinds of interactions it is not possible to avoid considering that quarks are strongly bound inside hadrons. This interplay is behind most of the research presented in the session "Heavy Quarks". We had the honour to have as a speaker Prof. Andrzej J. Buras, the recipient of the 2024 J.J. Sakurai prize of the American Physical Society awarded to recognize outstanding contributions to particle theory. In his talk, a new strategy in the search for physics beyond the Standard Model in the flavour sector has been discussed in details. The strategy, based on rare B and K decay modes, has the important feature of avoiding the use of the CKM matrix elements $|V_{cb}|$ and $|V_{ub}|$ elements, which are subject to tensions. A shopping list of desired measurements has been suggested, for the benefit of the programme of the experimental collaborations [20].

Experimental news in rare kaon decays obtained by the NA48/2 and NA62 Collaboration confirmed the importance of the kaon modes [21]. In particular, the branching ratio and the di-photon mass spectrum of the mode $K^+ \to \pi^+\gamma\gamma$, obtained by NA62 Collaboration, has been presented. Preliminary results for the branching fraction of $\pi^0 \to e^+e^-$ have been given. The first observation and measurement of the branching fraction of the $K_{\mu\nu}^{00}$ mode: $K^\pm \to \pi^0\pi^0\mu^\pm\nu$ obtained by the NA48/2 Collaboration at CERN have also been reported.

Rare decay modes of heavy mesons and baryons have been discussed, namely the radiative modes $B_c \to (B^*, B_1)\gamma$ [22] and the FCNC transitions of spin 1/2 double heavy baryons [23]: they are included in the physics programmes of the LHC experiments. The presentation by the Belle II Collaboration enriched the experimental view in flavour physics. New results have been presented on hadronic B decays, on the angle ϕ_3/γ, on time-dependent CP violating processes, on semileptonic B decays and the debated ratios $R(D)$ and $R(D^*)$. New analyses of electroweak and radiative penguin processes have been discussed, in particular for the very rare $B^+ \to K^+\nu\bar{\nu}$ mode. For the bottomonium spectroscopy, results have been quoted about $Y(10753)$ [24].

Developments have also been presented for the QCD effective theory suitable for heavy hadrons, the Heavy Quark Effective Theory. Improvements have been implemented in perturbative calculations and in the determination of nonperturbative quantities such as local and nonlocal matrix elements [25]. In this framework, the inclusive $\Lambda_b \to X_s\gamma$ decay, with the related spin effects, has been discussed, with a proposal of a novel expression for the shape function [26].

An important observable to test the Standard Model is the anomalous magnetic moment of the muon, the measurement of which is in tension with theoretical determinations. The uncertainty connected to the leading order hadronic vacuum polarization contribution to $(g-2)_\mu$ has been described in details, considering the methods currently used to determine such a term. The perspectives provided by the experiment MUonE, which is designed to collect data in the space-like region, have been discussed. The required improvements for the theoretical calculations of the elastic $e-\mu$ scattering have been described [27].

Searches for physics beyond the Standard Model include the dark photon. Systematic uncertainties related to the effects of the momentum distribution of the atomic electrons need to be precisely assessed, and they were discussed at the workshop [28]. They are relevant for dark matter searches (PADME, NA64, ...) and precision measurements (MUonE, ...) in fixed target experiments. In scenarios comprising vector dark matter, the electroweak phase transition can be characterized [29]. In this respect, new methods for computing the collision integrals involved in the description of cosmological phase transitions have been described [30].

Various formal studies related to the physics of strongly coupled systems have been presented at the workshop. They include the renormalized entanglement entropy [31], the quantum anomalies for chiral states of matter [32], the gravitational form factor of pion and proton [33], deterministic approaches to the dynamics of quantum fields [34]. The topology of the large N expansion in SU(N) Yang-Mills and the way to restore a puzzling violation of the spin-statistics theorem have been considered in details [35]. The fluctuations of a thermal field in a chiral Lagrangian with broken scale invariance have been discussed [36]. An example of synergy between particle physics and gravitational waves, represented by Feynman integrals, has been illustrated [37].

Insights on strong interactions come from heavy ion experiments and the related theoretical studies, presented in the session "QCD in extreme conditions of temperature and density". For the search of the critical point in the QCD phase diagram, data from the Beam Energy Scan program at the BNL Relativistic Heavy Ion Collider qualitatively suggest that the critical point should have a baryonic critical potential $\mu_B > 420$ MeV. This hint comes by looking at the plots of $n = 2, 3, 4$ factorial cumulants, which show a non-monotonic behavior expected in the presence of a critical point [38]. The QCD phase diagram at finite isospin chemical potential can be theoretically explored using AdS/CFT inspired methods. Using the Witten-Sakai-Sugimoto model, it is found in the baryon - isospin chemical potential plane at $T = 0$, that the pion mass increases with baryon density; at low baryon chemical potentials the baryons are absent. In the same model, the neutron star is constructed in agreement with experimental constraints for mass, radius, and tidal deformability [39].

At low T, for vanishing baryonic chemical potential and finite isospin chemical potential, a transition occurs from the hadronic phase to a superconducting one. In the Ginzburg-Landau theory, depending on the value of the coupling, a type I or type II superconducting phase can be realized. At the transition, a BPS bound is saturated by multi-vortex configurations. The BPS bound has been computed in chiral perturbation theory minimally coupled to electrodynamics, obtaining a critical value of the isospin chemical potential [40].

Heavy ion collisions provide a laboratory to study exotic charmonia, due to the large number of produced heavy quarks. A multiquark state of two charm ($c\bar{c}$) and two light quarks ($q\bar{q}'$) can have a molecular or compact structure. This environment can help shedding light on the structure of the $X(3872)$ comparing the production yield to that of $\psi(2S)$. Using an effective Lagrangian approach, the time evolution of the ratio $\mathcal{R}$ between the yields of $X(3872)$ and $\psi(2S)$ has been computed. $\mathcal{R}$ is different if $X(3872)$ is a molecule or a compact tetraquark. The ratio can be measured at LHC [41].

News from the ALICE Collaboration have been presented at the workshop, in particular concerning the production of nuclei and hypernuclei in Pb-Pb collisions [42], the heavy flavour production in pp collisions [43] and the quark-gluon plasma (QGP) features in small colliding systems [44]. Technological developments foreseen for the future runs have been described [45].

Spin polarization can be used to study the QGP. A numerical calculation in a hydrodynamic model, simulating Au-Au collisions at $\sqrt{s} = 200$ GeV and Pb-Pb collisions at $\sqrt{s} = 5.02$ TeV, shows that the spin polarization of Λ hyperons is very sensitive to initial conditions and bulk viscosity of the plasma. Hence, the azimuthal dependence of spin polarization can constrain the initial hydrodynamic conditions and the transport coefficients of the QGP [46]. Tensions are found for vector mesons, since the measured spin alignment is larger than expected. During the workshop, it has been proposed that local spin density fluctuations of quarks due to axial-vector and tensor interactions between quarks can induce the spin alignment of ϕ and K^{*0} mesons [47].

In many cases, dynamical calculations are needed to compute features of the QGP to be compared to data. To study the evolution of the electromagnetic fields produced in Ru+Ru and Zr+Zr collisions at $\sqrt{s_{NN}} = 200$ GeV, the effect of the electric and magnetic conductivities must be taken into account [48]. Moreover, to connect the nuclear structure to heavy-ion measurements, the elliptic shape of the initial entropy density distribution, which depends on the density of the colliding nuclei, has been studied through a cluster expansion decomposition, assuming scale invariance for the initial state [49].

Data from heavy ion collision experiments can constrain the Hagedorn temperature T_H. For a strongly-coupled confining gauge theory, T_H has been computed in a holographic model as a perturbative expansion in the coupling $1/g$. A remarkable agreement with an independent result is achieved, a non-trivial test of the holographic correspondence. Extrapolating the results to $SU(3)$ Yang-Mills, the prediction $T_H/T_c \sim 1.19$ has been obtained [50]. Other features of the QCD phases in a holographic approach have been investigated [51].

Looking at the future facilities, the perspectives for ePIC at the Electron Ion Collider have been discussed [52] together with some technological developments for the detectors [53, 54]. Various options for new high-energy colliders have been described, potentially operational after the conclusion of the LHC and High Luminosity-LHC programmes. The physics potential of various projects has been animately discussed, together with the technical requirements and various kinds of constraints [55]. The projects include lepton colliders, mainly dedicated to precision physics, proton-proton and multi-TeV muon colliders to explore the high-energy frontier. The process for defining the European Strategy for Particle

Physics has just started. It will involve the entire community in shaping the future of our science, which is foreseen to be bright.

References

[1] G. Ridolfi, Stefano and Marcello @ Work. A personal selection of results obtained by Stefano Catani and Marcello Ciafaloni, `these proceedings`

[2] A. Papa, Forward Higgs production: a challenge in high-energy perturbative QCD, `https://agenda.infn.it/event/37243/contributions/178263/`

[3] S. Mishra, Event shape distribution and the NLP corrections, `these proceedings`

[4] A. Srivastav, Cwebs in multiparton scattering amplitude: Structures at four loops, `these proceedings`

[5] G. Corcella, Top-mass determination from leptonic final states, `these proceedings`

[6] M. Guzzi, General Mass treatment for Z boson production in association with a heavy quark at hadron colliders, `these proceedings`

[7] G. Chirilli, Rapidity-only evolution of TMDs, `these proceedings`

[8] F. Murgia, Transverse Λ polarization in semi-inclusive DIS, e^+e^- and pp collisions and the TMD polarizing fragmentation, `https://agenda.infn.it/event/37243/contributions/178308/`

[9] E. Conroy, QCD measurements with ATLAS, `https://agenda.infn.it/event/37243/contributions/178144/`

[10] F. Anulli, ATLAS measurements of Drell Yan processes, `https://agenda.infn.it/event/37243/contributions/178727/`

[11] M. Mukhejee, CMS experimental results, `these proceedings`

[12] I. Garzia, Highlights from the BESIII experiment, `these proceedings`

[13] J. Zhang, Hyperon Physics at BESIII, `these proceedings`

[14] A. Palano, Study of charmonium decays to $K^0_S K\pi$ in the $B \to (K^0_S K\pi)K$ channels, `these proceedings`

[15] V. Mastrapasqua, Recent results in hadron spectroscopy from CMS, `these proceedings`

[16] W. Lucha, Various Illustrative Brief Glances at the Flavour-Cryptoexotic Tetraquark States, `these proceedings`

[17] H. Sazdjian, Scattering of mesons and emergence of tetraquarks in two-dimensional QCD, `these proceedings`

[18] M.A. Martin Contreras, A holographic approach to heavy non $q\bar{q}$-states using configurational entropy, `these proceedings`

[19] A. Cotrone, Low Energy Description of Single Flavor Baryons, `these proceedings`

[20] A.J. Buras, Hunting New Animalcula with Rare K and B Decays, `these proceedings`

[21] A. Bizzeti, Precision measurements with kaon and pion decays at CERN, `these proceedings`

[22] N. Losacco, Exclusive $c \to u\gamma$ transitions of B_c meson, `these proceedings`

[23] E. Askan, FCNC Decays of Spin-1/2 Double Heavy Baryons to Spin-3/2 Single Heavy Baryons, `these proceedings`

[24] A. Passeri, Recent results from Belle II experiment, `https://agenda.infn.it/event/37243/contributions/178193/`

[25] G. Paz, Recent developments in HQET, `these proceedings`

[26] F. Loparco, Inclusive rare Λ_b decays to photon, `these proceedings`

[27] F. Piccinini, Hadronic contribution to the muon $g-2$: theoretical challenges with the MUonE experiment, `these proceedings`

[28] G. Grilli di Cortona, Effects of atomic electron momentum distribution on resonant dark sector production, `these proceedings`

[29] M. Razzaq, Electroweak phase transition in a vector dark matter scenario, `these proceedings`

[30] L. Delle Rose, New calculation of collision integrals for cosmological phase transitions, `these proceedings`

[31] M. Fujita, Renormalized Entanglement Entropy and general entropic $c-$function, `these proceedings`

[32] S. Lionetti, Quantum Anomalies and Parity-odd CFT Correlators for Chiral States of Matter, `these proceedings`

[33] M. Tommasi, The Gravitational Form Factor of the Pion and Proton and the Conformal Anomaly, `these proceedings`

[34] M. Giachello, Symplectic quantization: a new deterministic approach to the dynamics of quantum fields, `these proceedings`

[35] M. Bochicchio, Topology of the large-N expansion in SU(N) Yang-Mills theory and spin-statistics theorem, `these proceedings`

[36] L. Passarella, Thermal field fluctuations in a chiral lagrangian with broken scale invariance, `these proceedings`

[37] M.K. Mandal, Feynman Integrals: Synergies between Particle Physics and Gravitational Waves, `these proceedings`

[38] M. Stephanov, QCD critical point: recent developments, `these proceedings`

[39] A. Schmitt, Dense QCD matter and neutron stars from holography, `https://agenda.infn.it/event/37243/contributions/228979/`

[40] A. Vera, BPS multi-vortices in chiral perturbation theory, `these proceedings`

[41] F.S. Navarra, How the hadron gas affects $X(3872)$ and $\psi(2S)$ production in heavy ion collision, `these proceedings`

[42] A. Mastroserio, Insights in nuclei and hypernuclei production in Pb-Pb collisions with ALICE, `https://agenda.infn.it/event/37243/contributions/178928/`

[43] L. Gyulai, Heavy-flavour production as a function of event activity in pp collisions with the ALICE experiment, `these proceedings`

[44] R.N. Patra, Exploring QGP Signature in Small System: Insights from ALICE in p–Pb and pp Collisions, `these proceedings`

[45] N. Nicassio, Physics programme of ALICE 3: a next-generation heavy-ion detector for LHC Run 5 and beyond, `these proceedings`

[46] F. Becattini, Polarization in relativistic heavy ion collisions, `https://agenda.infn.it/event/37243/contributions/233542/`

[47] K. Xu, Spin alignment of vector mesons induced by quantum spin density fluctuations, `https://agenda.infn.it/event/37243/contributions/228069/`

[48] I. Siddique, The effect of electric and chiral magnetic conductivities on azimuthally fluctuating electromagneti fields and observables, `https://agenda.infn.it/event/37243/contributions/228687/`

[49] S.F. Taghavi, Revealing nucleus properties by exploringultra-central symmetric heavy-ion collisions with scale-invariant initial states, `https://agenda.infn.it/event/37243/contributions/228685/`

[50] F. Bigazzi, Hagedorn temperature in confining gauge theories from holography, `these proceedings`

[51] A. Singh, Holographic phases of QCD, `these proceedings`
[52] S. Dalla Torre, Exploring the QCD frontier with ePIC at EIC, `https://agenda.infn.it/event/37243/contributions/178303/`
[53] M.R. Altamura, Silica aerogel characterization for the ePIC dRICH detector, `these proceedings`
[54] A. Colelli, Characterization of Monolithic Active Pixel Sensors for future collider experiments , `these proceedings`
[55] B. Mele, Physics at Future Accelerators, `these proceedings`

AUTHOR INDEX